Methods in Enzymology

Volume 402
BIOLOGICAL MASS SPECTROMETRY

METHODS IN ENZYMOLOGY

EDITORS-IN-CHIEF

John N. Abelson   Melvin I. Simon

DIVISION OF BIOLOGY
CALIFORNIA INSTITUTE OF TECHNOLOGY
PASADENA, CALIFORNIA

FOUNDING EDITORS

Sidney P. Colowick and Nathan O. Kaplan

Methods in Enzymology

Volume 402

Biological Mass Spectrometry

EDITED BY

A. L. Burlingame

DEPARTMENT OF PHARMACEUTICAL CHEMISTRY
UNIVERSITY OF CALIFORNIA
SAN FRANCISCO, CALIFORNIA

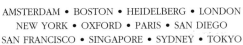

AMSTERDAM • BOSTON • HEIDELBERG • LONDON
NEW YORK • OXFORD • PARIS • SAN DIEGO
SAN FRANCISCO • SINGAPORE • SYDNEY • TOKYO
Academic Press is an imprint of Elsevier

ELSEVIER

Elsevier Academic Press
525 B Street, Suite 1900, San Diego, California 92101-4495, USA
84 Theobald's Road, London WC1X 8RR, UK

This book is printed on acid-free paper. ∞

For all information on all Elsevier Academic Press publications
visit our Web site at www.books.elsevier.com

ISBN-13: 978-0-12-182807-3
ISBN-10: 0-12-182807-7

PRINTED IN THE UNITED STATES OF AMERICA
05 06 07 08 09 9 8 7 6 5 4 3 2 1

Table of Contents

Biological Mass Spectrometry

Contributors to Volume 402

Article numbers are in parantheses following the name of Contributors.
Affiliations listed are current.

DAVID ARNOTT (8), *Department of Protein Chemistry, Genetech, Inc., South San Francisco, California*

MICHAEL A. BALDWIN (1, 9), *Mass Spectrometry Research Resource, Department of Pharmaceutical Chemistry, University of California, San Francisco, San Francisco, California*

KAREN BROWN* (14), *Lawrence Livermore National Laboratory, Livermore, California*

JENNIFER M. CAMPBELL (3), *Applied Biosystems, Framingham, Massachusetts*

ROBERT J. CHALKLEY (9), *Department of Pharmaceutical Chemistry, University of California, San Francisco, San Francisco, California*

GUILLAUME CHEVREUX (11), *Laboratoire de Spéctrometrie de Masse Bio-Organique (LSMBO), Strasbourg, France*

MAX L. DEINZER (10), *Department of Chemistry, Oregon State University, Corvallis, Oregon*

KAREN H. DINGLEY (14), *Lawrence Livermore National Laboratory, Livermore, California*

WERNER ENS (2), *Department of Physics and Astronomy, University of Manitoba, Winnipeg, Manitoba, Canada*

GARY L. GLISH (4), *Department of Chemistry, University of North Carolina at Chapel Hill, Chapel Hill, North Carolina*

KIRK C. HANSEN (9), *School of Medicine, University of Colorado Health Sciences Center, Aurora, Colorado*

JUSTIN M. HETTICK (6), *National Institute for Occupational Safety and Health, Health Effects Laboratory Division, Allergy and Clinical Immunology Research, Washington, D.C.*

ADAM H. LOVE (13), *Center for Accelerator Mass Spectrometry, Lawrence Livermore National Laboratory, Livermore, California*

CLAUDIA S. MAIER (10), *Department of Chemistry, Oregon State University, Corvallis, Oregon*

SCOTT A. MCLUCKEY (5), *Department of Chemistry, Purdue University, West Lafayette, Indiana*

KATALIN F. MEDZIHRADSZKY (7), *Department of Pharmaceutical Chemistry, School of Pharmacy, University of California, San Francisco, San Francisco, California; Proteomics Research Group, Biological Research Center, Szeged, Hungary*

JOSEPH W. MORGAN (6), *Department of Chemistry, Texas A&M University, College Station, Texas*

ANNE H. PAYNE (4), *Department of Chemistry, University of North Carolina at Chapel Hill, Chapel Hill, North Carolina*

NOELLE POTIER (11), *Laboratoire de Spéctrometrie de Masse Bio-Organique, Strasbourg, France*

*Current address: Cancer Biomarkers and Prevention Group, The Biocentre, University of Leicester, Leicester, United Kingdom.

CAROL V. ROBINSON (12), *Department of Chemistry, University of Cambridge, Cambridge, United Kingdom*

HÉLÈNE ROGNIAUX (11), *INRA URPVI – Plate-forme de Spectrométrie de Masse, Nantes, France*

DAVID H. RUSSELL (6), *Department of Chemistry, Texas A&M University, College Station, Texas*

KENNETH G. STANDING (2), *Department of Physics and Astronomy, University of Manitoba, Winnipeg, Manitoba, Canada*

JOHN T. STULTS (8), *Predicant Biosciences, Inc., South San Francisco, California*

PAULA TITO (12), *Department of Chemistry, University of Cambridge, Cambridge, United Kingdom*

KENNETH W. TURTELTAUB (14), *Biology and Biotechnology Research Program, Lawrence Livermore National Laboratory, Livermore, California*

ALAIN VAN DORSSELAER (11), *Laboratoire de Spéctrometrie de Masse Bio-Organique, Strasbourg, France*

MARVIN L. VESTAL (3), *Applied Biosystems, Framingham, Massachusetts*

JOHN S. VOGEL (13), *Center for Accelerator Mass Spectrometry, Lawrence Livermore National Laboratory, Livermore, California*

MITCHELL J. WELLS (5), *Department of Chemistry, Purdue University, West Lafayette, Indiana*

ZHONG-PING YAO (12), *Department of Chemistry, University of Cambridge, Cambridge, United Kingdom*

Preface

Mass spectrometry deals with the formation, manipulation, and measurement of charged substances in order to detect and identify them. Since its previous overview was published in this series (McCloskey, 1990), the Nobel Prize in Chemistry was awarded in 2002 to John B. Fenn and Koichi Tanaka for the discovery of two new methods for producing charged biomacromolecules from liquid and solid solution. These can be thought of as ways to isolate charged molecules in the gas phase that are formed simply from "normal acid-base protonation-deprotonation reactions" from volatile liquid buffers and solid matrices. Over the past two decades these techniques, electrospray (ESI) and matrix-assisted laser desorption (MALDI), have provided the remarkable window we needed to "see" into the machinery of cell biology and view its true molecular complexity for the first time.

These ways of producing ions work efficiently for virtually all biomacromolecules, so it is left to our scientific ingenuity to design ways to manipulate these charged molecules to elicit information that reveals their molecular structural nature. Hence, several generations of ion-optical and energy deposition strategies have emerged that make up the current tools of the trade—commercial mass spectrometers. It should be noted that the design and discovery of better strategies remains a vibrant, young pursuit.

Finally, advanced computational capabilities have evolved to record, process, and manage mass spectral information and provide interfaces with DNA and protein sequence repositories. The tools of bioinformatics are also being adapted and refined to provide visualization into our existing knowledge of biology.

But these developments represent just the beginning of positioning the kind of ingredients that will be employed to gain an understanding of human biology. This volume and its companion (Burlingame, 2005) are intended to describe the astounding strides that have brought us to our current methodological toolbox and also provide the foundation of knowledge indispensable to understanding the current practice of mass spectrometry, as well as to appreciate the rapidly expanding and accelerating horizons in this field.

Thus, this work is focused at the forefront of proteins and their complexities, including descriptions of the techniques and instrumentation being used, their sequence and structural identification based on interpretation of their tandem mass spectra, the strategies and issues in proteomics, studies of solution structures and interactions using isotope exchange, studies of non-covalent complexes with metal ions and ligands, and use of sub-attomole isotopic bio-tracers using accelerator mass spectrometry. All of these contributions are written by authorities who have made seminal contributions to their respective topics.

These foundations provide insight into the forefront of the experimental and technological platforms necessary to pursue a variety of major research themes surrounding protein biology, including proteomics, protein-protein interactions, glycobiology, epi-genetics, and systems biology.

I am indebted to all of my colleagues who have participated in this work, to Candy Stoner for her assistance and talents during the preparation phase, and to Raisa Talroze for the completion of both volumes. I would like to acknowledge the NIH, National Center for Research Resources, for generous financial support (Grant RR 01614).

A. L. BURLINGAME

References

Burlingame, A. L. (2005). "Mass Spectrometry: Modified Proteins and Glycoconjugates." *Meth. Enz.* **405**.

McCloskey, J. A. (1990). Mass spectrometry. *Meth. Enz.* **193,** 960.

METHODS IN ENZYMOLOGY

VOLUME 90. Carbohydrate Metabolism (Part E)
Edited by WILLIS A. WOOD

VOLUME 91. Enzyme Structure (Part I)
Edited by C. H. W. HIRS AND SERGE N. TIMASHEFF

VOLUME 92. Immunochemical Techniques (Part E: Monoclonal Antibodies and General Immunoassay Methods)
Edited by JOHN J. LANGONE AND HELEN VAN VUNAKIS

VOLUME 93. Immunochemical Techniques (Part F: Conventional Antibodies, Fc Receptors, and Cytotoxicity)
Edited by JOHN J. LANGONE AND HELEN VAN VUNAKIS

VOLUME 94. Polyamines
Edited by HERBERT TABOR AND CELIA WHITE TABOR

VOLUME 95. Cumulative Subject Index Volumes 61–74, 76–80
Edited by EDWARD A. DENNIS AND MARTHA G. DENNIS

VOLUME 96. Biomembranes [Part J: Membrane Biogenesis: Assembly and Targeting (General Methods; Eukaryotes)]
Edited by SIDNEY FLEISCHER AND BECCA FLEISCHER

VOLUME 97. Biomembranes [Part K: Membrane Biogenesis: Assembly and Targeting (Prokaryotes, Mitochondria, and Chloroplasts)]
Edited by SIDNEY FLEISCHER AND BECCA FLEISCHER

VOLUME 98. Biomembranes (Part L: Membrane Biogenesis: Processing and Recycling)
Edited by SIDNEY FLEISCHER AND BECCA FLEISCHER

VOLUME 99. Hormone Action (Part F: Protein Kinases)
Edited by JACKIE D. CORBIN AND JOEL G. HARDMAN

VOLUME 100. Recombinant DNA (Part B)
Edited by RAY WU, LAWRENCE GROSSMAN, AND KIVIE MOLDAVE

VOLUME 101. Recombinant DNA (Part C)
Edited by RAY WU, LAWRENCE GROSSMAN, AND KIVIE MOLDAVE

VOLUME 102. Hormone Action (Part G: Calmodulin and Calcium-Binding Proteins)
Edited by ANTHONY R. MEANS AND BERT W. O'MALLEY

VOLUME 103. Hormone Action (Part H: Neuroendocrine Peptides)
Edited by P. MICHAEL CONN

VOLUME 104. Enzyme Purification and Related Techniques (Part C)
Edited by WILLIAM B. JAKOBY

VOLUME 105. Oxygen Radicals in Biological Systems
Edited by LESTER PACKER

VOLUME 106. Posttranslational Modifications (Part A)
Edited by FINN WOLD AND KIVIE MOLDAVE

VOLUME 226. Metallobiochemistry (Part C: Spectroscopic and Physical Methods for Probing Metal Ion Environments in Metalloenzymes and Metalloproteins)
Edited by JAMES F. RIORDAN AND BERT L. VALLEE

VOLUME 227. Metallobiochemistry (Part D: Physical and Spectroscopic Methods for Probing Metal Ion Environments in Metalloproteins)
Edited by JAMES F. RIORDAN AND BERT L. VALLEE

VOLUME 228. Aqueous Two-Phase Systems
Edited by HARRY WALTER AND GÖTE JOHANSSON

VOLUME 229. Cumulative Subject Index Volumes 195–198, 200–227

VOLUME 230. Guide to Techniques in Glycobiology
Edited by WILLIAM J. LENNARZ AND GERALD W. HART

VOLUME 231. Hemoglobins (Part B: Biochemical and Analytical Methods)
Edited by JOHANNES EVERSE, KIM D. VANDEGRIFF, AND ROBERT M. WINSLOW

VOLUME 232. Hemoglobins (Part C: Biophysical Methods)
Edited by JOHANNES EVERSE, KIM D. VANDEGRIFF, AND ROBERT M. WINSLOW

VOLUME 233. Oxygen Radicals in Biological Systems (Part C)
Edited by LESTER PACKER

VOLUME 234. Oxygen Radicals in Biological Systems (Part D)
Edited by LESTER PACKER

VOLUME 235. Bacterial Pathogenesis (Part A: Identification and Regulation of Virulence Factors)
Edited by VIRGINIA L. CLARK AND PATRIK M. BAVOIL

VOLUME 236. Bacterial Pathogenesis (Part B: Integration of Pathogenic Bacteria with Host Cells)
Edited by VIRGINIA L. CLARK AND PATRIK M. BAVOIL

VOLUME 237. Heterotrimeric G Proteins
Edited by RAVI IYENGAR

VOLUME 238. Heterotrimeric G-Protein Effectors
Edited by RAVI IYENGAR

VOLUME 239. Nuclear Magnetic Resonance (Part C)
Edited by THOMAS L. JAMES AND NORMAN J. OPPENHEIMER

VOLUME 240. Numerical Computer Methods (Part B)
Edited by MICHAEL L. JOHNSON AND LUDWIG BRAND

VOLUME 241. Retroviral Proteases
Edited by LAWRENCE C. KUO AND JULES A. SHAFER

VOLUME 242. Neoglycoconjugates (Part A)
Edited by Y. C. LEE AND REIKO T. LEE

VOLUME 243. Inorganic Microbial Sulfur Metabolism
Edited by HARRY D. PECK, JR., AND JEAN LEGALL

Biological Mass Spectrometry

[1] Mass Spectrometers for the Analysis of Biomolecules

By MICHAEL A. BALDWIN

Abstract

Mass spectrometry (MS) has become a vital enabling technology in the life sciences. This chapter summarizes the fundamental aspects of MS, with reference to topics such as isotopic abundance and accurate mass and resolution. A broad and comprehensive overview of the instrumentation, techniques, and methods required for the analysis of biomolecules is presented. Emphasis is placed on describing the soft ionization methods and separation techniques employed in current state-of-the-art mass spectrometers.

As defined in a publication from the International Union of Pure and Applied Chemistry (IUPAC), MS (or mass spectroscopy) is "the study of systems by the formation of gaseous ions, with or without fragmentation, which are then characterized by their mass-to-charge ratios and relative abundances" (Todd, 1991). Since the publication of the last volume in *Methods in Enzymology* reviewing MS of biomolecules (McCloskey, 1990), there has been a revolution in the field. Two promising novel soft ionization methods emerging at that time were not generally available, partly because both were largely incompatible with the typical commercial sector mass spectrometers that were in widespread use. Although the particle bombardment/desorption techniques of plasma desorption MS (PDMS), fast atom bombardment (FAB), and liquid secondary ion MS (LSIMS), invented a decade earlier, had been making valuable contributions to the analysis of peptides, oligosaccharides, and other polar and involatile compounds, they were largely limited to the picomole range and thus lacked the sensitivity needed to tackle the most challenging problems. During that period when analysis of intact biological molecules such as small proteins first became possible, much research was focused on attempts to ionize ever larger molecules, many of which were standards purchased from commercial suppliers. With hindsight, simply measuring the molecular weight of a large molecule is often of limited utility, whereas digesting it chemically or enzymatically to smaller moieties and measuring the masses of even a subset of these can be very informative. Today, thanks to the maturation of soft ionization methods and new developments in mass analyzers optimized for these new ionization methods, MS is

METHODS IN ENZYMOLOGY, VOL. 402
0076-6879/05 $35.00
DOI: 10.1016/S0076-6879(05)02001-X

established as a fundamental technology in the biological sciences in routine use in numerous laboratories worldwide. There is no doubt that it is contributing to the solution of very many fundamental problems in biology and medicine (Burlingame *et al.*, 2000; Weston and Hood, 2004).

The selection of an optimal mass spectrometric method to tackle a particular task is rarely a straightforward consideration. As an example, MS is the enabling technology in the field of proteomics (deHoog and Mann, 2004), which involves protein identification in very complex mixtures such as cell lysates, tissues, or other biological samples, as well as the identification of interacting partners (Deshaies *et al.*, 2002), characterization of modifications, quantitation of expression levels, and studies of non-covalent protein complexes. In most cases, this involves an initial separation step, usually by one-dimensional (1D) or two-dimensional (2D) gel electrophoresis, perhaps labeling with an affinity-tagged ligand, followed by proteolysis with a site-specific protease such as trypsin to generate peptides, possibly multiple further separation/enrichment steps, then the mass spectrometric analysis of the peptide mixtures. Some of the questions that arise in choosing a mass spectrometer to carry out various aspects of these tasks are as follows:

- What level of sensitivity is required?
- How accurate must molecular weight measurements be, and is there a limit to the accuracy that is required or even useful?
- Will the determination of peptide molecular weight values be sufficient or will sequence data be necessary, and if so, what are the optimal techniques?
- Is it better to separate the peptide mixtures before MS analysis, for example, by high-performance liquid chromatography (HPLC)-MS, or can sufficient information be obtained on unseparated mixtures?
- If HPLC is not used, how will the selected method be affected by impurities that may be difficult to remove?
- What is the sample throughput, and is the technique amenable to automation?
- What will it cost?

Although such questions help to narrow the choices, ultimately there will be a number of alternative solutions, each of which has individual strengths and weaknesses.

Some Definitions and Principles

In addition to the 1991 IUPAC recommendations on nomenclature and definitions (Todd, 1991), in the same year the Committee on Measurements and Standards of the American Society for Mass Spectrometry also made recommendations on standard definitions of terms in use in MS (Price, 1991). The term MS encompasses a wide range of very different methods of analysis, each with its own unique characteristics. However, understanding some basic principles applicable to all these methods is essential and will allow the strengths and weaknesses of each to be evaluated and compared. All mass spectrometers have certain features in common: a sample introduction system, a method of creating ions from neutral sample molecules, a means of separating them and determining their mass-to-charge ratios, and an ion detector that may also provide a quantitative measure of the number of ions being detected within a given mass window or time period. Because the functions are frequently interrelated, these are represented in Fig. 1 as a series of interconnected boxes rather than as separate components. Thus, the method of sample introduction is intimately linked to the ionization step, and the choice of mass analyzers may also be predicated on the choice of ionization technique. Therefore, it is difficult to discuss any of these components in isolation.

The primary function of a mass spectrometer is to separate charge-bearing molecules according to molecular mass and to measure their mass numbers. The species studied must be ions rather than molecules because mass separation relies on the properties of charged particles moving under the influence of electric and magnetic fields, usually in high vacuum where the mean free path is sufficient to ensure that they mostly travel without collisions. Ions may be positively or negatively charged, although most mass spectrometric experiments are performed on positive ions. Positive ions can be formed either by removal of one or more electrons from each molecule or by the addition of one or more cations. A charged species that is formed by the removal or attachment of an electron is referred to as a *molecular ion,* whereas ionization that is achieved by addition or removal of a charged atom or molecule such as H^+ or NH_4^+ gives a species that is termed a *quasi-molecular ion.* The use of an earlier expression, *pseudo-molecular ion,* is not recommended (Todd, 1991). The addition of protons is the favored method for ion production in biological MS, so technically these are quasi-molecular rather than molecular ions. However, in common usage, they, too, tend to be referred to as *molecular ions.*

Ions are separated either spatially, as by a magnetic field, or temporally, as in time of flight (TOF). Even though the underlying physical

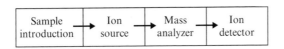

Fig. 1. The basic components of a mass spectrometer.

characteristics of these separation methods are well understood, because of potential deviations from ideal behavior, ion mass numbers are invariably obtained by comparison and calibration of the mass scale with known standard substances rather than by calculation from first principles. The degree of separation achieved is dependent on both the mass and the magnitude of the charge. By convention, the abscissa of the mass spectrum is calibrated in m/z units, where m is the mass and z is the number of electronic charges. The Thomson (Th) has been used to describe m/z units, although it has not achieved wide acceptance. Atomic and molecular weights are dimensionless because they are derived from a ratio of elemental mass values multiplied by 12.0000, the denominator being the mass of a ^{12}C atom. Nevertheless, they are frequently ascribed the symbol u, as recommended by the IUPAC (Todd, 1991), or Daltons (Da). An earlier symbol *amu*, an acronym for *atomic mass unit*, is no longer recommended. Although it is quite common to see the abscissa of a mass spectrum labeled with *Da*, technically this is incorrect because multiply charged ions appear not at m but at m/z.

Resolving Power and Resolution

The degree of separation achievable is determined by the resolving power (RP) of the mass analyzer. When two peaks of mass m and $m + \Delta m$ are just resolved according a specified criterion, the RP is defined as $m/\Delta m$, (e.g., if ions of mass 100.0 and 100.1 can be separated, then RP = 100/0.1 = 1000). This measurement can be made on a single peak by dividing the mass of the peak by its width without the need to observe the degree of separation. However, as shown in Fig. 2, there are different criteria for ion separation: originally Δm was measured at 5% above the baseline, but more recently it has come to be measured at 50%. The first criterion is commonly called the 10% *valley definition,* because two equal height peaks intersecting at the 5% point overlap to give a valley equal to 10% of their height. The second, based on the *full width* of a peak at *half* its *maximum* height (FWHM), is now the most widely adopted definition and is used in this chapter. However, because of the addition of the ion currents where adjacent peaks overlap, the FWHM effectively defines the point at which ions of different mass can no longer be separated.

The separation of the peaks within a mass spectrum obtained by an instrument of a specific RP is expressed in terms of resolution. This is usually reported as the reciprocal of RP and is expressed in parts per million (ppm). Thus, an instrument of RP of 10,000 gives a mass spectrum in which the peaks show a resolution of 1/10,000 (i.e., 100 ppm). Many mass analyzers such as a magnetic sector or a TOF have an approximately

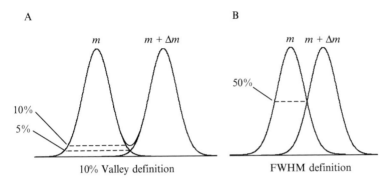

FIG. 2. Criteria for alternative definitions of resolving power.

constant RP over their mass range, which means that the peak shapes become broader with increasing mass. In such an instrument, the ability to resolve ions depends only on m and is independent of m/z. Thus, if RP = 1000, ions of $m = 1500.0$ and 1501.0 will not be separated, even when they are doubly charged and appear in the mass spectrum at m/z 750.0 and 750.5, respectively. Other instruments such as the quadrupole analyzer have an approximately constant peak width over the full mass range, in which case the definition RP $= m/\Delta m$ is mass dependent, so the RP is usually expressed in m/z units instead. Thus, a quadrupole mass analyzer that just separates peaks differing by 1 m/z unit is said to give a resolution of 1 mass unit. All multiply charged ions have peak spacing less than 1, so they would not be resolvable with such an analyzer, independent of mass.

Isotopes, Peak Shapes, and Mass Measurements

The stable *heavy* isotopes of the elements that make up most biological molecules (^{13}C, ^{2}H, ^{15}N, ^{18}O, etc.) have a major impact on the appearance of ions in a mass spectrum and on the way masses are defined and measured. The molecular ions of a pure organic compound will contain a mixture of species of different isotopic compositions, the distribution of which is determined by the natural abundances of the various stable isotopes of each element. Thus, even a pure compound gives a cluster of multiple molecular peaks in the mass spectrum, each peak separated by approximately 1 Da. The first peak in the isotopic cluster (that of lowest m/z) is made up of the lightest isotopes of each element respectively (^{12}C, ^{1}H, ^{14}N, ^{16}O, etc.) and is referred to as the *monoisotopic peak*. As illustrated in Fig. 3A for singly charged ions of the peptide hormone bradykinin, this gives the most intense peak for compounds of relatively low

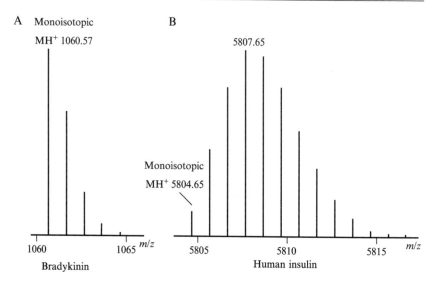

FIG. 3. Calculated isotopic peak intensities for (A) bradykinin and (B) human insulin. The m/z scales use nominal mass numbers, whereas the peak positions are defined by accurate mass numbers.

molecular weight. However, as shown in Fig. 3B for the small protein insulin, as the molecular weight increases, the height of the second and subsequent isotopic peaks become greater than that of the monoisotopic peak. For bradykinin, the monoisotopic ion represents 52% of the total ensemble of molecular ions, whereas for insulin it represents only 2.6%. At the molecular weight of medium-sized proteins (25–50 kDa), the cluster of peaks becomes broader and eventually the monoisotopic peak becomes vanishingly small. Many types of mass spectrometers can resolve the adjacent peaks in the clusters at low mass (e.g., for typical peptides of mass 500–2000 derived from a protein by tryptic digestion). However, at the molecular weight of proteins larger than insulin, very few instruments can resolve the isotopic clusters and they appear in the spectrum as a single broad peak.

Because of mass defects that arise from differences in nuclear binding energies, atomic masses do not precisely match the sum of their subatomic components, so the atomic weights of elements and their isotopes differ from integer values by small amounts (Rutherford, 1929). Ions of the same nominal mass but different atomic composition, termed *isobars,* have subtly different mass values. If these can be measured accurately, the atomic composition of the ions may be defined (Beynon, 1960). A simple example is given by ions of nominal mass 16, the compositions of which could be

CH_4 (16.0313), NH_2 (16.0187), or O (15.9949). The first two of these differ in mass by 1 part in 1269, so a mass spectrometer would need an RP higher than that to separate them if both ions occurred in the same spectrum. To define the composition of one of these ions would require any mass measurement error to be less half the reciprocal of the mass difference (i.e., <394 ppm). Most modern mass spectrometers would easily meet and exceed this performance. However, although such an analysis is straightforward for ions of low mass, it becomes increasingly difficult as the mass increases, partly because the absolute errors in mass measurement increase, but more so because the number of possible compositions of ions of any given mass increases nonlinearly with increasing mass.

Molecular masses may be either "nominal" or "accurate." The nominal masses are integer values based on H = 1, C = 12, N = 14, O = 16, and so on, which take no account of the fractional parts of the actual atomic masses. By contrast, the "accurate" masses include the fractional part, although whether experimentally measured values are truly accurate depends on several factors including the calibration of the m/z scale, which may be "external" or "internal." External calibration is carried out by comparison with the positions of peaks in a separate spectrum of a standard compound or mixture, recorded either before or after the spectrum of the sample of interest. This relies on the stability of the various components of the mass spectrometer over the time period between the recording of the calibration and the sample. This is unlikely to be as accurate as an internal calibration that is based on the positions of certain ions of known composition within the mass spectrum of the sample. Known trypsin autolysis fragments that commonly occur in unseparated protein digests are frequently used for internal calibration for the analysis of peptide mixtures. There is a connection between RP and the accuracy with which mass numbers can be measured, but they are not linked by any simple relationship. Mass measurement is based on defining the position of a peak on the m/z scale. In general, mass spectrometers of high RP enable this to be done more accurately; for example, with internal calibration, an instrument of RP $\geq$ 10,000 (resolution $\leq$ 100 ppm) should be able to give mass measurements with errors <10 ppm. High RP also makes it more likely that a peak will correspond to ions of a single atomic composition, rather than multiple overlapping components of different compositions. Note that mass measurements made on unseparated ions of different compositions and unknown relative abundance will be uninformative or misleading.

The existence of stable isotopes of most commonly occurring elements gives rise to two scales for atomic masses (i.e., monoisotopic and average chemical masses). The latter is derived from weighted averages based on the masses and natural abundances of the various isotopes of each element

(Rosman and Taylor, 1998; Vocke, 1999) Thus, the monoisotopic atomic mass of carbon is defined as 12.0000 precisely, corresponding to the atomic mass of a ^{12}C atom, whereas the average mass is approximately 12.011, calculated from the natural occurrence of ^{12}C and ^{13}C in the environment, the abundance of ^{13}C ranging from 0.98% to 1.15%, with a typical value of approximately 1.1% (Rosman and Taylor, 1998). If the individual isotopic peaks are resolved for a molecular ion, monoisotopic atomic weights should be used to calculate the mass value of the first peak in each cluster, giving a molecular weight referred to as the *monoisotopic value*. When a mass spectrometer is unable to resolve the separate isotopes, average atomic masses are used to calculate molecular weights. For typical proteins, the difference between monoisotopic and average atomic weights is approximately 0.6/1000. Theoretically, the masses of noise-free, well-resolved symmetrical peaks can be defined by the positions of their peak tops. However, the typical situation is less ideal and a peak position should be defined by its centroid, giving an average mass for all of the ions contributing to the peak. This is particularly important for unresolved isotopic peaks, because their profiles will be asymmetrical because of uneven isotopic distributions.

Some of these points are illustrated in Fig. 4, which shows simulated peak shapes for singly protonated bradykinin ions of monoisotopic mass 1060.56 at different FWHM resolutions. The simulations are based on Gaussian peak shapes, which are characteristic of many types of mass spectrometers. At the highest RP shown here (4000), the individual components are completely separated and the peak top coincides with the monoisotopic mass. At RP = 2000, the tails of the peaks overlap and it

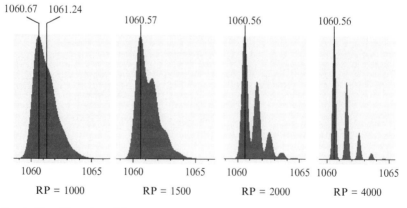

FIG. 4. The effect of resolving power on simulated stable isotopic clusters for bradykinin.

can be seen that this is close to the 10% valley definition for the separation of peaks. However, the peaks continue to be well defined and the overlap does not affect the positions of the peak tops. At RP = 1500, the peaks are extensively merged and there is a small shift of the monoisotopic peak top by 0.01 Da toward higher mass, equivalent to an error of 10 ppm. At RP 1000, the peaks are not resolved at all, although sufficient structure can be seen to indicate the underlying presence of adjacent unresolved ions. Here, the peak top is shifted to higher mass by 0.11 Da, which compared with the monoisotopic mass is an error of 100 ppm. However, comparing the peak top with the average mass of 1061.26 gives a much larger error, emphasizing that it is essential to use the centroid rather than the peak top for average mass assignments.

Soft Ionization

Until 30 years ago, mass spectrometers used in the analysis of organic compounds employed ionization by electron impact (EI) upon isolated analyte molecules that had been evaporated into a vacuum from the liquid or solid state by heating. EI causes electronic excitation with the expulsion of an electron, giving predominantly singly charged positive ions as in Fig. 5A. Such ions are often described as *radical cations* and *odd-electron ions,* because the removal of a single electron leaves an unpaired electron. The efficiency of ionization is low (1 in $\sim 10^4$), but this not a problem if sufficient material is available. Heating is likely to cause decomposition of biomolecules, and EI transfers sufficient energy that many molecular ions undergo unimolecular dissociation to smaller fragments before analysis.

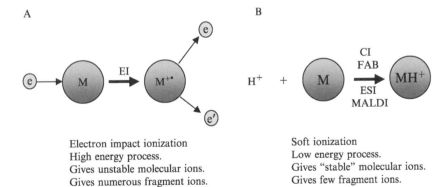

A

Electron impact ionization
High energy process.
Gives unstable molecular ions.
Gives numerous fragment ions.

B

Soft ionization
Low energy process.
Gives "stable" molecular ions.
Gives few fragment ions.

FIG. 5. Mass spectrometric ionization methods: (A) electron impact ionization, (B) soft ionization.

Fragment ions can be valuable because they give structural information, but compounds that contain particularly labile bonds are likely to dissociate to such an extent that no molecular ions are observed. Such an approach is not effective for the analysis of most polar, involatile, and thermally unstable biological molecules, although their properties may be improved by derivatization to reduce polarity, increase volatility, and perhaps induce some characteristic fragmentation reactions (Knapp, 1990).

Since about 1970, a series of alternative ionization methods provided improvements for biological analysis, including chemical ionization (CI) and FAB. Compared with EI, these are classified as *soft ionization methods* because they transfer less vibrational energy and form ions having lower internal energies. As illustrated in Fig. 5B, they give predominantly protonated ions that have no unpaired electrons and are inherently more stable than the radical cations formed by loss of an electron. Two soft ionization techniques developed in their modern form in the late 1980s proved so superior that they now account for virtually all applications of MS for the analysis of polar, involatile, and thermally labile biomolecules, namely electrospray ionization (ESI) (Fenn *et al.*, 1989, 1990) and matrix-assisted laser desorption ionization (MALDI) (Karas and Hillenkamp, 1988). The fundamental characteristics of each meant that the choice of ionization method largely dictated the nature of the mass analyzer. ESI (illustrated in Fig. 6A) involves spraying a continuous stream of dilute solution of analyte from a needle held at a high potential into a chamber at atmospheric pressure. For positive ion analysis, a typical solvent is 50% acetonitrile or methanol acidified with 1% acetic acid or 0.1% formic acid, conditions that ensure the unfolding of proteins and extensive protonation of most basic sites. Formation of droplets and evaporation of the solvent produces a continuous stream of ions, and initially ESI was applied exclusively to mass analyzers that work in a continuous mode, such as the quadrupole. The continuous source of ions could also be applied to magnetic sector analyzers but their narrow ion beam acceptance angle and requirement for high accelerating potential made such instruments less than ideal. Although some analysis of small molecules such as the study of drug metabolites employs atmospheric pressure chemical ionization, ESI has become the predominant ionization method for HPLC-MS in biomolecular analysis.

MALDI is a solid state sputtering/desorption method that produces ions by laser bombardment of crystals containing a small amount of analyte dispersed in a large amount of matrix. When carried out in high vacuum and at high accelerating voltage, the pulsed nature of laser radiation produces ions in pulses (as in Fig. 6B) that are best suited to TOF analysis. Because the samples are solids, MALDI is less easily adapted to chromatographic interfacing, although there have been preliminary experiments

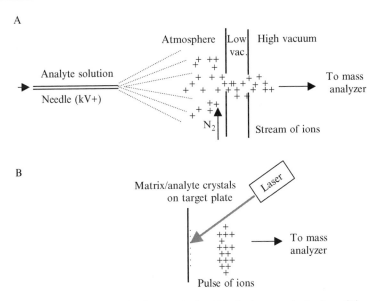

FIG. 6. Alternative soft ionization methods: (A) electrospray ionization, (B) matrix-assisted laser desorption ionization.

with MALDI from liquid droplets (Wattenberg *et al.*, 1997). Nanoflow HPLC or capillary electrophoresis can be interfaced to MALDI using vacuum deposition of the column eluent directly onto either a moving mylar tape within the TOF mass spectrometer (Preisler *et al.*, 2000, 2002), or drop-by-drop deposition onto a conventional target plate for off-line processing, or even onto a compact disc for insertion into a purpose-built ion source (Krutchinsky *et al.*, 2001). An advantage of off-line processing of HPLC fractions by MALDI compared with on-line liquid chromatography (LC)-ESIMS is the freedom to analyze the same spot several times, obtaining further levels of information with successive analyses.

Although both ESI and MALDI are in very widespread use, for neither technique is there full agreement on the mechanisms of ion formation and transfer to the gas phase. Overall, ESI is the simpler technique to deal with because the initial formation of ions in solution constitutes a typical example of acid–base equilibria, with pH-dependent protonation and dissociation reactions determining the nature of any charged species in the liquid droplets before evaporation (Cole, 2000; Kebarle, 2000). Most positive ion ESI-MS analyses are carried out in acidic solution, and the basic sites in an analyte such as a peptide effectively compete for the available protons. Peptide molecules having multiple basic sites are generally observed in the gas phase to have acquired multiple protons, suggesting that the efficiency of ion production is high. The first step in the creation of isolated ions is the formation of a charged cone of liquid (Taylor cone) from which positively charged droplets are ejected, particularly if the liquid includes an organic component that reduces its surface tension. As liquid evaporates, the droplets become smaller, and at the Rayleigh limit when the charge density becomes too great, they break up because of excessive repulsive forces between the positive charges. High-speed photography has established that the droplets tend to be elongated rather than spherical, and each primary droplet can emit a stream of smaller droplets that remove a disproportionate amount of charge. In the charge residue model (CRM), further evaporation causes this process to be repeated to the point at which each final droplet contains only one charged species. This is most favorable for ions of very high m/z values. By contrast, the ion evaporation model (IEM) that favors ions of lower m/z values proposes that individual ions are lost from the surface of the droplets *before* they reach the Rayleigh limit, thus reducing the net charge and preventing their coulombic destruction.

Despite the numerous applications of MALDI, the mechanisms of desorption and ionization are unclear. Most commercially available MALDI instruments employ the readily available nitrogen laser that gives 3-ns length pulses at a frequency of 337.1 nm. The use of other ultraviolet (UV) lasers has been explored, but apart from differences in the absorption coefficients of potential matrix compounds, the spectra are largely insensitive to wavelength. UV absorption is predominantly electronic, but at the low laser flux typically employed in MALDI, there is no realistic likelihood of multiphoton ionization occurring. Therefore, it must be concluded that the major function of the laser is to supply heat to the matrix. This is consistent with infrared radiation also being able to generate MALDI spectra that are similar to UV-MALDI spectra. Heating the matrix somehow effects the transfer of a portion of it into the vapor phase in the form of an expanding plume, presumably containing neutrals and ions and carrying the analyte. Whether the analyte was preionized in the solid state or becomes ionized in the plume, perhaps by excited state proton transfer, has yet to be established (Zenobi and Knochenmuss, 1998).

MALDI samples are co-crystallized with a very large molar excess of a matrix that acts as a chromophore for the laser radiation and protects even thermally sensitive analyte molecules from the direct effect of the laser irradiation. Effective matrix compounds in MALDI have been developed on a largely empirical basis, but the commonly used matrices such as 2,5-dihydroxybenzoic acid (DHB) and α-cyano-4-hydroxycinnamic acid (α-CHCA) popular for peptide analysis (Fig. 7) have several common features that are critical for their function. They must be physically

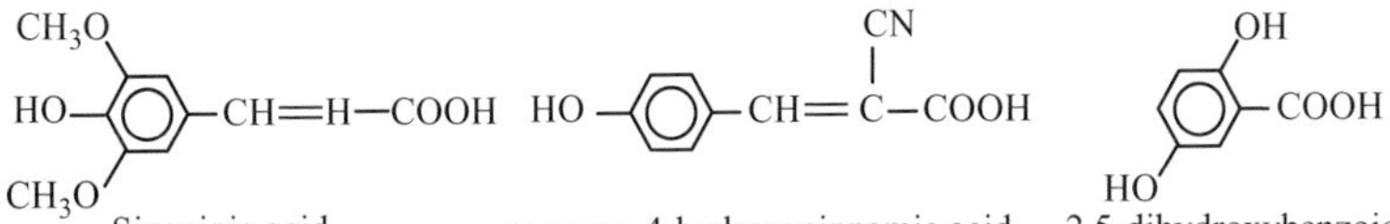

FIG. 7. Some matrices in common use for matrix-assisted laser desorption ionization of peptides and proteins.

compatible with the analyte to the degree that they can form mixed crystals, the analyte molecules being dispersed within the large excess of matrix. They must have an adequate coefficient for absorption of the UV radiation, which for peptides and proteins is largely achieved through the use of substituted and conjugated aromatics such as the cinnamic acid derivatives. The transfer of vibronic energy to the analyte ions is dependent on the nature of the matrix, allowing some "fine-tuning" of the spectra, for example, DHB imparts less internal energy than α-CHCA and gives molecular ions that are less prone to undergo unimolecular dissociation reactions (Karas *et al.*, 1995). For positive ion operation, they must act as a source of protons, so they are mostly acidic, whereas negative ion operation favors the use of basic matrices. For negative ion oligonucleotide analysis, basic co-matrices such as piperidine, imidazole, and triethylamine with proton affinities near or above the proton affinities of the nucleotide residues can serve as proton sinks during the desorption/ionization process to enhance the ionization (i.e., deprotonation) of the analyte, and they can also affect the nature and the extent of any fragmentation reactions (Simmons and Limbach, 1998). Matrix compounds must be relatively involatile and resistant to evaporation in high vacuum, a property of polar organic compounds that are extensively stabilized by hydrogen bonds. Some more or less desirable properties are relatively unpredictable; for example, the cinnamic acid derivative sinapinic acid was adopted early on as an effective matrix for peptides and proteins, but it gave rise to a photochemical adduct in which a sinapinic acid molecule becomes covalently attached to the analyte ion with loss of a water molecule. α-CHCA proved to be a superior alternative because it has no hydrogen on the α-carbon and cannot form an analogous adduct (Beavis and Chait, 1989).

Compared with ESI, MALDI ionization appears to be relatively inefficient. As stated earlier, ESI adds protons to all the basic sites in the analyte molecules, whereas the ions observed from MALDI are typically only singly charged. Even for a large protein containing numerous basic residues, it is rare to observe the attachment of more than two or perhaps three protons. Therefore, on statistical grounds, it is reasonable to conclude that either most analyte molecules remain uncharged or the ions become neutralized in MALDI and, therefore, do not contribute to the mass spectrum. However, in most applications, MALDI ions are formed in a high vacuum and with high kinetic energies, so they are transmitted through the mass spectrometer with high efficiency, whereas ESI ions are formed at atmospheric pressure and many are lost in the transfer to the vacuum system of the mass analyzer. In addition, MALDI is almost always used with TOF analysis, which maximizes the sensitivity of detection as it enjoys the Fellgett advantage (Fellgett, 1958), that is, as with interferometric detection of all wavelengths simultaneously in infrared spectroscopy, it allows detection of the entire ensemble of ions that are pulsed into the flight tube. This contrasts with a scanning instrument that detects only that fraction of ions passing through the mass analyzer at any given time and fails to detect the much larger number that are being discriminated against. This is an important consideration when only a limited amount of sample is available. On balance, the ultimate sensitivity of the two techniques in experiments optimized for small samples proves to be surprisingly similar; for example, routine measurements on peptide mixtures are relatively straightforward at the low femtomole level, and specific measurements can be made at the attomole level using either method. Note that at this level of sensitivity, careful sample handling to avoid losses by adsorption to surfaces is more likely to be the limiting factor for both analytical methods.

Because the technical aspects of ionization by MALDI and ESI are quite different, both offer advantages and disadvantages in certain situations. Unlike ESI, MALDI is relatively unaffected by modest amounts of salt and/or detergent. In fact, some detergent may be beneficial (Breaux *et al.*, 2000), although it may be necessary to remove excessive contaminants by cleanup of samples by solid phase extraction with "Zip-Tips" or similar (i.e., a pipette tip containing a reverse-phase adsorbent), allowing the peptides to be adsorbed, washed, and then eluted with a less polar solvent mixture (http://www.millipore.com). Also, ions formed by MALDI have higher internal energies than those from ESI, and the desorption process helps free the analyte ions from undesirable impurities, just as it frees them from the matrix. In the solution conditions of ESI, detergents may bind to and mask the sites on the analyte molecules that would normally carry a charge, and impurities are more likely to compete successfully for the available charge at the expense of the analyte.

It is important to note that even in the absence of impurities, analysis of peptide mixtures by either technique generally favors certain components, whereas others may be suppressed. Thus, the absence of a particular ion in the spectrum of a mixture is not proof of the absence of that peptide in the mixture. In the MALDI analysis of an unseparated tryptic digest, the peptides having less basic residues are suppressed (e.g., arginine-terminated peptides are favored over lysine-terminated peptides), although chemical derivatization as a means of reducing such discrimination has been investigated (Hale *et al.*, 2000). Ion suppression in ESI-MS is more dependent on polarity and surface activity. The more hydrophobic peptides that are less readily soluble in the liquid phase will migrate to the surface of the droplets and are more likely to form isolated ions in the gas phase than those that are readily soluble and remain in the bulk of the solution (Cech and Enke, 2000). This discriminatory effect in ESI can be reduced to some degree by coupling with HPLC so that at any given moment, the eluent is likely to contain no more than two or three peptides. Where possible, it is desirable to use both ionization methods because the information obtained may be complementary (Medzihradszky *et al.*, 2001). A significant advantage of using MALDI for limited quantities of sample is the ability to study the sample for as long as a sufficient portion remains on a target. As noted earlier in the context of LC-MALDI, in most instances, recording a typical MALDI spectrum consumes only a small fraction of a sample, even when the total amount deposited is only a few femtomoles, allowing the target to be removed from the instrument and retained for further analysis. By contrast, ESI is a flowing technique that consumes the entire amount of a sample within the component elution time.

Note that MALDI has been used less for quantitative applications because the sample preparations are quite heterogeneous and the components of a mixture may separate during crystallization. Thus, it is often necessary to identify a "sweet spot" among the crystals to obtain the best spectra. Although ESI spectra are generally more reproducible, ion currents are not necessarily proportional to relative concentrations in solution; therefore, quantitation requires suitable calibration and ideally the use of internal standards. Employing isotopically labeled analogs may resolve these problems for both MALDI and ESI. An example is the use of isotopically coded affinity tags (ICATs) developed by Aebersold (Gygi *et al.*, 1999). The original ICAT reagent was available in D_0 and D_8 forms to differentially label cysteine residues in two populations of proteins. These were then mixed and proteolyzed, and the tagged peptides were extracted by avidin affinity chromatography. MS analysis of the peptide mixtures using either MALDI or ESI revealed pairs of peaks separated by 8 Da, the relative intensities of which gave a direct measure of the relative proportions of the proteins from which they were derived. Newer ICAT reagents have employed ^{13}C as the heavy isotope because this does not cause the isotopically differentiated peptide pairs to separate in reversed phase HPLC, unlike deuterium, and chemically cleavable links may be incorporated to facilitate the removal of the heavy biotin affinity tag

(Hansen *et al.*, 2003). An alternative approach to chemical labeling is to carry out stable isotope labeling in cell culture (SILAC) using an isotopically labeled amino acid such as arginine, which will tag every Arg-containing peptide in a tryptic digest (Ong *et al.*, 2003). Although MALDI has been used to monitor biophysical characteristics of proteins, solution-based ESI is generally favored to observe the formation of non-covalent complexes of proteins with small ligands, metal ions, DNA, and even very large multiple protein complexes (Loo, 1997; Sobott *et al.*, 2005).

MALDI-TOF Mass Spectrometers

MALDI forms predominantly singly charged ions, so for the analysis of large molecules it requires a mass separation method having a high m/z range, such as a TOF analyzer. Samples are dissolved in water and mixed with a matrix dissolved in water alone or water mixed with an organic solvent such as methanol or acetonitrile, and sometimes acidified with formic acid or trifluoroacetic acid. A small drop of 1 μl or less is deposited on a multiposition sample plate and allowed to dry. Each sample spot will contain 1–5 μg of matrix, and for a typical unseparated digest of a protein, something between a femtomole and a picomole per peptide (i.e., a molar range of peptide:matrix in the region of $1:10^4$–10^7. The sample plate is loaded into the instrument through a vacuum lock without breaking vacuum. This can be viewed inside the vacuum chamber via a video camera, and its position can be manipulated as each sample is irradiated. The sample spot is normally larger than the focused laser beam and different regions of the spot can be selected manually. For heterogeneous sample preparations, it may be possible to select the regions of the target yielding higher ion current (i.e., the sweet spots). However, protocols are available that give relatively homogeneous sample preparations (Dai *et al.*, 1999; Vorm *et al.*, 1994) that are preferable for automated data collection, in which the laser is moved around the surface, either systematically or randomly.

Mass analysis by TOF has a long history, a commercial instrument being manufactured by the Bendix Corporation more than 40 years ago. The principle of linear TOF is straightforward, ions being pulsed into a flight tube by an accelerating potential, resulting in ion velocities that are inversely proportional to the square root of m/z. Thus, they become separated in the flight tube and their time of arrival at a detector situated at the opposite end can be converted to mass. However, the performance of the early instruments was very modest, because of variations in initial ion energies and ion positions before pulsing, compounded by the limitations of the electronics and timing circuitry. TOF analysis enjoyed a renaissance with the advent of MALDI, in which ions are formed in a pulse, the length of which is limited by the short pulse of the ionizing laser. Furthermore, by holding the MALDI target plate at an elevated potential, the ions are instantaneously accelerated toward a grounded aperture, so all ions are pulsed into the flight tube in a narrow burst (Fig. 6B). This development occurring in the late 1980s coincided with the availability of much faster and more reliable digital electronics for pulsing and timing, allowing the flight times to be measured with a precision and accuracy previously unattainable. For a nitrogen laser, the pulses have a half-width of 3 ns, so all ions are formed within this short period. Typical flight times are tens of microseconds, so the length of the initial pulse is relatively insignificant.

MALDI made possible the ionization of large biopolymers such as proteins and DNA or RNA fragments up to 100-mers, with TOF analysis being particularly suited to heavier species because it has no theoretical upper mass limit. Nevertheless, the typical RP of a few hundred for early MALDI-TOF mass spectrometers was inadequate and prevented accurate measurement of ion mass values. Two developments led to enhanced performance, namely delayed extraction (DE), described by a number of authors in 1995 (Brown and Lennon, 1995; King *et al.*, 1995; Vestal *et al.*, 1995; Whittal and Li, 1995) and the general adoption for MALDI-TOF instruments of the reflectron or electrostatic ion mirror that had been described many years earlier (Mamyrin and Shmikk, 1979). In the initial MALDI-TOF ion sources without DE, ions were accelerated as soon as they were desorbed by the laser pulse, with this acceleration occurring in a relatively high-pressure plume of desorbed ions and neutrals. This resulted in scattering and dissociation of many ions during acceleration, with consequent broadening of the mass spectral peaks. DE is achieved by placing a grid in front of the MALDI target and applying a high potential to both the target and the grid so the initial desorption occurs in a field-free region. In general, the ions of interest are heavy and slow moving, whereas most of the neutrals are relatively light molecules formed by thermolysis of the matrix and can diffuse away rapidly. After a delay period of perhaps 100 ns, the grid voltage is instantaneously reduced, causing the ions to be accelerated into the flight tube of the mass spectrometer, experiencing many fewer collisions. The improvement in resolution obtained from the use of DE is illustrated in Fig. 8 for the molecular ion of angiotensin, recorded in linear mode.

The DE source also has a focusing effect for ions with different initial velocities, as does the reflectron or ion mirror. In a linear TOF, ions of a single mass accelerated through the same potential should all have the same flight time and should arrive at the detector simultaneously. In practice, the kinetic energy imparted by the accelerating voltage is superimposed upon a range of energies arising from the laser desorption process.

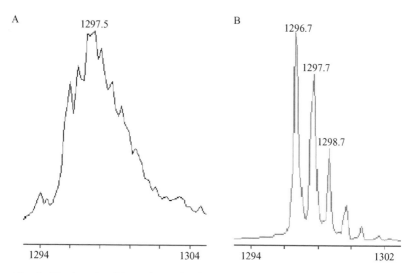

Fig. 8. Matrix-assisted laser desorption ionization–time-of-flight analysis of angiotensin ions recorded in linear mode using (A) continuous extraction and (B) delayed extraction.

Consequently, ions of a single mass have a distribution of arrival times at the detector. A reflectron situated at the far end of the flight tube provides a linear potential gradient from ground to just above the accelerating voltage. This slows the ions and then reverses their direction of flight, accelerating them back toward a detector situated at a slightly offset angle. The fastest ions penetrate further into the reflectron and travel a longer distance before reaching the detector than the slowest ions. Through a suitable choice of geometry, it is possible to have all ions of each unique mass arriving at the detector simultaneously, even though they have slightly different velocities. With the combination of the DE source and the reflectron, modern high-performance MALDI-TOF instruments easily attain an RP of 10,000 or more, which with suitable calibration allows mass measurements to be made with an accuracy of better than 10 ppm.

The layout of a modern high-performance instrument is illustrated in Fig. 9. Most such instruments have two detectors; with the reflectron voltages turned off the so-called *linear detector* may be used to detect heavier species such as intact protein ions. Such species may be successfully transmitted by the reflectron, but in general, they do not benefit from higher resolution measurements but may benefit from the higher sensitivity that can be achieved without the reflectron. The reflectron is predominantly used to analyze smaller species such as peptides, for which increased

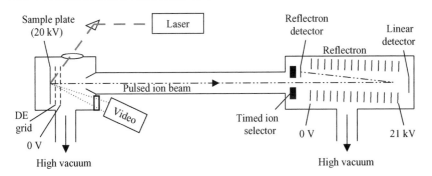

Fig. 9. A high-performance matrix-assisted laser desorption ionization–time-of-flight mass spectrometer.

resolution and more accurate mass measurements represent a substantial advantage (Clauser *et al.*, 1999). To optimize the efficiency of ion transmission, there may be a guidewire running along the center of the flight path. As ions enter the flight tube, a substantial fraction is likely to be diverging and potentially could be lost from the beam before detection. A small potential of opposite polarity to the charge on the ions applied to the guidewire makes these ions spiral around the wire as they travel down the tube, thereby maximizing the fraction of ions detected. However, capturing the more divergent ions tends to degrade the RP, so there is a trade-off between sensitivity and resolution, consequently, the guidewire may be replaced by focusing lenses. The low mass ions formed by the matrix (i.e., the fastest moving ions) are often abundant and they may saturate and temporarily degrade the detection system, but they provide little useful information. A timed ion selector can be used to deflect these out of the beam so that no ions are detected or recorded below a preset mass.

Mass Spectrometers Optimized for ESI

From both a semantic and mechanistic point of view, the classification of ESI as an ionization method for MS is questionable. More accurately, it is a means of isolating preexisting ions in solution and transferring them to the vacuum system of the mass spectrometer. Because it merely requires the evaporation of the solvent from droplets of liquid that are sprayed from a needle at atmospheric pressure, there is no requirement to impart significant amounts of internal energy to the ions. Evaporative cooling of the shrinking droplets is compensated for by gentle collisions with the high-pressure gas, which in some source designs is heated. However, as the ions are drawn through the mass spectrometer interface into regions of lower pressure, they are accelerated by potentials that may cause more energetic collisions with the remaining air molecules, converting kinetic energy into vibrational energy. The exploitation of this feature for structural analysis and compound class identification is described below.

Various forms of ESI allow operation over a wide range of flow rates, from a few nanoliters per minute up to hundreds of microliters per minute. The sample solution may be delivered via a syringe pump connected directly to a needle through which the liquid is pumped. Alternatively, the samples may be delivered directly in the eluent from an HPLC or capillary zone electrophoresis (CZE). For a solution containing a fixed concentration of analyte, the observed ion currents are largely independent of flow rate (Cole, 2000). Most biomolecular analyses are carried out with limited amounts of sample, so lower flow rates are favored, for example, nanobore HPLC interfaced directly with ESI requires flow rates of approximately 100–300 nl/min (Moseley, 2000). A version of ESI known as *nanospray* that is optimized for very small amounts of sample uses flow rates as low as 20–50 nl/min (Wilm and Mann, 1994). Here, 1–3 μl of sample solution is introduced into a quartz capillary, one end of which has been drawn out to a very narrow tip. The application of a high potential to the tip and a small positive pressure of gas from the other end causes the liquid to spray from the tip. Using micromanipulators and video cameras, the operator adjusts the position of the tip with respect to the entrance aperture of the mass spectrometer. The flow rate cannot be controlled directly, but with experience, an operator may obtain good-quality spectra from 1 μl of sample for as long as 1 h.

ESI is carried out at atmospheric pressure, and the ions are then transferred into the vacuum system, an arrangement that is better suited to certain instrument types than others. When it was first introduced, ESI was most often used in conjunction with a quadrupole mass analyzer. Quadrupole field mass separation technology, first developed in the 1950s by Wolfgang Paul, was well established, and single and triple quadrupole instruments were in widespread use. ESI has also become the commonly used ionization method for the 3D quadrupole ion trap (QIT), although MALDI has been employed as an ionization method for the QIT (Krutchinsky *et al.*, 2001) and the hybrid quadrupole-orthogonal acceleration TOF, described later in this chapter, has employed MALDI ionization for protein discovery, identification, and structural analysis (Baldwin *et al.*, 2001). Compared with a sector mass spectrometer, an advantage of both the linear quadrupole and the QIT for an atmospheric pressure ionization method is the low ion kinetic energy requirement; thus, there are no high accelerating voltages. Sector instruments use high voltages and high vacuum, both of which introduce complications in interfacing with spraying liquids at atmospheric pressure. Although quadrupoles were originally developed for analysis of small organic molecules and the m/z range was usually limited to about 2000, an upper limit of 3000–4000 is readily achievable, which is adequate even for ESI of proteins, as proteins sprayed from acid solution commonly give a distribution of multiply charged ions within the m/z range of about 700–2000. Despite the convenience and simplicity of linear quadrupoles, the resolution of such instruments is low. QITs can give superior resolution over a narrow mass range, but when these instruments are scanned over their full mass range, the resolution is also limited.

Much higher resolution can be achieved with a modified version of TOF analysis. In a conventional MALDI-TOF instruments, the ions generated from a static solid sample within the vacuum system are accelerated into the flight tube within nanoseconds of their formation. For an alternative ionization method such as ESI, ions are generated externally and transferred into the vacuum system in a procedure that takes milliseconds. If these ions were to be accelerated along the central axis of the ion source by application of a pulsed potential, the new velocity would be added to their previous component of velocity in the same direction. Furthermore, at the moment of acceleration, the spatial distribution of ions along this axis would further broaden the distribution of arrival times for ions of a given mass, degrading the RP and making accurate mass determination impossible. However, the ion beam from an external source such as ESI can be focused and constrained such that it has very little component of motion in any direction orthogonal to its primary axis. At a point along the ion path, a section of the beam passes between two parallel electrodes, one being solid (the pusher) and the other a grid (the puller). The application of a pulsed potential between these two electrodes accelerates the ions orthogonally into a reflectron TOF (Dawson and Guilhaus, 1989; Dodonov *et al.*, 1991; Guilhaus *et al.*, 2000; Verentchikov *et al.*, 1994). Although the accelerating voltage is typically much lower than in a MALDI-TOF (e.g., 4 kV compared with 20 kV), the resolution and mass measurement accuracy are comparable. To achieve high sensitivity in an oaTOF, it is necessary to maximize the fraction of the ions pulsed into the TOF. This requires that the ion velocity into the orthogonal acceleration region be matched to the duty cycle of the ion pulses, ensuring that most ions are pulsed into the TOF and minimizing any fraction that passes through the accelerator without experiencing acceleration. This is difficult to achieve for the whole m/z range because ion velocities in the beam entering the accelerator are likely to vary inversely with mass. The maximum pulsing rate is limited by the flight time of the slowest ions through the TOF (highest m/z), otherwise, the fastest ions from the subsequent pulse (lowest m/z) might overtake them before they reach the detector. However, ion trapping before pulsing into the oaTOF can increase the sensitivity for the low mass ions without compromising the performance for the high mass ions (Chernushevich, 2000). Typically, an oaTOF is operated with a repetition rate of about 10 kHz.

ESI of analyte molecules that contain multiple basic sites gives multiply charged ions, so molecular weights are not revealed directly. However, there is usually a distribution of charge states giving rise to a series of ions in the mass spectrum. Because these are spaced nonlinearly, the charge states can be determined by inspection and the molecular weight can be calculated (Covey *et al.*, 1988). All manufacturers of ESI instruments provide software that does this automatically by a process known as *deconvolution,* which displays a profile of the molecular species present. This is shown in Fig. 10 for a recombinant insulin-like growth factor (IGF)–binding protein (IGFBP4). In this case, the deconvoluted spectrum reveals the presence of two separate species, the lower mass component being within 0.4 Da of the calculated molecular weight for IGFBP4 of 25,735.3 Da. The higher mass component is 16 Da heavier, most likely attributable

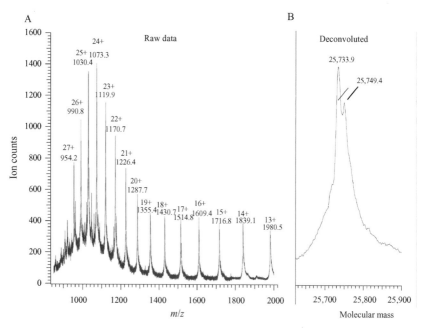

FIG. 10. The electrospray ionization mass spectrum of a recombinant protein, insulin-like growth factor–binding protein-4.

to partial oxidation of methionine to methionine sulfoxide. Note that in this example the isotopic peaks are unresolved, which is typical of most mass spectra of proteins, with Fourier transform MS (FTMS) being the only technique able to resolve such species.

Ionization by ESI is very soft and does not usually result in fragment ions, but as mentioned earlier, suitable strategies to increase internal energy to induce fragmentation are possible. Within an intermediate pressure region of the ESI interface is a plate with an aperture commonly referred to as the *skimmer* or *nozzle*. A voltage applied to this plate can be used to increase the ion kinetic energy. Collisions with neutral gas molecules result in a fraction of the kinetic energy being converted to vibronic energy, whereby otherwise stable molecular ions may be energized and may undergo unimolecular dissociation. Thus, in an HPLC-MS experiment, it is possible to use a single quadrupole to observe fragment ions that are characteristic of particular compound classes. For example, by monitoring m/z 79 for PO_3^- anions throughout the HPLC separation, phosphopeptides may be identified in a protein digest (Carr *et al.*, 1996; Huddleston *et al.*. 1993a). Similarly, sugar oxonium ions such as m/z 204 for N-acetylhexosamine may be monitored to characterize glycopeptides (Huddleston *et al.*, 1993b). More sophisticated methods for the study of fragment ions providing substantially superior performance have been developed, whatever ionization method is adopted. These are considered later in the separate section on tandem MS.

Biophysical Studies by ESI-MS

Because the multiply charged ions typical of ESI can be separated by an analyzer designed for ions of relatively low m/z, it is not necessary to employ a "high-mass" instrument. An exception to this general rule is the study of large protein complexes under native conditions by ESI, when relatively few sites are accessible for proton attachment; consequently, the required m/z range is significantly higher and may substantially exceed 10,000. The softness of ESI is a major advantage when studying noncovalent complexes, because these can be quite labile. Although the typical ESI conditions are strongly denaturing, it is possible to use pH-neutral solutions and to include some salts and buffers in low concentration, although these are likely to degrade the performance of the instrument and necessitate frequent cleaning of the ESI interface. It is advantageous to use volatile buffers if possible, such as ammonium formate or bicarbonate, depending on the pH level required.

Using a mass spectrometer to monitor complexes formed in solution requires that they can be transferred to the solvent-free vacuum without dissociation. Non-covalent interactions can be subdivided into those that may be enhanced by the removal of the dielectric, water, and those that will be weakened (Loo, 1997). Hydrophobic interactions are stabilized within an ionic environment (i.e., in aqueous salt solution) and thus are generally the most labile in the solvent-free vacuum of the mass spectrometer. In contrast, interactions between metal ions and peptides or proteins are electrostatic in nature and are strengthened upon removal of water. This makes the study of such interactions relatively straightforward, although the occurrence of nonspecific interactions of no biological relevance must be minimized by the use of suitable controls. It has also been demonstrated that some very large protein complexes can be stabilized in the gas phase by the presence of multiple solution-derived small molecules and ions (Sobott et al., 2005).

Hydrogen–deuterium exchange reactions occurring in solution and monitored by ESI-MS can be the basis for other biophysical studies. Of the many hydrogen atoms in a protein, those that are bonded to carbon are generally stable, whereas OH, NH, and SH bonds are polar and can exchange in aqueous solution. However, the amide NH groups involved in maintaining secondary structure are more stable and less prone to exchange. For example, the geometry of the α-helix requires each backbone carbonyl group to form a hydrogen bond with the spatially adjacent NH group of the amino acid on the next turn of the helix, four residues toward the C-terminus. Such effects are revealed by dissolving the protein in D_2O rather than H_2O, the rate at which deuterium exchanges with the hydrogen atoms in the amide bonds of the protein backbone being indicative of secondary structure maintained by hydrogen bonding. Furthermore, deeply buried hydrophobic sections of a protein are less accessible to water, so the rate of H/D exchange in such regions will be slower than for solvent-exposed regions. Consequently, increased or decreased rates of isotopic exchange may correlate with tertiary structural differences.

Using nuclear magnetic resonance (NMR) imaging, pioneering work in this area has been carried out by Walter Englander for more than 20 years, and this method continues to enjoy widespread use (Wand and Englander, 1996). However, an alternative is to employ MS to observe changes in molecular mass caused by the introduction of the heavy isotope. Not only can this technique monitor changes in the molecular weight of the intact protein, but the protein can also be rapidly digested to smaller peptides that are then analyzed by MS, usually using HPLC-MS, and possibly tandem MS. HPLC-MS is viable because H/D exchange is catalyzed by hydrogen ions at low pH (pH < 2) but by hydroxide ions at higher pH levels (>3). Thus, for the pH region most likely to be of interest in protein folding, pH 3–8, the rate of H/D exchange increases by one order of magnitude per pH unit. For example, the rate of exchange in an experiment conducted at pH 7 can be slowed by a factor of 10,000 by reducing the pH to 3, which is close to optimum for digestion by the nonspecific protease pepsin and is the pH used for reversed phase HPLC separations. The rate may be reduced further if the temperature is lowered from ambient to close to 0°. Thus, further exchange is effectively "frozen" as soon as the pH and temperature are reduced. By digesting the protein into relatively small peptides, it is possible to isolate and identify the regions that show the slowest exchange and thereby determine those residues involved in maintaining the structure (Smith et al., 1997).

ESI-MS requires only extremely small amounts of material for such experiments (e.g., a few microliters of micromolar solution), whereas NMR typically requires 0.5 ml of millimolar solution. Other advantages over NMR include the relatively simple assignment of peptide masses to defined regions of the sequence, whereas NMR requires a detailed assignment of each of the amide protons. Further, the mass spectrometric method allows experiments to be carried out rapidly over a wide range of pH levels. In a refinement of the mass spectrometric method, it is possible to use collision-induced dissociation (CID) and tandem MS to define the degree of exchange at individual residues within a peptide (Deng et al., 1999).

The Need to Obtain Structural Information: Tandem Mass Spectrometry

Early success with protein identification by automated database searching was based on mass mapping of peptides derived by tryptic digestion of the protein, usually separated by 2D sodium dodecylsulfate (SDS) polyacrylamide gel electrophoresis (PAGE) and analyzed by MALDI-TOF. In 1993, several groups independently demonstrated that a MALDI mass map of the peptides in an unseparated protein digest could correctly identify a protein that was present in a database, even when only a subset of the peptides was ionized and detected. Because every protein can potentially give a unique pattern of peptides, this technique is often referred to as *peptide mass fingerprinting* (PMF). However, the uniqueness of the match depends on the fraction of potential peptides that is actually observed and the accuracy with which their masses can be measured. Because of the high level of activity in genomic sequencing and the completion of several genomes, databases have grown dramatically larger, enforcing a requirement for more accurate mass measurement to increase the selectivity and specificity of mass mapping. So now PMF is rarely adequate and peptide sequence determination is virtually essential. This is certainly the case for protein identification in complex mixtures when there are too many peptide signals and PMF returns an unacceptably high number of candidates; to obtain protein sequence when no match is found in a database; or to identify and localize posttranslational modifications. Sequence information can be obtained readily by tandem MS, frequently known as MS/MS, which is illustrated schematically in Fig. 11.

Conventional MS produces ions that are separated by m/z and analyzed directly. If a soft ionization method is used (i.e., one that causes minimal fragmentation), the mass spectrum will yield the molecular weight values of compounds present in the analyte but little or no structural information. In a tandem mass spectrometer, ions of a particular mass number selected by the first mass analyzer (MS1) can be subjected to collisions with neutral gas atoms or molecules that cause excitation of the ions, resulting in unimolecular decomposition. This process is known as CID. The ionic fragments are then separated in a second analyzer (MS2), yielding structural information such as the amino acid sequence of a peptide. Collision conditions can be defined for two different regimens of CID, so either the high-velocity ions typical of sector instruments and MALDI-TOF mostly undergo a single collision that converts sufficient kinetic energy into internal energy for reaction to occur or the slower ions typical of quadrupoles, QITs, and ion cyclotron resonance experience multiple collisions in a higher pressure gas cell, each of which transfers a smaller amount of energy but the aggregate effect of all collisions gives sufficient energy for reaction. It is also possible to transfer sufficient energy in the ionization process for a substantial fraction of ions to dissociate, even using a soft ionization technique such as MALDI, by using higher than normal laser power and using a relatively "hot" matrix such as α-CHCA. High-energy CID can

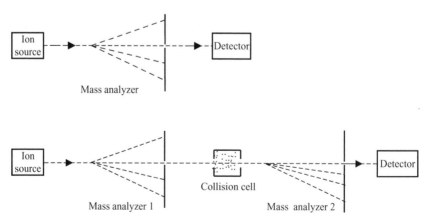

FIG. 11. The principles of (A) mass spectrometry and (B) tandem mass spectrometry.

also be achieved in a MALDI-TOF instrument equipped with a collision cell (Stimson *et al.*, 1997).

Although MALDI and ESI are described as soft ionization methods, each will yield ions with a distribution of internal energies. Those with the highest energies may dissociate before acceleration to appear as normal fragments in the spectrum or in the time spent between acceleration and analysis, in which case they are generally referred to as *metastable ions*. The initial internal energy also determines how much additional energy is needed from the collision events for the dissociation of otherwise stable ions to occur rapidly enough to yield fragment ions in the CID spectrum. It is instructive to consider the factors affecting unimolecular ion fragmentation in tandem MS, in which the selected parent or precursor ion dissociates to yield a new lighter ion with the loss of a neutral fragment. Such reactions are typical of all chemical processes in that they are controlled by kinetic and thermodynamic properties. To express this in its simplest form, the amount of internal energy of the primary ions, either with or without collisional excitation, determines whether they can undergo dissociative reactions within a certain time period, as defined by the geometry of the mass analyzer (Levsen, 1978). In the positive ion mode, soft ionization methods all form ions by cationization, giving inherently stable even-electron ions. This contrasts with the earlier method of ionization by electron impact that removed an electron from each molecule to form an odd-electron radical cation. Most of the reactions of odd-electron ions involve homolytic bond breaking, in which a neutral radical is lost to form an even electron fragment ion. This is illustrated in reaction 1 of Fig. 12, for a process that is sometimes described as α-cleavage (McLafferty and Turecek, 1993). (By convention, the movement of a single electron is indicated by a fishhook arrow.) For those ions possessing sufficient internal energy, these reactions that involve the dissociation of a single bond are rapid. An alternative group of reactions involves the breaking of more than one bond concurrent with the making of new bonds, thereby causing a rearrangement in which a neutral molecule is eliminated, as in reaction 2. An example of such a reaction that proceeds via a six-membered transition state is the McLafferty rearrangement, reaction 2b. The energy requirements for such reactions are often lower, but the entropic demands are higher, so the reactions are slower and may be seen to give metastable ions rather than conventional fragment ions. Because soft ionization gives even electron ions, homolytic bond dissociation is unfavorable and most reactions of ions formed by ESI or MALDI involve heterolytic processes, frequently accompanied by hydrogen rearrangements, as shown in reaction 3. Unless the ions gain excess internal energy by a mechanism such as collisional activation, these tend to be slow processes. The rates of fast

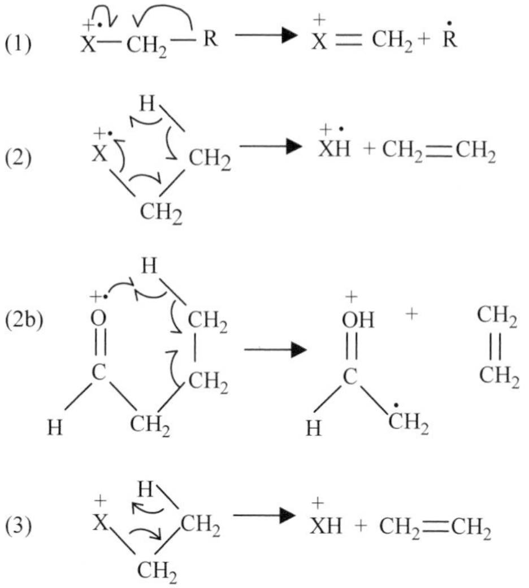

Fig. 12. Some reaction schemes for the unimolecular dissociation of ions.

mass spectrometric reactions have been analyzed using the technique of field ionization kinetics (FIK). Field ionization occurs in the vacuum in close proximity to a very sharp point or edge of a metal emitter, and FIK enables the study of fragmentation reactions that occur within the potential gradient derived from such an emitter (Derrick, 1995).

The maximum amount of kinetic energy that can be converted to internal energy in a single collision is defined by the center of mass energy (E_{CM}). Assuming the velocity of the target gas atom or molecule is negligible, the E_{CM} depends on the original ion kinetic energy (KE) and the masses of the ion (m_i) and neutral target (m_n).

$$E_{CM} = KE.m_n/(m_i + m_n)$$

Thus, the mass of the neutral target plays a major role (e.g., for an ion of mass 1000 and KE of 1 keV): A neutral of mass 4 (helium) gives a theoretical maximum E_{CM} of ~4 eV, whereas mass 40 (argon) gives ~40 eV. Consequently, the nature of the dissociation reactions will be affected significantly by the choice of neutral gas, and any dissociation that requires higher internal energy will benefit from a heavier target gas. In practice, the average amount of kinetic energy converted to internal energy will be less than the theoretical maximum, a phenomenon known as *inelastic scattering*, so a distribution of energies will be transferred by the collisions. Note also that conversion of kinetic energy to internal energy will result in the incident ions traveling more slowly after collisions. For ions that undergo CID, this will be seen as energy loss for the fragments, whereas fragment ions formed by metastable dissociation will not experience this energy loss. Ion fragmentation may also be accompanied by conversion of internal energy into kinetic energy, giving rise to a broadening of the kinetic energy distribution for the fragment ions, which will apply to both metastable and CID fragments. This phenomenon was well documented for sector instruments in experiments such as mass analyzed ion kinetic energy spectroscopy (MIKES), in which energy loss apparently shifts the peaks to lower mass as slower ions experience greater deflection, a phenomenon that correlates with the energy requirements of the dissociation reactions (Vachet *et al.*, 1996). Slowing the ions down in a linear TOF would appear to make the ions heavier, but in theory a reflectron should compensate for this and the net effect is likely to be modest.

As stated earlier, collisions between ions and neutrals cause excitation of the ions with a distribution of elevated internal energies. The higher the average internal energy, the greater the number of channels that will be available for dissociation reactions, many possible reactions being accessible only to ions that are activated by high KE collisions. The differences between high- and low-energy CID of peptides are detailed elsewhere in this volume (see Chapter 7). Briefly, fragmentation at an amide (or "peptide") bond giving a *b* ion (N-terminal fragment) or a *y* ion (C-terminal fragment) is a relatively low-energy process involving hydrogen rearrangements such as shown in reaction 3. Higher energy processes lead to cleavage elsewhere along the backbone, for example, between an α-C and a carbonyl-C to give an *a* ion or simultaneous backbone and side-chain cleavage to give *d* or *w* ions that can distinguish leucine from isoleucine, or formation of the low mass immonium ions that can characterize the amino acids present. Thus, metastable ion spectra and typical low-energy CID spectra show sequence information in the form of *b* and *y* series ions, but they are devoid of some of the more structurally informative ions. Also, any particularly favorable amide cleavage such as at the N-terminus of a proline residue, where the secondary amino nitrogen can stabilize the positive charge, may dominate the unimolecular spectrum at the expense of other sequence ions. Consequently, if sufficient fragmentation is desired to derive the sequence from first principles, high-energy CID may be essential. On the other hand, if fragment ions are being used to search the database for a peptide sequence "tag" (Mann and Wilm, 1994), which is a series of fragment peaks that defines a sequence of amino acids within a peptide, low-energy CID will often suffice.

Previously, there were very few instruments other than multiple sectors capable of providing high-energy CID spectra, whereas there are numerous alternatives for low-energy CID, mostly using ESI interfaced to the triple quadrupole, the QIT, the quadrupole interfaced to an oaTOF, or the FT-ion cyclotron resonance (FT-ICR) spectrometer. QITs in particular have achieved wide acceptance for ESI with HPLC-MS and HPLC-MS/MS, taking advantage of their modest cost, fast scanning, ease of computer control, and tandem MS and MS^n capabilities. However, recent developments have blurred the differentiation between instruments optimized for MALDI and ESI. MALDI ionization at relatively high pressure (0.01–760 Torr) and low accelerating voltage (5–10 V) gives ions that become thermally equilibrated before being drawn into the higher vacuum region required for any mass analyzer. Thus, the pulse nature of the ions illustrated in Fig. 6B becomes smeared out and more equivalent to a continuous stream of ions. The overall result is that the ESI and MALDI methods have become more interchangeable, and a number of instruments previously associated mainly with ESI are also available with MALDI as an option (Baldwin *et al.*, 2001; Krutchinsky *et al.*, 1998, 2001; Laiko *et al.*, 2000; O'Connor *et al.*, 2002).

Quadrupoles and QITs

Both QITs and linear quadrupoles constrain ions to circular orbits under the influence of combined direct current (DC) and radiofrequency (RF) potentials (see Chapters 2 and 4) (Todd and March, 1995). For any combination of DC and RF fields, only ions of a selected m/z follow stable orbits. The fundamental difference is that ions in a 3D QIT have virtually no component of linear translation, whereas in a linear quadrupole made up of four parallel rods, ions travel along spiral paths, emerging to be detected or diverted into a further mass analyzer. Figure 13A illustrates a simple linear single stage quadrupole analyzer with a channeltron detector. Here, the four rods are cylindrical, although the theoretical optimum cross-section is hyperbolic. The detector is shown mounted on the central axis of the analyzer, but improved signal/noise ratios can be achieved by deflecting the ions into an offset detector. There is virtually no mass selectivity in a linear quadrupole if the DC voltage is set to zero, but the RF voltage constrains ions to spiral orbits within the ion beam, so an RF-only quadrupole can make a very effective collision cell. This is used in the triple quadrupole and in a hybrid tandem mass spectrometer that combines a linear quadrupole for MS1, an RF-only collision cell, and an oaTOF for MS2, referred to here as a *QqTOF*.

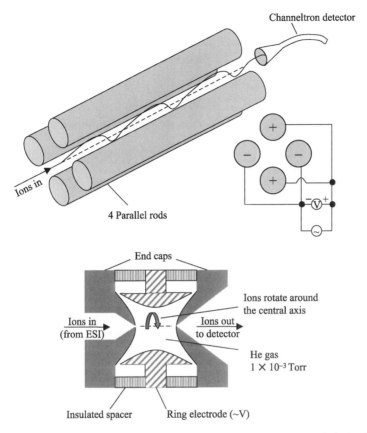

FIG. 13. Two alternative designs of quadrupole mass spectrometers: (A) the linear quadrupole and (B) the quadrupole ion trap.

Although QITs were first described about the same time as linear quadrupole analyzers, for many years their commercial development lagged behind. Both linear quadrupoles and 3D QITs are relatively simple, compact, and potentially inexpensive devices that are easily controlled by computer. However, only in the last decade have QITs come to enjoy wide popularity. Now, thanks in part to a vigorous program of improvement and the realization that the QIT could be more versatile than either the linear quadrupole or the triple quadrupole, they are in widespread use for a variety of analytical applications. As seen in Fig. 13B, the 3D QIT has a doughnut-shaped ring electrode to which RF and DC voltages are applied, as well as two end-caps. Ions are usually formed externally and pulsed in through an aperture in one end-cap, being constrained from further linear motion by DC potentials applied to the end-caps. They can also be pulsed out through the opposite end-cap to be detected externally. By application of appropriate voltages, ions of a single m/z can be stored within the trap, where they can be subjected to processes such as multiple steps of CID, often referred to as MS^n. Thus, despite having only a single analyzer of modest proportions and cost, the QIT can be an extremely effective tandem mass spectrometer (Hao and March, 2001). The most common application of the ESI-QIT is as a mass selective detector for HPLC-MS and MS/MS, for which it was a pioneer in the development of automated CID analyses, often described as *data-dependent experiments* (DDEs) (Tiller *et al.*, 1998). The simple and robust nature of the QIT makes it ideal for unattended automated operation. There are a number of variations of this, but generally, in DDE mode the instrument carries out a survey scan of the peptides eluting from the HPLC and then selects the strongest peaks for CID. After each CID experiment, the next survey scan selects new target peptides. If the same peptides as before continue to give strong signals in adjacent scans, they are excluded from repetitive analysis in favor of weaker peaks. Thus, this ensures that the maximum possible number of peptides is sampled, potentially covering a wide dynamic range. Although the QIT enables such experiments to be performed at modest cost compared with the QqTOF, its performance is inferior. Most peptides ionized by ESI give multiply charged ions, the charge states of which can be determined only if the RP of the instrument is sufficient to separate the naturally occurring isotopes. Unlike the QqTOF, the 3D QIT is unable to achieve this in full-scan mode. The mass range for CID is also limited, low mass ions including immonium ions and internal dipeptides and tripeptides not being observed. Although it is possible for proteins to be identified in a database search for sequence tags based only on ions within the mid and upper mass ranges, the lower mass ions are valuable for peptide sequencing. Note that DDE techniques are equally applicable to much higher performance tandem mass spectrometers, potentially yielding peptide spectra suitable for *de novo* sequencing (Lingjun *et al.*, 2001).

A new addition to the QIT family is the linear 2D ion trap, which exploits elements of both the linear quadrupole and the 3D trap. A hybrid tandem version, the Q-trap, employs a linear quadrupole as MS1, an RF-only quadrupole as the collision cell, and the linear ion trap as MS2. Compared with the 3D QIT, this offers certain performance advantages, including less susceptibility to space charge effects and the ability to display CID fragment ions throughout the entire m/z range.

The QqTOF

Although the oaTOF has been employed as the sole mass analyzer in a number of successful home-built and commercial mass spectrometers, probably its true strength is as the second analyzer in a hybrid tandem mass spectrometer. A sector instrument with an oaTOF designed for ESI or MALDI operation allowed high-resolution separation in both MS1 and MS2 (Bateman *et al.*, 1995; Medzihradszky *et al.*, 1996), but the reduced sensitivity imposed by the narrow acceptance angle of the sector analyzers and technical limitations of combining ESI and MALDI with a sector instrument prevented this technique from achieving more general acceptance. However, the use of a quadrupole mass analyzer as MS1 and an RF-only quadrupole collision cell in combination with an oaTOF as MS2 presents a very powerful combination in terms of sensitivity, resolution, and mass range (see Chapter 2). Illustrated in Fig. 14, the QqTOF is considered by many users to be the instrument of choice for

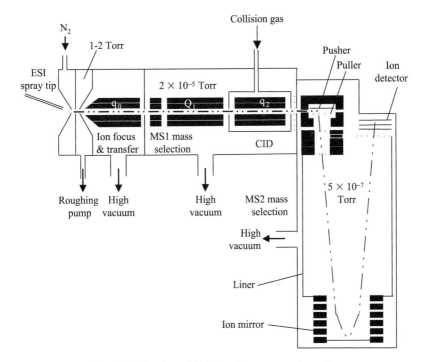

FIG. 14. A Qq-time-of-flight tandem mass spectrometer.

high-performance HPLC-MS/MS, giving resolution in the region of 10,000, mass range in MS mode of 12,000 or more (RF-only quadrupole), and mass measurement accuracy of 10 ppm or better (Morris *et al.*, 1996; Shevchenko *et al.*, 1997). For MS/MS operation, the mass range is limited by the maximum m/z selectable by the quadrupole, typically 3000, but because most CID experiments for biopolymer characterization are carried out on smaller fragments, such as peptides from a digested protein, this is generally adequate. As described earlier, to improve the duty cycle and, therefore, the sensitivity of the QqTOF for low mass ions by up to 20 times, one manufacturer has introduced a degree of mass-dependent trapping immediately before the accelerator. Ions are slowed down in a mass-selective fashion and then pulsed into the accelerator region immediately before the accelerating voltage pulse is applied (Chernushevich, 2000). A further innovation with some QqTOF instruments is the use of multiple reflections in the oaTOF, which is part of a more general application of multiple reflection MS (Casares *et al.*, 2001). Instead of collecting the ions after their first reflection, they are reflected back into the TOF for a second pass, each ion following a W-path rather than the V-path of the conventional oaTOF. This gives superior resolution of about 20,000 that starts to make it competitive with FT-ICR, though with an inevitable reduction in sensitivity compared with single-pass operation.

FT-ICR

Of the available commercial mass spectrometers, the FT-ICR is capable of the highest performance in terms of resolution and mass accuracy (see Chapter 4). It has been claimed that a mass measurement on a single peptide can give an accurate mass tag that is sufficient to identify a protein in a database (Conrads *et al.*, 2000), although there have been few demonstrated examples of this approach. For protein analysis, the performance specifications of most high-performance tandem mass spectrometers such as the QqTOF are relevant only to the analysis of fragments (peptides) but not to intact proteins. By contrast, the FT-ICR is capable of isotopic resolution of intact proteins. Ion activation can be induced by infrared laser radiation, resulting in multiphoton dissociation, or by pulsed admission of CID gases. However, it is difficult to introduce sufficient energy through collisions of large ions with light gases to achieve significant fragmentation and only the lowest energy pathways are populated, so conventional CID on large proteins is relatively uninformative. Electron capture–induced dissociation (ECD) was introduced to provide more comprehensive sequencing of small to mid-sized proteins, possibly with automated computer interpretation (Horn *et al.*, 2000). This relied on capture of low-energy electrons by multiply charged ions, which then fragment to give c-type ions (i.e., C-terminal fragment ions formed by backbone cleavage between an amide nitrogen and the adjacent α-carbon). Although initial studies employed thermal energy electrons, higher energy electrons (11 eV) produce radical cations such as z-type ions, further fragmentation of which gives w ions that, for example, can distinguish the isomeric amino acid residues of leucine and isoleucine (Kjeldsen et al., 2003). There are problems inherent in studies on larger ions, such as the difficulty of correctly assigning the monoisotopic mass due to the low abundance of this ion. Nevertheless, protein identification and substantial sequence information can be obtained in what have come to be known as "top-down" experiments on multiply charged intact protein ions (Sze et al., 2002), compared with the more commonly used "bottom-up" approach based on proteolysis and peptide analysis.

Although the FT-ICR spectrometer can itself be used for MS^n experiments, ions are generally prepared externally and pulsed in via another ion transmission device. This may another mass analyzer, such as a quadrupole, giving an extra dimension of tandem MS (Lingjun et al., 2001). Remote ionization allows different types of ionization techniques to be employed, but ESI is clearly the most popular. Although FT-ICR provides unique opportunities to manipulate ions in MS^n experiments, the electronic pulse sequences are quite complex to set up and operate, and until recently it had not been established that the FT-ICR could be operated successfully in a routine analytical environment without highly experienced and skilled operators. Thus, it went against one trend in MS that seeks to make sophisticated instrumentation more readily accessible to nonspecialist operators, an increasingly important aim considering the mushrooming demand for MS compared with the small number of people being trained in this area. However, the Thermo Finnigan LTQ FTMS is a new hybrid instrument that combines a linear ion trap with an FT-ICR. As this comes from a company that has supplied large numbers of ion traps with software noteworthy for its ease of use, this instrument with a specification of 2 ppm mass accuracy with external calibration and 500,000 RP may do much to popularize FT-ICR.

High-Energy MALDI Tandem Mass Spectrometry

Despite a wide choice of instrument types and substantial differences in their performance characteristics, all the instruments referred to in this chapter give only low-energy CID. By contrast, the ion kinetic energies in conventional TOFs are in the kiloelectron volt range and should be ideal for high-energy CID. Furthermore, the MALDI process imparts more energy than ESI and may give ions that have sufficient internal energy to undergo unimolecular decomposition without further collisional activation. Ions that decompose by unimolecular processes may give rise to "prompt" fragments and metastable ions. For prompt fragments, unimolecular decomposition occurs before acceleration (i.e., within the ~100 ns residence time in the DE source), and the fragment ions give peaks at the normal fragment m/z values. The metastable ions that dissociate later (i.e., after acceleration) will give fragment ions having the same velocity as the precursor ions and will arrive at the detector at the same time in a linear TOF but will be filtered out in a reflectron TOF. As a result of dissociation occurring during acceleration giving fragments with somewhat reduced velocities, some fragment ions are characterized by broader peaks occurring at higher m/z values than the expected fragment mass.

The sector oaTOF, referred to earlier in this chapter, gave high-resolution tandem mass spectra but with the sensitivity limit of 1 pmol. Until recently, post-source decay (PSD) with a reflectron TOF instrument was the only other widely available sequencing technique employing MALDI (Spengler *et al.*, 1992). This involves reducing the reflectron voltage in a series of steps to allow fragment ions formed after acceleration and, therefore, having reduced kinetic energies to be analyzed; thus, it is a means of studying what were described earlier as *metastable ions.* For such experiments, the laser is operated at higher intensity than usual to increase the internal energy deposition and thereby increase the fraction of ions having sufficient energy to dissociate within the time available, typically a few microseconds. This rather cumbersome method that gives poor resolution and low mass accuracy is becoming superseded by new MALDI-TOF instruments, including a tandem TOF/TOF design that gives high-energy CID spectra.

A MALDI-TOF/TOF based on a conventional high-performance DE-reflectron TOF has been described that can be operated in MS or MS/MS modes (see Chapter 3) (Medzihradszky *et al.*, 2000; Vestal *et al.*, 2000). This is illustrated in Fig. 15. If ionization is achieved with a high-frequency YAG laser operating at 200 Hz rather than the conventional nitrogen laser with a maximum frequency of about 20 Hz, spectral data can be acquired very rapidly. A typical YAG-laser wavelength is 354 nm rather than 337 nm, which has little effect on the spectra, although the laser power required for DHB matrix is greater because of low absorption at this wavelength. The resolution of the precursor ions in MS/MS mode (typically 0.5%) is defined by the opening and closing of a timed ion selector. The ions are then decelerated (e.g., to 1 keV) before entering an electrically floating collision chamber. The chamber contains a collision gas at a pressure designed to maximize the number of single collisions between ions and molecules but

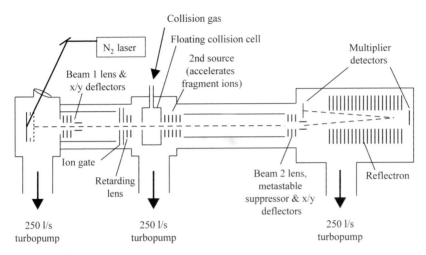

Fig. 15. A time-of-flight (TOF)-TOF tandem mass spectrometer.

to minimize the number of ions experiencing multiple collisions. After traversing the collision cell, any remaining precursor ions and the ionic products of dissociation reactions traveling at the same velocity are accelerated by a high-voltage pulse into MS2, which is a reflectron TOF. Unlike PSD, this results in all fragment ions having almost the same high kinetic energy, within ± 0.5 keV, allowing the whole mass range to be well focused by the reflectron. Consequently, the resolution of the fragment ion spectra is comparable to the separation of normal ions in a typical reflectron TOF. This instrument can be used with a range of collision gases and with different collision energies, although raising the collision energy results in a greater spread of final kinetic energies that may cause some loss of resolution in MS2.

For automated protein identification by the analysis of tryptic digests, the throughput of such an instrument is dramatically enhanced compared with manual operation of a more conventional tandem mass spectrometer. With a laser operating at 200 Hz and assuming 500 laser shots are sufficient to give a mass spectrum with good signal/noise ratios, an entire 96-spot plate can be scanned in 4 min. With real-time database searching, the majority of the proteins that have been digested to give multiple peptides can be identified by peptide mass fingerprinting during this time. The mass spectrometer can then be directed to carry out the further analysis of each spot by CID, first by making measurements on selected ions to confirm the previous assignments, and then to study peptides not assigned by mass mapping. The number of shots required per peptide and the total number of peptides to be studied will determine the time required for a complete analysis of all 96 spots, but it is likely to be completed within an hour or two. The provision of a multiplate cassette to hold 24 or more target plates enables unattended operation for as much as 24 h at a time. Thus, the comprehensive analysis of several thousand samples per day with a single instrument is becoming a realistic undertaking (Huang *et al.*, 2002).

Ion Detection

In an instrument that produces and transmits a continuous ion beam, the ions arriving at a detector represent an electrical current. This current can be amplified and recorded (e.g., as a function of m/z in a conventional scan or at selected m/z values in a multiple ion monitoring experiment). The first stage of amplification is usually achieved with an electron multiplier having either discrete dynodes, as in Fig. 16, or a continuous dynode. The principle is based on emission of secondary electrons caused by an ion or an electron striking a metal surface. The multiplier illustrated is set up for positive ion operation as the first dynode is held at a negative potential attractive to positive ions. Successive dynodes are then progressively less negative, so the emitted electrons are accelerated between them. The number of secondary electrons emitted per event depends on the bombarding velocity at each dynode, which is controlled by the potential between the dynodes. Rather than the small number of dynodes illustrated here, in reality a discrete dynode multiplier may have 20 or more. If every impact

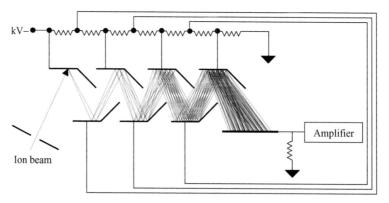

Fig. 16. The design and operating principles of a discrete dynode electron multiplier operating in positive ion mode.

releases an average of two electrons, the overall gain for a 20-dynode multiplier will be 2^{20} (i.e., $\sim 10^6$). A small increase in the voltage between the dynodes that increases the yield to an average of 2.25 electrons/impact will increase the gain to 2.25^{20} (i.e., 10^7). The typical electron energy required to achieve this multiplication factor is about 100 eV, so a 20-dynode multiplier might be operated with an overall potential drop of 2 kV. Because there is a maximum output current that a multiplier can provide, the overall gain is adjusted to prevent this from being exceeded for the most abundant ions in the spectrum, so the output current is proportional to the ion current. In this mode the response time or bandwidth of the electronic amplifier is selected to give a continuous current that does not reflect the pulsed arrival of the ions at the detector. This conventional approach to ion detection and measurement in MS is best suited to ion currents approximately within the range 1×10^{-12} to 1×10^{-16} A, but it is generally unsuitable for very weak or intermittent ion currents or for experiments in which the m/z values are changing very rapidly, such as in a TOF mass spectrometer in which an entire spectrum may be obtained within about 100 μs.

The alternative continuous dynode multiplier or channeltron made of glass is less expensive to produce and is generally smaller, so its use was pioneered in simpler and more compact instruments such as the quadrupole, as shown in Fig. 13A. A number of designs have been explored, but all involve a hollow tube, usually curved, coated on the inside with a high-resistance material, down which the electrons travel. A voltage gradient down the tube causes successive sections to act like the discrete dynodes in Fig. 16. The gain is dependent on the ratio of length to width. In general, the gain and lifetime of such a multiplier are likely to be less than those of the discrete dynode variety. It is also possible to have a 2D array of channels in a flat glass plate where the output electrons fall onto a phosphorescent screen and the signal is detected by a photodiode array. This is referred to as *multichannel plate* (mcp) detector. In an instrument that gives a spatial separation of the ions of different m/z, such as a magnetic sector, an mcp can be mounted at the focal plane of the analyzer. The position at which an ion falls on the mcp will define its m/z, so the signal from such a detector can give a direct representation of the spectrum over a limited mass range. The gain of a single plate is lower than the gain of a conventional multiplier, but the plates can be stacked so that electrons emerging from the first plate can be multiplied again by a second plate. The spacing of the channels within the plate, which is typically 5–25 μm, defines the maximum resolution that can be achieved. The advantage of such an arrangement is that ions of a range of m/z values are detected simultaneously, increasing the sensitivity compared with scanning the ion current across a slit.

In most mass spectrometers, the multiplier/detector system acts as a transient recorder, the output current varying with the ion current. For experiments that produce lower ion currents, it is possible to monitor the arrival at the detector of individual ions, counting the number of ions arriving within a certain time window. Here, the electron multiplier is operated close to saturation, each ion falling on the detector giving an isolated pulse that can be detected and stored. For example, in the MDS-Sciex QSTAR instrument, the ions fall onto any one of four anodes of a four-anode multichannel plate detector with a time-to-digital converter capable of detecting single ions (Chernushevich et al., 2001). Such a detector can record only one ion striking an individual anode within the counting period, but the presence of four anodes decreases the likelihood of two or more ions arriving at a single anode close enough in time to give only a single pulse. However, high ion currents that result in multiple ions striking a detector within the counting period will cause the stronger peaks to be truncated and distorted. The width of the detector pulse may also exceed the counting period, giving a dead-time after the arrival of an ion that can cause further signal distortion at high ion currents. The ion intensities are recorded as "counts" for ions arriving within each timing period, which in this instrument are 625-ps wide, and accumulated in "bins." The final mass spectrum is plotted as a histogram of the ion counts in each bin. Even at a resolution of about 10,000, each peak is several bins wide at low m/z values, the number of bins per peak increasing with mass as the square root of m/z. In normal MS mode the ion current is usually more than sufficient to determine the mass of each ion accurately, but it may be necessary to enhance the gain slightly for CID/MSMS experiments because the initial precursor ion current is likely to be distributed across many fragments. The threshold for ion detection is effectively adjusted by increasing the detector voltage for CID/MSMS, thereby increasing the average pulse height per ion. Thus, the "counts" recorded for a given peak do not necessarily constitute an absolute measure of the total number of ions of that mass transmitted and detected by the oaTOF.

MALDI-TOF is somewhat different as ion detection is intermediate between the analog techniques typical of quadrupoles and sector instruments and the pure pulse counting of oaTOFs. Relatively large numbers of ions are formed in isolated pulses in MALDI so the multiplier is not operated in saturated mode. Nevertheless, the response is nonlinear at higher ion currents as the multiplier is easily saturated by multiple ions arriving simultaneously (Westman et al., 1997). The output from the detector is approximately proportional to the number of ions arriving within a given time and is stored in a high-frequency digital oscilloscope. Thus, again the eventual output is a histogram of ion currents, which with a 500-MHz oscilloscope would be recorded every 2 ns. Ions of high m/z

values such as singly charged protein ions moving relatively slowly through the mass analyzer may not strike the first dynode of the multiplier with sufficient velocity to induce secondary electron emission. Thus, detection efficiency drops off at higher m/z values. The sensitivity is usually increased by the application of a conversion dynode held at a high potential of opposite polarity to the charge on the ions, toward which the ions are accelerated. This is referred to as *post-acceleration detection* and is essential for negative ion operation, because the first dynode of the electron multiplier is usually negative and would repel negative ions.

Only FT-ICR measures ion currents by a completely different method. At a fixed magnetic field, all ions of the same m/z value have the same cyclotron frequency and move together in a coherent ion packet. This cyclotron motion of ions is responsible for the induction of an electric current in coils surrounding the cell, m/z being determined from the cyclotron frequency. This is advantageous for weak currents because the signal can be monitored for longer times to improve the signal/noise ratio. Also the method is nondestructive, so the ions remain in the trap and are available for further experiments.

Conclusions

The properties of ESI and MALDI induced a quantum leap in biological MS, so that MS is now firmly established as an essential core technology in much biological research. When the last volume of *Methods in Enzymology* dealing with MS was published (McCloskey, 1990), ESI and MALDI were just emerging as the ionization methods that would finally allow MS to analyze virtually any biomolecule without chemical derivatization or transformation. Since then, the methods for ion separation and detection have been refined to more fully exploit the characteristics of these ionization methods. Thus, the tools available for biological MS are the most sophisticated and versatile ever, even though new methods continue to emerge, such as selective binding and differential analysis directly from protein chips (Fung *et al.*, 2001). In many respects, further improvements are now incremental and the current technology for biological MS is in a mature state, although absolute sensitivity still does and probably always will remain a challenge. Now MS is rapidly moving from a regimen in which samples are handled singly to the point at which automated systems are handling many thousands of samples per day. This is important for the application of MS to proteomics, which requires the identification of multiple proteins that are linked through their cellular actions, defining their levels of expression in different tissues in normal and abnormal states (e.g., during disease) and characterizing their posttranslational modifications and their binding partners. Despite some valuable first steps in developing reliable quantitative methods in biological MS, this still presents substantial difficulties. Thus, looking ahead the immediate challenges are to analyze larger numbers of samples within shorter periods of time, achieve even higher sensitivity of detection, and improve the ability to quantitate the molecules under investigation. This will require enhanced data processing and interpretation, with validated scoring methods that will allow automation to the degree that reliable results will be generated by computerized methods with little or no human input.

Acknowledgment

Financial support from the National Institutes of Health, No. NCRR 01614, is gratefully acknowledged.

References

Baldwin, M. A., Medzihradszky, K. F., Lock, C. M., Fisher, B., Settineri, C. A., and Burlingame, A. L. (2001). Matrix-assisted laser desorption/ionization coupled with quadrupole/orthogonal acceleration time-of-flight mass spectrometry for protein discovery, identification, and structural analysis. *Anal. Chem.* **73,** 1707–1720.

Bateman, R. H., Green, M. R., Scott, G., and Clayton, E. (1995). A combined magnetic sector-time-of-flight mass spectrometer for structural determination studies by tandem mass spectrometry. *Rapid Commun. Mass Spectrom.* **9,** 1277.

Beavis, R., and Chait, B. T. (1989). Cinnamic acid derivatives as matrices for ultraviolet laser desorption mass spectrometry of proteins. *Rapid Commun. Mass Spectrom.* **12,** 432–435.

Beynon, J. H. (1960). "Mass Spectrometry and Its Applications to Organic Chemistry." Van Nostrand, Princeton, NJ.

Breaux, G. A., Green-Church, K. B., France, A., and Limbach, P. A. (2000). Surfactant-aided, matrix-assisted laser desorption/ionization mass spectrometry of hydrophobic and hydrophilic peptides. *Anal. Chem.* **72,** 1169–1174.

Brown, R. S., and Lennon, J. J. (1995). Mass resolution improvement by incorporation of pulsed ion extraction in a matrix-assisted laser desorption/ionization linear time-of-flight mass spectrometer. *Anal. Chem.* **67,** 1998.

Burlingame, A. L., Carr, S. A., and Baldwin, M. A. (eds.) (2000). *In* "Mass Spectrometry in Biology and Medicine." Humana, Totowa, NJ.

Carr, S. A., Huddleston, M. J., and Annan, R. S. (1996). Selective detection and sequencing of phosphopeptides at the femtomole level by mass spectrometry. *Anal. Biochem.* **239,** 180–192.

Casares, A., Kholomeev, A., and Wollnik, H. (2001). Multipass time-of-flight mass spectrometers with high resolving powers. *Int. J. Mass Spectrom.* **206,** 267–273.

Cech, N. B., and Enke, C. G. (2000). Relating electrospray ionization response to nonpolar character of small peptides. *Anal. Chem.* **72,** 2717–2723.

Chernushevich, I. V. (2000). Duty cycle improvement for a quadrupole-time-of-flight mass spectrometer and its use for precursor ion scans. *Eur. J. Mass Spectrom.* **6,** 471–479.

Chernushevich, I. V., Loboda, A. V., and Thomson, B. A. (2001). An introduction to quadrupole-time-of-flight mass spectrometry. *J. Mass Spectrom.* **36,** 849–865.

Clauser, K. R., Baker, P. R., and Burlingame, A. L. (1999). Role of accurate mass measurement ($+/-10$ ppm) in protein identification strategies employing MS or MS/MS and database searching. *Anal. Chem.* **71**, 2871–2882.

Cole, R. B. (2000). Some tenets pertaining to electrospray ionization mass spectrometry. *J. Mass Spectrom.* **35**, 763–772.

Conrads, T. P., Anderson, G. A., Veenstra, T. D., Pasa-Tolic, L., and Smith, R. D. (2000). Utility of accurate mass tags for proteome-wide protein identification. *Anal. Chem.* **72**, 3349–3354.

Covey, T. R., Bonner, R. F., Shushan, B. I., and Henion, J. (1988). The determination of protein, oligonucleotide and peptide molecular weights by ion-spray mass spectrometry. *Rapid Commun. Mass Spectrom.* **2**, 249–256.

Dai, Y., Whittal, R. M., and Li, L. (1999). Two-layer sample preparation: A method for MALDI-MS analysis of complex peptide and protein mixtures. *Anal. Chem.* **71**, 1087–1091.

Dawson, J. H. J., and Guilhaus, M. (1989). Orthogonal-acceleration time-of-flight mass spectrometer. *Rapid Commun. Mass Spectrom.* **3**, 155–159.

deHoog, C. L., and Mann, M. (2004). Proteomics. *Annu. Rev. Genomics Hum. Genet.* **5**, 267.

Deng, Y., Pan, H., and Smith, D. L. (1999). Selective isotope labeling demonstrates that hydrogen exchange at individual peptide amide linkages can be determined by collision-induced dissociation mass spectrometry. *J. Amer. Chem. Soc.* **121**, 1966.

Deshaies, R. J., Seol, J. H., McDonald, W. H., Cope, G., Lyapina, S., Shevchenko, A., Verma, R., and Yates, J. R., III. (2002). Charting the protein complexome in yeast by mass spectrometry. *Mol. Cell. Proteomics* **1**, 3–10.

Dodonov, A. F., Chernushevich, I. V., and Laiko, V. V. (1991). Atmospheric pressure ionization time of flight mass spectrometer. *Proceedings of the 12th International Mass Spectrometry Conference,* Amsterdam.

Fellgett, P. B. (1958). Spectrométre interférential multiplex pour measures infrarouges sur les étoiles. *J. Phys. Radium* **19**, 237.

Fenn, J. B., Mann, M., Meng, C. K., Wong, S. F., and Whitehouse, C. M. (1989). Electrospray ionization for mass spectrometry of large biomolecules. *Science* **246**, 64–71.

Fenn, J. B., Mann, M., Meng, C. K., Wong, S. F., and Whitehouse, C. M. (1990). Electrospray ionization-principles and practice. *Mass Spectrom. Rev.* **9**, 37.

Fung, E. T., Thulasiraman, V., Weinberger, S. R., and Dalmasso, E. A. (2001). Protein biochips for differential profiling. *Curr. Opinion Biotechnol.* **12**, 65–69.

Guilhaus, M., Selby, D., and Mlynski, V. (2000). Orthogonal acceleration time-of-flight mass spectrometry. *Mass Spectrom. Rev.* **19**, 65–107.

Gygi, S. P., Rist, B., Gerber, S. A., Turecek, F., Gelb, M. H., and Aebersold, R. (1999). Quantitative analysis of complex protein mixtures using isotope-coded affinity tags. *Nat. Biotechnol.* **17**, 994–999.

Hale, J. E., Butler, J. P., Knierman, M. D., and Becker, G. W. (2000). Increased sensitivity of tryptic peptide detection by MALDI-TOF mass spectrometry is achieved by conversion of lysine to homoarginine. *Anal. Biochem.* **287**, 110–117.

Hansen, K. C., Schmitt-Ulms, G., Chalkley, R. J., Hirsch, J., Baldwin, M. A., and Burlingame, A. L. (2003). Mass spectrometric analysis of protein mixtures at low levels using cleavable ^{13}C-isotope–coded affinity tag and multidimensional chromatography. *Mol. Cell Proteomics* **2**, 299–314.

Hao, C., and March, R. E. (2001). A survey of recent research activity in quadrupole ion trap mass spectrometry. *Int. J. Mass Spectrom.* **212**, 337–357.

Horn, D. M., Zubarev, R. A., and Mclafferty, F. W. (2000). Automated *de novo* sequencing of proteins by tandem high-resolution mass spectrometry. *Proc. Natl. Acad. Sci. USA* **97**, 10313–10317.

Huang, L., Baldwin, M. A., Maltby, D. A., Medzihradszky, K. F., Baker, P. R., Allen, N., Rexach, M., Edmondson, R. D., Campbell, J., Juhasz, P., Martin, S. A., Vestal, M. L., and Burlingame, A. L. (2002). The identification of protein–protein interactions of the nuclear pore complex of *Saccharomyces cerevisiae* using high throughput matrix-assisted laser desorption ionization time-of-flight tandem mass spectrometry. *Mol. Cell Proteomics* **1**, 434–450.

Huddleston, M. J., Annan, R. S., Bean, M. F., and Carr, S. A. (1993a). A selective detection of phosphopeptides in complex mixtures by electrospray liquid chromatography mass spectrometry. *J. Am. Soc. Mass Spectrom.* **4**, 710–717.

Huddleston, M. J., Bean, M. F., and Carr, S. A. (1993b). Collisional fragmentation of glycopeptides by electrospray ionization LC/MS and LC/MS/MS: Methods for selective detection of glycopeptides in protein digests. *Anal. Chem.* **65**, 877–884.

Karas, M., and Hillenkamp, F. (1988). Laser desorption ionization of proteins with molecular masses exceeding 10,000 daltons. *Anal. Chem.* **60**, 2299–2301.

Karas, M., Bahr, U., Strupat, K., Hillenkamp, F., Tsarbopoulos, A., and Pramanik, B. N. (1995). Matrix dependence of metastable fragmentation of glycoproteins in MALDI TOF mass spectrometry. *Anal. Chem.* **67**, 675–679.

Kebarle, P. (2000). A brief overview of the present status of the mechanisms involved in electrospray mass spectrometry. *J. Mass Spectrom.* **35**, 804–817.

King, T. B., Colby, S. M., and Reilly, J. P. (1995). High resolution MALDI-TOF mass spectra of three proteins obtained using space velocity correlation focusing. *Int. J. Mass Spectrom. Ion Processes* **145**, L1–L7.

Kjeldsen, F., Haselmann, K. F., Sørensen, E. S., and Zubarev, R. A. (2003). Distinguishing of Ile/Leu amino acid residues in the PP3 protein by (hot) electron capture dissociation mass spectrometry. *Anal. Chem.* **75**, 1267.

Knapp, D. R. (1990). Chemical derivatization for mass spectrometry. *Methods Enzymol.* **193**, 314–329.

Krutchinsky, A. N., Kalkum, M., and Chait, B. T. (2001). Automatic identification of proteins with a MALDI-quadrupole ion trap mass spectrometer. *Anal. Chem.* **73**, 5066–5077.

Krutchinsky, A. N., Loboda, A. V., Spicer, V. L., Dworschak, R., Ens, W., and Standing, K. G. (1998). Orthogonal injection of matrix-assisted laser desorption/ionization ions into a time-of-flight spectrometer through a collisional damping interface. *J. Am. Soc. Mass Spectrom.* **9**, 508–518.

Laiko, V. V., Baldwin, M. A., and Burlingame, A. L. (2000). Atmospheric pressure matrix-assisted laser desorption/ionization mass spectrometry. *Anal. Chem.* **72**, 652–657.

Levsen, K. (1978). "Fundamental Aspects of Organic Mass Spectrometry." Verlag Chemie, Weinheim.

Li, L., Masselon, C.D., Anderson, G. A., Pasa-Tolic, L., Lee, S.-W., Shen, Y., Zhao, R., Lipton, M. S., Conrads, T. P., Tolic, N., and Smith, R. D. (2001). High-throughput peptide identification from protein digests using data-dependent multiplexed tandem FTICR mass spectrometry coupled with capillary liquid chromatography. *Anal. Chem.* **73**, 3312.

Loo, J. A. (1997). Studying noncovalent protein complexes by electrospray ionization mass spectrometry. *Mass Spectrom. Rev.* **16**, 1–23.

Mamyrin, B. A., and Shmikk, D. V. (1979). The linear mass reflectron. *Zh. Eksp. Teor. Fiz.* **76**, 1500.

Mann, M., and Wilm, M. (1994). Error-tolerant identification of peptides in sequence databases by peptide sequence tags. *Anal. Chem.* **66**, 4390–4399.

McCloskey, J. A. (ed.) (1990). *In* Mass Spectrometry, **193**.

McLafferty, F. W., and Turecek, F. (1993). "Interpretation of Mass Spectra," 4 Ed. University Science Books, Mill Valley, CA.

Medzihradszky, K. F., Adams, G. W., Burlingame, A. L., Bateman, R. H., and Green, M. R. (1996). Peptide sequence determination by matrix-assisted laser desorption ionization employing a tandem double focusing magnetic-orthogonal acceleration time-of-flight mass spectrometer. *J. Am. Soc. Mass Spectrom.* **7**, 1–10.

Medzihradszky, K. F., Campbell, J. M., Baldwin, M. A., Falick, A. M., Juhasz, P., Vestal, M. L., and Burlingame, A. L. (2000). The characteristics of peptide collision-induced dissociation using a high-performance MALDI-TOF/TOF tandem mass spectrometer. *Anal. Chem.* **72**, 552–558.

Medzihradszky, K. F., Leffler, H., Baldwin, M. A., and Burlingame, A. L. (2001). Protein identification by in-gel digestion, high-performance liquid chromatography, and mass spectrometry: Peptide analysis by complementary ionization techniques. *J. Am. Soc. Mass Spectrom.* **12**, 215–221.

Morris, H. R., Paxton, T., Dell, A., Langhorne, J., Berg, M., Bordoli, R. S., Hoyes, J., and Bateman, R. H. (1996). High sensitivity collisionally activated decomposition tandem mass spectrometry on a novel quadrupole/orthogonal-acceleration time-of-flight mass spectrometer. *Rapid Comun. Mass Spectrom.* **10**, 889–896.

O'Connor, P. B., Mirgorodskaya, E., and Costello, C. E. (2002). High pressure matrix-assisted laser desorption/ionization Fourier transform mass spectrometry for minimization of ganglioside fragmentation. *J. Am. Soc. Mass Spectrom.* **13**, 402–407.

Ong, S.-E., Kratchmarova, I., and Mann, M. (2003). Properties of ^{13}C-substituted arginine in stable isotope labeling by amino acids in cell culture (SILAC). *J. Proteome Res.* **2**, 173–181.

Preisler, J., Hu, P., Rejtar, T., and Karger, B. L. (2000). Capillary electrophoresis–matrix-assisted laser desorption/ionization time-of-flight mass spectrometry using a vacuum deposition interface. *Anal. Chem.* **72**, 4785–4795.

Preisler, J., Hu, P., Rejtar, T., Moskovets, E., and Karger, B. L. (2002). Capillary array electrophoresis-MALDI mass spectrometry using a vacuum deposition interface. *Anal. Chem.* **74**, 17.

Price, P. (1991). Standard definitions of terms relating to mass spectrometry. *J. Am. Soc. Mass Spectrom.* **2**, 336–348.

Rosman, K. J. R., and Taylor, P. D. P. (1998). Isotopic compositions of the elements 1997. *Pure Appl. Chem.* **70**, 217–235.

Rutherford, E. (1929). Discussion on the structure of atomic nuclei. *Proc. Roy. Soc. Lon.* **A123**, 373.

Shevchenko, A., Chernushevich, I., Ens, W., Standing, K. G., Thomson, B., Wilm, M., and Mann, M. (1997). Rapid *de novo* peptide sequencing by a combination of nanoelectrospray, isotopic labeling and a quadrupole/time-of-flight mass spectrometer. *Rapid Comun. Mass Spectrom.* **11**, 1015–1024.

Simmons, T. A., and Limbach, P. A. (1998). The influence of co-matrix proton affinity on oligonucleotide ion stability in matrix-assisted laser desorption/ionization time-of-flight mass spectrometry. *J. Am. Soc. Mass Spectrom.* **9**, 668–675.

Smith, D. L., Deng, Y., Yuzhong, and Zhang, Z. (1997). Probing the non-covalent structure of proteins by amide hydrogen exchange and mass spectrometry. *J. Mass Spectrom.* **32**, 135–146.

Sobott, F., McCammon, M. G., Hernandez, H., and Robinson, C. V. (2005). The flight of macromolecular complexes in a mass spectrometer: One contribution of 17 to a discussion meeting "Configurational energy landscapes and structural transitions in clusters, fluids and biomolecules." *Philos. Trans. R. Soc. A* **363**, 379–391.

Spengler, B., Kirsch, D., Kaufmann, R., and Jaeger, E. (1992). Peptide sequencing by matrix-assisted laser-desorption mass spectrometry. *Rapid Comun. Mass Spectrom.* **6**, 105–108.

Stimson, E., Truong, O., Richter, W. J., Waterfield, M. D., and Burlingame, A. L. (1997). Enhancement of charge remote fragmentation in protonated peptides by high-energy CID MALDI-TOF-MS using "cold" matrices. *Int. J. Mass Spectrom. Ion Processes* **169/170**, 231–240.

Sze, S. K., Siu, K., Ge, Y., Oh, H., and McLafferty, F. W. (2002). Top-down mass spectrometry of a 29-kda protein for characterization of any posttranslational modification to within one residue. *Proc. Natl. Acad. Sci. USA* **99**, 1774–1779.

Tiller, P. R., Land, A. P., Jardine, I., Murphy, D. M., Sozio, R., Ayrton, A., and Schaefer, W. H. J. J. (1998). Application of liquid chromatography-mass spectrometry(n) analyses to the characterization of novel glyburide metabolites formed *in vitro*. *Chromatogr. A* **794**, 15–25.

Todd, J. F. J. (1991). Recommendations for nomenclature and symbolism for mass spectroscopy. *Pure Appl. Chem.* **63**, 1541–1566.

Todd, J. F. J., and March, R. E. (eds.) (1995). *In* "Practical Aspects of Ion Trap Mass Spectrometry." CRC Press, Boca Raton, FL.

Vachet, R. W., Winders, A. D., and Glish, G. L. (1996). Correlation of kinetic energy losses in high-energy collision-induced dissociation with observed peptide product ions. *Anal. Chem.* **68**, 522–526.

Verentchikov, A. V., Ens, W., and Standing, K. G. (1994). Reflecting time-of-flight mass spectrometer with an electrospray ion source and orthogonal extraction. *Anal. Chem.* **66**, 126–133.

Vestal, M. L., Juhasz, P., and Martin, S. A. (1995). Delayed extraction matrix-assisted laser desorption time-of-flight mass spectrometry. *Rapid Commun. Mass Spectrom.* **9**, 1044–1050.

Vestal, M. L., Juhasz, P., Hines, W., and Martin, S. A. (2000). A new delayed extraction MALDI-TOF MS-MS for characterization of protein digests. *In* "Mass Spectrometry in Biology and Medicine" (A. L. Burlingame, S. A. Carr, and M. A. Baldwin, eds.), p. 1. Humana Press, Totowa, NJ.

Vocke, R. D., Jr. (1999). Atomic weights of the elements 1997. *Pure Appl. Chem.* **71**, 1593–1607.

Vorm, O., Roepstorff, P., and Mann, M. (1994). Improved resolution and very high sensitivity in MALDI-TOF of matrix surfaces made by fast evaporation. *Anal. Chem.* **66**, 3281–3287.

Wand, A. J., and Englander, S. W. (1996). Protein complexes studied by NMR spectroscopy. *Curr. Opin. Biotechnol.* **7**, 403–408.

Wattenberg, A., Barth, H.-D, and Brutschy, B. (1997). Copper-binding abilities of the tripeptide diglycylhistidine studied by laser-induced liquid beam ionization/desorption mass spectrometry in aqueous solution. *J. Mass Spectrom.* **32**, 1350–1355.

Westman, A., Brinkmalm, G., and Barofsky, D. F. (1997). MALDI induced saturation effects in chevron microchannelplate detectors. *Int. J. Mass Spectrom. Ion Processes* **169/170**, 79–87.

Weston, A. D., and Hood, L. (2004). Systems biology, proteomics, and the future of health care: Toward predictive, preventative, and personalized medicine. *J. Proteome Res.* **3**, 174–196.

Whittal, R. M., and Li, L. (1995). High-resolution matrix-assisted laser desorption/ionization in a linear time-of-flight mass spectrometer. *Anal. Chem.* **67**, 1950–1954.

Wilm, M. S., and Mann, M. (1994). Electrospray and Taylor Cone theory, Dole's beam of macromolecules at last? *Int. J. Mass Spectrom. Ion Processes* **136**, 167–180.

Zenobi, R., and Knochenmuss, R. (1998). Ion formation in MALDI mass spectrometry. *Mass Spectrom. Rev.* **17**, 337–366.

[2] Hybrid Quadrupole/Time-of-Flight Mass Spectrometers for Analysis of Biomolecules

By WERNER ENS and KENNETH G. STANDING

Abstract

The basic principles of quadrupole/time-of-flight (TOF) mass spectrometers are discussed. These instruments can be used for ions produced either by electrospray ionization (ESI) or by matrix-assisted laser desorption ionization (MALDI). In the most common configuration, the functions of collisional cooling, parent ion selection, and collision-induced dissociation are carried out successively in three separate quadrupoles. The ions are then injected orthogonally into a TOF spectrometer, which makes the m/z measurement. Thus, these hybrid instruments benefit from the versatile ability of quadrupoles to carry out various tasks and from the high performance of TOF spectrometers in both simple mass spectrometry (MS) and tandem (MS/MS) modes. Significantly, collisions in the initial quadrupole decouple the instrument almost completely from the ion production process, so the quadrupole/TOF spectrometer is a stable device that is relatively insensitive to variations in the ion source.

Introduction

In recent years MS has emerged as a major analytical tool in biotechnology and biochemistry. Tandem MS (MS/MS) has become an especially valuable means for determining biomolecular structure (Gross, 1990; Yost and Boyd, 1990). In this technique, a "parent ion" derived from the analyte of interest is selected in one mass spectrometer and then is broken up, usually by collisions with the ambient gas in a collision cell (Hayes and Gross, 1990). The resulting "daughter ions" are examined in a second mass analyzer, yielding information about the structure of the parent. The triple quadrupole spectrometer ($Q_1q_2Q_3$) has been the most popular instrument for such MS/MS measurements (Yost and Boyd, 1990). In this device, Q_1 and Q_3 are quadrupole mass filters, in which Q_1 serves to select the parent ion and Q_3 scans the daughter ion spectrum. The daughter ions are produced in the collision cell enclosed in a quadrupole ion guide q_2 (radiofrequency [RF] excitation only), which focuses the ions toward the axis.

A complete mass spectrum can be obtained only from a quadrupole mass filter by scanning; ion species in the spectrum are examined one at a

METHODS IN ENZYMOLOGY, VOL. 402
0076-6879/05 $35.00
DOI: 10.1016/S0076-6879(05)02002-1

time, discarding all others, which considerably reduces sensitivity (typically by a factor ~1000) if a complete mass spectrum is needed. In the most common mode of operation of the triple quadrupole instrument, the first quadrupole Q_1 merely selects a given *parent* ion, so no scanning is involved. Indeed, a quadrupole is well suited to this role, because it couples efficiently to the collision cell. However, the entire mass spectrum of *daughter* ions is often of interest, and it must be obtained by scanning. The consequent reduction of sensitivity is a significant handicap when only a limited amount of sample is available.

This suggests that it is worthwhile to replace the final quadrupole Q_3 by a TOF spectrometer, in which the complete daughter ion spectrum can be measured in parallel (without scanning), yielding a large increase in sensitivity. The TOF spectrometer also provides this maximum sensitivity at full resolution, in contrast to the quadrupole, in which an increase in window width, the usual method of increasing sensitivity, degrades the resolution. Moreover, TOF mass analyzers have a number of other features that are particularly useful for the high-efficiency analysis of biomolecules. First, their m/z range is effectively unlimited, apart from problems of ion production and detection. Second, a TOF instrument contains no narrow slits or similar restricting elements, which of course reduce sensitivity. Early TOF spectrometers suffered from poor resolution, but the use of electrostatic reflectors (Mamyrin *et al.*, 1973), and the rediscovery (Brown and Lennon., 1995; Colby *et al.*, 1994) of the benefits of delayed extraction (Wiley and McLaren, 1955) created substantial improvements in resolving power (Vestal *et al.*, 1995). Moreover, developments in fast electronics removed earlier limitations in the recording of TOF spectra, enabling the rapid response of the instrument to be exploited more fully. A TOF instrument, thus, provides in many cases an optimum combination of resolution, sensitivity, and fast response, particularly under conditions in which the whole mass spectrum is required (Cotter, 1997).

For these reasons, hybrid quadrupole/TOF instruments have been developed as alternative devices for MS/MS measurements (Chernushevich *et al.*, 2001). In one such configuration (QqTOF) (Shevchenko *et al.*, 1997), the first two sections (Qq) of a triple quadrupole spectrometer are coupled to a TOF spectrometer. In another configuration (QhTOF) (Morris *et al.*, 1996), an RF hexapole h encloses the collision cell instead of the RF quadrupole q. The functions of the first two sections of both the triple quadrupole and the hybrid instruments are the same—selection of the parent ion in Q and its dissociation in the collision cell within q or h. The difference appears in the final section, in which a TOF spectrometer replaces the final quadrupole mass filter. Figure 1 shows a schematic diagram of a QqTOF instrument containing a single-stage electrostatic mirror. In addition to

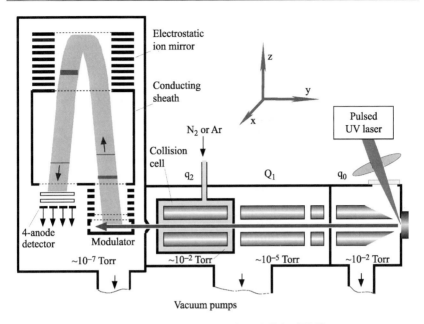

Fig. 1. Schematic diagram of a quadrupole/time-of-flight (TOF) mass spectrometer, showing the collisional damping RF quadrupole q_0, the mass-selecting quadrupole Q_1, and the collision cell enclosed in the RF quadrupole q_2. Although the instrument was originally configured with an electrospray ion source (Shevchenko *et al.*, 1997), it has been used most commonly with a matrix-assisted laser desorption ionization (MALDI) ion source, as shown. The TOF section has a single-stage mirror and a four-anode detector, as described in the text.

the mass-selecting quadrupole Q_1 and the collision cell q_2, this instrument includes another quadrupole q_0 to provide collisional cooling of the ion beam (see discussion later in this chapter). The mirror and the ion detector are similar to those used previously in single-MS measurements with pulsed ionization sources (e.g., MALDI or pulsed Cs^+ bombardment) (Tang *et al.*, 1988).

Ion Production

As in other mass spectrometric applications for the study of biomolecules, the most effective ion production methods are ESI (Aleksandrov *et al.*, 1985; Cole, 1997; Yamashita *et al.*, 1984) and MALDI (Hillenkamp *et al.*, 1991; Karas *et al.*, 1988). Using these techniques, intact gas phase molecular ions with molecular weights up to hundreds of kilodaltons can be formed.

Because the ion beams produced by ESI and MALDI have very different characteristics, it has been customary to examine them in different types of mass analyzer, but the hybrid quadrupole/TOF instruments have the advantage that they can be adapted to use either type of ion source. We first discuss the type of TOF instrument necessary for efficient use of an ESI source, and then we discuss the problems and benefits of interfacing a MALDI source to this device.

Coupling an ESI Source to a Quadrupole/TOF Instrument

Electrospray produces a continuous beam of ions. Like other continuous ion sources, it is most compatible with mass spectrometers that operate in a similarly continuous fashion, such as quadrupole mass filters. Indeed, the combination of an ESI source and a quadrupole mass filter has become a popular and satisfactory configuration, so there is no difficulty in coupling an electrospray source to the initial quadrupole sections of the hybrid quadrupole/TOF mass spectrometer. However, problems do arise in coupling the beam leaving the quadrupoles to the TOF section, because a TOF spectrometer requires a pulsed beam. The most straightforward coupling technique is to chop the continuous electrosprayed beam into short packets and to inject these along the spectrometer axis (the z axis in Fig. 1) as in MALDI/TOF instruments. However, this procedure involves a tremendous loss in sensitivity, because gating ions into the TOF analyzer in this way to produce well-separated ion packets imposes a serious reduction in the fraction of the total beam injected into the TOF analyzer. At best, this fraction then has the value ([length of packet]/[time between packets]). For example, Pinkston *et al.* (1986) found that the need to extract short ion packets with a small energy spread meant that only 0.0025% of the total ion current contributed to the recorded spectra when they interfaced a continuous chemical ionization source to a TOF mass analyzer.

Orthogonal Injection. Fortunately TOF instruments can tolerate a relatively large spatial or velocity spread in a plane perpendicular to the spectrometer axis, as illustrated by the large sources (up to ~1 cm diameter) typically used in fission fragment desorption (PDMS). This tolerance can be exploited by injecting electrospray ions into the TOF instrument perpendicular to the z axis (i.e., "orthogonal injection") (Chernushevich *et al.*, 1997, 1999; Guilhaus *et al.*, 2000). This geometry provides a high-efficiency interface for transferring ions from a continuous beam to a pulsed mode. Another advantage is the small velocity spread in the z direction that is usually observed, making high resolution easier to obtain.

Such a configuration is illustrated in the QqTOF instrument shown in Fig. 1. Here, the electric field direction in the single-stage electrostatic ion mirror defines the spectrometer axis (the z axis). After leaving the collision cell, ions are re-accelerated along the $-y$ axis to the desired energy. This energy is usually less than 10 eV, so the ions still have a relatively low velocity. They are then focused by ion optics into a parallel beam that enters continuously into a storage region in the ion modulator of the TOF analyzer, still perpendicular to the spectrometer axis. Initially, the storage region is field free, so ions continue to move in their original direction in the gap between a flat plate below and a grid above. When a voltage pulse is applied to the plate, a sausage-shaped ion packet is pushed out of the storage region into an accelerating column (the second stage of the modulator), in which the ions are accelerated to kiloelectron volt energies. The longest flight time of interest determines the maximum injection frequency, typically 4–10 kHz. Thus, a series of ion packets is injected into the accelerating column at that frequency and, from there, into a field-free drift region of the TOF instrument, providing a maximum flight time of about 100 μs or more between pulses.

Collisional Cooling

Both transmission and resolution in the TOF analyzer in QqTOF instruments are very sensitive to the "quality" of the ion beam entering it (much more so than the final q in triple quadrupoles). Thus, the application of "collisional cooling" to TOF measurements (Krutchinsky *et al.*, 1998) has been an important development, because this technique can produce significant improvements in beam quality. Such cooling was introduced previously in quadrupole mass spectrometers, which have commonly used an RF beam guide to couple the ion source to the mass selecting quadrupole q. This beam guide consists of a set of quadrupole (or other multipole) rods to focus the ion beam onto the axis, and in early instruments, the beam guide was operated at relatively high vacuum, based on the argument that any gas in that region would cause beam attenuation. However, it was demonstrated by Douglas and French (1992) and Xu *et al.* (1993) that a higher pressure (0.01–1.00 torr) in the beam guide leads to collisional damping of the ion beam, thus improving its beam quality and consequently increasing the ion transmission. For similar reasons such a beam guide provides a useful interface to a TOF mass spectrometer, and Krutchinsky *et al.* (1998) have reported calculations and measurements that illustrate the collisional cooling obtained in that case (Fig. 2).

In the collisional cooling ion guide, the RF field in the quadrupole focuses ions onto its axis while collisions with the molecules of the ambient gas reduce the ion velocities to near-thermal values (Krutchinsky *et al.*, 1998; Tolmachev *et al.*, 1997). Thus, the beam leaving the ion guide has a

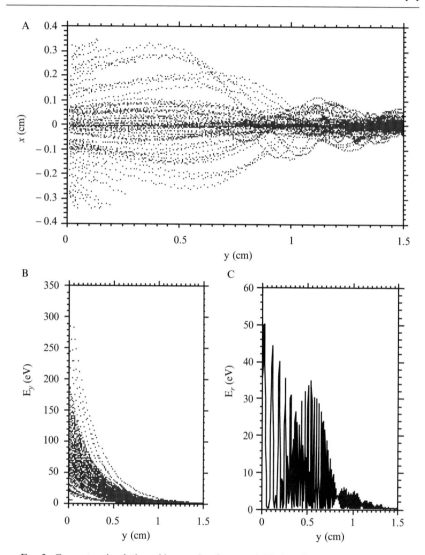

FIG. 2. Computer simulation of ion motion for myoglobin ions in a quadrupole ion guide at a pressure of 0.1 torr (Krutchinsky *et al.*, 1998). (A) Projection of ion trajectories on the *yz* plane. (B) "Energy" E_z of the ions as a function of position along the quadrupole *z* axis, where $E_z = mv_z^2/2$. (C) "Energy" E_r of the ions as a function of position along the quadrupole *z* axis, where $E_r = mv_r^2/2$.

greatly reduced spatial spread in the x-z plane, and much smaller velocity spreads in all three directions (x, y, and z), and these properties have no observable dependence on the original parameters of the beam delivered by the source. Indeed, collisional cooling decouples the mass measurement almost completely from processes occurring in the ion source, that is, the ions "forget" how they were formed.

The quadrupole/TOF instrument of Fig. 1 takes advantage of this technique by sampling ions from the ion source through the additional quadrupole q_0, in which the ions produced by the ion source (the parent ions) are cooled. In addition, radial and axial collisional damping of both parent and daughter ion motions occurs in the collision cell q_2, because this cell also is operated in the RF-only mode at a pressure of several millitorrs or more. In both cases, the cooling results in better transmission through the quadrupoles. Moreover, the ions retain their near-thermal energy *spreads* as they are reaccelerated to the desired energies after leaving the quadrupoles, so the benefits of cooling are retained for the ions entering the TOF analyzer.

Orthogonal Injection of MALDI Ions

The rationale for trying to couple MALDI to a TOF instrument by orthogonal injection is not immediately obvious, because axial injection of MALDI ions into a TOF spectrometer has often been described as an ideal marriage between compatible techniques. In the usual geometry, ions are ejected from the target along the axis of the TOF instrument (normal to the sample surface) by bombardment with a pulsed laser. A TOF start signal is supplied by the laser pulse or by the necessary extraction pulse if delayed extraction is used. The rather long time between pulses (usually ~100 ms) gives plenty of flight time, even for very heavy ions. The initiation of delayed extraction has provided excellent resolution, and equally excellent mass accuracy (~10 ppm) can be obtained in favorable cases with sufficiently careful calibration.

By contrast, direct orthogonal injection of MALDI ions into a TOF spectrometer suffers from considerably greater difficulties than in the ESI case, because of the larger velocity and angular spreads of the MALDI ions, and because of the pulsed nature of the MALDI beam. Consequently, attempts to provide this coupling by direct injection of MALDI ions have usually yielded unimpressive results in single-MS measurements (Krutchinsky *et al.*, 1998a; Spengler and Cotter, 1990). Additional problems arise for MS/MS measurements in a hybrid instrument with a mass-selecting quadrupole, because the high-velocity ions from MALDI may not spend enough time in a short quadrupole to permit efficient mass discrimination.

Nevertheless, axial injection of MALDI ions does have its own problems. Some peak broadening is produced by the energy spread of the ions in the plume ejected from the target by the laser pulse, even with delayed extraction. Delayed extraction itself complicates the mass calibration and can be optimized for only part of the mass range at a time, so spectra may have to be recorded in segments, and high mass accuracy may be difficult to attain. Moreover, optimal focusing conditions, as well as the mass calibration, depend to some extent on laser fluence, the type of sample matrix and support, the sample preparation method, and even the location of the laser spot on the sample. Finally, perhaps the most serious handicap of the axial MALDI/TOF configuration is the difficulty in obtaining structural information by MS/MS measurements. Such measurements have normally been carried out by the so-called *post-source decay technique* in which a parent ion is selected by a gate in the flight tube and the products of its metastable decay are observed after reflection (Chaurand *et al.*, 1999; Spengler, 1997). The capabilities of this method are seriously restricted by the limited resolution obtained for parent ion selection ($\sim$200) and the limited mass accuracy of the daughter ion measurement ($\sim$0.2 Da), as well as the difficulties of controlling the fragmentation process and interpreting the results.

The possibility of orthogonal injection of MALDI ions was revived by the realization that conditions for it can be made much more favorable by collisional cooling. Thus, a MALDI ion source can be coupled directly to a TOF spectrometer through an RF-quadrupole ion guide with collisional cooling (Krutchinsky *et al.*, 1998a), similar to the one used for electrospray. For example, the MALDI source shown in Fig. 1 replaced the electrospray source originally installed in this hybrid instrument (Loboda *et al.*, 2000a; Shevchenko *et al.*, 2000). When MALDI ions are injected into such a spectrometer, collisions in the ion guide damp the ion motion, improving the m/z range and the resolution. Measurements with this configuration give resolution and sensitivity comparable to axial MALDI performance and have several additional benefits:

1. The collisions in the ion guide also spread the ion pulse out along the quadrupole axis, producing a quasi-continuous beam (as shown in Fig. 3 [Krutchinsky *et al.*, 1998]), which can then be treated just like an electrosprayed beam. Although it might seem odd to produce a continuous beam from the pulsed MALDI beam, only to pulse it again for injection into the TOF instrument, the final beam has highly favorable properties very different from those of the original one. In particular, the start time for the TOF measurement no longer needs to be correlated with the laser

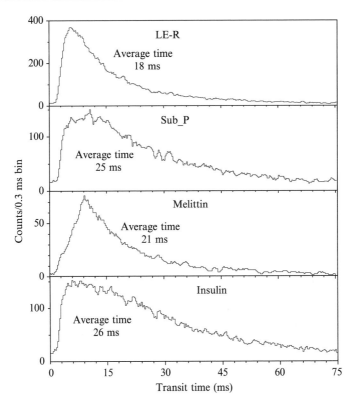

FIG. 3. Measured transit times through a collisional cooling quadrupole (Krutchinsky *et al.*, 1998). Note that for a 10-kHz injection, the time between injection pulses is 0.1 ms, so the beam entering the time-of-flight modulator is almost continuous.

pulse, so the beam can be injected into the spectrometer at a high repetition rate (up to 10 kHz) as in ESI/TOF, although the laser itself may run at about 20 Hz or less. This reduces the number of ions in a single time measurement by about the same factor ($\sim$500 $=$ 10,000/20) and, thus, eliminates problems of peak saturation and detector shadowing from intense matrix signals. As a result, the count rates obtained are compatible with digital pulse-counting methods (i.e., time-to-digital converters).

2. The residence time of ions in the guide is typically a few milliseconds or more, as shown in Fig. 3 (Krutchinsky *et al.*, 1998), so the plume of ions has more time to expand and cool before it is injected into the flight tube, eliminating much of the metastable decay otherwise taking place in this region and, thus, simplifying the spectrum.

3. The performance of the instrument is independent of ion source conditions, as a result of the decoupling of the ion desorption event from the flight time measurement by the ion guide and the collision cell. Thus,there is much greater flexibility in the choice of different sample-preparation methods and different laser wavelengths, pulse widths, and fluences. Even insulating target supports can sometimes be tolerated.

4. The optimum conditions for orthogonal injection are independent of m/z, so the detailed adjustments required in axial MALDI are unnecessary. The detector is set for single-ion counting and the laser is simply set to any convenient fluence up to the maximum. The laser spot can be moved to a different spot on the target without any readjustment. The acquisition of MALDI spectra is, therefore, extremely simple, requiring very little operator expertise. The measured flight time has a simple quadratic dependence on mass; a simple two-point calibration gives optimum mass accuracy over the entire mass range. We have found that such an external calibration gives mass accuracies better than about 10 ppm in the instrument of Fig. 1 if performed within an hour of the measurement, provided that the room temperature is fairly stable.

5. Both ESI and MALDI sources can be used on the same instrument with minimal changes in configuration (Krutchinsky *et al.*, 1998a, 2000), because the ion beam entering the spectrometer has similar properties in both cases.

6. Perhaps most significant is the ability to carry out MS/MS measurements on MALDI ions in the quadrupole/TOF instruments just as on ESI ions, with similar high values of sensitivity (fmole) and control over conditions for collision-induced dissociation. Parent ions (up to a few thousand daltons) can be selected with unit mass resolution in the mass-selecting quadrupole; high resolving power ($\sim$10,000) and mass accuracy ($\sim$10 ppm or better) for the daughter ions are provided by the TOF analyzer. For structural determination, both MS measurements (mass mapping) and MS/MS measurements can be carried out interactively on the same sample and in the same instrument, without breaking vacuum.

On the other hand, an orthogonal injection instrument with collisional cooling is more complex than an axial-injection MALDI instrument, even though the quality of the data in the single-MS mode is rather similar. More seriously, there are losses associated with the orthogonal injection process that are not found with axial injection, as discussed later, so sensitivity is expected to be somewhat lower. Moreover, most existing orthogonal injection instruments use small accelerating voltages ($\sim$4 kV), so they have low efficiency for large singly charged ions (as produced by MALDI). The report of an orthogonal injection TOF spectrometer with 20 kV acceleration voltage (Ackloo and Loboda, 2005; Loboda *et al.*, 2003) is encouraging, but such a configuration is not yet available in a quadrupole/TOF instrument. In addition, there still may be difficulties in transmitting high m/z ions efficiently through the quadrupole ion guide, and problems related to fragmentation and adduct-formation for large singly charged ions as a result of their long transit times.

Features of the Orthogonal Injection/TOF Configuration

Flight Tube High Voltage

In any TOF spectrometer, the need for DC acceleration of the ions means that either the flight tube or the previous stage of the instrument must be elevated to the DC acceleration potential, several kilovolts or more. In axial MALDI/TOF spectrometers (and in an early ESI/TOF instrument [Verentchikov *et al.*, 1994]), the latter solution is adopted, that is, the previous stage (the source itself) is run at high voltage. However, this is not practical in quadrupole/TOF instruments because of the need to couple the TOF section to the quadrupoles, which normally operate close to ground potential. Thus, the former solution is chosen, raising the flight tube to the accelerating potential. In this case, the ions are accelerated to their final energy by the DC field produced by the potential V_z applied across the acceleration column, and positive ions are accelerated from ground to high negative voltages, rather than from high positive voltages to ground as in axial MALDI/TOF mass spectrometers. The subsequent motion is similar in both cases; ions of different m/z values are separated in time as they pass down the flight tube and are reflected by the mirror.

However, it is usually impractical to put high voltage on the whole vacuum chamber, so the flight tube must be built as a separate structure within the vacuum chamber (i.e., the "conducting sheath;" shown in Fig. 1). This structure is typically formed from a mesh because of the need for efficient evacuation of the interior, and particular care must be taken to avoid penetration of external electric fields through the mesh, which would degrade the performance of the instrument. The difficulty of keeping this rather large structure at high voltage is one reason that orthogonal injection TOF instruments have run at fairly modest accelerating voltages until the model mentioned earlier was developed (Ackloo and Loboda, 2005; Loboda *et al.*, 2003).

Direction of Entry and m/z *Range*

When the electric field in the mirror has the correct value, the total flight time t is $2L/v_z$, where L is the z-component of the effective flight path (Tang *et al.*, 1998). The distance that the ions move in the y direction during this time is $v_y t = 2L\, v_y/v_z$, so the ratio v_y/v_z must have the value D/2L if the ions are to strike the detector, where D is the separation in the y direction between the modulator and the detector. Here, the axial velocity component v_z is determined by the energy provided by the accelerating column ($mv_z^2/2 = q\,V_z$), where V_z is the voltage across the accelerating column, m is the ion mass, and q is the charge, that is, v_z is proportional to $\sqrt{q/m}$. However, in general the *initial* value of the y component of velocity v_y will have a different form, usually some fraction of the velocity of the supersonic jet in which the ions enter the vacuum system (Beavis and Chait, 1991; Fenn, 2000). Unless this is modified, the ratio v_y/v_z, will depend on mass and charge, resulting in a corresponding restriction in the m/z ratio observable at a given setting of V_z.

Again collisional cooling makes an important contribution. If the energy of an ion is reduced to near-thermal values by collisional damping, its approximate energy $mv_y^2/2$ on entering the storage region will be $q\,V_y$, where V_y is the voltage difference between the collision cell and the modulator. The ratio of velocities in the two orthogonal directions is then $v_y/v_z = \sqrt{V_y/V_z}$, (independent of m and q), so the ratio of voltages can be adjusted to satisfy the required relation $(V_y/V_z) = (D/2L)^2$ for ions of all m/z values. Thus, all ions enter the field-free drift region in a direction making a small angle with the z axis (set by the ratio of accelerating voltages), and in principle they all strike the detector. For example, collisional cooling has made it possible to observe ions over a mass range from about 1300 Da to more than 1 MDa at the same time (m/z range from ~600 to more than 12,000) in an orthogonal injection TOF spectrometer (Krutchinsky *et al.*, 1998), as shown in Fig. 4. Note that this requires no additional deflection in the drift region, which could degrade the mass resolution (Dodonov *et al.*, 1994; Mlynski and Guilhaus, 1996).

Time Focusing, Resolution, and Mass Accuracy

Factors limiting the resolution of orthogonal injection TOF spectrometers have been discussed in detail by Dodonov *et al.* (1994, 2000). Even if the mechanical and electrical tolerances in the instrument are controlled precisely, the TOF resolution is still limited by the initial spatial and velocity spreads of the ions (z and v_z) in the storage region of the modulator, emphasizing the importance of collisional cooling before the ions enter this region. For example, a spread in the z coordinates of the

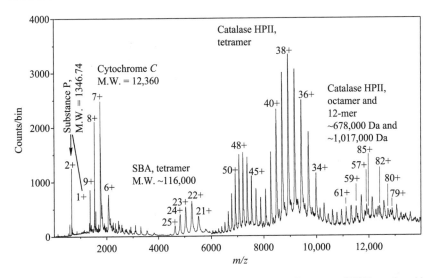

FIG. 4. The m/z spectrum of a mixture of four components in a water/0.08% acetic acid solution (Krutchinsky *et al.*, 1998): Substance P (1347 Da, 1 μM concentration), cytochrome *C* (12,360 Da, 1 μM), soy bean agglutinin tetramer (116 kDa, 3 μM), and the tetramer (339 kDa), octamer (678 kDa), and 12-mer of catalase HPII (5 μM).

ions (typically a few millimeters), leads to variations in energy (and consequently in velocity) of the ions leaving the modulator. The effect of these variations on the flight time can be reduced in a linear spectrometer by adjusting the fields in the modulator so that ions originating from different z positions in the storage region arrive at the plane of the detector at about the same time (Wiley and McLaren, 1955), thus improving the resolution; the ions exchange their original spatial spread for a velocity distribution on this plane. However, considerably better results can be obtained in a reflecting spectrometer, like the one in Fig. 1. Here, the modulator fields can be adjusted to place the time-focus plane fairly close to the source (Verentchikov *et al.*, 1994), thus minimizing the axial dimension of the ion packet. The plane then acts as an "object plane" for the mirror, and the electric field in the mirror can be adjusted to correct the ion velocity spread, as first shown by Mamyrin (1973). This greatly increases the flight time and, therefore, the mass dispersion, without increasing the width of an individual peak. A single-stage mirror, such as the one shown in Fig. 1, corrects the velocity spread in the object plane to first order, which is good enough to give a resolving power of about 10,000 as long as the energy spread in the plane is less than about 5% of the total energy.

By using a two-stage mirror (Dodonov *et al.*, 2000; Loboda *et al.*, 2000; Mamyrin *et al.*, 1973), it is possible to adjust voltages to obtain second or higher-order time focusing; spectrometer designs have achieved a resolving power between 15,000 and 20,000 (Dodonov *et al.*, 2000; Loboda *et al.*, 2000). Improvements in resolving power can also be obtained by increasing the length of the flight path (with a corresponding loss of sensitivity), in which case the main factor limiting the resolution is the residual energy spread in the beam. The flight path may be lengthened either directly (Vestal *et al.*, 1995) or by introducing additional reflectors (Piyadasa *et al.*, 1999; Wollnik and Przwloka, 1990). For example, a resolving power of 20,000 has been specified for a commercial instrument that uses a second mirror to increase the flight path (Hoyes *et al.*, 2000).

Nevertheless, the combination of collisional cooling with the single-stage mirror and effective flight path approximately 2.8 m in the instrument of Fig. 1 yields a resolving power ($M/\Delta M_{FWHM}$) between 8000 and 10,000 for peptides with masses 1000–6000 Da, as shown in Fig. 5A for a MALDI UV source with a few nanoseconds pulse length. The same resolution is achieved for other lasers and matrices, as shown in Fig. 5B for the molecular ion of insulin, using a glycerol matrix and an infrared laser with a 200-ns pulse width. The highest resolution obtained with a similar orthogonal-injection TOF geometry, but without collisional cooling (using an ESI source) was about 5000 (Verentchikov, 1994). Note, however, that the peak shape is determined mostly by the isotopic envelope for heavy ions produced by electrospray ionization (m > 10,000 Da).

The approximate 10,000 resolution is sufficient to distinguish separate isotopic peaks in ESI spectra for most peptides and, hence, determine their individual charge states. Such capability is particularly useful when interpreting complicated spectra from mixtures such as tryptic digests or spectra resulting from fragmentation of multiply charged ions. For smaller ions, the resolution decreases, but it remains sufficiently high to partially resolve two or more peaks within the same nominal mass, as shown in Fig. 5C (Chernushevich *et al.*, 1997). This is an advantage of TOF mass spectrometers over quadrupoles, where the resolution usually decreases linearly with decreasing mass in this range.

The high resolving power delivered by the TOF section of the QqTOF instrument is an important advantage of the quadrupole/TOF instruments over triple quadrupoles, in which high resolution is difficult to obtain along with adequate sensitivity. An even more important advantage is the high mass accuracy (better than 10 ppm or 10 mDa) that can be obtained, resulting from the combination of the high resolution and the decoupling

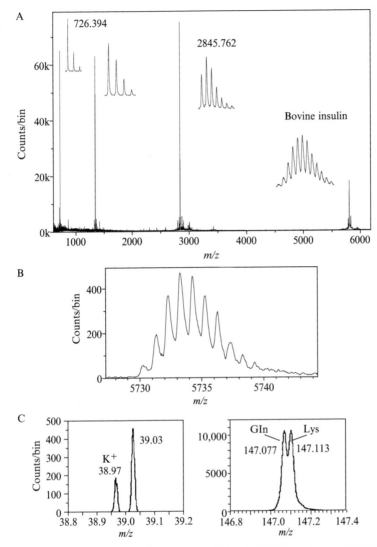

FIG. 5. (A) Positive ion spectrum of a mixture of dalargin (726 Da), substance P (1348 Da), melittin (2846 Da), and bovine insulin (5733 Da) (Loboda *et al.*, 2000a). (B) Spectrum of bovine insulin with an infrared laser. (C) Small molecule mass spectra observed in an electrospray/time-of-flight instrument (Chernushevich *et al.*, 1997).

of the ion production process from the mass measurement by collisional cooling, which makes the ion beam insensitive to changes in ion source conditions. For example, such accuracy has been crucial in identifying a mutation in the coat protein of a particular brome mosaic virus isolate (She *et al.*, 2001). In that case, the measured mass of a particular tryptic fragment differed from the predicted value by 99.074 Da, most obviously achieved by deletion of valine 27 (99.068 Da) in the predicted sequence. However, the differences Δ between measured and calculated values for the b ions are all negative on this assumption, with an average $\Delta = -11$ mDa (Table I), suggesting a search for other possibilities. Instead, replacement of Arg 26 by Gly (genetic code change AGG to GGG) gives almost the same calculated mass change (-99.080 Da), but considerably better agreement for the b ions ($\Delta = -0.4$ mDa), as shown in Table II. This stimulated a more careful search of the MS/MS spectrum (Fig. 6), which showed that there were very small peaks (initially not assigned) corresponding to b_3 and y_{15} in the latter peptide, confirming the substitution of Arg by Gly.

TABLE I

COMPARISON OF CID FRAGMENTATION OF THE SINGLY-CHARGED ION (m/z 1734.98) FROM THE BMV-W PEPTIDE (RESIDUES 24–41) WITH MASS VALUES CALCULATED ASSUMING ^{27}VAL DELETION (I.E., FROM TARQPVIVEPLAAGQGK)[a]

b ion	m/z (Obs.)	MH$^+$ (Calc.)	Δ (mDa)
b1	—	102.056	—
b2	173.091	173.093	−2
b3	329.180	329.194	−14
b4	457.246	457.252	−6
b5	—	554.305	—
b6	653.361	653.373	−12
b7	766.441	766.458	−17
b8	865.511	865.526	−15
b9	991.566	994.569	3
b10	—	1091.621	—
b11	1204.700	1204.705	−5
b12	—	1275.743	—
b13	1346.764	1346.780	−16
b14	—	1403.801	—
b15	1531.843	1531.860	−17
b16	1588.857	1588.881	−24
b17	1716.973	1716.976	−3

[a] Values listed are monoisotopic masses (MH$^+$).

TABLE II

COMPARISON OF CID FRAGMENTATION OF THE SINGLY-CHARGED ION (m/z 1734.98) FROM THE
BMV-W PEPTIDE (RESIDUES 24–41) WITH MASS VALUES CALCULATED ASSUMING A
MUTATION OF R TO G AT POSITION 26 (i.e., FROM TAGVQPVIVEPLAAGQGK)

b ion	m/z (Obs.)	MH$^+$ (Calc.)	Δ (mDa)
b1	—	102.056	—
b2	173.091	173.093	−2
b3	230.118	230.114	+4
b4	329.180	329.182	−2
b5	457.246	457.241	+5
b6	—	554.294	—
b7	653.361	653.362	−1
b8	766.441	766.446	−5
b9	865.511	865.515	−4
b10	994.566	994.557	+9
b11	—	1091.610	—
b12	1204.700	1204.694	+6
b13	—	1275.731	—
b14	1346.764	1346.768	−4
b15	—	1403.790	—
b16	1531.843	1531.849	−6
b17	1588.857	1588.870	−13
b18	1716.973	1716.965	+8

Duty Cycle and m/z Discrimination

As remarked earlier, all ions are recorded in parallel in a TOF analyzer, so the overall efficiency is much larger than in scanning instruments. This advantage is possessed by both orthogonal injection TOF instruments and axial TOF spectrometers. However, the axial spectrometers are usually more efficient, because all of the output of their pulsed ion sources can be injected into the flight tube.

On the other hand, orthogonal injection normally involves losses and some m/z dependence associated with the duty cycle, defined here as the fraction of the beam entering the TOF spectrometer that is available for acceleration into the flight tube. To avoid spectral overlap, an injection pulse cannot be applied until the slowest ion from the previous pulse has reached the detector. For instruments without deflection, the velocity component v_y of an ion in the flight tube is the same as its velocity in the storage region. Thus, the slowest ions (the ions of highest mass) entering the storage region would cover the distance between the modulator and the detector (D) by the time the next injection pulse arrives, if the storage region were long enough. However, only a finite slice of this beam of length

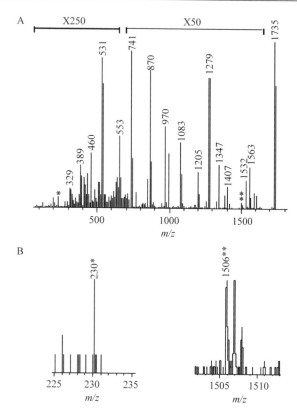

Fɪɢ. 6. MS/MS spectrum of the parent ion with $m/z = 1734.98$ from a tryptic digest of brome mosaic virus. (A) overall spectrum, (B) expanded views of the above spectrum near 230 and 1506 Da.

Δl can be accelerated and detected, where Δl is determined by the size of the apertures in the detector and/or the modulator. This sets an upper limit on the duty cycle for this m/z value as $\Delta l/D$.

If the incoming beam is monoenergetic, v_y is proportional to $1/\sqrt{m/z}$, so lighter ions will cover a distance larger than D between TOF pulses. Thus, the duty cycle is proportional to $\sqrt{m/z}$ and there is systematic discrimination against low-m/z ions:

$$\text{DutyCycle}(m/z) = \frac{\Delta l}{D}\sqrt{\frac{m/z}{(m/z)_{\text{max}}}}, \tag{1}$$

where $(m/z)_{max}$ corresponds to the ion in the spectrum with the largest m/z value to be measured. Nevertheless, this discrimination is predictable, and signals at low m/z values are normally stronger than at high m/z values, so in most cases this defect is preferable to the decrease in transmission at high m/z that is characteristic of quadrupole mass filters.

In most reflecting TOF instruments with orthogonal injection, the duty cycle is between 5% and 30%, depending on the m/z value of the ion and the instrumental parameters. Whitehouse et al. (1998) demonstrated an improvement in the duty cycle by trapping ions in a two-dimensional ion trap (an RF octopole with electrostatic apertures at each end), and then gating them in short bursts into the TOF spectrometer. These results have been extended by Chernushevich (2001); by varying trapping and gating parameters, different modes of operation were achievable, from 100% duty cycle over a narrow m/z range to <50% duty cycle over a wider mass range, although ions outside of these ranges were lost partially or completely. In spite of the reduced m/z range, this technique can be very useful when applied in certain MS/MS modes of operation, where only selected fragment ions have to be monitored, as discussed later in this chapter.

Ion Detection

The TOF detector normally consists of two large area microchannel plates in a chevron configuration. Signals may be recorded by either digital or analog techniques, each of which has advantages, depending mainly on the maximum intensity of the ion signal (Ens et al., 1993). In the digital technique used in the instrument of Fig. 1, the detector signal passes to a preamplifier and a constant-fraction discriminator and is then registered in single ion counting mode by a time-to-digital converter (TDC). This has the advantage that the time measurement is independent of the detector pulse width, so this parameter does not limit the resolution.

The TDC can record hundreds of stops for each injection pulse, but only one stop per injection pulse for a given m/z species (i.e., within the "dead time" of the TDC, about a few nanoseconds). To avoid spectral distortion, the count rate for an individual ion species has an effective limit somewhat less than the frequency of the injection pulse. For axial MALDI, this frequency is the laser repetition rate, usually less than 20 Hz, so TDCs are not very useful for such measurements. In the configuration of Fig. 1, on the other hand, collisional cooling spreads the ions from a single laser pulse over tens of milliseconds, so typical injection frequencies of 4–10 kHz can be used. Thus, the limitation is much less severe than in axial MALDI, and in most cases, spectral distortion is negligible, even for the highest laser fluence obtainable from the usual 20 Hz laser. However, for higher *average* power lasers, and especially for simple spectra, where the total ion current is concentrated in a few peaks, distortion will be observed, and it limits ultimate throughput for orthogonal MALDI.

The problem is more severe for electrospray because the average ion currents can be considerably higher and are more difficult to control. The problem is particularly acute in LC/MS experiments, where the spectra *are* simple, and the maximum count rate determines the available dynamic range.

The most straightforward solution to the distortion problem, when using a TDC, is to divide the anode into a number (n) of sections, each connected to an independent TDC channel. Assuming a uniform distribution of ions over the anode, this gives an *n*-fold increase in the allowed data rate. For example, the instrument of Fig. 1 employs a four-segment anode and a four-channel TDC (both from Ionwerks, Inc., Houston, Texas). In principle, *n* can be increased to any value needed, and up to 256 segments have been reported (Bouneau *et al.*, 2003), but cross-talk becomes a serious problem beyond about four or eight channels, and the expense of the TDCs soon becomes significant. An alternative solution is to use anode segments of different sizes (Gonin *et al.*, 2000). In this case, only the small anode is used for high count rates, and the intensity is scaled up according to the geometry.

For count rates that are not significantly greater than one ion per injection pulse per species per anode segment, it is possible to correct for the distortion using simple software algorithms (Chernushevich *et al.*, 2001). Other strategies are also possible, such as measuring weaker peaks (e.g., a ^{13}C isotope peak) to scale intense peaks or introducing dynamic current limiting procedures.

Ions may also be recorded by an analog device such as an integrating transient recorder, which is better suited to high-intensity signals, and can run at the maximum repetition rate (reciprocal of the maximum flight time). However, such devices are limited in resolution by the width of the detector pulse (Coles and Guilhaus, 1994), and they require much more careful matching of the detector and recorder gains, which depend on signal intensity. Moreover, because of inherent noise in the analog-to-digital conversion, they are poorly suited to low count rates, such as the measurement of pulses caused by single ions striking the detector. Note that existing transient recorders are well suited for conventional (axial injection) MALDI because that technique typically operates at a much lower repetition rate (a few hertz), and high instantaneous current, whereas a high repetition rate in the range of several kilohertz is critical for obtaining a reasonable duty cycle in TOF instruments with orthogonal injection of continuous beams.

Sensitivity

The sensitivity of the QqTOF geometry for electrospray ions has been considered already. In single-MS mode, or for daughter ion scanning, the parallel detection of the TOF system gives a 10–100-fold improvement in sensitivity over a triple quadrupole instrument, in spite of the duty-cycle losses associated with orthogonal TOF. For parent-ion scans, using appropriate trapping strategies, the QqTOF sensitivity approaches that of the triple quadrupole, as discussed later.

The situation is somewhat different when comparing orthogonal MALDI with conventional axial-injection MALDI. In axial injection MALDI, essentially all desorbed ions are injected directly into the TOF analyzer. In orthogonal MALDI, there are intrinsic duty cycle losses, as well as losses associated with the transmission of the quadrupole ion guides. Moreover, the TOF analyzer in orthogonal TOF uses at least one additional grid, and the voltage is typically lower, although this is not a disadvantage in principle.

In spite of these intrinsic losses associated with orthogonal MALDI, there are some features that provide compensation. The single-ion counting method is more sensitive than analog detection, particularly when weak signals are present together with strong signals. In addition, because the quality of the spectrum in orthogonal MALDI is independent of the target conditions, it is possible to make more efficient use of the deposited sample by increasing the laser fluence and irradiating the sample until it is essentially all removed from the target.

Thus, in practice, the sensitivity of orthogonal MALDI approaches that of axial MALDI. Using the simplest type of dried droplet sample preparation with 0.5 μl sample deposited on the target, about 70 amole substance P yields a usable mass spectrum (Loboda *et al.*, 2000). Significantly cleaner spectra can be achieved by using smaller spots deposited from smaller volumes. In experiments with Karger *et al.* to test an off-line interface for CE MALDI (Ens *et al.*, 2001), we deposited a mixture of four angiotensin peptides as a 100-μm wide strip. A single portion of this strip corresponding to the width of the laser beam ($\sim$100 μm) contained about 80 amoles of each of the peptides. Figure 7 shows a spectrum obtained from such a spot using 40 laser shots and with one laser shot. These spectra still provide excellent signal/noise ratios, suggesting the lower limit of sensitivity is well below 80 attomoles.

Throughput

The QqTOF mass spectrometer does not itself determine the throughput in most electrospray experiments, because the throughput is usually limited by the flow rate of the ion source. There has been some development of

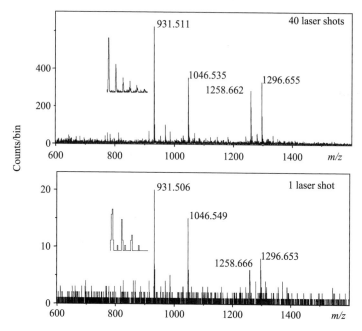

Fig. 7. Spectra of a mixture of four angiotensin peptides deposited as a 100-μm wide strip, using 40 laser shots (top) and a single laser shot (lower), incident on a single portion of this strip corresponding to the width of the laser (about 100 μm) containing about 80 attomoles of each of the peptides.

multi-spraying technologies, but these have not yet gone beyond the existing capability of the QqTOF instrument.

The situation is different in orthogonal MALDI, because there is usually no direct coupling between the target preparation and the MS measurement. If separation is carried out off-line, as in the example mentioned earlier (Ens *et al.*, 2001), the time scale of the separation is independent of the time scale of the MS measurement, and multiple separations can be carried out at the same time. Under such conditions, particularly in the high volume applications that have become the focus of several proteomics projects, the mass spectrometer itself may well be the bottleneck.

As mentioned earlier, the ultimate limit in throughput for a QqTOF instrument with single-ion counting is presented by the data system. For a repetition rate of 10 kHz and a 4-anode detector, the maximum count rate for a single ion species is about 40,000 ions/s, making use of software corrections for spectral distortion. Then for a dynamic range of 1000, assuming 40 ions is sufficient to define a peak (see Loboda *et al.*, 2000, for examples in which this is more than adequate), spectra can be acquired at the rate of 1/s. For a dynamic range of 100, the rate is 10 spectra/s.

In principle, these rates can be reached in any experiment by simply increasing the repetition rate of the laser, assuming sufficient sample is available to sustain the count rate for the duration of the acquisition. In practice, for single-MS experiments, the limit is often reached using a simple nitrogen laser with 150 μJ pulses up to 20 Hz. Indeed, even for weak signals such as those shown in Fig. 5, a spectrum like the one in Fig. 5A can be acquired in 2 s, and the one in Fig. 5B at the rate of 20/s. For these data, the extraction pulse rate was 4 kHz, so the highest count rate was about 10% of the maximum.

However, optimum acquisition of MS/MS spectra requires much more initial ion signal, because only one parent ion is selected, and the daughter ions are divided over all the peaks in the final spectrum. As a result, a 20-Hz nitrogen laser will not normally reach the limiting count rate, and good-quality MS/MS spectra typically take a minute or more to obtain using this laser. Again, the time required can be reduced to the aforementioned limits by using higher repetition rate lasers. Reasonably inexpensive kHz Nd-YAG lasers are now available, although the energy per pulse is only in the tens of microjoules range, so the laser beam must be focused to a smaller spot on the target. We have demonstrated that such a laser, appropriately focused, brings the required time to acquire good MS/MS spectra in the instrument of Fig. 1 into the range of a few seconds (Loboda *et al.*, 2000a). At such high repetition rates, the sample is consumed rather quickly, so sample rastering becomes necessary.

Quadrupole Operation

When used as a mass filter, quadrupole rods have both RF- and DC-voltage components applied, and the reduced Mathieu parameters q_M and a_M characterize the amplitudes of both components (Dawson, 1995):

$$q_M \cong \frac{4eV}{(m/z)\omega^2 r_0^2} \quad a_M \cong \frac{8eU}{(m/z)\omega^2 r_0^2}, \tag{2}$$

where e is the charge of the electron, V and ω are the amplitude and angular frequency of the RF voltage, U is the value of the DC voltage, and r_0 is the inscribed radius of the quadrupole. When $U = 0$, the quadrupole acts simply as an ion guide.

Single MS Mode

As just mentioned, the simplest mode of operation of the QqTOF instrument (single MS or TOF MS) occurs when the mass filter Q is operated in the RF-only mode ($U = 0$, $a_M = 0$), so the quadrupole sections simply act as transmission elements that deliver ions to the TOF spectrometer. This mode of operation is applicable, for example, to the measurement of the proteolytic fragments produced from enzymatic digestion of a protein (peptide mass mapping). The resulting spectra benefit from the high resolution and mass accuracy of the TOF instrument, as well as its ability to record all ions in parallel without scanning, as pointed out earlier. Such single-MS spectra can be acquired either with or without a collision gas in q. In the former case, the collision energy is kept below 10 eV to avoid fragmentation, and both sensitivity and resolution benefit from the additional collisional damping provided by the gas in the collision cell.

In this RF-only operational mode ($a_M = 0$), quadrupoles serve as high-mass filters: Ions are rejected if they have m/z values below a cutoff value corresponding to $q_M = 0.908$. Although there is no such sharp cutoff for ions with high m/z values, their transmission suffers because of poorer focusing, because the depth of the effective potential well is inversely proportional to m/z (Gerlich, 1992). Ions can be lost in any one of the three multipoles on their way to the TOF section. The resulting effective bandpass within which ions are transmitted simultaneously into the TOF instrument normally covers approximately an order of magnitude in m/z. If a wider m/z range is required, the RF voltage can be switched between two or more RF levels during spectrum acquisition, but in this case, light ions are lost when the RF voltage is "high," and heavy ions are poorly focused in a shallow potential well when the RF voltage is "low."

Tandem MS Mode

For tandem MS in the instrument shown in Fig. 1, Q is operated in the mass filter mode, so ions are transmitted only within a narrow m/z window, corresponding to $q_M = 0.706$ and $a_M = 0.237$. A monoisotopic parent ion of interest may be selected or the window may be widened to increase the sensitivity by transmitting the corresponding full isotopic cluster. Note that a measure of the distribution of isotopes selected by Q is provided by the distribution actually observed for the undissociated ions in the MS/MS spectrum.

After selection in Q, the parent ion is accelerated to an energy between 20 and 200 eV before it enters the collision cell q, where it undergoes collision-induced dissociation during the first few collisions with neutral gas molecules (usually argon or nitrogen). The resulting daughter ions (as well as the remaining parent ions) are collisionally cooled and focused as they pass through the rest of the collision cell, thus improving the quality of the ion beam entering the TOF spectrometer.

The maximum m/z transmitted through a quadrupole mass filter with reasonable efficiency depends on the choice of parameters in Eq. (1). In most commercial triple quadrupole or quadrupole/TOF spectrometers, these parameters have been chosen to give a maximum m/z value of 4000 or less. Thus, tandem measurements can be carried out in these instruments on ions of very high mass produced by electrospray (in correspondingly large charge states), but only up to masses of a few thousand daltons on the singly charged ions produced predominantly by MALDI. This mass range is usually adequate for sequencing applications, though in several instruments, the quadrupoles have been modified to give a larger m/z range for examination of ions with higher values of m/z (Collings and Douglas, 1997; Sobott et al., 2002; Whitehouse et al., 1998), and the most recent quadrupole/TOF commercial instruments also have an extended m/z range as a result of the need to analyze MALDI ions.

In general, the problems of tuning the quadrupoles in the quadrupole/TOF instruments are similar to those encountered with triple quadrupoles, as described in some detail by Yost and Boyd (1990). However, the higher pressures normally used for collisional cooling lead to long residence times in the collision cell, which tends to produce adducts and make it difficult to switch rapidly from one parent ion to another, as required in some types of measurement. Consequently, a weak axial electric field in the collision cell has been found to be beneficial, and various methods of producing such a field have been implemented (Loboda et al., 2000b; Thomson et al., 1996). The presence of the axial field means also that optimal collisional cooling can be attained for ions of all m/z values with the same buffer gas pressure. However, it is important that the method of producing the axial field does not restrict the m/z range transmitted, because the quadrupole/TOF instrument detects ions of all m/z values in the same measurement. The most recent design (Loboda et al., 2000b) appears to provide a satisfactory solution to this problem.

Parent Ion Scans in the Quadrupole

The most common mode of measurement in triple quadrupole mass spectrometers is the daughter (or product) ion scan that measures the daughter ion spectrum that results from collision-induced dissociation of a selected parent ion. As discussed earlier, quadrupole/TOF hybrid instruments have considerable advantages over triple quadrupoles for this type of measurement.

However, in sequencing unknown samples, it is helpful to be able to identify relevant parent ion peaks that may be obscured by chemical noise. For this purpose, an alternative mode of triple quadrupole operation can be used. This is the parent (or precursor) ion scan, in which the final quadrupole setting is fixed and the first quadrupole is scanned. This measures the spectrum of parent ions, giving rise to selected daughters, such as the immonium ions that characterize particular amino acid residues (Papayannopoulos, 1995) or phosphotyrosine (Steen *et al.*, 2001). Such parent ion scans are useful tools for identifying components of a mixture that lose a particular diagnostic fragment. Here, the simple arguments for preferring a TOF spectrometer as a final stage in a composite instrument do not apply, because only a single daughter is being monitored, and a similar scanning function is carried out in the first quadrupole in both the triple quadrupole and the quadrupole/TOF hybrid instrument. Moreover, the efficiency for measuring the products of collisional-induced dissociation is between one and two orders of magnitude lower in the quadrupole/TOF hybrid than in the triple quadrupole, because of the reduced duty cycle in the former instrument, as discussed earlier, and because of losses at grids and in the detector. Thus, there are advantages in doing parent ion scans in triple quadrupoles.

Nevertheless, the duty cycle losses in the quadrupole/TOF instrument can be largely removed by trapping ions in the collision cell (Chernushevich, 2001), as described earlier. This reduces the m/z range, but in this case the range often needs only to encompass the desired daughter ion, so there is no disadvantage to trapping. The higher resolution obtainable in the TOF spectrometer is also helpful in discriminating between the diagnostic daughter ions and background. For example, there are 21 possible ions from unmodified peptides that are within 0.1 Da of the phosphotyrosine immonium ion at 216.043 Da (Steen *et al.*, 2001). The final quadrupole in a triple quadrupole instrument will not likely resolve them, resulting in false-positive identifications of phosphotyrosine, but the TOF instrument easily discriminates between the possibilities (Steen *et al.*, 2001).

Another report describes a second important application of trapping in the collision cell—"peptide end sequencing" (Nielsen *et al.*, 2002). If MALDI mass mapping fails to identify a protein unambiguously after a proteolytic digestion, an MS/MS measurement is carried out in which the intensity of low mass daughters (e.g., 120–500 m/z) is accentuated by trapping. Typically the N-terminal b_2 and b_3 ions, as well as the y_1 and y_2 C-terminal daughters, can be observed, and such end sequencing of one or two tryptic peptides is usually sufficient to provide unique identification of any protein included in the database.

In addition, the TOF instrument can check for the presence of multiple diagnostic fragments at the same time, provided they are within the m/z range trapped. Thus, quadrupole/TOF instruments are becoming increasingly useful for parent ion scans.

Acknowledgments

This work was supported by grants from the U.S. National Institutes of Health (GM59240), from the Natural Sciences and Engineering Research Council of Canada, and from MDS Sciex. We thank our collaborators at Manitoba, particularly Igor Chernushevich, Andrew Krutchinsky, Alexandre Loboda, Victor Spicer, and Anatoli Verentchikov, who all made important contributions to the work reported here. We are also grateful to Andrej Shevchenko and to Bruce Thomson for informative discussions.

References

Ackloo, S., and Loboda, A. (2005). Applications of a matrix-assisted laser desorption/ionization orthogonal time-of-flight mass spectrometer. 1. Metastable decay and collision-induced dissociation for sequencing peptides. *Rapid Commun. Mass Spectrom.* **19,** 213–220.

Aleksandrov, M. L., Gall, L. N., Krasnov, N. V., Nikolayev, V. I., Pavlenko, V. A., and Shkurov, V. A. (1985). Ion extraction from solutions at atmospheric pressure. *Dokl. Akad. Nauk* **277,** 379–383, *Dokl. Phys. Chem.* **277** 572–576.

Beavis, R. C., and Chait, B. T. (1991). Velocity distributions of intact high mass polypeptide molecule ions produced by matrix assisted laser desorption. *Chem. Phys. Lett.* **181,** 479–484.

Bouneau, S., Cohen, P., Della Negra, S., Jacquet, D., Le Beyec, Y., Le Bris, J., Pautrat, M., and Sellem, R. (2003). 256-anode channel plate device for simultaneous ion detection in time of flight measurements. *Rev. Sci. Instrum.* **74,** 57–67.

Brown, R. S., and Lennon, J. J. (1995). Mass resolution improvement by incorporation of pulsed ion extraction in a linear time-of-flight mass spectrometer. *Anal. Chem.* **67,** 1998–2003.

Chaurand, P., Luetzenkirchen, F., and Spengler, B. (1999). Peptide and protein identification by MALDI and MALDI-PSD time-of-flight mass spectrometry. *J. Am. Soc. Mass Spectrom.* **10,** 91–103.

Chernushevich, I. V. (2001). Duty cycle improvement for a quadrupole time-of-flight mass spectrometer and its use for precursor ion scans. *Eur. J. Mass Spectrom.* **6,** 471–479.

Chernushevich, I. V., Ens, W., and Standing, K. G. (1997). Electrospray ionization time-of-flight mass spectrometry. *In* "Electrospray Ionization Mass Spectrometry" (R. B. Cole, ed.), pp. 204–234. John Wiley & Sons, New York.

Chernushevich, I. V., Ens, W., and Standing, K. G. (1999). Orthogonal-injection time-of-flight mass spectrometry for the analysis of biomolecules. *Anal. Chem.* **71,** 452A–461A.

Chernushevich, I.V, Loboda, A., and Thomson, B. (2001). An introduction to quadrupole/time-of-flight mass spectrometry. *J. Mass Spectrom.* **36,** 849–865.

Colby, S. M., King, T. B., and Reilly, J. P. (1994). Improving the resolution of matrix-assisted laser desorption/ionization time-of-flight mass spectrometry by exploiting the correlation between ion position and velocity. *Rapid Commun. Mass Spectrom.* **8,** 865–868.

Cole, R. B. (ed.) (1997). *In* "Electrospray Ionization Mass Spectrometry." John Wiley & Sons, New York.

Coles, J., and Guilhaus, M. (1994). Resolution limitations from detector pulse-width and jitter in a linear orthogonal-acceleration time-of-flight mass spectrometer. *J. Am. Soc. Mass Spectrom.* **5,** 772–778.

Collings, B., and Douglas, D. (1997). An extended mass range quadrupole for electrospray mass spectrometry. *Int. J. Mass Spectrom. Ion Processes* **162,** 121–127.

Cotter, R. J. (1997). *In* "Time-of-Flight Mass Spectrometry." American Chemical Society, Washington, DC.

Dawson, P. H. (ed.) (1995). *In* "Quadrupole Mass Spectrometry and Its Applications." American Institute of Physics, Woodbury, NY.

Dodonov, A. F., Chernushevich, I. V., and Laiko, V. V. (1994). Electrospray ionization on a reflecting time-of-flight mass spectrometer. *In* "Time-of-Flight Mass Spectrometry" (R. J. Cotter, ed.), pp. 108–123. American Chemical Society Symposium Series. American Chemical Society, Washington, DC.

Dodonov, A. F., Loboda, A. V., Kozlovski, V. I., Soulimenkov, I. V., Raznikov, V. V., Zhen, Z., Horwath, T., and Wollnik, H. (2000). High-resolution electrospray ionization orthogonal-injection time-of-flight mass spectrometer. *Eur. J. Mass Spectrom.* **6,** 481–490.

Douglas, D. J., and French, J. B. (1992). Collisional focusing effects in radio frequency quadrupoles. *J. Am. Soc. Mass Spectrom.* **3,** 398–407.

Ens, W., Krokhin, O., Bromirski, M., Standing, K. G., Hu, P., Rejtar, T., and Karger, B. L. (2001). Off-line coupling of CE with a MALDI QqTOF mass spectrometer. *In* "Proceedings of the 49th ASMS Conference on Mass Spectrometry and Allied Topics." Chicago, IL.

Ens, W., Standing, K. G., and Verentchikov, A. (1993). Digital and analog methods for time-of-flight mass spectrometry. *In* "Proceedings of the International Conference on Instrumentation for Time-of-Flight Mass Spectrometry" (G. J. Blanar and R. J. Cotter, eds.). LeCroy Corp., Chestnut Ridge, NY.

Gerlich, D. (1992). Inhomogeneous RF fields: A versatile tool for the study of processes with slow ions. *Adv. Chem. Phys.* **82,** 1–176.

Gonin, M., Fuhrer, K., and Schultz, J. A. (2000). A new concept to increase dynamic range using multi-anode detectors. *In* "Proceedings of the 48th ASMS Conference on Mass Spectrometry and Allied Topics." Long Beach CA.

Guilhaus, M., Selby, D., and Mlynski, V. (2000). Orthogonal acceleration time-of-flight mass spectrometry. *Mass Spectrom. Rev.* **19,** 65–107.

Gross, M. L. (1990). Tandem mass spectrometry: Multisector magnetic instruments. *Methods Enzymol.* **193,** 131–153.

Hayes, R. N., and Gross, M. L. (1990). Collision-induced dissociation. *Methods Enzymol.* **193,** 237–263.

Hillenkamp, F., Karas, M., Beavis, R. C., and Chait, B. T. (1991). Matrix-assisted laser desorption/ionization mass spectrometry of biopolymers. *Anal. Chem.* **63,** 1193A–1203A.

Hoyes, J. B., Bateman, R. H., and Wildgoose, J. L. (2000). A high resolution time-of-flight mass spectrometer with selectable drift length. *In* "Proceedings of the 48th ASMS Conference on Mass Spectrometry and Allied Topics." Long Beach, CA.

Karas, M., and Hillenkamp, F. (1988). Laser desorption ionization of proteins with molecular masses exceeding 10.000 daltons. *Anal. Chem.* **60,** 2299–2301.

Krutchinsky, A. N., Chernushevich, I. V., Spicer, V., Ens, W., and Standing, K. G. (1998). A collisional damping interface for an electrospray ionization TOF mass spectrometer. *J. Am. Soc. Mass Spectrom.* **9,** 569–579.

Krutchinsky, A. N., Loboda, A. V., Spicer, V. L., Dworschak, R., Ens, W., and Standing, K. G. (1998a). Orthogonal injection of MALDI ions into a time-of-flight spectrometer through a collisional damping interface. *Rapid Commun. Mass Spectrom.* **12,** 508–518.

Krutchinsky, A. N., Zhang, W., and Chait, B. T. (2000). Rapidly switchable matrix-assisted laser desorption/ionization and electrospray quadrupole time-of-flight mass spectrometry for protein identification. *J. Am. Soc. Mass Spectrom.* **11,** 493–504.

Loboda, A. V., Ackloo, S., and Chernushevich, I. V. (2003). A high-performance matrix-assisted laser desorption/ionization orthogonal time-of-flight mass spectrometer with collisional cooling. *Rapid Commun. Mass Spectrom.* **17,** 2508–2516.

Loboda, A. V., Krutchinsky, A. N., Bromirski, M., Ens, W., and Standing, K. G. (2000a). A tandem quadrupole/time-of-flight mass spectrometer with a matrix-assisted laser desorption/ionization source: Design and performance. *Rapid Commun. Mass Spectrom.* **14,** 1047–1057.

Loboda, A., Krutchinsky, A., Loboda, O., McNabb, J., Spicer, V., Ens, W., and Standing, K. G. (2000b). Novel linac II electrode geometry for creating an axial field in a multipole ion guide. *Eur. J. Mass Spectrom.* **6,** 531–536.

Mamyrin, B. A., Karataev, V. I., Shmikk, D. V., and Zagulin, V. A. (1973). The mass-reflectron, a new nonmagnetic time-of-flight mass spectrometer with high resolution. *Sov. Phys. JETP* **37,** 45–48.

Mlynski, V., and Guilhaus, M. (1996). Matrix-assisted laser/desorption ionization time-of-flight mass spectrometer with orthogonal acceleration geometry: Preliminary results. *Rapid Commun. Mass Spectrom.* **10,** 1524–1530.

Morris, H. R., Paxton, T., Dell, A., Langhorne, J., Berg, M., Bordoli, R. S., Hoyes, J., and Bateman, R. H. (1996). High-sensitivity collisionally-activated decomposition tandem mass spectrometry on a novel quadrupole/orthogonal-acceleration time-of-flight mass spectrometer. *Rapid Commun. Mass Spectrom.* **10,** 889–896.

Nielsen, M. L., Bennett, K. L., Larsen, B., Moniatte, M., and Mann, M. (2002). Peptide end sequencing by orthogonal MALDI tandem mass spectrometry. *J. Proteome Res.* **1,** 63–71.

Papayannopoulos, I. A. (1995). The interpretation of collision-induced dissociation tandem mass spectra of peptides. *Mass Spectrom. Rev.* **14,** 49–73.

Pinkston, J. D., Rabb, M., Watson, I. T., and Allison, I. (1986). New time-of-flight mass spectrometer for improved mass resolution, versatility, and mass spectrometry/mass spectrometry studies. *Rev. Sci. Instrum.* **57,** 583–589.

Piyadasa, C. K. G., Håkansson, P., and Ariyaratne, T. R. (1999). A high resolving power multiple reflection MALDI time-of-flight mass spectrometer. *Rapid Commun. Mass Spectrom.* **13,** 620–624.

She, Y.-M., Haber, S., Seifers, D. L., Loboda, A., Chernushevich, I., Perreault, H., Ens, W., and Standing, K. G. (2001). Determination of the complete amino acid sequence for the coat protein of brome mosaic virus by time-of-flight mass spectrometry: Evidence for mutations associated with change of propagation host. *J. Biol. Chem.* **276,** 20039–20047.

Shevchenko, A. A., Chernushevich, I. V., Ens, W., Standing, K. G., Thomson, B., Wilm, M., and Mann, M. (1997). Rapid *de novo* peptide sequencing by a combination of nanoelectrospray, isotopic labeling and a quadrupole/time-of-flight mass spectrometer. *Rapid Commun. Mass Spectrom.* **11,** 1015–1024.

Shevchenko, A., Loboda, A., Ens, W., and Standing, K. G. (2000). MALDI quadrupole time-of-flight mass spectrometry: A powerful tool for proteomic research. *Anal. Chem.* **72,** 2132–2141.

Sobott, F., Hernandez, H., McCammon, M. G., Tito, M. A., and Robinson, C. V. (2002). A tandem mass spectrometer for improved transmission and analysis of large supramolecular assemblies. *Anal. Chem.* **74,** 1402–1407.

Spengler, B. (1997). Post-source decay analysis in matrix-assisted laser desorption/ionization mass spectrometry of biomolecules. *J. Mass Spectrom.* **32,** 1019–1036.

Spengler, B., and Cotter, R. J. (1990). Ultraviolet laser desorption/ionization mass spectrometry of proteins above 100,000 daltons by pulsed ion extraction time-of-flight analysis. *Anal. Chem.* **62,** 793–796.

Steen, H., Küster, B., Fernandez, M., Pandey, A., and Mann, M. (2001). Detection of tyrosine phosphorylated peptides by precursor ion scanning quadrupole TOF mass spectrometry in positive ion mode. *Anal. Chem.* **73,** 1440–1448.

Tang, X., Beavis, R., Ens, W., Lafortune, F., Schueler, B., and Standing, K. G. (1988). A secondary ion time-of-flight mass spectrometer with an ion mirror. *Int. J. Mass Spectrom. Ion Processes* **85,** 43–67.

Thomson, B., Jolliffe, C., and Javahery, R. (1996). RF-only quadrupoles with axial fields. *In* "Proceedings of the 44th ASMS Conference on Mass Spectrometry and Allied Topics." Portland, OR.

Tolmachev, A. V., Chernushevich, I. V., Dodonov, A. F., and Standing, K. G. (1997). A collisional focusing ion guide for coupling an atmospheric pressure ion source to a mass spectrometer. *Nucl. Instr. Meth.* **B124,** 112–119.

Verentchikov, A. N., Ens, W., and Standing, K. G. (1994). Reflecting time-of-flight mass spectrometer with an electrospray ion source and orthogonal extraction. *Anal. Chem.* **66,** 126–133.

Vestal, M. L., Juhasz, P., and Martin, S. A. (1995). Delayed extraction matrix-assisted laser desorption time-of-flight mass spectrometry. *Rapid Commun. Mass Spectrom.* **9,** 1044–1050.

Whitehouse, C., Gulcicek, E., Banks, J. F., and Andrien, B. (1998). A two-dimensional ion trap API TOF mass spectrometer. *In* "Proceedings of the 46th ASMS Conference on Mass Spectrometry and Allied Topics." Orlando, FL.

Wiley, W. C., and McLaren, I. H. (1955). Time-of-flight mass spectrometer with improved resolution. *Rev. Sci. Instrum.* **26,** 1150–1157.

Wollnik, H., and Przewloka, M. (1990). Time-of-flight mass spectrometers with multiply-reflected ion trajectories. *Int. J. Mass Spectrom. Ion Processes* **96,** 267–274.

Xu, H. J., Wada, M., Tanaka, J., Kawakami, H., Katayama, I., and Ohtani, S. (1993). A new cooling and focusing device for ion guide. *Nucl. Instr. Meth.* **A333,** 274–281.

Yamashita, M., and Fenn, J. B. (1984). Electrospray ion source: Another variation on the free-jet theme. *J. Chem. Phys.* **88,** 4451–4459.

Yost, R. A., and Boyd, R. K. (1990). Tandem mass spectrometry: Quadrupole and hybrid instruments. *Methods Enzymol.* **193,** 154–200.

Further Reading

Fenn, J. B. (2000). Mass spectrometric implications of high-pressure ion sources. *Int. J. Mass Spectrom. Ion Processes* **200,** 459–478.

[3] Tandem Time-of-Flight Mass Spectrometry

By MARVIN L. VESTAL and JENNIFER M. CAMPBELL

Abstract

A new tandem time-of-flight (TOF–TOF) instrument has been developed by modifying a standard matrix-assisted laser desorption ionization (MALDI)-TOF instrument to make high-performance, high-energy collision-induced dissociation (CID) MALDI tandem mass spectrometry (MS) a practical reality. To optimize fragment spectra quality, the selected precursor ion is decelerated before entering a floating collision cell and the potential difference between the source and the collision cell defines the collision energy of the ions. Standard operating conditions for tandem MS use a 1-kV collision energy with single-collision conditions and increased laser power for ion formation. Hence, both high- and low-energy fragments are observed in MALDI TOF-TOF spectra. On standard peptides, sensitivities down to 1 fmol are demonstrated. On a mixture of two solution tryptic digests at the 25-fmol level, 23 spectra were sufficient to result in proper database identification.

Introduction

It is generally accepted that MS is essential for protein identification and characterization. MALDI-TOF is typically the first method employed for protein identification, used for the determination of accurate masses of peptides formed by enzymatic digestion, and MS-MS in various forms is used both as a more definitive method for identification and as the principal means for protein characterization (Eng *et al.*, 1994; Mann and Wilm, 1994). Two-dimensional (2D) gel electrophoresis is by far the most widely accepted technique for high-resolution separation of protein mixtures, and alternatives such as micro–high-performance liquid chromatography (HPLC) and capillary electrophoresis have only recently been seriously considered (Davis *et al.*, 1995; Olivares *et al.*, 1987). Advances in MALDI-TOF MS combined with advances in 2D gel electrophoresis and other separation techniques promise to revolutionize the speed and sensitivity of the separation, quantitation, identification, and characterization of proteins in complex mixtures (Shevchenko *et al.*, 1996).

Tandem MS is now established as the method of choice for characterizing proteins, although no single MS-MS instrument or technique has established dominance (Ducret *et al.*, 1998; Shevchenko *et al.*, 1997; Wilm

METHODS IN ENZYMOLOGY, VOL. 402   0076-6879/05 $35.00
DOI: 10.1016/S0076-6879(05)02003-3

et al., 1996). In these techniques, peptide mixtures are introduced into the mass spectrometer either as a continuous flow of a liquid solution, such as in nanospray, or as described later for MALDI-TOF. A molecular ion of interest is selected by the first MS, ions are caused to fragment, usually by collision with a neutral gas, and the fragment ion masses and intensities are measured using the second MS. Most MS-MS applications employ triple quadrupoles (Yost and Enke, 1978), hybrid quadrupole-TOF systems (Morris *et al.*, 1996) or ion traps, either quadrupole (Louris *et al.*, 1987) or magnetic (as in Fourier transform–ion cyclotron resonance [FT-ICR]) (Cody and Frieser, 1982). These techniques employ low-energy CID in which the ions are fragmented by a large number of relatively low-energy collisions. An alternative technique is high-energy CID in which the collision energy is sufficient to cause fragmentation as the result of a single collision, and the possible number of collisions that the ions undergo is small (i.e., <10). Before the present work, high-energy CID was available only on tandem magnetic sector instruments (Sato *et al.*, 1987) or a hybrid of a magnetic sector with TOF (Medzihradszky *et al.*, 1996). These instruments are complex and expensive and are not readily interfaced with sensitive ionization techniques such as MALDI and electrospray.

Description of the Instrument

The new TOF-TOF instrument is derived from a standard high-performance MALDI-TOF instrument (Vestal *et al.*, 1995) that has been modified to make high-performance, high-energy CID MALDI-TOF-TOF MS-MS a practical reality. An earlier version of the TOF-TOF system was based on a very large MALDI-TOF instrument (Vestal *et al.*, 2000a). The instrument described here is based on concepts similar to those described earlier, but several modifications have been made to improve sensitivity and to achieve better resolution and mass accuracy in a smaller format. The ion source optics and electronics are essentially unmodified from those employed on the commercial instrument, except that the high-voltage pulse circuit providing the delayed extraction pulsed voltage has been modified to operate at higher repetition rates. The construction of the ion mirror is similar to that used earlier except that the single-stage mirror has been replaced with a two-stage mirror to provide higher resolution over a broad energy range. The additional elements required for MS-MS operation are installed in the field-free region between the ion source and the ion mirror. These include a set of deflection electrodes for precursor selection, a deceleration lens, an electrically isolated collision cell, a second pulsed ion accelerator, a set of deflection electrodes for metastable suppression, and additional focusing and beam-steering electrodes. An additional 250-L/s turbo molecular pump was added to the system to evacuate the housing surrounding the collision cell. Similar pumps are employed on the source and mirror housing as in the original MALDI-TOF instrument.

A schematic diagram of the TOF-TOF instrument is shown in Fig. 1. For peptide mass fingerprinting or other applications of MALDI-TOF, no voltage is applied to the elements added for MS-MS, and the system performance in both linear and reflector modes is equal to or better than the conventional system described earlier (Vestal *et al.*, 1995). By applying appropriate direct current (DC) and pulsed voltages to the added elements, the system is converted to a tandem combination of a linear delayed extraction MALDI-TOF system for producing and selecting a precursor ion, a collision cell for energizing the selected ions to fragment, and a second reflecting TOF system for focusing and detecting the fragment ions. A field-free space is provided between the collision cell and the second analyzer to allow time for the ions to fragment before they are reaccelerated.

Ions produced in the source are focused in space and time at the timed-ion-selector (*TIS* in Fig. 1). A time delay generator is programmed to open the gate of the selector at the time the lightest mass of interest reaches the gate, and the gate is closed when the highest mass of interest passes through the gate. Ions outside the selected mass window are deflected away from the collision cell. Selected ions are retarded by the deceleration lens and enter the collision cell with the selected laboratory collision energy, defined by the potential difference between the source and the collision cell. The laboratory collision energy may be varied by adjusting the ion

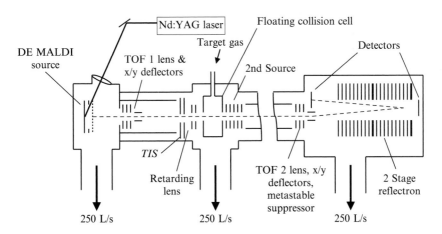

FIG. 1. Instrumental apparatus. See text for details of operation.

source potential relative to that of the collision cell. Fragment ions produced in the collision cell or in the adjacent field-free region continue to travel with essentially the same velocity as the precursor ions until they reach the second pulsed accelerator. After the ions enter the pulsed accelerator, a high-voltage pulse is applied to accelerate the selected precursor and fragment ions, and the amplitude of the pulse and the voltages applied to the ion mirror are adjusted to produce optimum resolution at the detector. With the apparatus illustrated in Fig. 1, excellent resolution is obtained over the full mass range of the fragment spectrum with a single set of parameters.

Fragmentation may also occur in the field-free region between the source and the selector or in the region between the pulsed accelerator and the mirror. Fragments produced in the first region are rejected by the deceleration lens, but fragments occurring in the second region may produce broad unfocused peaks similar to those observed in conventional MALDI–post-source decay (PSD) (Kaufmann et al., 1994). These "metastable" peaks may be removed by applying a deflection pulse to the second ion deflector (labeled "metastable suppressor" in Fig. 1) at the time that the remaining precursor ions and fragments produced in this field-free region reach the vicinity of the deflector. This deflector can be adjusted to completely remove the precursor ion, but generally the applied voltage is set to allow about 3–4% of the precursor ions to be transmitted. This lower voltage reduces the noise due to metastables to an acceptable level while still allowing the transmitted precursor ions to be used as reference peaks for calibration of the fragment spectra.

The TOF-TOF system may employ any ultraviolet (UV) laser appropriate for conventional MALDI. Our earlier work employed a nitrogen laser (337 nm) at pulse rates between 2 and 20 Hz (Medzihradszky et al., 2000). A diode-pumped Nd:YAG laser operating at 355 nm has been used at rates between 20 Hz and 2 kHz (Vestal et al., 2000b). Most of the results presented here were obtained at a laser rate of 200 Hz. Although operating parameters may need to be slightly altered as the laser rate is changed, there is no effect of the acquisition speed on the performance of the instrument. It should be noted, however, that higher laser rates allow data to be obtained more rapidly, and the ability to store and process the data may become limiting.

Performance Data

Precursor Selection

The selected ions, along with fragments produced (either by unimolecular dissociation or CID) in this field-free region, pass through the collision cell, enter the pulsed ion accelerator (labeled "2nd Source" in Fig. 1), and are accelerated into the second MS by applying a high-voltage pulse (typically 7 kV). The timed ion selector is located in the field-free space between the ion source and the collision cell, and the time that the selector is opened and the later time when the acceleration pulse is applied are calculated, as has been described previously (Vestal and Juhasz, 1998). If the timed ion selector is not used, then ions arriving at the accelerator after the pulse is applied do not enter the second MS. A spectrum obtained with the selector off and the accelerator pulse timed to focus the matrix trimer ion, nominal mass 568, is shown in Fig. 2A. The sharp peaks at lower mass are due to fragments formed both in the collision cell and in the field-free region between the collision cell and the accelerator, and the broad peaks are due to lower mass matrix ions, which are formed in the source and are transmitted, with relatively low efficiency, through the second source and into the second MS before application of the acceleration pulse. With the selector off or set to transmit a relatively wide window, the mass range for precursor ions that can be reaccelerated and focused spans about 5% of the nominal mass chosen. Precursor ions and their fragments within this mass range are focused and the masses of the fragments of all transmitted precursors can be accurately determined. Thus, it is possible to determine the relative intensities of fragment ions from nearby precursors. This feature may be useful for quantitation using isotopic labeling techniques such as isotope-coded affinity tags (ICATs) (Gygi et al., 1999).

The spectrum obtained with precursor set to transmit only the matrix trimer ion is shown in Fig. 2B. Laser intensity and other parameters are identical to those used in Fig. 2A. As can be seen from comparing Fig. 2A and B, the m/z 568 is transmitted unattenuated, the isotope peak at m/z 569 is attenuated by about an order of magnitude, and only the focused fragment ions are detected at lower mass. This resolving power for precursor selection is about 500 using a 10% valley criterion or about 1000 FWHM. The former value accurately represents the ability to select a particular precursor with minimal contamination by adjacent precursor masses. This resolution is sufficient to allow selection of the isotopic envelope of peptides throughout the mass range of interest but is not adequate for selecting a single isotope at higher mass. A single isotope can be selected up to m/z 1000, but the sensitivity is significantly reduced. The current design of the timed ion selector provides higher resolving power for fragments than for the precursor; this ensures that the relative contribution from fragment ions resulting from precursors with masses similar to the selected ion that are not completely suppressed is always less than that of the selected ion itself. As is discussed below, in many cases the ability to transmit and focus neighboring ions allows more than one peptide to be identified even if they

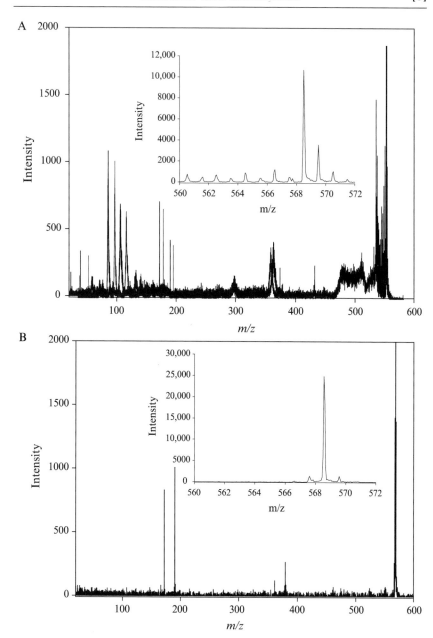

FIG. 2. MS-MS spectrum of the trimer ions of CHCA recorded (A) without and (B) with the timed ion selector being operated. The insets highlight the precursor region of the spectra.

have the same nominal mass or differ by an amount that is insufficient to allow complete separation by the precursor selector.

Sensitivity

Under ideal conditions, the ionization efficiency of MALDI is very high. In peptide mass fingerprinting, absolute ionization efficiency is not a primary concern because chemical noise from matrix components and impurities is usually the major factor limiting sensitivity. In MS-MS, the quality of the fragment spectrum is determined by the total number of ions accumulated in that spectrum; only at very low sample concentrations is the chemical noise the limiting factor. Using α-cyano-4-hydroxycinnamic acid (CHCA) matrix, we have measured overall efficiencies of about 1% for peptides loaded at low concentrations by desorbing all of the sample loaded and integrating the total number of ions detected. These efficiency measurements include the transfer efficiency of the analyzer (estimated to be ~40% when operated in MS-MS mode). Efficiencies for other popular matrices (e.g., 2,5-dihydroxybenzoic acid [DHB]) may be as much as an order of magnitude lower. Examples of some results showing the sensitivity in MS-MS mode that is routinely achieved on simple mixtures of peptides are shown in Figs. 3–5. These results were obtained from a mixture of three peptides (des-Arg[1] bradykinin, angiotensin I, and Glu[1] fibrinopeptide B) at equimolar concentrations. These samples were loaded at total amounts ranging from 1 pmol to 1 fmol each in 3-mm diameter sample spots using a standard dried droplet method. The resulting average surface concentrations range from approximately 1 fmol/100 micron diameter laser spot down to 1 attomole/laser spot. At the laser intensities used for these measurements, about 1000 laser shots can be accumulated from each laser spot before that portion of the sample spot is depleted, and the relative response from each laser spot within the sample spot varies by about a factor of five.

The result shown in Fig. 3B represents the accumulation of 1000 shots on a "good" laser spot from the 10 fmol sample. Figure 3C shows the result of 10,000 shots accumulated from several laser spots from the 1 fmol sample, and Fig. 3A is the accumulated spectrum from 5000 shots from the 1 pmol sample. In each case, a spectra from all laser shots fired were averaged into the final result; selective averaging was not employed. Approximately 10 attomoles of sample was consumed in obtaining the spectra shown in Fig. 3B and C, and 5 fmol was consumed to obtain spectrum A. In all spectra, internal fragment ions are indicated by asterisks, and neutral losses from the labeled ions are indicated by plus signs. These spectra were obtained with the metastable suppressor in operation, thereby

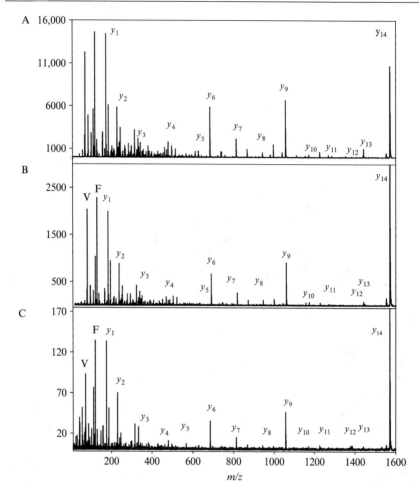

Fɪɢ. 3. MS-MS spectra of Glu[1] fibrinopeptide B (EGVNDNEEGFFSAR) recorded with the laser operated 20% above threshold, N_2 as a collision gas, and single-collision conditions for (A) 5000 shots from a 1 pmol sample spot, (B) 1000 shots from a 10 fmol sample spot, and (C) 10,000 shots from a 1 fmol sample spot.

reducing the precursor ion intensity by about a factor of 30. The total integrated ion intensity of the precursor ion (before attenuation) is approximately equal to the total integrated fragment ion intensity in these spectra. The total number of fragment ions integrated in Fig. 3C is approximately 50,000; about 100,000 in Fig. 3B; and 5,000,000 in Fig. 3A. At the highest concentration, the ionization efficiency is lower by about a factor of 10

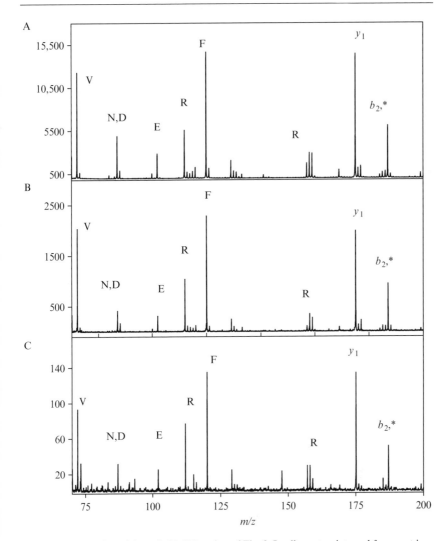

Fig. 4. Expansion of the m/z 70–200 region of Fig. 3. In all spectra, internal fragment ions are indicated by asterisks, and neutral losses from the labeled ions are indicated by plus signs.

because of depletion of the matrix ions. Reliable detection and mass measurement of a fragment ion requires collection of at least 10 ions in the two or three digitizer channels that define the peak. The relative intensities of several of the structurally significant y and w ions (e.g., 8, 10, 11, and 12) in this particular case are in the range of 1–3% of the most

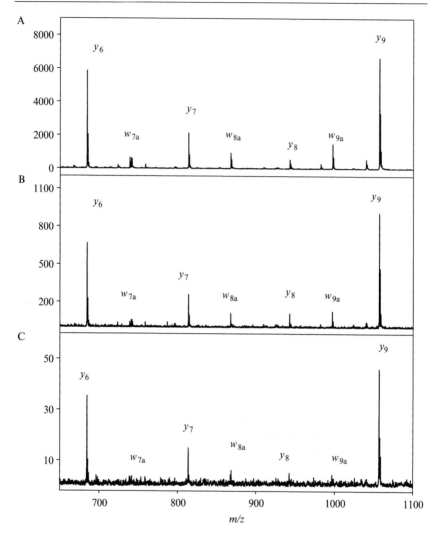

FIG. 5. Expansion of the m/z 650–1100 region of Fig. 3.

intense fragment peak. Nevertheless, these ions are detectable and the dominant fragment masses can be determined with an uncertainty of less than 50 ppm. Portions of the spectra are expanded in Figs. 4 and 5 to better illustrate the quality of the spectra. A representative set of masses, intensities, and mass errors is summarized in Table I.

TABLE I
MASS ACCURACIES FOR THE GLU[1] FIBRINOPEPTIDE B SPECTRUM SHOWN IN FIGS. 3–5, PANEL A[a]

	m/z	Observed	Height	Δ (Da)	Δ (ppm)
V	72.0813	72.0973	2517	−0.0160	−221.7
L	87.0558	87.0708	908	−0.0150	−172.3
F	120.0813	120.0946	3735	−0.0133	−110.9
R	129.1140	129.1190	1078	−0.0050	−38.8
y_1-NH$_3$	158.0930	158.1003	1067	−0.0073	−46.4
y_1	175.1195	175.1194	8039	0.0001	0.9
y_2-NH$_3$	229.1301	229.1310	833	−0.0009	−4.0
y_2	246.1566	246.1540	1639	0.0026	10.6
y_3-NH$_3$	316.1621	316.1512	1411	0.0109	34.5
y_3	333.1886	333.1801	1171	0.0085	25.4
y_4	480.2571	480.2445	1168	0.0126	26.3
y_6	684.3469	684.3455	4810	0.0014	2.0
y_7	813.3895	813.3967	2122	−0.0072	−8.9
w_8	867.4001	867.4621	856	−0.0620	−71.4
w_9	996.4427	996.5135	1195	−0.0708	−71.0
y_9-NH$_3$	1039.4485	1039.4786	830	−0.0301	−29.0
y_9	1056.4750	1056.4893	7817	−0.0143	−13.5
y_{13}	1441.6348	1441.7113	1411	−0.0765	−53.1

[a] Columns denote the identity of the fragment, the theoretical mass of the peptide, the observed mass using default instrumental calibration, the height of the peak, and mass errors in m/z and in ppm.

The minimum measurable peak in these examples ranges from 0.02% of total fragment intensity in Fig. 3C, 0.01% in Fig. 3B, and 0.0002% in Fig. 3A. The more intense peaks in the fragment spectra are typically 1–2% of the total fragment intensity; thus, the dynamic range in the spectra obtained at low sample levels is about 100. By accumulating a large number of laser shots, the dynamic range may exceed 10,000 at sample concentrations above approximately 1 fmol/laser spot. Unfortunately, as the dynamic range exceeds about 1000, a peak is observed at virtually every possible fragment mass, and these additional peaks may be of little utility in elucidating structure. At very low concentrations, increasing the number of ions integrated beyond about 100,000 may not improve the quality of the spectra significantly, because ions due to chemical noise from matrix components and impurities within the selected precursor mass window may be the limiting factor.

The total number of laser shots that must be accumulated to achieve a particular result depends on several factors. If the goal is to measure the masses of only the most intense peaks for peptide identification by database searching, then accumulation of a few thousand ions may be sufficient. For samples containing fewer than 10 peptides at surface concentrations higher than about 10 attomoles/laser spot, this requires only about 100 laser shots. On the other hand, acquiring the same number of ions from a particular component present at this concentration in a mixture of 100 peptides requires 10 times as many laser shots. If the goal is *de novo* sequencing or determination of posttranslational modifications, then a larger number of ions and laser shots is required. Accumulation of 10,000 shots is usually sufficient for even the most demanding case. As several hundred thousand shots are generally required to significantly deplete the sample with standard dried droplet sample preparation, there is no problem recording several high-quality MS-MS spectra from each sample spot.

Fragmentation

The fragmentation processes occurring in the TOF-TOF instrument are closely related to those occurring in the PSD process (Kaufmann *et al.*, 1994) employed in most MALDI-TOF mass spectrometers equipped with reflecting analyzers. Ions may be excited sufficiently in the MALDI source to spontaneously decay after extraction, or they may fragment as the result of an energetic collision with a neutral molecule. This technique has been useful, but its utility has been limited by a number of instrumental factors. As discussed earlier in this chapter, these limitations on speed, sensitivity, resolution, and mass accuracy, as well as ease of use, have been overcome in the TOF-TOF system.

Fragmentation of peptide ions in various MS-MS instruments has been extensively reviewed (Shulka and Futrell, 2000). The energy imparted to the ions during the formation process, which leads to unimolecular fragmentation, is strongly dependent on the ionization method. In MALDI, both the laser intensity and the properties of the matrix affect the distribution of internal energy in the peptide ions. Near threshold for ionization, little fragmentation is observed, but as the laser intensity is increased above threshold, significant unimolecular fragmentation may occur. More fragmentation is generally observed for "hot" matrices such as CHCA than for "cooler" matrices such as DHB. At modest laser intensities (~10–20% above threshold), the unimolecular fragmentation is dominated by simple peptide bond cleavages to yield the y and b ions observed in low-energy CID and in PSD. At high laser intensities, the fragments due to small neutral losses (e.g., ammonia and water) from these y and b ions predominate, and a ions may also increase with laser intensity. In general, immonium ions and internal fragments are weak in the unimolecular spectra even at high laser intensity. Increasing the laser intensity beyond 20–30%

above threshold rapidly increases the rate of sample consumption, but the observed ion intensities tend to approach a saturation value, indicating reduced ionization efficiency under these conditions. Spectra recorded in the TOF-TOF when no gas is added to the collision cell are very similar to those obtained in low-energy CID experiments on singly charged ions in other MS-MS instruments such as hybrid quadrupole TOF (Anderson *et al.*, 1998) or quadrupole ion traps (Doroshenko and Cotter, 1996).

Fragmentation in the TOF-TOF instrument may also be produced by high-energy collisions with a neutral gas. This approach is similar to that employed in the tandem magnetic sector instruments, and the results obtained are comparable even though fast atom bombardment ionization was employed in the earlier work (Johnson *et al.*, 1988). The major differences appear to be that the TOF-TOF provides much higher detection efficiency and resolution over the entire mass range of the fragments, while the sector instruments detected the low mass ions with rather low efficiency. In high-energy collisions, the possible internal energy gain after the collision is determined by the laboratory kinetic energy of the ions, the mass of the neutral partner, and the number of collisions that occur. This energy is added to that imparted by the ionization process. The effect from all of these variables has been studied over limited ranges, but most of the routine work has employed a laboratory collision energy of 1 kV, argon, neon, or nitrogen as collision gas, and collision gas pressures corresponding to average collision numbers between zero and three. These conditions appear to be satisfactory for a wide range of peptides observed in protein digests, but there are some indications that either higher laboratory energy or higher mass collision partners may be desirable for producing more extensive fragmentation of peptides with masses above m/z 3000. A few selected examples of results of these studies on fragmentation are given below; the complete results are published elsewhere (Campbell, 2003).

The fragment spectra obtained at different collision gas pressures for Glu[1] fibrinopeptide B (EGVNDNEEGFFSAR), a typical pseudo-tryptic peptide with a C-terminal arginine and no other basic residues, are shown in Figs. 6–10. The spectrum in Fig. 6A was obtained with no collision gas added, Fig. 6B at a pressure corresponding to about one collision, on average, and Fig. 6C to about two collisions. Because Poisson statistics apply, for the condition used for recording Fig. 6B, about 37% of the ions undergo one collision, and 26% two or more. In Fig. 6C, 27% of the ions undergo one collision, and 59% two or more. With no collision gas, the observed intensities of immonium ions and internal fragments are very weak, and no v and w ions due to side-chain cleavages are observed. As collision gas is added, the pattern of the y ions is only slightly affected, with the low mass ions increasing in intensity relative to the higher masses.

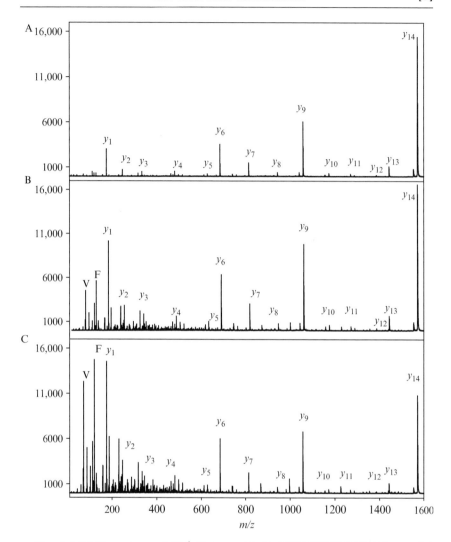

FIG. 6. MS-MS spectrum of Glu[1] fibrinopeptide B (EGVNDNEEGFFSAR) recorded with the laser operated 20% above threshold and with (A) no gas added to the collision cell, and N_2 added to the collision cell such that the average number of collisions was (B) one and (C) two.

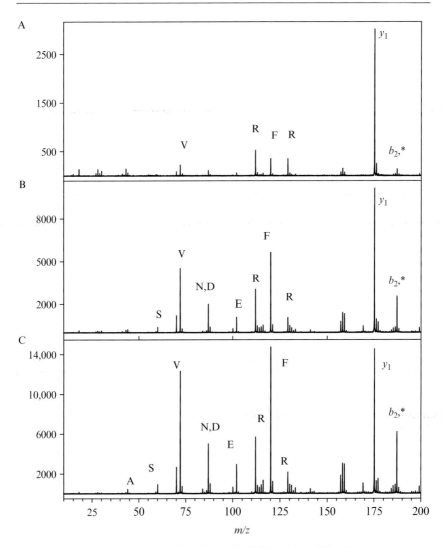

FIG. 7. Expansion of the m/z 10–200 region of Fig. 6.

In contrast, the immonium ions, internal fragment ions, and side-chain cleavage ions all increase approximately in proportion to the average collision number. This behavior appears to be independent of the peptide sequence and can be used to determine whether a proposed assignment of a fragment peak is plausible. In these measurements, the dynamic range is

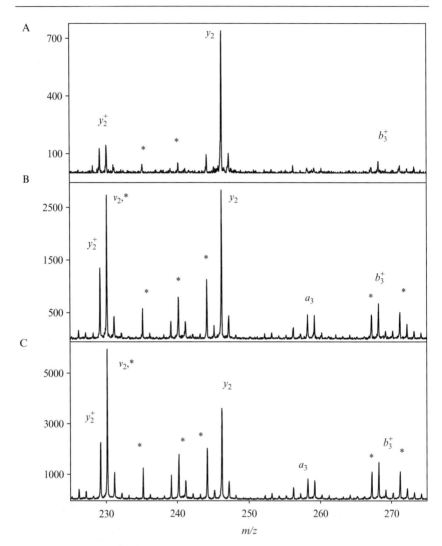

FIG. 8. Expansion of the m/z 225–275 region of Fig. 6.

greater than 1000, and in the range where dipeptide and tripeptide internal fragments are observed (Figs. 8 and 9), a detectable peak is observed at each nominal mass. These peaks of low relative intensity are generally not useful for determining structure. On the other hand, the intensities of several of the y, v, and w ions in the high mass region (Fig. 10) are 1% or

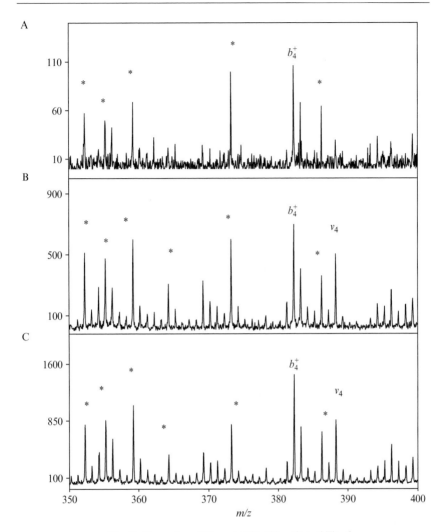

FIG. 9. Expansion of the m/z 350–400 region of Fig. 6.

less of the most intense fragment peaks, and these peaks, which are essential for determining the sequence, may not be reliably measured if the dynamic range is less than 100.

Substance P (RPKPQQFLM-NH$_2$) has been studied using a wide variety of ion-activation methods, including surface-induced (Nair *et al.*, 1996), photo-induced (Barbacci and Russell, 1999), and electron capture–induced

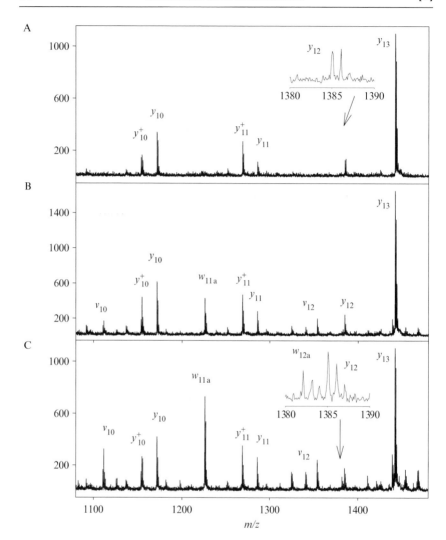

FIG. 10. Expansion of the m/z 1075–1500 region of Fig. 6.

(Axelsson *et al.*, 1999) dissociations. Additionally, Substance P has been widely adopted as a "test molecule" for CID, and several studies describing the fragmentation of substance P in sector instruments are available (Hayes and Gross, 1990). With arginine at the N-terminus, lysine at residue 3, and no basic residue near the C-terminus, the positive charge tends to reside on the N-terminal fragments, in contrast to the previous example, and the fragment spectrum is dominated by *a, b,* and *d* ions. An extensive study of the fragmentation of substance P has been completed to quantitatively determine the dependence of fragmentation in TOF-TOF on operating parameters (Campbell, 2003). Results from that study are selected to demonstrate some specific attributes.

Total fragmentation efficiency as a function of gas pressure in the housing surrounding the collision cell is shown for laboratory collision energies of 1 and 2 kV in Fig. 11. The pressure in the collision cell is estimated to be about three orders of magnitude higher but is not measured directly, and the ionization gauge readings are uncorrected for its response to a particular gas relative to air. In this example, neon was the collision gas, but qualitatively similar results were obtained for all of the gases studied including He, Ar, and Kr. These results were obtained by integrating the total fragment ion intensity, and separately, the total precursor ion intensity over the isotopic envelope. For all the experiments, the laser was operated near threshold; thus, the contribution of unimolecular reactions to the total fragment ion intensity is minimal. Although there is considerable scatter in these data, it is clear that fragmentation efficiency

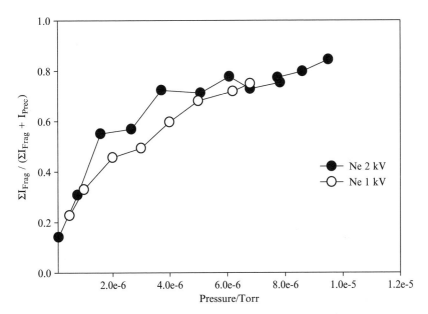

FIG. 11. Fragmentation efficiency as a function of collision cell housing pressure for MS-MS of substance P using near-threshold laser intensity, Ne as a collision gas, and 1 kV (○) and 2 ktV (●) laboratory frame collision energies.

in the 1–2 collision pressure regimen is enhanced at the higher collision energy. Inspection of the spectra shows that most added fragment ion intensity is observed as immonium ions and internal fragments. This trend is further illustrated in Fig. 12, where the total immonium and the total precursor ion intensities relative to total ion intensity are plotted as a function of housing pressure for neon at 2 kV. In this case, the total ion intensity is confined to those peaks that are assigned to expected fragments of the peptides. As a result, the low-intensity peaks that are not assigned are neglected in this example.

A portion of the substance P fragment spectrum obtained under three operating conditions is shown in Fig. 13. In Fig. 13A, a sample spectrum from the data of Fig. 11, the laser was operated near threshold, and the gas pressure was adjusted so that the precursor was attenuated by about 70%. The data in Fig. 13B were obtained under identical conditions except that the laser intensity was increased by about 20%, and the data in Fig. 13C used the same high laser intensity, but the collision gas was removed. Except for the higher relative intensity of the a-17 ions, the spectrum in Fig. 13A is virtually identical to that obtained by high-energy CID in a four-sector instrument (Hayes and Gross, 1990). Clearly, the b ions are

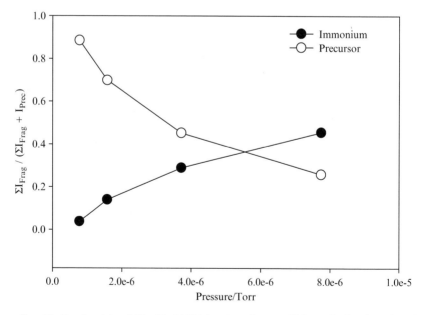

FIG. 12. For the data of Fig. 11, 2-kV laboratory frame collisions, the fraction of total assigned ion intensity that is observed as immonium (●) and precursor (○) ions.

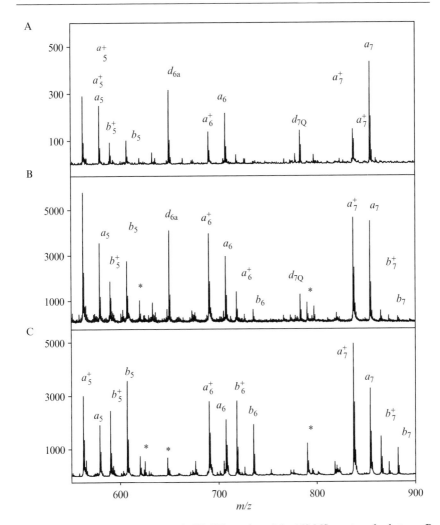

FIG. 13. The expansion of the m/z 550–900 portion of the MS-MS spectra of substance P recorded with (A) a threshold laser and gas pressures operating such that 70% of the precursor ion was suppressed, (B) laser intensity 20% above threshold with the same gas pressure as (A), and (C) laser intensity as in (B), with no gas in the collision cell.

primarily due to excitation in the ion source, whereas the d ions are only produced by high-energy collisions. The a ions are produced by both processes, as are some of the internal fragments, whereas the immonium ions are primarily produced by high-energy collisions.

Applications to Protein Identification and Characterization

Although the TOF-TOF system is relatively new, applications to identification and characterization of proteins have been reported (Juhasz *et al.*, 2002). One example is presented here to demonstrate the current performance of the technique. For this experiment, 5 pm each of β-galactosidase (*Escherichia coli*) and glycogen phosphorylase (rabbit muscle) were digested with bovine trypsin and lyophilized. One-hundred microliters of CHCA was added to the separate dry samples, and 0.5 μl of each resulting solution were mixed on a 3-mm diameter sample spot. Thus, the nominal sample loading on the MALDI plate, assuming no losses and complete digestion, is 50 fmol total, or about 50 attomoles/laser spot. The peptide mass fingerprint obtained from this sample is shown in Fig. 14. At least 100 peptides are observed that can be assigned to these two proteins. As noted earlier, the ionization efficiency for any component in such a complex mixture is substantially reduced by competition for the available protons;

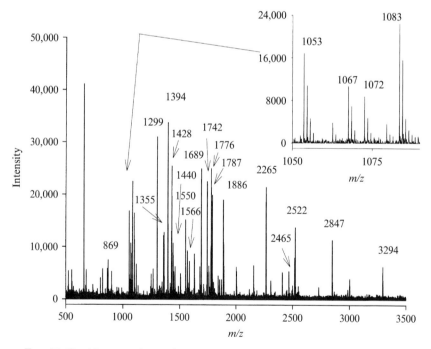

FIG. 14. Peptide mass fingerprint of a mixture of tryptic digests of 25 fmol of β-galactosidase (*Escherichia coli*) and 25 fmol of glycogen phosphorylase (rabbit muscle) on a 3-mm sample spot.

nevertheless, 50 fmol of sample loaded onto a single 3-mm spot is sufficient to allow excellent fragment spectra to be obtained on a large number of these peptides. For this example, 10,000 laser shots were accumulated for each of 22 selected mass peptides, as indicated by the mass numbers in Fig. 14. The region between m/z 1050 and 1090 is expanded in the inset showing the presence of at least six peptides in this mass range; the four labeled peaks were selected for MS-MS analysis. Spectra obtained for these peaks are shown in Fig. 15. These spectra were adequate for unambiguous identification of the peptide via database searching. Only the most intense peaks are labeled in Fig. 15, and the number of fragment peaks matching expected fragments of the peptide found is indicated in Table II. No fragment peaks from adjacent precursors were observed, and the unassigned masses were plausible low mass fragments not included in the set used in the database searching program.

All of the more intense peaks in the region below m/z 2500 provided MS-MS of sufficient quality to allow *de novo* sequencing, although in this

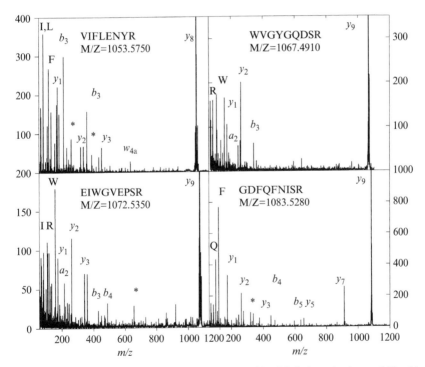

FIG. 15. MS-MS spectra for each of the four peptides labeled on the inset of Fig. 14. Results of the database search for these spectra are shown in Table II.

TABLE II

DATABASE SEARCHING RESULTS FROM THE PEPTIDES NOTED ON FIG. 14[a]

m/z	Unmatched/ submitted	Immonium ions?	Rank	Sequence of peptide
869.45	1/11	Yes	1	FAAYLER
1053.57	0/15	Yes	1	VIFLENYR
1067.49	3/9	Yes	1	WVGYGQDSR
1072.54	1/9	Yes	1	EIWGVEPSR
1083.52	1/10	No	1	GDFQFNISR
1299.62	9/35	Yes	1	ELNYGPHQWR
1355.67	0/16	Yes	1	DYYFALAHTVR
1394.73	2/36	Yes	1	LPSEFDLSAFLR
1428.68	10/32	No	1	DWENPGVTQLNR
1440.73	9/21	No	1	LLSYVDDEAFIR
1442.70	1/9*	No	1	VLYPNDNFFEGK
1550.77	0/13	Yes	1	IGEEYISDLDQLR
1566.79	5/13	No	2	DFNVGGYIQAVLDR
1689.88	2/20	Yes	1	ARPEFTLPVHFYGR
1742.90	1/33	Yes	1	LSGQTIEVTSEYLFR
1776.11	8/30	No	1	IENGLLLLNGKPLLIR
1787.73	2/10	No	1	HSDYAFSGNGLMFADR
1886.90	6/24	No	1	GYNAQEYYDRIPELR
2265.20	2/26	Yes	1	DVSLLHKPTTQISDFHVATR
2465.19	0/13	Yes	1	IDGSGQMAITVDVEVASDTPHPAR
2522.23	2/20	Yes	1	VVQPNATAWSEAGHISAWQQWR
2847.41	0/26	Yes	1	VTVSLWQGETQVASGTAPFGGEIIDER
3292.61	2/40	Yes	1	GVNRHEHHPLHGQVMDEQTMVQDILLMK

[a] The MS-Tag (version 3.6) portion of the Protein Prospector package was used to search the Swiss Prot. 5.10.2001 protein database. The columns denote the calibrated mass of the peptide in MS mode, the portion of unmatched to submitted fragment ions (with $m/z >$ 140), whether or not the detected immonium ions were used in the search, and what rank MS-Tag gave to the known sequence of the ions (shown in the last column).

case, the fragment masses alone were sufficient to tentatively assign the sequence. One example is shown in Figs. 16 and 17. The masses submitted the database search are shown in Table III. Of the 36 fragment peaks with $m\backslash z$ values more than 150 that were submitted, 34 matched known fragments of the tryptic peptide LPSEFDLSAFLR from β-galactosidase within the specified mass error of 100 millimass units. It should be noted that both unmatched peaks were of relatively low intensity and that the peak at m/z 1184.68 can be assigned to y_{11}, although its mass error is greater than 100 millimass units. The fragment ions provide complete coverage of the amino acid sequence of the peptide, so it would be possible to use the spectrum for *de novo* sequencing. At higher mass (e.g., m/z 3292 shown in Fig. 18), sequence coverage is incomplete, but the spectrum is sufficient to un-ambiguously identify the peptide from the database. The arginine at position 4 due to the missed cleavage causes the spectrum to be dominated by *a* and *b* ions, with particularly intense ions due to cleavage at the

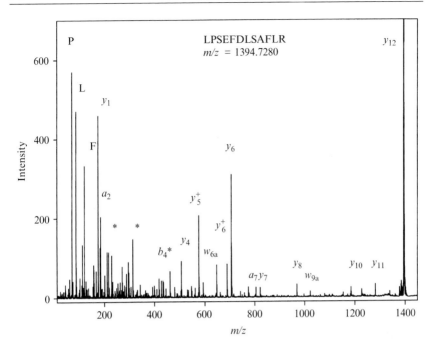

FIG. 16. Exemplary MS-MS spectrum of the peptide of mass 1394 from Fig. 14.

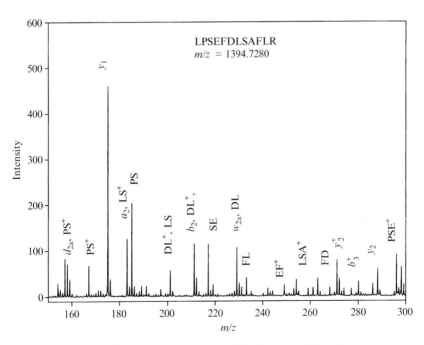

FIG. 17. Expansion of the m/z 150–300 region of Fig. 16.

TABLE III
DETAILS OF THE MS-TAG DATABASE SEARCH FOR THE SPECTRUM OF THE PEPTIDE
LPSEFDLSAFLR (m/z 1394) SHOWN IN FIG. 16 AND 17[a]

Assignment	m/z	Δ (millimass units)
PS-28	157.1	−2.4
PS-18	167.08	2.5
y_1	175.12	−2.6
a_2	183.14	−9.9
LS	201.11	−11.3
b_2	211.15	6.8
w_{a2}	229.13	−0.4
FL-28	233.16	−8.8
FD	263.11	10.6
y_2-17	271.18	2.9
LSA	272.16	0.9
y_2	288.19	−11.4
b_3	298.16	−14.7
SAF	306.14	−0.8
PSE	314.15	13
v_3	343.22	7.1
a_4	399.24	16.1
y_3-17	418.24	−7.5
b_4	427.23	9.2
y_3-17	433.25	−10.2
PSEF-28	435.27	37.4
PSEF	461.22	15.2
y_4	506.33	19.4
a_5	546.34	51.3
y_5-17	576.26	−52.8
y_5	593.33	−8
w_{a6}	647.39	40.1
b_6	689.36	−38.8
y_6	706.46	33.7
a_7	774.43	29.2
y_7-17	804.46	35.8
y_7	821.49	33.7
y_8	968.59	71.9
y_{10}	1184.68	80.8
	1227.75	
	1281.76	

[a] Column denotes the identity of the fragment ion, the submitted mass, and the calculated mass error in millimass units.

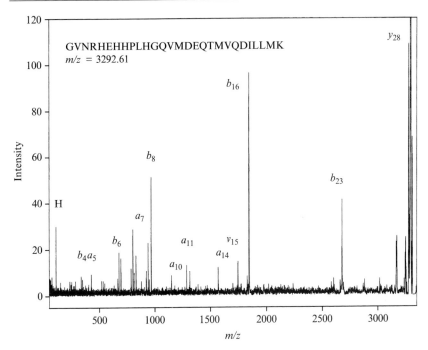

FIG. 18. Sample high-mass MS-MS spectrum of peak with m/z 3292 from Fig. 14.

C-terminal side of aspartic acid (b_{16} and b_{23}) and at the N-terminal side of proline (b_8).

One of the more difficult examples is illustrated in Fig. 19, corresponding to a region of the peptide mass fingerprint containing a number of relatively weak peaks close together as indicated in Fig. 14. It was not possible to separate the two components at nominal masses 1440 and 1442 in the precursor selection without sacrificing too much intensity; nevertheless, it was possible to identify both peptides by database searching as indicated in Table II. The major fragments corresponding to each are indicated in Fig. 19. A total of 21 fragments were submitted for the search, and all except 9 matched the component at m/z 1440. Eight of the nine remaining fragments matched the peptide with molecular weight 1442.

The results of the database search on the 23 peptides are summarized in Table II. At least one MS-MS spectrum was obtained on each of these masses, and 10,000 laser shots were summed for each. For the more intense peaks, a few hundred laser shots were sufficient for identifying the peptide

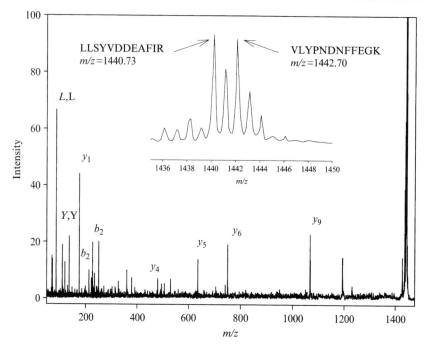

Fig. 19. MS-MS spectrum recorded for m/z 1440 from Fig. 14. Inset of precursor region demonstrates that fragments are formed from two equally abundant precursor ions. Fragments from LLSYVDDEAFIR are shown in bold type.

by searching the database, but the signal/noise ratio and dynamic range is improved by summing more shots. In most cases, all of the more intense peaks in the spectra were assigned to expected fragments from the peptide found. In cases in which a significant number of unassigned fragments are observed, the peptide mass fingerprint is rather congested and it appears that several relatively low-intensity peptide peaks may also be present within the selected precursor window.

Conclusion

Work has established the MALDI-TOF-TOF system as a useful tool for protein identification using database searching, as well as for protein characterization and *de novo* sequencing. This system is now available commercially. Further work on software for automatic interpretation of the spectra is required to make the high-throughput promise a reality.

References

Anderson, U. N., Colburn, A. W., Makarov, A. A., Papakis, E. N., Davis, S. C., Hoffman, A. D., and Thomson, S. (1998). In-series combination of a magnetic-sector mass spectrometer with a time-of-flight quadratic-field ion mirror. *Rev. Sci. Instrum.* **69,** 1650–1660.

Axelsson, J., Palmblad, M., Hakansson, K., and Hakansson, P. (1999). Electron capture dissociation of substance P using a commercially available Fourier transform ion cyclotron resonance mass spectrometer. *Rap. Comm. Mass. Spectrom.* **13,** 474–477.

Barbacci, D. C., and Russell, D. H. (1999). Sequence and side-chain specific photofragment (193 nm) ions from protonated substance P by matrix assisted laser desorption ionization time-of-flight mass spectrometry. *J. Am. Soc. Mass Spectrom.* **10,** 1038–1040.

Campbell, J. M. (2003). Mapping the properties of center of mass collision energy on a MALDI TOF/TOF mass spectrometer—fundamentals and applications. *In* "Proceedings of the 51st ASMS Conference." Montreal, P.Q., June 8–12.

Cody, R. B., and Frieser, B. S. (1982). High-resolution detection of collision-induced dissociation fragments by Fourier transform mass spectrometry. *Anal. Chem.* **54,** 1431–1433.

Davis, M. T., Stahl, D. C., Hefta, S. A., and Lee, T. D. (1995). A microscale electrospray interface for on-line capillary liquid chromatography/tandem mass spectrometry of complex peptide mixtures. *Anal. Chem.* **67,** 4549–4556.

Doroshenko, V. M., and Cotter, R. J. (1996). Pulsed gas introduction for increasing CID efficiency in a MALDI/quadrupole ion trap mass spectrometer. *Anal. Chem.* **68,** 463–472.

Ducret, A., Van Oostveen, I., Eng, J. K., Yates, J. R., III, and Aebersold, R. (1998). High throughput protein characterization by automated reverse-phase chromatography/electrospray tandem mass spectrometry. *Protein Sci.* **7,** 706–719.

Eng, J. K., McCormack, A. L., and Yates, J. R., III (1994). An approach to correlate tandem mass spectral data of peptides with amino acid sequences in a protein database. *J. Am. Soc. Mass Spectrom.* **5,** 976–989.

Gygi, S. P., Rist, B., Gerber, S. A., Turacek, F., Gelb, M. H., and Aebersold, R. (1999). Quantitative analysis of complex protein mixtures using isotope-coded affinity tags. *Nat. Biotechnol.* **17,** 994–999.

Hayes, R. N., and Gross, M. L. (1990). Collision-induced dissociation. *Methods Enzymol.* **193,** 237–263.

Johnson, R. S., Martin, S. A., and Biemann, K. (1988). Collision-induced fragmentation of $(M + H)^+$ ions of peptides. Side chain specific sequence ions. *Int. J. Ion Proc. Mass Spectrom.* **86,** 137–154.

Juhasz, P. J., Campbell, J. M., and Vestal, M. L. (2002). MALDI TOF-TOF technology for peptide sequencing and protein identification. *In* "Mass Spectrometry and Hyphenated Techniques in Neuropeptide Research" (J. Sibberring, R. Ekman, D. M. Desiderio, and N. M. Nibbering, eds.). Wiley, Hoboken, NJ.

Kaufmann, R., Kirch, D., and Spengler, B. (1994). Sequencing of peptides in a time-of-flight mass spectrometer: Evaluation of postsource decay following matrix-assisted laser desorption ionisation (MALDI). *Int. J. Mass Spectrom. Ion Proc.* **131,** 355–385.

Louris, J. N., Cooks, R. G., Syka, J. R. P., Kelley, P. E., Stafford, G. C., Jr., and Todd, J. F. J. (1987). Instrumentation, applications, and energy deposition in quadrupole ion-trap tandem mass spectrometry. *Anal. Chem.* **59,** 1677–1685.

Mann, M., and Wilm, M. (1994). Error-tolerant identification of peptides in sequence databases by peptide sequence tags. *Anal. Chem.* **66,** 4390–4399.

Medzihradszky, K. F., Adams, G. W., Bateman, R. H., Green, M. R., and Burlingame, A. L. (1996). Peptide sequence determination by matrix-assisted laser desorption ionization employing a tandem double focusing magnetic-orthogonal acceleration time-of-flight mass spectrometer. *J. Am. Soc. Mass Spectrom.* **7**, 1–10.

Medzihradszky, K. F., Campbell, J. M., Baldwin, M. A., Falick, A. M., Juhasz, P., Vestal, M. L., and Burlingame, A. (2000). The characteristics of peptide collision induced dissociation using a high-performance MALDI TOF/TOF tandem mass spectrometer. *Anal. Chem.* **72**, 552–558.

Morris, H. R., Paxton, T., Dell, A., Langhorne, J., Berg, M., Bordoli, R. S., Hoyes, J., and Bateman, R. H. (1996). High sensitivity collisionally-activated decomposition tandem mass spectrometry on a novel quadrupole/orthogonal-acceleration time-of-flight mass spectrometer. *Rap. Comm. Mass Spectrom.* **10**, 889–896.

Nair, H., Somogyi, A., and Wysocki, V. H. (1996). Effect of alkyl substitution at the amide nitrogen on amide bond cleavage: electrospray ionization/surface–induced dissociation fragmentation of substance P and two alkylated analogs. *J. Mass Spectrom.* **31**, 1141–1148.

Olivares, J. A., Nguyen, N. T., Yonker, C. R., and Smith, R. D. (1987). On-line mass spectrometric detection for capillary zone electrophoresis. *Anal. Chem.* **59**, 1230–1232.

Sato, K., Asada, T., Ishihara, M., Kunihiro, F., Kammei, Y., Kubota, E., Costello, C. E., Martin, S. A., Scoble, H. A., and Biemann, K. (1987). High-performance tandem mass spectrometry: Calibration and performance of linked scans of a four sector instrument. *Anal. Chem.* **59**, 1652–1659.

Shevchenko, A., Wilm, M., Vorm, O., and Mann, M. (1996). Mass spectrometric sequencing of proteins from silver stained polyacrylamide gels. *Anal. Chem.* **68**, 850–858.

Shulka, A. K., and Futrell, J. H. (2000). Tandem mass spectrometry: Dissociation of ions by collisional activation. *J. Mass Spectrom.* **35**, 1069–1090.

Vestal, M. L., Juhasz, P., and Martin, S. A. (1995). Delayed extraction matrix assisted laser desorption time-of-flight mass spectrometry. *Rapid Comm. Mass Spectrom.* **9**, 1044–1050.

Vestal, M., and Juhasz, P. (1998). Resolution and mass accuracy in matrix assisted laser desorption ionization time-of-flight. *J. Am. Soc. Mass Spectrom.* **9**, 892–911.

Vestal, M. L., Juhasz, P., Hines, W., and Martin, S. A. (2000a). A new delayed extraction MALDI TOF MS-MS for characterization of protein digests. *In* "Mass Spectrometry in Biology and Medicine" (A. L. Burlingame, S. A. Carr, and M. A. Baldwin, eds.), pp. 1–16. Humana Press, Totowa, NJ.

Vestal, M. L., Campbell, J. M., Hayden, K. M., and Juhasz, P. J. (2000b). Performance Evaluation of Improved MALDI-TOF MS-MS System. *In* "Proceedings of the 48th ASMS Conference." Palm Beach, CA. June 11–15.

Wilm, M., Shevchenko, A., Houthaeve, T., Briet, S., Schweigrer, L., Fotsis, T., and Mann, M. (1996). Femtomole sequencing of proteins from polyacrylamide gels by nano-electrospray mass spectrometry. *Nature* **379**, 466–469.

Yost, R. A., and Enke, C. E. (1978). Selected ion fragmentation with a tandem quadrupole mass spectrometer. *J. Am. Chem. Soc.* **100**, 2274–2275.

Further Reading

Shevchenko, A., Chernushevich, I., Ens, W., Standing, K. F., Thomson, B., Wilm, M., and Mann, M. (1998). Rapid *"de novo"* peptide sequencing by a combination of nanoelectrospray, isotopic labeling and a quadrupole/time-of-flight mass spectrometer. *Rapid Comm. Mass Spectrom.* **11**, 1015–1024.

[4] Tandem Mass Spectrometry in Quadrupole Ion Trap and Ion Cyclotron Resonance Mass Spectrometers

By Anne H. Payne and Gary L. Glish

Abstract

Instruments that trap ions in a magnetic and/or electric field play a very important role in the analysis of biomolecules. The two predominant instruments in the category of trapping instrument are the quadrupole ion trap mass spectrometer (QIT-MS) and the ion cyclotron resonance (ICR) MS. The latter is also commonly called *Fourier transform MS* (FT-MS). The QIT is an inexpensive, simple, and rugged MS used for various routine applications. The ICR-MS is an expensive, high-performance instrument with figures of merit for resolution and mass accuracy surpassing all other mass spectrometers. This chapter covers the basic principles of operation of these instruments, including the trapping/manipulation/detection of ions and various approaches used to activate ions to perform tandem mass spectrometry experiments.

Introduction

The world of mass spectrometers can be divided into two general types of instruments: ion-trapping instruments and beam instruments. The major distinguishing characteristic of ion-trapping instruments versus other types of MSs is that tandem mass spectrometry (MS/MS) is performed via a tandem-in-time method, rather than tandem in space. This means that each stage of mass spectrometry is performed in the same analyzer, sequentially in time. In contrast, a beam instrument, such as a triple quadrupole, has each step of MS/MS performed in different analyzers that are sequentially separated in space, one analyzer for each stage of MS/MS. In beam instruments, each subsequent stage requires the addition of another reaction region and mass analyzer. An immediately obvious advantage of trapping instruments is that multiple stages of MS/MS (MS^n) can be performed without instrumental modifications. The number of MS/MS stages possible in a trapping instrument is limited only by the ion intensity (Louris *et al.*, 1990). An added benefit of trapping instruments is that 80–90% of MS/MS product ions can be trapped, whereas in linear quadrupole instruments, MS/MS efficiencies are typically an order of magnitude lower (Johnson

METHODS IN ENZYMOLOGY, VOL. 402
0076-6879/05 $35.00
DOI: 10.1016/S0076-6879(05)02004-5

et al., 1990), and efficiencies in other beam instruments are even worse (Yost *et al.*, 1979). The high MS/MS efficiency allows ion traps to perform even more stages of MS/MS. In our laboratory, for example, up to MS^8 has been performed to provide a great deal of information about the sequence of peptides (Lin and Glish, 1998).

There are two common types of ion-trapping mass analyzers: the QIT-MS and the ICR instrument. These two instruments operate using very different principles but have some similar characteristics. This chapter discusses the basic principles of operation of these two instruments and the variety of approaches to MS/MS that are available. It will start with the QIT, which is much less expensive and, thus, much more commonly found in bioanalytical laboratories. It is worth noting, however, that the ICR is a much more mature technology and that many of the methods used with the QIT have been adapted from ICR experiments.

Quadrupole Ion Trap Mass Spectrometer

Overview

Although the QIT was patented in 1960 (Paul and Steinwedel, 1960), it was not widely used until the 1980s. Today, interest in and applications for the QIT-MS have dramatically expanded. The versatility and availability of this instrument can make the QIT-MS the workhorse of a laboratory. Typical operation of commercial ion traps gives a resolution of about 2000 and a mass range of 4000–6000 Da/charge, although the QIT-MS is capable of resolutions of more than 10^6 and a range of 70,000 Da/charge (Kaiser *et al.*, 1991; Londry *et al.*, 1993). Whereas the figures of merit of commercial QITs are significantly lower than those for an ICR, the QIT-MS is operationally and mechanically much simpler. In addition to the inherent advantages of a trapping instrument, the QIT-MS also boasts relatively simple vacuum requirements, a small footprint, fast analysis, and ruggedness, and it is relatively inexpensive. The QIT-MS can be interfaced with a wide variety of ionization methods, including continuous sources such as electrospray ionization (ESI) (VanBerkel *et al.*, 1990) and secondary ion mass spectrometry (SIMS) (Kaiser *et al.*, 1989), as well as pulsed methods like laser desorption (LD)(Glish *et al.*, 1989) and matrix-assisted laser desorption ionization (MALDI) (Chambers *et al.*, 1993; Jonscher *et al.*, 1993). The QIT-MS also has the greatest sensitivity of any MS (Cooks *et al.*, 1991). These features make the QIT-MS a valuable tool for analysis of biological molecules.

Trapping Ions in a QIT-MS

Apparatus. The QIT-MS operates analogously to a linear quadrupole, but in three dimensions rather than two. A QIT-MS is composed of three hyperbolic electrodes: two end-caps and one ring (Fig. 1). In a linear quadrupole mass filter, certain combinations of alternating current (AC) and direct current (DC) voltages allow an ion to have a stable trajectory, which means they pass through the filter and strike a detector. In the QIT-MS, those ions with stable trajectories are trapped in the volume encompassed by the electrodes. The donut-shaped ring electrode takes the place of one pair of poles. The end-caps replace the other pair of poles. The dimensions of the QIT-MS are described in terms of the distance from the center of the trap to the closest point on each of these electrodes. The ring electrode defines the radial direction, r_o. The end-cap electrodes define the axial direction, z_o. One end-cap electrode will usually have a hole in it to allow ions to enter the trapping volume. In rare instances, ions may also be injected through the ring, in which case the ring electrode would have an entrance hole. Ions are typically detected by ejecting them so they will strike a standard electron multiplier detector. The detector is located

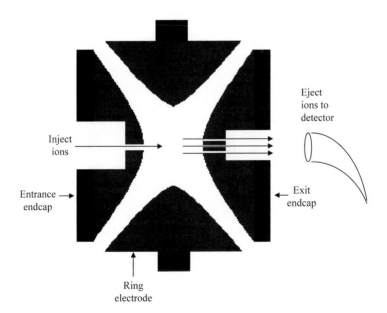

FIG. 1. Cross-sectional view of the quadrupole ion trap. The entrance end-cap has one larger hole for ion injection of focused ions while the exit end-cap has several smaller holes to allow ejection.

beyond the end-cap opposite of where the ions are injected, and thus, this end-cap also has holes, to allow the ions egress.

To generate the trapping field, an AC voltage is applied to the ring electrode, the end-cap electrodes or both the ring and the end-caps. In the latter arrangement, the voltage applied to the ring is 180 degrees phase shifted relative to the voltage applied to the end-cap. An optional DC voltage may be applied to either the ring or the end-caps. The movement of an ion in the trapping field is described by a second-order differential equation. A general solution to this type of differential equation was discovered by Mathieu more than a century ago. From this solution, the combinations of AC and DC voltages that result in a stable trajectory for an ion can be found (March and Londry, 1995). From Mathieu's solution, the Mathieu parameters, a_u and q_u, are used to describe the stability of an ion in both the axial (z) and the radial (r) direction, where

$$a_z = -2a_r = \frac{-16eU}{m(r_0^2 + 2z_0^2)\Omega^2} \tag{1}$$

and

$$q_z = -2q_r = \frac{8eU}{m(r_0^2 + 2z_0^2)\Omega^2}. \tag{2}$$

These equations incorporate the ions' characteristics of mass (m) and charge (e), the trap dimensions both radial (r_o) and axial (z_o), the AC frequency (Ω), and the AC (V) and DC (U) amplitudes. The AC voltage used is in the radiofrequency (RF) range and, thus, is commonly referred to as *RF*. The combinations of a_u and q_u values that result in a stable trajectory in both the axial and the radial direction are shown in the stability diagram in terms of the axial (z) direction (Fig. 2). For practical reasons, r_o, z_o, and Ω are fixed, and ion stability is controlled by changing the magnitudes of the RF and DC voltages.

Higher Order Fields. Electric trapping fields are generated by the applied RF voltage, which creates a three-dimensional (3D) trapping field. With prescribed dimensions of r_0 and z_0 ($r_0^2 = 2z_0^2$), and a specific hyperbolic shape of the electrodes, the trapping field exerts a restoring force that increases linearly as ions move away from the center of the trap. This force pushes the ions back to the center of the ion trap when they move away and results in their trapping. However, because of field imperfections such as those caused by the entrance and exit holes and intentional distortions in z_o, higher order fields are also present (Doroshenko and Cotter, 1997b; Wang et al., 1993).

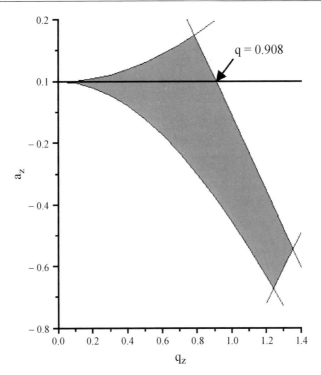

Fig. 2. Stability diagram. Ions with a_z and q_z values that fall in the shaded region will have a stable trajectory and, thus, be trapped within the quadrupole ion trap mass spectrometer.

The electric field is the first derivative of the electric potential. Thus, the quadrupole potential, which is quadratic, provides a linear field, whereas a hexapole potential, which is cubic, provides a quadratic field as shown in Fig. 3. These fields affect the motion and stability of the ions in the trap. Higher order fields are weakest at the center of the trap but become more significant toward the edges (greater slope). Therefore, ion motion can most easily be described when the ions are close to the center of the trap, where they experience mostly linear fields.

Resonance. An ion is trapped in the QIT-MS by maintaining stable periodic motion induced by the applied voltages. Each *m/z* ion has a unique periodicity of its motion, known as the *secular frequency*. The frequency in the axial direction (ω_z) can be approximated ($q_z < 0.4$) in terms of the ion's a_z and q_z parameters by

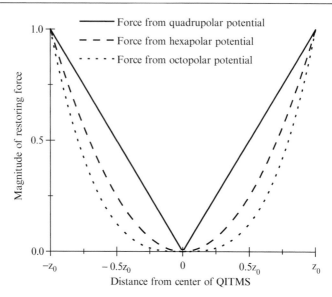

FIG. 3. Magnitude of restoring forces from higher order fields as a function of distance from the center of the ion trap.

$$\omega_z = \frac{(a_z^2 + \frac{q_z^2}{2})^{1/2}\Omega}{2}.$$ (3)

This characteristic frequency can be used to increase the kinetic energy of selected ions via resonance with a supplementary AC voltage applied across the end-cap electrodes. An ion moving at the frequency of the supplementary AC voltage will gain kinetic energy in the axial direction.

Bath Gas. In contrast to the ICR and most other mass analyzers, the QIT-MS benefits from a somewhat higher background gas pressure than is typically used for mass spectrometry (Stafford *et al.*, 1984). This is another advantage to the QIT-MS in that pumping requirements are not as stringent, and bath gas molecules are already present for use in CID. Bath gas molecules provide collisions with the trapped ions. These low-energy collisions damp the kinetic energy of the ions and restrict their movement to closer to the center of the trap. This is referred to as *collisional cooling.* At the center of the trap, the ions form a smaller cloud, and contributions of higher order fields are less. By condensing the ions to the center of the trapping volume, ion loss is reduced, and ions are ejected more coherently. Both sensitivity and resolution are improved by the presence of the bath gas (Louris *et al.*, 1989). Generally, helium is the gas used, as collisional scattering is lower than with more massive gas molecules. The helium is typically added to a level of approximately 1 m Torr.

Ion-Trapping Capacity and Space Charge. The QIT-MS electrodes define a finite trapping volume. Because like charges repel each other, a finite number of ions can be trapped. It has been estimated that 10^5 ions can be trapped at once (Cooks *et al.*, 1991). Whereas this capacity is more than sufficient for detection of an analyte, problems can arise if the trap "fills up" with extraneous ions. If the trapping limit is reached by extraneous ions, the dynamic range for analyte detection will be reduced, and in the worst case, analyte ions may not be observed at all.

The presence of a large number of ions results in an effective DC charge, known as *space charge*. Space charging is the result of overfilling the ion trap and begins to have an effect on performance long before the ion-trapping capacity is reached. Beyond a certain point, not only will increasing the ion accumulation time *not* increase the signal, but an overfilled trap also detrimentally affects the resolution of the QIT-MS. Resolution in the QIT-MS depends on ejecting all ions of the same *m/z* as a coherent packet to a detector. Ions in a space-charged trap will be forced by coulombic repulsions to spread out into a larger cloud. As the ion packet becomes more spread out, the resolution suffers. Because the *m/z* is determined by the RF level at the time the ions strike the detector, if the ions are spread out over time, the signal intensity of a single *m/z* is read over a longer time. This results not only in degradation in resolution but also as reduced signal current at the detector. The observation of broader than normal peaks is often indicative of a space-charged trap. Solutions include lowering ionization times or ejecting extraneous ions, either resonantly or by increasing the RF amplitude before mass analysis.

Ion Injection

All discussion up to this point has involved analysis and manipulation of ions. However, ions must first be formed, injected, and successfully trapped before they can be analyzed (Doroshenko and Cotter, 1997a; Louris *et al.*, 1989). Previous chapters have discussed ionization techniques. Pairing these with the QIT-MS, however, is not entirely straightforward. The complication is that these ions must have sufficient kinetic energy to enter the trap, overcoming the fringing fields created by the RF voltage that obscure the entrance. However, when ions have sufficient kinetic energy to enter the trap, they have sufficient energy to also exit the trap.

Trapping Ions. There are two possible solutions to this problem: increasing the trapping field strength or decreasing the ions' kinetic energies, once the ions have entered the trapping volume. Ions can be injected through a hole in either an end-cap electrode (axially) or the ring electrode (radially). For methods in which the RF is increased, injection through the ring is superior (O and Schuessler, 1981c). When the RF is constant, injection through an end-cap is preferred (O and Schuessler, 1981a).

If the RF is changed as the ions are injected, it can either be pulsed on (Kishore and Ghosh, 1979; O and Schuessler, 1981b,c) or ramped to higher voltages once the ions are inside the trap (Doroshenko and Cotter, 1997a). For practical reasons, ramping up the RF voltage is the easier technique to implement. However, only those ions in the trapping volume at the instant the RF is raised will be trapped. Therefore, these methods are ill-suited for pairing with continuous ion sources such as ESI because the duty cycle is very poor. However, these methods have seen use with MALDI, a pulsed ionization source. By timing the laser pulse with the RF, significant ion trapping can be achieved.

The more common method for trapping ions is to remove kinetic energy from ions in the trap so that the trapping fields are sufficient to hold them. Although the trapping process is not well understood, significant increases in trapping efficiencies are seen when a bath gas is present in the ion trap. In this case, ions are given sufficient kinetic energy to enter the trap. These ions are then thought to lose energy due to collisions with the gas molecules in the trapping volume (Quarmby and Yost, 1999). Sufficient kinetic energy is removed from the ions so that they are effectively trapped. This method requires a balance so that the RF level is low enough to admit the ions but high enough that the field can trap the ions once they have lost some of their kinetic energy through collisions. With a helium pressure of 1 mTorr, an RF level corresponding to a lower trapping limit of 40–50 Da/e is appropriate.

A potential problem with ion injection is that the population of ions trapped can be biased (Louris *et al.*, 1989). There appears to be some correlation between the injected ions' *m/z* and the RF level of the trap during injection. By using a higher RF level during ion injection, the lighter ions are either no longer stable in the trap or not able to pass the fields at the trap entrance while the heavier ions are more effectively trapped by the stronger field. The ion population can be shifted to favor heavier ions in this way. In general, this effect is small and a single RF level during injection is generally acceptable, but caution should be used when deducing the relative solution concentrations of species, keeping in mind this potential *m/z* bias.

Selective Accumulation. One strength of the QIT-MS is that ions can be accumulated over long periods of time to increase the number of ions present when they are ejected to the detector. However, once the trap is full, extending accumulation times will not improve the signal and can even be detrimental (see the section "Ion-Trapping Capacity and Space Charge," earlier in this chapter). If the analyte of interest is a small component of a mixture, this can be problematic, as the trap will fill up with the extraneous ions. However, if the extraneous ions could be ejected *during* ion accumulation, the trap would not fill up, and the analyte population could be increased. This can be done by resonantly ejecting the unwanted ions.

Selective accumulation can most easily be achieved using stored waveform inverse FT (SWIFT) (see the section "Ion Manipulation," in the section "Ion Cyclotron Resonance" section) because a large band of frequencies can be used to eject the unwanted ions. The waveform does not include those frequencies corresponding to the analytes so that these ions are not ejected. This technique can be particularly useful with MALDI, which produces a great excess of matrix ions that can quickly fill up the QIT-MS. Selective accumulation works best when the analyte ions are significantly different in m/z than the unwanted ions because at the RF levels used for injection, the higher mass ions have closely spaced secular frequencies. When the secular frequencies are too closely spaced, it is difficult to eject ions of unwanted m/z that are near the analyte m/z without also ejecting analyte ions. In such cases, ions can be injected for a period of time, the RF level raised to increase the spacing of the secular frequencies, and then the unwanted ions ejected. After ejection of the unwanted ions, another accumulation period can be used and the cycle repeated.

Ion Detection

Methods of Detection. In the first couple of decades after the invention of the QIT, the main mode of operation was a method termed *mass selective stability.* In this mode, the QIT-MS is operated analogous to the linear quadrupole; the parameters are set so that one m/z at a time has a stable trajectory. The trapped m/z is then ejected to the external detector. The next m/z is then trapped and ejected, and the process continues until the mass spectrum is acquired. This mode of operation offers little or no advantage over a linear quadrupole. Alternative modes of operation offer advantages over linear quadrupoles. When these modes became routine, the use of the QIT-MS was greatly increased.

Mass Selective Instability. The first commercial QIT-MSs were operated in a mode known as *mass selective instability.* In this mode of operation, the DC = 0 V ($a_z = 0$) and an AC voltage is applied to the ring electrode while the end-caps are grounded, so all ions lie along the q_z axis (Kelley *et al.*, 1985) (Fig. 4). All ions with $q_z < 0.908$ (the boundary of the stability diagram) are trapped. Because $m/z \propto 1/q_z$ (Eq. [2]), smaller

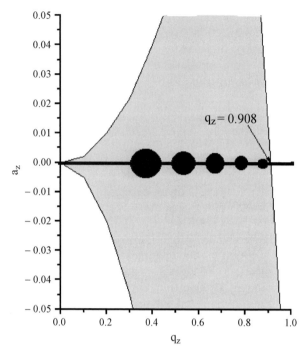

FIG. 4. Mass-selective instability mode. All ions lie along the q_z axis with smallest ions at the highest q_z values. By raising the radiofrequency level, each ion's q_z value will shift up. The ions will sequentially become unstable at $q_z = 0.908$ from smallest to largest.

m/z ions have a larger q_z and are closer to the boundary of the stability diagram. The m/z which corresponds to $q_z = 0.908$ is the lowest mass trapped and is typically referred to as the *cutoff mass*. The amplitude of the RF determines this cutoff mass, so often the RF level is given in terms of the cutoff mass rather than by the voltage amplitude. Mass analysis can be achieved by increasing the amplitude of the RF. As the amplitude increases, each ion's q_z value increases. When an ion's q_z value becomes just greater than 0.908, that ion's trajectory becomes unstable, and it is ejected from the trap. Low m/z ions will be ejected first, followed by higher m/z ions. An exit hole in one end-cap allows the ions to escape from the trap and strike the detector. By correlating the resulting signals at the detector to the RF amplitude, the m/z of the ions can be determined. This is known as the *mass selective instability* mode of operation (Stafford *et al.*, 1984).

RESONANCE EJECTION. The operation mode known as *resonance ejection* has made the QIT-MS a practical instrument for the analysis of biological molecules. In the mass selective instability mode of operation, the mass range is limited to about 650 Da/e because of practical limitations in the amplitude of the RF achievable without electrical breakdown. However, the resonance ejection mode of detection can circumvent this problem (Kaiser *et al.*, 1989). By applying a sufficiently large or long supplemental voltage at the resonant frequency, the ion gains enough kinetic energy to exit the trap. The first use of resonance ejection involved scanning the resonance frequency and holding the RF voltage constant (Ensberg and Jefferts, 1975). The most common implementation now involves applying a supplemental voltage at a fixed frequency and ramping the RF amplitude. Because secular frequencies are dependent upon the RF amplitude (Eqs. [2] and [3]), the secular frequencies of the trapped ions change as the RF amplitude changes. As the ions' frequencies come into resonance with the supplemental voltage, from lower *m/z* to higher *m/z,* they will be ejected. Resonance ejection can provide superior resolution and sensitivity compared to mass selective instability (as discussed in the section "Resolution"). More importantly, resonance ejection can be used to increase the mass range. Because of these features, resonance ejection is used on most current commercial QITs.

Nonlinear Resonance Ejection. As mentioned previously, higher order fields are naturally present in the QIT, and they are also intentionally induced by changes in the geometry from the mathematically ideal geometry. The higher order fields induce nonlinear resonances at specific locations in the stability diagram. One such location, due to hexapole resonance, is at a secular frequency, ω_z, equal to one-third the RF trapping frequency, Ω. The strength of this resonance can be increased by increasing the hexapolar contribution. This is done by distorting the geometry of the trapping electrodes in a nonsymmetrical manner.

The nonlinear resonances have little effect on the ions if they are at the center of the trapping volume; however, as the ion motion moves further from the center, the nonlinear resonances can cause rapid ejection of the ions from the trap (Franzen *et al.*, 1995). Thus, by applying a supplementary voltage at a frequency of $\omega_z, = \{1/3\}\Omega$, very rapid ejection of the ions can be achieved. This nonlinear resonance ejection provides improved performance such as faster scan rates and better resolution. At least one commercially available QIT operates using nonlinear resonance ejection.

Image Current Detection. Another mode of detection is also possible in the QIT-MS. Image current detection is the type of detection used in ICRs (see the section "Ion Detection" in the "Ion Cyclotron Resonance" section for details). Because each *m/z* has a characteristic secular frequency, the image current is a function of the secular frequencies of the trapped ions. The secular frequencies (and thus the *m/z*) in the image current can be determined via FT. This type of detection has particular advantages in that it is nondestructive, can detect very high m/zs, and ion populations can be remeasured to either detect a change in the ions or achieve better sensitivity.

Image current detection is not currently used in commercial QIT-MS instruments but has been demonstrated (Badman *et al.*, 1999; Goeringer *et al.*, 1995; Soni *et al.*, 1996). A complication of this mode of detection is isolating the detector electrodes from the RF on the ring electrode. The trapping RF is large enough to obscure the ions' image currents if not properly shielded. Ions can be excited either resonantly (Goeringer *et al.*, 1995) or by a high-voltage DC pulse (100 V, 1 μs) applied to one end-cap (Badman *et al.*, 1999; Soni *et al.*, 1996) to cause them to move coherently. Both of these methods function analogously to the broadband excitation used before detection in an FT-ICR. FT detection enables ICRs to provide vastly improved resolution and sensitivity, and this detection method could potentially enhance these performance characteristics in the QIT-MS as well. However, the best resolutions achieved thus far have been approximately 1000 (Badman *et al.*, 1999). Complications such as RF interference and space-charge–induced frequency shifts (covered in the section "Space Charge") will have to be overcome before image current detection will see widespread use.

Resolution. Resolution in the QIT-MS in either mass selective instability or resonance ejection mode depends on how quickly and coherently a packet of ions can be ejected to the detector. In both of these modes, the RF amplitude is increased at a constant rate. As the ions are ejected and strike the detector, the resulting signals can be correlated to an RF level at which the ions were ejected and the mass can be determined. Resonance ejection not only increases the mass range of the QIT-MS but can also provide higher resolution spectra (Goeringer *et al.*, 1992).

Scan rate is also related to resolution in resonance ejection. A faster scan rate results in lower resolution. A small amount of time is necessary for ions to become unstable and be ejected as they come into resonance with the ejection frequency. If the RF amplitude increases too quickly, all ions of a given m/z do not all have time to become destabilized equally and are not ejected together. Normal scan rates for unit mass resolution are 5000–13,000 Da/s. Slower scan rates are used to obtain higher resolution.

Ion Manipulation

Once ions have been trapped, analysis via MS/MS can follow. This is accomplished in three steps: isolation of the desired parent ion, activation of that ion, and detection of the product ions. Isolation of an ion is achieved using resonance ejection, which has already been discussed as a mode of ejection and detection. To isolate a parent ion, ions of both lower and higher m/z must be ejected. Lower m/z ions can be ejected by simply ramping the RF to a sufficient voltage so low m/z ions are ejected at $q_z = 0.908$. However, higher mass ions can be simultaneously ejected by applying a supplemental frequency corresponding to an m/z just higher than the parent ion's. As the RF amplitude increases, sequentially higher m/z ions will be resonantly ejected while those masses lower than the parent ion are ejected via mass selective instability. Ions can also be isolated using SWIFT in the same manner described in the section "Selective Accumulation."

Ion Activation

Collisional Activation. Several methods can be used for ion activation, each with different advantages and limitations. The most common form of activation is by collision with a neutral target gas as part of the overall process known as collision-induced dissociation (CID), which is discussed in more detail in Chapter 5. There are several methods to cause ions to undergo energetic collisions where kinetic energy is converted into internal energy. The excess internal energy then induces the parent ions to dissociate to product ions, from which structural information can be deduced.

The type of dissociation products seen with these collisional activation methods in ion traps can be different from those seen in other mass analyzers. There is a dramatic difference in the collision energies accessed in a sector instrument, typically in the 3–10 keV range, versus the 10s of electron volts in a QIT-MS. (It should be noted that the collision energy in a QIT-MS is ill-defined because the ions' kinetic energy is constantly changing due to the dynamic nature of the electric trapping field.) This difference in magnitude of the collision energy in a sector versus a QIT-MS leads to substantial differences in the amount of internal energy deposited and the internal energy distribution, with the result being notable differences in MS/MS spectra.

Although the collision energies in triple quadrupole instruments can be similar in magnitude to those in the QIT-MS, two other differences often lead to different MS/MS spectra being observed. First, helium is typically the collision gas in the QIT-MS, whereas argon is more commonly used in triple quadrupoles. With argon as the collision gas, much more kinetic energy can be converted to internal energy per collision, again leading to a different internal energy distribution. The second difference between the QIT-MS and triple quadrupole (and sector instruments) is the time frame for the reaction. Ions must dissociate (react) in 100 ms or less in the triple quadrupole instrument to be observed in the next stage of analysis.

However, in the ion trap, the times are two to three orders of magnitude longer. This longer time frame allows low energy, but kinetically slow reactions to occur, and can significantly favor such reactions.

RESONANCE EXCITATION. The first and most common method for collisional activation is resonance excitation (Charles *et al.*, 1994; March *et al.*, 1990; Splendore *et al.*, 1996; Williams *et al.*, 1993). This method employs the same phenomenon of resonance as the resonance ejection method discussed previously. When a supplemental AC voltage is applied across the end-cap electrodes at an ion's secular frequency, that ion's kinetic energy increases. The difference between excitation and ejection is one of degrees. By judiciously choosing the amplitude of the supplementary voltage and the time it is applied, the ion's kinetic energy can be increased without supplying sufficient energy for ejection. A typical excitation voltage would be 300–500 mV$_{p-p}$ and 10–40 ms long. Because the ions are manipulated by changing the amplitude of the RF, the sequence of events, or scan function, can be shown in terms of the RF amplitude. A scan function including isolation and resonance excitation is shown in Fig. 5.

Unfortunately, determining the frequency needed for resonance excitation is not always straightforward. The equation for determining the secular frequency of an ion (Eq. [3]) is not exact. The resonant frequency can be affected by higher order fields and can change as the ions gain kinetic

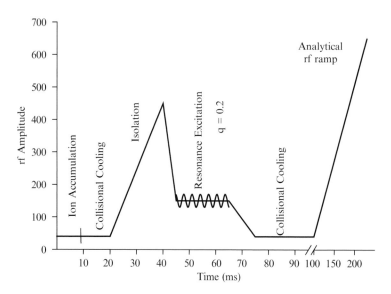

FIG. 5. Scan function for MS/MS using resonance excitation collision-induced dissociation.

energy (Franzen, 1991). The number of trapped ions also affects the secular frequencies of those ions (Vedel and André, 1984). The space charge of the ions acts as a DC potential. The effective a_z value no longer equals zero, and the secular frequency changes (Eq. [3]). To address this problem, some commercial QITs try to carefully control the number of ions trapped. Another approach is to use a narrow range of frequencies rather than a single frequency for resonance excitation.

Controlling the number of trapped ions is especially problematic if the ion source output fluctuates significantly, as can be the case with MALDI. This may be one reason that no commercial QITs offer MALDI as an ionization option. Frequency shifts caused by the number of trapped ions are part of the motivation for finding other means of dissociation, although resonance excitation CID is still the most common method for ion dissociation.

Once the appropriate frequency is determined, the ion's kinetic energy can be increased. However, one must be cautious of the competition between ejection and excitation (Charles et al., 1994). If too much energy is supplied, a significant fraction of the parent ion population can be lost through ejection. The MS/MS efficiency is then decreased.

$$\text{MS/MS Efficiency} = \frac{\sum(\text{production ion intensities})}{(\text{initial parent ion intensity})}. \qquad (4)$$

This not only hampers the identification of the product ions but also limits subsequent stages of MS/MS due to an insufficient number of ions.

To weight the experiment in favor of excitation over ejection, the parent ion should be trapped as strongly as possible. The ions can be thought of as being held in an energy well, called the *pseudo-potential well*. The depth of this well is the maximum kinetic energy the ion can have and still remain trapped. The deeper this well, the more kinetic energy the ion can gain before it is ejected. Ions in deeper wells can, thus, undergo more energetic collisions and a greater number of collisions. This allows the ion to gain enough internal energy to dissociate before it is ejected. The depth of the well, D, is related to the q_z parameter and can be approximated by the Dehmelt pseudo-potential well model for q_z values less than 0.4:

$$D = \frac{(q_z^2 m r_0^2 \Omega^2)}{32e} \qquad (5)$$

This equation shows that ions trapped with higher q_z values reside in a deeper well and can be excited to higher kinetic energies without being ejected. By increasing the RF amplitude, an ion's q_z value can be increased

(Eq. [2]) However, at higher q_z values, lower m/z ions are not stable, so the smaller product ions are not trapped. To balance between the well depth and trapping the product ions, a q_z of 0.2–0.4 is generally chosen. Thus, product ions less than about one-third the m/z of the parent ion are generally not observed.

HEAVY GAS CID. A variation on typical resonant excitation CID involves the use of heavy gases (Ar, Xe, Kr) as the collision gas. Heavy gases offer the benefit of higher energy deposition per collision. The maximum amount of kinetic energy (E_k) that can be converted to internal energy through a collision (E_{com}) is given by

$$E_{com} = E_k \left(\frac{M_n}{M_n + M_p} \right), \tag{6}$$

where M_p is the mass of the parent ion and M_n is the mass of the neutral molecule. For a parent ion with a given E_k, a heavy gas provides a higher M_n and a greater percentage of energy deposition than helium. The addition of heavy gas molecules to the QIT-MS allows higher energy deposition and a greater degree of dissociation via CID. Also, the frequency range for efficient excitation is wider with heavy gases, reducing the required precision of the secular frequency (Vachet and Glish, 1996). In addition, q_z values as low as 0.05 can be used, allowing low m/z products to be detected (Doroshenko and Cotter, 1996). However, the heavy gas also deteriorates the resolution and sensitivity of the QIT-MS due to greater scattering upon collision. Pulsing in the heavy gas for CID and allowing it to pump away before detection can alleviate these problems at the cost of increased time per scan (reduced duty cycle).

BOUNDARY-ACTIVATED DISSOCIATION. Methods of activation other than resonance excitation have also been explored to circumvent some of the problems. One such technique is boundary-activated dissociation (BAD) (March et al., 1993; Paradisi et al., 1992; Vachet and Glish, 1998). In this technique, a DC pulse is applied to the end-caps instead of an AC voltage. This causes a change in ions' a_z values. As the a_z value becomes large enough, the ion's a and q values approach the boundary of the stability diagram (Fig. 6). The a_z value is chosen so that the ion is not ejected, but its trajectory increases in amplitude. As the ions' motions become larger in amplitude, there is greater force acting upon the ions, causing their kinetic energy to increase. Collisions with the bath gas molecules can then induce dissociation just as with resonance excitation. Only the magnitude of the DC must be adjusted to optimize the dissociation. As in resonance excitation, the ion must reside in a sufficiently deep well that the large-amplitude oscillations do not result in ejection. So, again, the q_z values used are

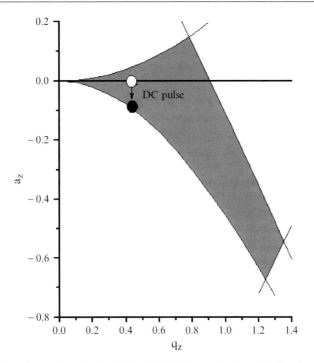

FIG. 6. Boundary-activated dissociation (BAD). The application of a direct current (DC) pulse changes the ion's a_z value. As the a_z value approaches the boundary, the ion gains kinetic energy.

typically between 0.2 and 0.4, and low m/z product ions are not observed. However, the secular frequency is irrelevant in BAD, so no tuning is required, and any fluctuations in frequencies are inconsequential. Therefore, BAD is particularly useful for MS/MS when the ion intensity fluctuates significantly from scan to scan, as is common with MALDI.

RED-SHIFTED OFF-RESONANCE LARGE-AMPLITUDE EXCITATION. Another activation technique that has been shown to be particularly useful for larger ions ($m/z > 1000$) is called red-shifted off-resonance large-amplitude excitation (RSORLAE) (Qin and Chait, 1996a,b). Larger ions often require a large amount of energy before dissociation is induced due to the number of degrees of freedom. Because of the competition between ejection and excitation, it can be difficult to increase the ions' kinetic energy to allow more energetic collisions without sacrificing MS/MS efficiency due to ion ejection. RSORLAE involves first increasing the RF amplitude over a period of about 10 ms, immediately dropping the RF level, and applying an excitation voltage. This voltage has a very large amplitude (21 V_{p-p} compared to $\sim$400 mV in resonance excitation). However, the ions are not ejected because the frequency is shifted from the secular frequency by approximately 5% to lower frequencies (red shifted). The excitation times are variable from 30 ms to as long as 1 s.

The reason this technique can deposit larger amounts of energy in an ion without ejection lies in the higher order fields in the trapping field. These higher order fields get stronger closer to the edges of the trapping volume. The secular frequencies are affected by these higher order fields by being shifted to higher frequencies (blue shifted) as the ion moves out from the center of the trap. The initial ramping and drop in the RF amplitude serves to first compress the ions to the center of the trap under the stronger trapping field. Then, when the amplitude is dropped, the cloud expands, and the higher order fields have more effect on the ions. The excitation frequency is red shifted, so the ions are further excited to the edges of the trap, the blue-shifting secular frequencies move further off resonance. This allows the ions to experience a more sustained excitation period, building up internal energy, without gaining enough kinetic energy to be ejected.

RSORLAE results not only in a greater degree of dissociation than resonance excitation, but also a slightly lower q_z value can be used ($q_z \approx$ 0.15). Thus, somewhat smaller product ions can be trapped than those seen with resonance excitation or BAD. RSORLAE is of particular value when paired with MALDI. Because MALDI produces predominantly singly charged ions, ions of m/z values more than 1000 are common. Also, because the excitation frequency does not have to exactly match the secular frequency, fluctuations in the secular frequency are not as problematic. RSORLAE has been used to dissociate MALDI-produced ions up to m/z values 3500 with significant improvements over resonance excitation.

NONRESONANCE EXCITATION. One other activation technique using CID also involves a non-resonant approach (Wang *et al.*, 1996). Non-resonance excitation is capable of increasing an ion's kinetic energy to 40 eV and can do so in only a few microseconds. Like RSORLAE, the ion remains trapped even with a higher kinetic energy than can be achieved via resonance excitation. Thus, the internal energy deposition can also be greater. This non-resonance excitation involves applying a low-frequency square wave (50–500 Hz) to the end-caps. Because this frequency is so low in comparison to the secular frequencies of the ions ($\sim$100,000 Hz), this square wave can more reasonably be thought of as a series of DC pulses. The square wave causes the trapping field to change instantaneously. The ions will accelerate quickly to compensate for this change. While doing so, collisions with the bath gas convert the kinetic energy to internal energy. This process is repeated as the square wave cycles. Non-resonance excitation does not result in ejection because the kinetic energy added to the ions does not continuously increase the ions' periodic stable motions as each of the previously discussed methods does. Instead, the ions change from one periodic motion to another very quickly. The kinetic energy is gained as the ions adjust from one motion to the other, not as a periodic motion is increased in amplitude.

Non-resonance excitation differs from resonant excitation in that it is not selective for any one species of ions. All trapped ions will be excited simultaneously. This can be desirable or not, depending on the application. Also, because more energy can be deposited quickly by non-resonance techniques, higher energy dissociation channels can be accessed. Thus, non-resonance excitation can form a different group of product ions from those seen with resonance excitation.

SURFACE-INDUCED DISSOCIATION. A final type of ion activation that uses collisions to increase the internal energy of ions is surface-induced dissociation (SID) (Lammert and Cooks, 1991). However, this method differs from CID in that the collisions do not involve gas molecules. Rather, the collision is between the ion and a surface such as the ring electrode. SID has several advantages including larger energy deposition and the non-resonant nature of the excitation. Resonance excitation has been estimated to increase the average kinetic energy to around 9 eV (Williams *et al.*, 1993), although depending on the experimental conditions, this can vary somewhat. RSORLAE and non-resonance excitation techniques can increase the kinetic energy somewhat more, to tens of electron volts. In contrast, SID kinetic energies are estimated to be in the hundreds of electron volts. The combination of such large kinetic energies and collisions with a massive surface, as opposed to light He atoms, allows SID to form products with very high threshold energy levels. The excitation is achieved via a very large DC pulse, typically 300–400 V, lasting 1–4 μs, applied to the end-caps. This pulse causes the ions to strike the ring electrode. The DC pulse must be carefully timed with the phase of the trapping RF frequency. The position of the ions upon acceleration must be such that the product ions are reflected into the trap and can subsequently be detected. Although this technique does show the aforementioned advantages, it also has drawbacks. Besides the experimental difficulty of implementing the fast, large, carefully timed DC pulse, SID has shown very poor efficiencies. The most abundant product ions observed are less than 5% that of the parent ion. And, as in most of the previously mentioned techniques, q_z values must be raised somewhat, to 0.2–0.4 in this case, which prevents the observation of low m/z product ions.

Infrared Multiphoton Dissociation. Another method for inducing dissociation differs from all the previously discussed techniques in that it does not use collisions. Instead of increasing the kinetic energy of ions and then using collisions to convert that to internal energy, infrared multiphoton dissociation (IRMPD) uses photons to deposit energy into the ions (Boué *et al.,* 2000; Colorado *et al.,* 1996; Stephenson *et al.,* 1994). IR wavelength photons are sent into the trapping volume where they can be absorbed by the trapped ions. An IR-transparent window in the vacuum housing allows access to the trap. A laser beam can then be directed into a hole drilled through the ring electrode. An intense laser beam is used because multiple photons must be absorbed before an ion gains enough energy to dissociate. Either a 25- or a 50-watt CO_2 laser can be used, and irradiation times of approximately 100 ms are usually sufficient. IRMPD offers several advantages over other dissociation techniques. First, it is not a resonant technique, and any ion in the path of the laser will be excited. This can be used, for example, to dissociate multiple analytes simultaneously. The laser beam will also excite product ions. These ions will subsequently dissociate and can provide a richer product ion spectrum. However, the spectrum can also be more complicated and hide the genealogy of the ions. The order in which product ions are formed in this type of MS/MS is not as clear as in stepwise MS^n. These relationships can be revealed but require additional steps to the experiment (Colorado *et al.,* 1996).

One very important advantage of IRMPD is that low q_z values can be used. All other excitation methods described rely on increasing the kinetic energy to increase the internal energy of ions. Therefore, the ions must be able to absorb that kinetic energy without being ejected, which requires higher q_z values. IRMPD has no such restrictions, and low m/z product ions are easily trapped and observed. However, a complication with IRMPD in the QIT-MS is that collisions with the bath gas molecules between absorption events can remove energy from the ions. If collisions occur quickly enough compared to the rate of photon absorption, the ions will not reach the critical energy for dissociation. As the bath gas pressure is increased from 1 to 4×10^{-5} Torr, dissociation rates decrease. Almost no dissociation is seen at bath gas pressures above 8×10^{-5} Torr, even with a 50-Watt laser pulse lasting hundreds of milliseconds. Heating the ion trap/bath gas to 160° allows typical bath gas pressures to be used (Payne and Glish, 2001). This technique, termed *thermally assisted IRMPD,* can improve sensitivity by more than an order of magnitude versus IRMPD at ambient temperature and reduced bath gas pressure.

Gas-Phase Reactions. Another type of MS/MS reaction for which trapping instruments are particularly useful involves gas-phase ion–molecule and ion–ion reactions. The QIT-MS is particularly versatile for these types of analyses because of the combination of controlled timing of the reactions, mass-selecting capabilities for both reactants and products, and MS^n for both the formation of the reactants (McLuckey et al., 1991) and the examination of the products (Donovan and Brodbelt, 1992). Often, these reactions are between the trapped ion and a volatile neutral introduced to the vacuum chamber. However, reactions between two ions have also been reported (Herron et al., 1996a,b; Stephenson and McLuckey, 1996a,b, 1997a). Ion traps allow the reaction to be controlled by holding the reactants together for a variable time before ejection and detection. Therefore, studying the kinetics is straightforward, and the extent of the reaction can be well controlled.

ION–MOLECULE REACTIONS. One very useful type of ion–molecule reaction for biological molecules involves hydrogen-deuterium exchange. A protonated analyte reacts with a deuterated species such as CH_3OD, D_2O, or ND_3. As active hydrogens are exchanged for deuteriums, the m/z of the analyte is seen to increase because deuteriums are more massive than hydrogens. The number of exchangeable hydrogens can then be determined. This is of particular interest in evaluating the conformation of the analyte. For example, if a protein or peptide is folded in the gas phase, some hydrogens will be protected from exchange within the interior of the protein. However, if the same species is unfolded, more hydrogens are exposed, and the degree of exchange will be greater. H/D exchange can also be used to determine when exchangeable hydrogens are obscured by an adduct. This structural information is then used to deduce the analyte's nature in solution and how the structure is affected upon transfer to the gas phase.

Ion–molecule reactions that are specific for a certain functional group or structure can also be used to provide information about an analyte. For example, multiply charged peptides can be reacted with HI (Stephenson and McLuckey, 1997b). The basic groups of the N-terminus, Arg, His, and Lys, will each gain an HI adduct. By monitoring the change in m/z, the number of HI adducts can be determined and the number of basic sites deduced.

Deprotonation reactions have been used with multiply protonated protein ions. These ions can be held in an ion trap in the presence of a basic reagent such as dimethylamine. The dimethylamine molecules will extract protons from the protein, reducing its charge. The rate of deprotonation changes as the charge state changes. The change in rate reflects the different structures of the multiply charged proteins (McLuckey et al., 1990). Further, deprotonation reactions between 1,6-diaminohexane and CID product ions have been used to determine the charge states of these ions by observing the m/z shift as z is changed (McLuckey et al., 1991).

Ion–Ion Reactions. Reactions between two oppositely charged species can also be used. A QIT-MS is particularly useful for these types of reactions, as both positively and negatively charged ions can be trapped simultaneously under normal operating conditions (this is not true for ICR instruments). Ion–ion reactions provide greater control in that ions are injected as needed, while neutral reactant molecules are usually present at a constant pressure. As with the ion–molecule reactions, deprotonation can also be achieved via ion–ion reactions. Ion–ion reactions have an advantage over deprotonating ion–molecule reactions in that anions are much stronger bases than neutral compounds. Anions can completely deprotonate multiply charged proteins and peptides in ion–ion reactions, whereas neutral compounds used in ion–molecule reactions typically cannot. Ion–ion deprotonation reactions have been useful to determine charge states and to de-clutter charge-state convoluted spectra. By decreasing the charge on trapped ions, the m/z values shift. Where multiple charge states of proteins overlap, shifting to a lower charge state for each protein can separate the overlapping signals. Ion–ion deprotonation of multiply charged proteins can be done using fluorocarbon anions (Stephenson and McLuckey, 1996b). A similar reaction involving protonated pyridine ions allowed determination of the charge states of negative CID product ions (Herron *et al.*, 1996b). Electron-transfer reactions have also been used to these ends (Stephenson and McLuckey, 1997a).

Ion Cyclotron Resonance

Overview

The modern era of ICR has its beginning in the work of Marshall and Comisarow (1974). By applying FT techniques, improved performance in numerous aspects of the experiment became possible. Today the ICR may be the most powerful mass spectrometer available. It is renowned for its unparalleled mass resolving power and provides accurate mass measurements as good as or better than any other type of mass spectrometer. The performance capabilities of the ICR have been steadily improving over the years as computers have become more powerful and as stronger magnetic fields have been obtained. Many of the performance characteristics, such as mass range and mass resolving power, are related to the strength of the magnetic field. As magnet technology has evolved, ICR instruments have gone from magnetic field strengths on the order of 1 Tesla, generated by electromagnets, to 3.0, 4.7, 7.0, and now 9.4 Tesla fields available with superconducting magnets. The desire for high magnetic fields provides two of the major contrasts between ICR and QIT-MS instruments—size and cost. Whereas the QIT-MS is a benchtop instrument, an ICR requires significant laboratory floor space.

As mentioned, the performance characteristics are a function of the magnetic field strength. Current state-of-the-art instruments with 9.4-Tesla magnets can reach resolving powers in excess of 10^6 for ions of m/z of 1000. The theoretical mass range exceeds 10^5 Da/e and mass accuracies can be in the sub–part-per-million range. Like the QIT, most any type of ionization technique can be coupled with an ICR. In many ways, the QIT-MS and ICR complement each other, and although QIT-MS instruments are found in a wide variety of laboratories, ICR is becoming an indispensable tool in state-of-the-art MS facilities involved in biotechnology-related areas.

Trapping Ions in Ion Cyclotron Resonance

Principles. The main trapping force in ICR is the magnetic field. However, unlike the QIT in which the RF electric field traps ions in all three dimensions, the magnetic field traps the ions only in two dimensions, the plane perpendicular to the magnetic field. The ion motion in this plane is the cyclotron motion and is simply described by the basic equation of motion of a charged particle in a magnetic field:

$$\frac{mv^2}{r} = Bev. \tag{7}$$

in which m is the mass of the ion (kilograms), r is the radius of the ion motion (meters), v is the ion velocity (meters/s), e is the charge on the ion (coulombs), and B is the magnetic field strength (Tesla). Equation (7) can be rearranged to give:

$$\frac{m}{e} = \frac{Br}{v} = B\omega_c, \tag{8}$$

in which ω_c is the angular velocity (radians/s) at which the ion is orbiting in the magnetic field. The angular velocity is related to cyclotron frequency by

$$f_c = 2\pi\omega_c. \tag{9}$$

In most ICR instruments, the magnetic field is fixed, so each mass-to-charge ratio has a unique cyclotron frequency (f_c). It is worth noting that the cyclotron frequency of a given mass-to-charge ratio is independent of the velocity (kinetic energy) of the ions.

Because the magnetic field traps ions only in two dimensions, an electric field is used to trap the ions in the third dimension, the axis of the magnetic field. A potential of the same polarity as the charge of the ions being trapped is applied to trapping plates, which are mounted perpendicular to the magnetic field. Ions with axial kinetic energies less than the potential applied to the trapping plates move in a potential as well as simple harmonic oscillators. This trapping motion is independent of the cyclotron motion.

In addition to cyclotron and trapping motion of the ions in an ICR instrument, there is also a motion termed the *magnetron motion*. The magnetron motion results from the combined effect of the magnetic and electric fields. Like the cyclotron motion, the magnetron motion has a frequency associated with it. However, the magnetron frequency is independent of the mass-to-charge ratio of the ions. Also, whereas the cyclotron frequency is typically in the range of a few kilohertz to a few megahertz, the magnetron motion frequency is typically less than 100 Hz and is given by

$$f_m = \frac{\alpha V}{\pi a^2 B}, \tag{10}$$

where V is the magnitude of the trapping potential, B is the magnetic field strength, a is the distance between the trapping plates, and α is a constant related to the analyzer cell geometry. Thus, the magnetron motion depends not only on the magnetic and electric fields, but also on the analyzer cell geometry. The magnetron motion can be thought of as coupling with the cyclotron motion and displacing the center of the cyclotron motion.

Apparatus. Although many analyzer cell geometries have been used in ICR, the most common is the cubic cell (McIver, 1970). This cell, as the name implies, is a six-sided cube. The two sides that are orthogonal to the magnetic field are the trapping plates discussed previously. The plates on the other four sides of the cube are not important in the trapping of the ions but are used for the ion manipulation and detection processes, which are discussed below. While the cube is the simplest arrangement, a cylinder cut in quarters along the magnetic field axis is also a common geometry. Again, the main purpose of the individual sections of the cylinder is to provide the means to manipulate and detect the ions that are trapped in the cell. There have been many variations of these two basic designs, typically implemented to shim the electric field to improve various aspects of performance. Schematics of these two basic cells are shown in Fig. 7.

Ion-Trapping Capacity and Space Charge. As with QIT-MS, there is a limit to the number of charges that can be trapped in an ICR analyzer cell due to the coulombic repulsion between the ions. The QIT-MS can hold slightly more ions than an ICR, but the difference is not great. And just like the QIT-MS, performance is degraded in the ICR before the space charge limit is reached. Shifts in the cyclotron frequency along with peak

A

Detect

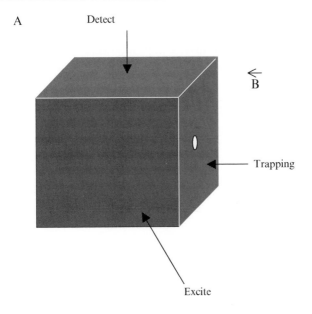

B

Trapping

Excite

B

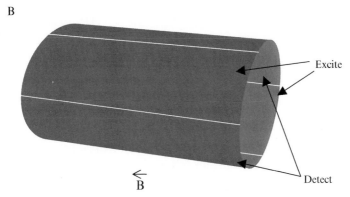

Excite

B

Detect

FIG. 7. Schematics of (A) cubic and (B) cylindrical ion cyclotron resonance (ICR) cells.

broadening can occur. If ions of similar mass are present, the two peaks can coalesce into a single peak.

Ion Injection

Ion Sources. Since the early days of ICR, a common means to get ions into the trapping region is to form them by electron ionization within the trapping cell. A filament to generate the ionizing electrons is located just outside the cell, within the magnetic field. This arrangement allows efficient ionization and trapping of volatile compounds. Most biological compounds, however, are involatile and require a desorption ionization technique. Desorption ionization techniques require that the sample be external to the cell. Thus, the ions formed must be injected into the cell.

For MALDI, it is possible to locate a sample probe just outside the analyzer cell and let the ions diffuse into the cell after being formed by the laser pulse (Hettich and Buchanan, 1991; Koster *et al.*, 1992). However, the kinetic energy of the MALDI-desorbed ions, which increases with mass, limits the trapping efficiency of this approach and can cause discrimination based on the ion mass.

Coupling ESI with ICR has increased challenges due to the vast pressure differences between the source and the analyzer. ESI is an atmospheric pressure technique and optimum ICR performance is at very low pressures (e.g., 10^{-9} Torr or less), a pressure differential of 11–12 orders of magnitude. This can only be achieved through a number of stages of differential pumping. Although there has been one example of an ESI probe with multiple stages of differential pumping that can be inserted into the magnetic field near the cell (Hofstadler and Laude, 1992), all commercial instruments have the source located outside of the magnetic field, typically at least a meter away from the cell (Henry *et al.*, 1989).

Trapping Ions. Injecting ions into the analyzer cell is not a trivial task. Not only do ions have to penetrate a large magnetic field, but they also must be slowed down enough to be trapped in the axial dimension. Several approaches have been used to help the ions traverse the magnetic field gradient. The earliest and probably most common approach is use of an RF-only ion guide (McIver *et al.*, 1985). In addition to RF-only ion guides, DC wire ion guides have also been used (Limbach *et al.*, 1993).

After traversing the magnetic field, the next issue is trapping the ions in the cell. To get the ions into the cell, the potential on the front trapping plate (closest to the ion source) must be lower than the kinetic energy of the ions. To keep the ions from going straight through the cell, there must be either a higher voltage on the rear trapping plate (i.e., the one farthest from the ion source), or the axial kinetic energy must be decreased after the ions enter the cell. By having the axial trapping plates at different voltages, ions that have kinetic energies lower than the potential on the rear trapping plate will be turned around and head back in the direction of the source. If there is no change in the ion kinetic energy, the front trapping plate potential must be raised to a level greater than the kinetic energy of the ions. When this is done, no more ions can enter the cell. The time it takes a given ion to travel from the front plate to the maximum distance toward the rear plate (depending on the potential applied to the rear plate) and back to the front plate will be a function of the mass of the ion (smaller ions will have larger velocities). For a continuous beam of ions, this limits the number of ions that can be trapped but will not discriminate based on mass. However, for pulsed beams of ions, mass discrimination may occur. The gating period (i.e., the length of time that the trapping potential is lowered on the front plate to allow ions in) can be varied to optimize the trapping of a particular mass-to-charge ratio relative to the rest of the species present (Dey *et al.*, 1995).

Another method for trapping ions is to decrease the kinetic energy of the ions through collisions with a background gas. This can allow longer ion accumulation times. However, the analyzer cell pressure is increased so that there is loss of resolution (Beu and Laude, 1991). A gas can be pulsed into the analyzer cell, but this substantially increases the duty cycle because of long pump-out times (20–120 s) to return to the low pressure needed for optimal resolution (Beu *et al.*, 1993). As an alternative to the gated trapping or the use of collisions to trap ions, ions can be deflected off-axis in a technique known as "sidekick" (Caravatti and Allemann, 1991). The sidekick method provides the ability to accumulate ions in the analyzer cell for longer periods with reduced mass discrimination, increasing the number of ions that are trapped.

Whereas MALDI naturally forms a pulse of ions, ESI generates a continuous beam of ions. However, a common mode of operation of ESI is to accumulate the ions in an RF-only multipole external to the magnetic field and then inject the ions as a pulse (Senko *et al.*, 1997). Because the ion injection portion of the ICR experiment is typically only a small fraction of the overall cycle time, accumulating the ions externally and injecting them in a pulse can greatly increase the effective duty cycle. One point of caution concerning external ion accumulation in an RF-only multipole is that it has been shown that above a certain space-charge value, dissociation of the ions can occur. This leads to a mass spectrum that is not representative of the ion species being formed in the ion source (Sannes-Lowery and Hofstadler, 2000).

The ability to inject ions from external ionization (EI) sources makes the combination of almost any ionization technique with ICR feasible. Although MALDI and EI can readily be performed in the magnetic field near the analyzer cell, it is now common to have these ionization sources external to the field. This allows ready access to many ionization techniques on a given instrument. In some designs, it is possible to rather quickly switch from one ionization method to another. This ability is facilitated by the fact that mass calibration is dependent only on the magnetic field, and thus, the same calibration applies regardless of the ionization technique, in contrast to many types of mass spectrometers. It is also possible to form ions via EI with a filament within the magnetic field and, at the same time, inject ions from an external source. This allows even more accurate mass measurements to be obtained.

Ion Detection

Image Current. ICR is unique among MS techniques in that the ions can be readily detected without impinging them on an electron multiplier detector or any other type of detector external to the mass analysis region. They are detected inside the cell without destroying them, and therefore, after the ions have been detected, they can be manipulated further to perform more sophisticated mass spectrometric experiments. Two steps are involved in the ion-detection process: ion excitation and subsequent detection of the current induced on the analyzer cell plates by the ions (image current).

In the basic scheme, two of the four plates parallel with the magnetic field are used to excite the ions (excitation plates), and the other two are used to detect the ions (detection plates). The excitation involves applying a broadband signal to the excitation plates that spans the cyclotron frequency range of ions that are trapped. This excitation signal will cause the ions to gain kinetic energy. Because the cyclotron frequency remains constant, the ions will increase the radius of their orbit. By judicious choice of the excitation signal parameters, the orbit radius can be increased to just slightly less than the dimension of the cell. The excited ions are coherent in their motion (i.e., ions of each individual mass-to-charge move together in a packet).

If the ions are positively charged, as they approach a detection plate, electrons are attracted to that plate. As the packet of ions continues its orbit and approaches the opposite detection plate, electrons are attracted to that plate. A current that oscillates at the same frequency of the cyclotron motion of the ions can be detected in an external circuit between the two detection plates. This current is called the *image current* and is detected as a function of time. The current can then be mathematically processed using the Fourier equation to change from a time-domain signal to a frequency-domain signal. This process gives the technique its commonly used name, *FT-ICR* or *FT-MS,* although it should be noted that Fourier processes have been used with other types of mass spectrometers and are not unique to ICR. After the cyclotron frequency of the trapped ions has been determined, the mass-to-charge ratio can be obtained from Eqs. (8) and (9). Figure 8 shows theoretical time-domain signals and the corresponding frequency-domain spectra.

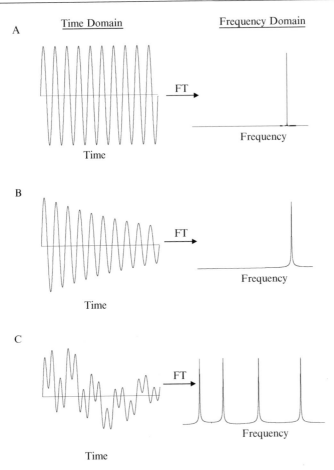

FIG. 8. Examples of Fourier Transform. (A) Long time domain signal gives high resolution in frequency domain. (B) Time domain is exponentially damped (as by collisions) to give a lower frequency resolution. (C) Several frequencies convoluted in time domain are separated into frequencies by Fourier transform.

Resolution. The high-resolution capabilities of the ICR are based on the capability to measure the image current for a relatively long period (seconds). The longer the time-domain signal can be measured, the more precisely the cyclotron frequency can be determined. The higher the precision is, the better the resolution (i.e., the separation between two peaks) unless peak coalescence occurs due to space charge. In addition to scaling with the acquisition time, the resolution increases linearly with increasing magnetic field and decreases linearly as a function of m/z.

The main limit on the resolution is the pressure in the system. The image current is damped (Fig. 8B) as ions undergo collisions with neutral molecules in the vacuum system. Several effects are responsible for the decreased signal due to ion–molecule collisions. First, the ions' velocity is decreased, reducing their cyclotron radius, moving them farther from the detection plate. Thus, there is less charge induced on the detection plate. Also, as the cyclotron radius decreases, the magnetron radius increases, which can cause the ions to be lost from the cell. Finally, scattering resulting from the collisions dephases the coherent ion motion, again reducing the charge induced on the detection plates.

The effect of pressure on resolution (higher pressure, poorer resolution) leads to one of the biggest contrasts between the operation of an ICR and a QIT-MS. In an ICR experiment, pressures in the analyzer cell are typically in the 10^{-9} Torr pressure range (or lower), whereas the QIT-MS is typically operated in the range of 10^{-3} Torr. Thus, as noted earlier, the ICR requires significantly more vacuum pumping, especially when ionization techniques at atmospheric pressure are used.

Ion Manipulation

Stored Waveform Inverse Fourier Transform. A unique characteristic of the ion-trapping instruments is the ability to selectively manipulate the ions that are trapped. This is a result of the fact that each different mass-to-charge ratio has a unique frequency of motion in the trapping field. In the ICR, this motion is the cyclotron motion, discussed earlier. By appropriate application of an RF potential to the analyzer cell, the cyclotron motion can be resonantly excited, analogous to resonant excitation in a QIT-MS. The cyclotron motion is excited for ion detection but can also be excited for other purposes. A standard means for manipulating ions is known as *SWIFT* (Marshall *et al.*, 1985). In SWIFT, a frequency-domain waveform is generated to achieve the desired purpose of the ion manipulation (e.g., to eject all ions except those of a selected m/z that will be the parent ions for an MS/MS experiment) (Fig. 9). Then, an inverse FT of the frequency domain spectrum is performed to obtain a time-domain signal that can be applied to the appropriate (excitation) cell plates. This time-domain signal will then cause all the ions in the analyzer cell to increase in radius to the point at which they strike a plate and are lost. Very complex ion manipulation can readily be performed by this method.

Quadrupolar Axialization. Another technique used for ion manipulation is quadrupolar axialization (Bollen *et al.*, 1990). This technique involves converting magnetron motion to cyclotron motion by electronic manipulation of the ions. The conversion of magnetron to cyclotron motion

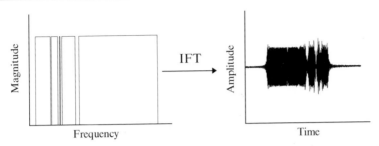

FIG. 9. Stored Waveform Inverse Fourier Transform (SWIFT). The frequency domain waveform shown here will excite all ions except those with frequencies corresponding to those notched out. An inverse Fourier transform is performed to obtain a time domain signal to be applied to the excitation cell plates. The phases are modulated to stagger the phases of the frequencies to spread the power more evenly over the waveform.

decreases the magnetron radius while increasing the cyclotron radius. The cyclotron radius can simultaneously be decreased by collisions, as described previously. The net effect is to focus the ions to the center of the analyzer cell. This improves the performance of an ICR instrument for a number of different parameters and experiments, such as mass resolution and accuracy, MS/MS, and ion remeasurement (Speir *et al.*, 1993).

Ion remeasurement is another unique feature of ion-trapping instruments. Because the ions are not destroyed when they are detected via image current measurement, they can be refocused to the center of the analyzer cell using quadrupolar axialization. Once in the center of the cell, the ions can be efficiently excited and detected again. This cycle, detection followed by axialization, can be done many times. This process can substantially improve the signal-to-noise ratio of a spectrum and the MS^n efficiency.

Ion Activation

Like the QIT-MS, there are numerous methods to increase the internal energy of ions to cause them to dissociate. Some of the techniques are identical, and others are similar in principle. In addition, just like the QIT, multiple stages of MS/MS can readily be performed.

Collisional Activation. As with all other instruments capable of performing MS/MS experiments, collisions of the selected parent ions with neutral gas molecules is by far the most common means of ion activation in ICR. However, a major drawback to this approach is the conflict between the need for a collision gas to be present to effect collisional activation and the need to have ultralow pressures to obtain the high mass resolution characteristic of ICR analysis. Two approaches have been taken to address this conflict. The simplest approach involves pulsing the collision gas into the analyzer cell during the time period of the collisional activation and then delaying the detection sequence several seconds to allow the collision gas to pump out and the pressure to return to a level allowing the desired resolution to be obtained (Carlin and Freiser, 1983).

As an alternative to pulsing a collision gas into the vacuum system, a differentially pumped dual cell geometry can be used (Littlejohn and Ghaderi, 1986). In this arrangement, two cubic-trapping cells are used with a common trapping plate between them. This common plate separates regions of the vacuum system with a conductance-limiting aperture in the center of the plate. This aperture allows ions to be transferred back and forth between the two cells while keeping a pressure differential of over two orders of magnitude between the two cells. Thus, collisional activation can be performed at a higher pressure and then the product ions transferred to the low-pressure cell for high-resolution mass analysis.

The physical aspects of collisional activation in an ICR are the same as in the QIT-MS (see the section "Surface-Induced Dissociation"). One difference in the experiment implementation is the nature of the collision gas. Typically, helium is used in the QIT-MS, mostly because it is already present; for other purposes, argon is more commonly used in the ICR experiment, and thus, a greater fraction of kinetic energy can be converted to internal energy in the ICR. On some occasions, this may lead to different MS/MS spectra being obtained from the two instruments.

GAS-PHASE COLLISIONAL ACTIVATION. As in the QIT-MS, the ions trapped in the ICR analyzer cell have low kinetic energies. Thus, some means to increase their kinetic energy to effect collisional activation must be used. There are a number of ways in which this can be done. The original collisional activation involved dipolar resonant excitation of the cyclotron motion (Cody et al., 1982), analogous to the resonant excitation that is the most common method used with the QIT-MS. The maximum kinetic energy that the ions can obtain during excitation is a function of the magnetic field strength and can be hundreds of electron volts. The actual kinetic energy can be calculated for this and other forms of resonant excitation (Schweikhard and Marshall, 1993).

A problem with resonant excitation in ICR is that product ions are formed with significant cyclotron radii, which complicates detection (and transfer between cells in a dual-cell system). There are alternative methods to standard resonant excitation that are currently used: sustained off-resonance irradiation (SORI) (Gauthier et al., 1991), very low energy (VLE) excitation (Boering et al., 1992), and multiple excitation collisional activation (MECA) (Lee *et al.*, 1993). The most common of these methods is SORI. This is a non-resonant technique in which a frequency slightly different than the ion cyclotron frequency is used to excite the ions. This difference in frequency causes the ion motion to be alternatively accelerated and then decelerated. This allows long excitation periods while keeping the ions near the center of the cell.

VLE also provides alternating periods of acceleration and deceleration. Resonant excitation is used to accelerate the ions for a period of time, after which the resonant frequency is phase shifted 180°, causing the ions to be decelerated. This sequence can be repeated many times, providing a very similar result as SORI. MECA is similar to VLE except collisional cooling is used to decelerate the ions rather than doing it electronically.

SURFACE-INDUCED DISSOCIATION. Another activation method that avoids the use of a collision gas is SID. In this technique, the ions are forced to collide with a surface in the mass spectrometer. In the ICR, that surface is typically one of the trapping plates (Ijames and Wilkins, 1990; Williams *et al.*, 1990). A voltage pulse is applied to one trapping plate, accelerating the ions so that they strike the opposite plate. The amplitude and duration of this voltage pulse are the important parameters. Although in theory this technique is probably the easiest of all the techniques to implement, requiring the least amount of modification or extra equipment, it is not commonly used. This is probably because of the relatively low efficiency with which product ions are collected and because the modified surfaces, which can enhance SID, are difficult to use. However, work has been focused toward these issues (Danell and Glish, 2000; Zhong *et al.*, 1997), so SID may become a more widely used technique.

Photodissociation. Photodissociation is a logical combination with ICR. It obviates the need to add a collision gas to the instrument, and because the ions are trapped in a well-defined region and at low energy, relatively good overlap between the laser beam and ions can be achieved. Both UV and IR photodissociation have been used to dissociate biological ions in ICR experiments. UV photodissociation was the first method tried for biomolecules (Bowers *et al.*, 1984; Hunt *et al.*, 1987). Extensive dissociation is observed, in fact so much that it is difficult to interpret the resulting MS/ MS spectra. As an alternative, infrared multiphoton photodissociation (IRMPD) has been explored (Little *et al.*, 1994). This activation method is somewhat similar to the SORI method in that the ions gain internal energy relatively slowly, so the lower energy dissociation pathways predominate.

Blackbody Infrared Radiative Dissociation. A relatively new method to dissociate ions that also relies on photons to energize the ions is termed *blackbody infrared radiative dissociation* (BIRD) (Price *et al.*, 1996). This method basically involves heating the trapped ions by absorption of IR photons emitted (blackbody radiation) from the analyzer cell plates/ vacuum chamber. By heating the vacuum system up significantly above room temperature (typically 150–200°), trapped ions can be energized to levels sufficient for dissociation to occur at observable rates. The IR photon absorption (heating) process of the ion is in competition with IR emission, which cools the ions. It often takes tens to hundreds of seconds to effect dissociation by the BIRD technique. Obviously, BIRD is not a high-throughput method. However, fundamental properties of biomolecules, such as activation energies, can be determined under appropriate experimental conditions.

Ion–Molecule Reactions. Though sometimes not thought of as such, isolation of ions of a specific m/z and reaction of those ions with a neutral gas to form products is a classic MS/MS experiment. One of the main strengths of ICR from the outset has been the ability to use ion–molecule reactions. For many years, the main application of ICR was the study of gas phase bimolecular chemistry. As the instrumentation has improved and matured, ion–molecule chemistry has played a reduced role behind the high-resolution and accurate mass measurement capabilities. However, it is becoming apparent that using ion–molecule reactions for fundamental studies of biomolecules will be a great strength of ICR because of its resolution capabilities. In particular, H/D exchange of peptides and proteins to probe gas-phase conformations of these species is emerging as a major tool (McLafferty *et al.*, 1998). Insight into the effect of parameters such as charge state can be obtained from comparing the number of exchangeable protons and the rate of exchange. A related experiment involves the stripping of protons from multiply charged peptides and proteins (Gross and Williams, 1995). Being able to control the reaction time allows kinetic measurements to be made, which provide yet another piece of data in the analysis of biomolecules.

Electron-Capture Dissociation. The newest activation technique that is unique to the ICR is electron-capture dissociation (Zubarev *et al.*, 1998). In this method, low-energy electrons (<0.2 eV) are captured by multiply charged ions (typically proteins) in the analyzer cell. An intriguing feature of the electron-capture–induced dissociation is that the product ions formed are c and z ions that results from cleavage of the backbone amine bond, not b and y ions typically observed with other methods as a result of the peptide bond cleavage. In addition, dissociation is observed at more sites along the protein backbone relative to other activation methods.

Acknowledgment

This work was supported by National Institutes of Health grant No. GM49852.

References

Badman, E. R., Patterson, G. E., Wells, J. M., Santini, R. E., and Cooks, R. G. (1999). Differential non-destructive image current detection in a Fourier transform quadrupole ion trap. *J. Mass Spectrom.* **34,** 889–894.

Beu, S. C., and Laude, D. A. (1991). Ion trapping and manipulation in a tandem time-of-flight-Fourier transform mass-spectrometer. *Int. J. Mass Spectrom. Ion Processes* **104,** 109–127.

Beu, S. C., Senko, M. W., Quinn, J. P., Wampler, F. M., and McLafferty, F. W. (1993). Fourier-transform electrospray instrumentation for tandem high-resolution mass-spectrometry of large molecules. *J. Am. Soc. Mass Spectrom.* **4,** 557–565.

Boering, K. A., Rolfe, J., and Brauman, J. I. (1992). Control of ion kinetic-energy in ion-cyclotron resonance spectrometry—very-low-energy collision-induced dissociation. *Rapid Commun. Mass Spectrom.* **6,** 303–305.

Bollen, G., Moore, R. B., Savard, G., and Stolzenberg, H. (1990). The accuracy of heavy-ion mass measurements using time of flight-ion cyclotron-resonance in a Penning trap. *J. Applied Phys.* **68,** 4355–4374.

Boué, S. M., Stephenson, J. L., and Yost, R. A. (2000). Pulsed helium introduction into a quadrupole ion trap for reduced collisional quenching during infrared multiphoton dissociation of electrosprayed ions. *Rapid Commun. Mass Spectrom.* **14,** 1391–1397.

Bowers, W. D., Delbert, S. S., Hunter, R. L., and McIver, R. T. (1984). Fragmentation of oligopeptide ions using ultraviolet-laser radiation and Fourier-transform mass-spectrometry. *J. Am. Chem. Soc.* **106,** 7288–7289.

Caravatti, P., and Allemann, M. (1991). The infinity cell—a new trapped-ion cell with radiofrequency covered trapping electrodes for fourier-transform ion-cyclotron resonance mass-spectrometry. *Organic Mass Spectrom.* **26,** 514–518.

Carlin, T. J., and Freiser, B. S. (1983). Pulsed valve addition of collision and reagent gases in Fourier-transform mass-spectrometry. *Anal. Chem.* **55,** 571–574.

Chambers, D. M., Goeringer, D. E., McLuckey, S. A., and Glish, G. L. (1993). Matrix-assisted laser desorption of biological molecules in the quadrupole ion trap mass spectrometer. *Anal. Chem.* **65,** 14–20.

Charles, M. J., McLuckey, S. A., and Glish, G. L. (1994). Competition between resonance ejection and ion dissociation during resonant excitation in a quadrupole ion trap. *J. Am. Soc. Mass Spectrom.* **5,** 1031–1041.

Cody, R. B., Burnier, R. C., and Freiser, B. S. (1982). Collision-induced dissociation with fourier-transform mass-spectrometry. *Anal. Chem.* **54,** 96–101.

Colorado, A., Shen, J. X., Vartanian, V. H., and Brodbelt, J. (1996). Use of infrared multiphoton photodissociation with SWIFT for electrospray ionization and laser desorption applications in a quadrupole ion trap mass spectrometer. *Anal. Chem.* **68,** 4033–4043.

Cooks, R. G., Glish, G. L., McLuckey, S. A., and Kaiser, R. E. (1991). Ion trap mass spectrometry. *Chem. Eng. News* **69,** 26–41.

Danell, R. M., and Glish, G. L. (2000). A new approach for effecting surface-induced dissociation in an ion cyclotron resonance mass spectrometer: A modeling study. *J. Am. Soc. Mass Spectrom.* **11,** 1107–1117.

Dey, M., Castoro, J. A., and Wilkins, C. L. (1995). Determination of molecular weight distributions of polymers by MALDI-FTMS. *Anal. Chem.* **67,** 1575–1579.

Donovan, T., and Brodbelt, J. (1992). Characterization of the dissociation behavior of gasphase protonated and methylated lactones. *J. Am. Soc. Mass Spectrom.* **3,** 47–59.

Doroshenko, V. M., and Cotter, R. J. (1996). Pulsed gas introduction for increasing peptide CID efficiency in a MALDI/quadrupole ion trap mass spectrometer. *Anal. Chem.* **68,** 463–472.

Doroshenko, V. M., and Cotter, R. J. (1997a). Injection of externally generated ions into an increasing trapping field of a quadrupole ion trap mass spectrometer. *J. Mass Spectrom.* **31,** 602–615.

Doroshenko, V. M., and Cotter, R. J. (1997b). Losses of ions during forward and reverse scans in a quadrupole ion trap mass spectrometer and how to reverse them. *J. Am. Soc. Mass Spectrom.* **8,** 1141–1146.

Ensberg, E. S., and Jefferts, K. B. (1975). The visible photodissociation spectrum of ionized methane. *Astrophys. J.* **195,** L89–L91.

Franzen, J. (1991). Simulation study of an ion cage with superimposed multipole fields. *Int. J. Mass Spectrom. Ion Processes* **106,** 63–78.

Franzen, J., Gabling, R.-H., Schubert, M., and Wang, Y. (1995). Nonlinear ion traps. *In* "Practical Aspects of Ion Trap Mass Spectrometry" (R. E. March and J. F. J. Todd, eds.), Vol. I, "Fundamentals of Ion Trap Mass Spectrometry," pp. 49–167. CRC Press, Boca Raton, FL.

Gauthier, J. W., Trautman, T. R., and Jacobson, D. B. (1991). Sustained off-resonance irradiation for collision-activated dissociation involving Fourier transform mass spectrometry. Collision-activated dissociation technique that emulates infrared multiphoton dissociation. *Analytica Chimica Acta* **246,** 211–225.

Glish, G. L., Goeringer, D. E., Asano, K. G., and McLuckey, S. A. (1989). Laser desorption mass spectrometry and MS/MS with a three-dimensional quadrupole ion trap. *Int. J. Mass Spectrom. Ion Processes* **94,** 15–24.

Goeringer, D. E., Crutcher, R. I., and McLuckey, S. A. (1995). Ion remeasurement in the radio frequency quadrupole ion trap. *Anal. Chem.* **67,** 4164–4169.

Goeringer, D. E., Whitten, W. B., Ramsey, J. M., McLuckey, S. A., and Glish, G. L. (1992). Theory of high-resolution mass spectrometry achieved via resonance ejection in the quadrupole ion trap. *Anal. Chem.* **64,** 1434–1439.

Gross, D. S., and Williams, E. R. (1995). Experimental-measurement of Coulomb energy and intrinsic dielectric polarizability of a multiply protonated peptide ion using electrospray-ionization Fourier-transform mass-spectrometry. *J. Am. Chem. Soc.* **117,** 883–890.

Henry, K. D., Williams, E. R., Wang, B. H., McLafferty, F. W., Shabanowitz, J., and Hunt, D. F. (1989). Fourier-transform mass-spectrometry of large molecules by electrospray ionization. *Proc. Natl. Acad. Sci. USA* **86,** 9075–9078.

Herron, W. J., Goeringer, D. E., and McLuckey, S. A. (1996a). Ion-ion reactions in the gas phase: Proton transfer reactions of protonated pyridine with multiply charged oligonucleotide anions. *J. Am. Soc. Mass Spectrom.* **6,** 529–532.

Herron, W. J., Goeringer, D. E., and McLuckey, S. A. (1996b). Product ion charge state determination via ion/ion proton transfer reactions. *Anal. Chem.* **68,** 257–262.

Hettich, R. L., and Buchanan, M. V. (1991). Matrix-assisted laser desorption Fourier-transform mass-spectrometry for the structural examination of modified nucleic-acid constituents. *Int. J. Mass Spectrom. Ion Processes* **111,** 365–380.

Hofstadler, S. A., and Laude, D. A. (1992). Trapping and detection of ions generated in a high magnetic-field electrospray ionization Fourier-transform ion-cyclotron resonance mass-spectrometer. *J. Am. Soc. Mass Spectrom.* **3,** 615–623.

Hunt, D. F., Shabanowitz, J., and Yates, J. R. (1987). Peptide sequence analysis by laser photodissociation Fourier transform mass spectrometry. *J. Chem. Soc. Chem. Commun.* 548–550.

Ijames, C. F., and Wilkins, C. L. (1990). Surface-induced dissociation by Fourier transform mass spectrometry. *Anal. Chem.* **62,** 1295–1299.

Johnson, J. V., Yost, R. A., Kelley, P. E., and Bradford, D. C. (1990). Tandem-in-space and tandem-in-time mass spectrometry: Triple quadrupoles and quadrupole ion traps. *Anal. Chem.* **62,** 2162–2172.

Jonscher, K., Currie, G., McCormack, A. L., and Yates, J. R. III (1993). Matrix-assisted laser desorption of peptides and proteins on a quadrupole ion trap mass spectrometer. *Rapid Commun. Mass Spectrom.* **7,** 20–26.

Kaiser, R. E. Jr., Cooks, R. G., Stafford, G. C. Jr., Syka, J. E. P., and Hemberger, P. E. (1991). Operation of a quadrupole ion trap mass spectrometer to achieve high mass/charge ratios. *Int. J. Mass Spectrom. Ion Processes* **106,** 79–115.

Kaiser, R. E. Jr., Louris, J. N., Amy, J. W., and Cooks, R. G. (1989). Extending the mass range of the quadrupole ion trap using axial modulation. *Rapid Commun. Mass Spectrom.* **3,** 225–229.

Kelley, P. E., Stafford, G. C., and Stephens, D. R. (1985). United States Patent No. 4,540,884 .

Kishore, M. N., and Ghosh, P. K. (1979). Trapping of ions injected from an external source into a three-dimensional r.f. quadrupole field. *Int. J. Mass Spectrom. Ion Processes* **29,** 345–350.

Koster, C., Castoro, J. A., and Wilkins, C. L. (1992). High-resolution matrix-assisted laser desorption ionization of biomolecules by Fourier-transform mass-spectrometry. *J. Am. Chem. Soc.* **114,** 7572–7574.

Lammert, S. A., and Cooks, R. G. (1991). Surface-induced dissociation of molecular ions in a quadrupole ion trap mass spectrometer. *J. Am. Soc. Mass Spectrom.* **2,** 487–491.

Lee, S. A., Jiao, C. Q., Huang, Y. Q., and Freiser, B. S. (1993). Multiple excitation collisional activation in Fourier-transform mass-spectrometry. *Rapid Commun. Mass Spectrom.* **7,** 819–821.

Limbach, P. A., Marshall, A. G., and Wang, M. (1993). An electrostatic ion guide for efficient transmission of low-energy externally formed ions into a Fourier-transform ion-cyclotron resonance mass-spectrometer. *Int. J. Mass Spectrom. Ion Processes* **125,** 135–143.

Lin, T., and Glish, G. L. (1998). C-terminal peptide sequencing via multistage mass spectrometry. *Anal. Chem.* **70,** 5162–5165.

Littlejohn, D. P., and Ghaderi, S. (1986). United States Patent No. 4,581,533 .

Little, D. P., Speir, J. P., Senko, M. W., O'Connor, P. B., and McLafferty, F. W. (1994). Infrared multiphoton dissociation of large multiply charged ions for biomolecule sequencing. *Anal. Chem.* **66,** 2809–2815.

Londry, F. A., Wells, G. J., and March, R. E. (1993). Enhanced mass resolution in a quadrupole ion trap. *Rapid Commun. Mass Spectrom.* **7,** 43–45.

Louris, J. N., Amy, J. W., Ridley, T. Y., and Cooks, R. G. (1989). Injection of ions into a quadrupole ion trap mass spectrometer. *Int. J. Mass Spectrom. Ion Processes* **88,** 97–111.

Louris, J. N., Brodbelt-Lustig, J. S., Cooks, R. G., Glish, G. L., Van Berkel, G. J., and McLuckey, S. A. (1990). Ion isolation and sequential stages of mass spectrometry in a quadrupole ion trap mass spectrometer. *Int. J. Mass Spectrom. Ion Processes* **96**, 117–137.

March, R. E., and Londry, F. A. (1995). Theory of quadrupole mass spectrometry. *In* "Practical Aspects of Ion Trap Mass Spectrometry" (R. E. March and J. F. J. Todd, eds.), Vol. I, pp. 25–48. CRC Press, Boca Raton, FL.

March, R. E., McMahon, A. W., Allinson, E. T., Londry, F. A., Alfred, R. L., Todd, J. F. J., and Vedel, F. (1990). Resonance excitation of ions stored in a quadrupole ion trap. Part II. Further simulation studies. *Int. J. Mass Spectrom. Ion Processes* **99**, 109–124.

March, R. E., Weir, M. R., Londry, F. A., Catinella, S., Traldi, P., Stone, J. A., and Jacobs, B. (1993). Controlled variation of boundary-activated ion fragmentation processes in a quadrupole ion trap. *Can. J. Chem.* **72**, 966–976.

Marshall, A. G., and Comisarow, M. (1974). Fourier transform ion cyclotron resonance [FT-ICR] spectroscopy. *Chem. Phys. Lett.* **25**, 282–283.

Marshall, A. G., Wang, T.-C. L., and Ricca, T. L. (1985). Tailored excitation for Fourier transform ion cyclotron resonance mass spectrometry. *J. Am. Chem. Soc.* **107**, 7893–7897.

McIver, R. T. (1970). A trapped ion analyzer cell for ion cyclotron resonance spectroscopy. *Rev. Sci. Instrum.* **41**, 555–558.

McIver, R. T., Hunter, R. L., and Bowers, W. D. (1985). Coupling a quadrupole mass-spectrometer and a Fourier-transform mass-spectrometer. *Int. J. Mass Spectrom. Ion Processes* **64**, 67–77.

McLafferty, F. W., Guan, Z., Haupts, U., Wood, T. D., and Kelleher, N. L. (1998). Gaseous conformational structures of cytochrome *c. J. Am. Chem. Soc.* **120**, 4732–4740.

McLuckey, S. A., Glish, G. L., and Berkel, G. J. V. (1991). Charge determination of product ions formed from collision-induced dissociation of multiply protonated molecules via ion/molecule reactions. *Anal. Chem.* **63**, 1971–1977.

McLuckey, S. A., Van Berkel, G. J., and Glish, G. L. (1990). Reactions of dimethylamine with multiply charged ions of cytochrome. *c. J. Am. Chem. Soc.* **112**, 5668–5670.

O, C. S., and Schuessler, H. A. (1981a). Confinement of ions created externally in a radio-frequency ion trap. *J. Applied Phys.* **52**, 1157–1166.

O, C.-S., and Schuessler, H. A. (1981b). Confinement of ions injected into a radiofrequency quadrupole ion trap: Pulsed ion beams of different energies. *Int. J. Mass Spectrom. Ion Processes* **40**, 67–75.

O, C.-S., and Schuessler, H. A. (1981c). Confinement of pulse-injected external ions in a radiofrequency quadrupole ion trap. *Int. J. Mass Spectrom. Ion Processes* **40**, 53–66.

Paradisi, C., Todd, J. F. J., Traldi, P., and Vettori, U. (1992). Boundary effects and collisional activation in a quadrupole ion trap. *Organic Mass Spectrom.* **27**, 251–254.

Paul, W., and Steinwedel, H. (1960). United States Patent No. 2,939,952.

Payne, A. H., and Glish, G. L. (2001). Thermally assisted infrared multiphoton photodissociation in a quadrupole ion trap. *Anal. Chem.* **73**, 3542–3548.

Price, W. D., Schnier, P. D., and Williams, E. R. (1996). Tandem mass spectrometry of large biomolecule ions by blackbody infrared radiative dissociation. *Anal. Chem.* **68**, 859–866.

Qin, J., and Chait, B. T. (1996a). Matrix-assisted laser desorption ion trap mass spectrometry: Efficient isolation and effective fragmentation of peptide ions. *Anal. Chem.* **68**, 2108–2112.

Qin, J., and Chait, B. T. (1996b). Matrix-assisted laser desorption ion trap mass spectrometry: Efficient trapping and ejection of ions. *Anal. Chem.* **68**, 2102–2107.

Quarmby, S. T., and Yost, R. A. (1999). Fundamental studies of ion injection and trapping of electrosprayed ions on a quadrupole ion trap. *International J. Mass Spectrom.* **190/191**, 81–102.

Sannes-Lowery, K. A., and Hofstadler, S. A. (2000). Characterization of multipole storage assisted dissociation: Implications for electrospray ionization mass spectrometry characterization of biomolecules. *J. Am. Soc. Mass Spectrom.* **11**, 1–9.

Schweikhard, L., and Marshall, A. G. (1993). Excitation modes for fourier transform-ion cyclotron-resonance mass-spectrometry. *J. Am. Soc. Mass Spectrom.* **4**, 433–452.

Senko, M. W., Hendrickson, C. L., Emmett, M. R., Shi, S. D. H., and Marshall, A. G. (1997). External accumulation of ions for enhanced electrospray ionization Fourier transform ion cyclotron resonance mass spectrometry. *J. Am. Soc. Mass Spectrom.* **8**, 970–976.

Soni, M., Frankevich, V., Nappi, M., Santini, R. E., Amy, J. W., and Cooks, R. G. (1996). Broad-band Fourier transform quadrupole ion trap mass spectrometry. *Anal. Chem.* **68**, 3314–3320.

Speir, J. P., Gorman, G. S., Pitsenberger, C. C., Turner, C. A., Wang, P. P., and Amster, I. J. (1993). Remeasurement of ions using quadrupolar excitation Fourier-transform ion-cyclotron resonance spectrometry. *Anal. Chem.* **65**, 1746–1752.

Splendore, M., Londry, F. A., March, R. E., Morrison, R. J. S., Perrier, P., and André, J. (1996). A simulation study of ion kinetic energies during resonant excitation in a stretched ion trap. *Int. J. Mass Spectrom. Ion Processes* **156**, 11–29.

Stafford, G. C. J., Kelley, P. E., Syka, J. E. P., Reynolds, W. E., and Todd, J. F. J. (1984). Recent improvements in and analytical applications of advanced ion trap technology. *Int. J. Mass Spectrom. Ion Processes* **60**, 85–98.

Stephenson, J. L., Booth, M. M., Shalosky, J. A., Eyler, J. R., and Yost, R. A. (1994). Infrared multiple photon dissociation in the quadrupole ion trap via a multipass optical arrangement. *J. Am. Soc. Mass Spectrom.* **5**, 886–893.

Stephenson, J. L., and McLuckey, S. A. (1996a). Ion/ion reactions in the gas phase: Proton transfer reactions involving multiply-charged proteins. *J. Am. Chem. Soc.* **118**, 7390–7397.

Stephenson, J. L., and McLuckey, S. A. (1996b). Ion-ion proton transfer reactions for protein mixture analysis. *Anal. Chem.* **68**, 4026–4032.

Stephenson, J. L., and McLuckey, S. A. (1997a). Adaptation of the Paul trap for study of the reaction of multiply charged cations with singly charged anions. *Int. J. Mass Spectrom. Ion Processes* **162**, 89–106.

Stephenson, J. L., and McLuckey, S. A. (1997b). Counting basic sites in oligopeptides via gas-phase ion chemistry. *Anal. Chem.* **69**, 281–285.

Vachet, R. W., and Glish, G. L. (1996). Effects of heavy gases on the tandem mass spectra of peptide ions in the quadrupole ion trap. *J. Am. Soc. Mass Spectrom.* **7**, 1194–1202.

Vachet, R. W., and Glish, G. L. (1998). Boundary-activated dissociation of peptide ions in a quadrupole ion trap. *Anal. Chem.* **70**, 340–346.

VanBerkel, G. J., Glish, G. L., and McLuckey, S. A. (1990). Electrospray ionization combined with ion trap mass spectrometry. *Anal. Chem.* **62**, 1284–1295.

Vedel, F., and André, J. (1984). Influence of space charge on the computed statistical properties of stored ions cooled by a buffer gas in a quadrupole RF trap. *Phys. Rev. A* **29**, 2098–2101.

Wang, M., Schachterle, S., and Wells, G. (1996). Application of nonresonance excitation to ion trap tandem mass spectrometry and selected ejection chemical ionization. *J. Am. Soc. Mass Spectrom.* **7**, 668–676.

Wang, Y., Franzen, J., and Wanczek, K. P. (1993). The non-linear resonance ion trap. Part 2. A general theoretical analysis. *Int. J. Mass Spectrom. Ion Processes* **124**, 125–144.

Williams, E. R., Henry, K. D., McLafferty, F. W., Shabanowitz, J., and Hunt, D. F. (1990). Surface-induced dissociation of peptide ions in Fourier-transform mass spectrometry. *J. Am. Soc. Mass Spectrom.* **1**, 413–416.

Williams, J. D., Cooks, R. G., Syka, J. E. P., Hemberger, P. H., and Nogar, N. S. (1993). Determination of positions, velocities, and kinetic energies of resonantly excited ions in the quadrupole ion trap mass spectrometer by laser photodissociation. *J. Am. Soc. Mass Spectrom.* **4,** 792–797.

Yost, R. A., Enke, C. G., McGilvery, D. C., Smith, D., and Morrison, J. D. (1979). High efficiency collision-induced dissociation in an RF-only quadrupole. *Int. J. Mass Spectrom Ion Physics* **30,** 127–136.

Zhong, W., Nikolaev, E. N., Futrell, J. H., and Wysocki, V. H. (1997). Tandem FTMS studies of surface-induced dissociation of benzene monomer and dimer ions on a self-assembled fluorinated alkanethiolate monolayer surface. *Anal. Chem.* **69,** 2496–2503.

Zubarev, R. A., Kelleher, N. L., and McLafferty, F. W. (1998). Electron capture dissociation of multiply charged protein cations. A nonergodic process. *J. Am. Chem. Soc.* **120,** 3265–3266.

[5] Collision-Induced Dissociation (CID) of Peptides and Proteins

By J. Mitchell Wells and Scott A. McLuckey

Abstract

The most commonly used activation method in the tandem mass spectrometry (MS) of peptides and proteins is energetic collisions with a neutral target gas. The overall process of collisional activation followed by fragmentation of the ion is commonly referred to as *collision-induced dissociation* (CID). The structural information that results from CID of a peptide or protein ion is highly dependent on the conditions used to effect CID. These include, for example, the relative translational energy of the ion and target, the nature of the target, the number of collisions that is likely to take place, and the observation window of the apparatus. This chapter summarizes the key experimental parameters in the CID of peptide and protein ions, as well as the conditions that tend to prevail in the most commonly employed tandem mass spectrometers.

Introduction

Evidence for the gas-phase CID of molecular ions is apparent in the first mass spectra recorded by Sir J. J. Thomson with his parabola mass spectrograph, and the phenomenon was a subject of study throughout the development of MS during the first half of the twentieth century. A summary of the early work on CID has been published (Cooks, 1995). The modern application of CID to the detection, identification, and structural

METHODS IN ENZYMOLOGY, VOL. 402
0076-6879/05 $35.00
DOI: 10.1016/S0076-6879(05)02005-7

analysis of organic molecules, to complex mixture analysis, and to biopolymer sequencing can be traced to multiples works (Beynon *et al.*, 1973; Futrell and Tiernan, 1972; Jennings, 1968; McLafferty *et al.*, 1973; Wachs *et al.*, 1972) on CID itself and on the closely allied topic of metastable ion dissociation (Cooks *et al.*, 1973). Detailed study of CID during the latter half of the twentieth century has resulted in an understanding of the energy transfer mechanisms and dissociation chemistry of small ions (<500 Da) (Cooks, 1978; McLafferty, 1983, 1992; Shukla and Futrell, 2000; Turecek and McLafferty, 1996). Tandem mass spectrometers, which use gas-phase collision regions to produce product ions from precursor ions, have proven extremely useful for the identification and characterization of ions and for complex mixture analysis. This usefulness is due to several factors, including the ease of implementation of CID (relative to alternative methods such as photo- or surface-induced dissociation [SID]), the fact that CID is universal (i.e., all ions have a collision cross section), and the fact that CID cross sections are typically high.

The last 20 years has seen a revolution in the application of MS and tandem MS to biological problems, due to the advent of ionization methods such as matrix-assisted laser desorption ionization (MALDI) (Hillenkamp and Karas, 2000; Karas *et al.*, 1985) and electrospray ionization (ESI) (Cole, 1997; Whitehouse *et al.*, 1985), capable of producing ions from biologically derived samples such as peptides, proteins, and nucleic acids. Given the popularity and wide application of CID for use in tandem MS of small ions, it is not surprising that CID has been extended to the tandem MS of these larger biological ions (Biemann and Papayannopoulos, 1994; Hunt *et al.*, 1986). Indeed, tandem MS with CID has become an important tool in the growing field of proteomics, the effort to elucidate the protein complement for a given species as a function of physiological conditions (Aebersold and Goodlet, 2001; Larsen and Roepstorff, 2000; Yates, 1998). As is often the case in science, the application of the technique has outpaced the understanding of the underlying phenomena that dictate the appearance of the resulting data. The energy-transfer mechanisms that operate during collisions of large (>1000 Da) possibly multiply charged ions with small target gas atoms or molecules are not as well understood as they are for smaller ions. The dissociation chemistry of large multiply charged biological ions upon activation by collisions is also the subject of continuing study. The intent of this chapter is to describe the important experimental variables that affect the CID of biologically derived macroions and to describe what is known about how such ions behave as a function of those variables, with reference to examples from the literature. The discussion is largely focused on the behavior of peptide and protein ions, as these species constitute by far the most widely studied biological ions by CID. The ranges of the relevant variables that can be accessed with available instrumentation is discussed. Note that in terms of instrumentation, the discussion is kept narrowly focused on CID behavior; other characteristics of tandem mass spectrometers, such as precursor and product mass resolution, efficiency of ion transfer between mass analysis stages, sensitivity, and others, are not discussed. The reader should be aware that although a given instrument may access a range of CID variables, which yields useful tandem mass spectra in terms of the dissociation that occurs, other instrument criteria must also be considered when evaluating the applicability of a tandem mass spectrometer to a particular problem.

Experimental Variables that Influence the CID Behavior of
   Biological Ions

A wide variety of conditions has been used to effect CID of biological ions as new types of tandem mass spectrometers have been developed or as existing types have been fitted with ion sources capable of producing biological ions. Some activation methods have been developed specifically to improve the quality of tandem MS data for biological ions. A set of figures of merit for CID are outlined here to aid in the discussion of how peptides and proteins dissociate as a function of the CID conditions used:

1. The time over which the activation occurs relative to the time for unimolecular dissociation or rearrangement.
2. The amount of energy that can be deposited during the activation.
3. The distribution of the deposited energy.
4. The variability of the deposited energy.
5. The form of the deposited energy (vibrational vs. electronic).
6. The time scale of the instrument used (the kinetic window within which dissociation reactions must occur in order to be observed).
7. The efficiency of the CID process, in terms of cross section or rate constant.

The first figure of merit, the time scale of the activation, refers to how fast energy is deposited in the ion relative to how fast energy equilibrates through the degrees of freedom of the ion, how fast the ion dissociates or rearranges, and how fast energy is removed by cooling processes such as deactivating collisions or photon emission. McLuckey and Goeringer (1997) have distinguished three time regimens for activation methods used in tandem MS, including collisional methods: fast, slow, and very slow (slow heating). Figure 1 illustrates typical ranges of activation times for a number of activation methods, with the collision-based methods of interest here highlighted.

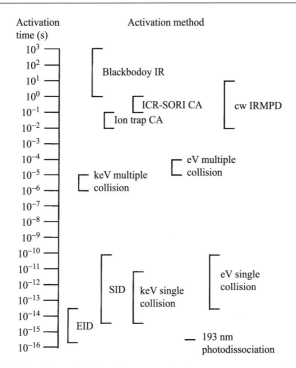

FIG. 1. Activation time scale (time between activating events, e.g., gas-phase collision or photon absorption) for a wide variety of methods used to dissociate biological ions, with the gas-phase collision–based methods highlighted in bold. (Reproduced with permission from McLuckey and Goeringer, 1997.)

Fast methods are those in which the input of energy occurs rapidly relative to unimolecular dissociation, such as single-collision CID at kiloelectron volts or electron volts (in the laboratory frame of reference) kinetic energy. Slow CID methods are those in which multiple collisions can occur, with long (tens to hundreds of microseconds) intervals between individual collisions. In this regimen, chemistry can occur between collisions, meaning that ions can dissociate or isomerize during the activation. This type of CID is effected in collision cells, such as the central collision quadrupole of triple quadrupole tandem mass spectrometers and other instruments that use multipole collision regions, which are operated at pressures of a few millitorr to as high as a torr (Dodonov et al., 1997). Very slow activation methods, or slow heating methods, are distinguished from slow methods by the fact that deactivating processes, such as cooling collisions or photon emission, can occur between activation events. In this regimen, a steady state between activation and deactivation can be achieved. If the dissociation rate is low relative to the activation/deactivation rate (i.e., if the ions are in the so-called "rapid energy exchange" condition), then the ion internal energy can be described with Boltzmann statistics. To a good approximation, the ions can be considered to have an internal temperature, and ion dissociation can be described by the Arrhenius theory. Quadrupole ion trap collisional activation in approximately 1 millitorr of helium bath gas (Goeringer et al., 1999; Louris et al., 1987) and sustained off-resonance irradiation (SORI) (Gauthier et al., 1991) in Fourier transform–ion cyclotron resonance (FT-ICR) MS are examples of slow heating CID methods. Direct heating of the helium bath gas is another form of slow heating that can be used for CID in quadrupole ion traps (Asano et al., 1999a; Butcher et al., 1999). Multiple excitation collisional activation (MECA) (Williams and McLafferty, 1990) and very-low-energy collisional activation (VLE-CA) (Boering et al., 1992a,b) are other forms of slow heating that have been used in FT-ICR. Slow heating methods lead to similar protein and peptide dissociation as slow methods, although an important difference is that for slow heating methods, the product ions are generally not subject to further activation, and hence less sequential dissociation is observed.

The amount of energy that can be deposited in an ion depends on the relative collision energy of the colliding ion/neutral pair. Most CID experiments can be roughly categorized into one of two energy regimens: "high-energy" CID with kiloelectron volt kinetic energy in the laboratory frame of reference and "low-energy" CID with 1–100 eV kinetic energy in the laboratory frame. It bears mentioning here that as discussed throughout this section, there are many variables that determine the amount, distribution, form, and so on, of internal energy deposited into the ion, and further, there are a number of variables in addition to those pertaining to the internal energy that determine the appearance of a CID spectrum. Hence, the historic distinction between "high"- and "low"-energy CID based only on the laboratory kinetic energy is an oversimplification that should probably be avoided. Some statement regarding the collision energy, together with an indication of the number of collisions, the mass of the collision partner, and the activation time is probably a minimum in describing a particular CID experiment.

The kinematics for collisions of small ions with neutrals have been carefully described (Cooks, 1978; McLuckey, 1992; Shukla and Futrell, 2000) and show that it is the relative kinetic energy of the collision partners that is available for transfer to internal energy of the ion. This center-of-mass collision energy, KE_{com}, is a fraction of the laboratory kinetic energy, KE_{lab}, as given by Eq. (1) if the velocity of the neutral is ignored:

$$KE_{com} = \left(\frac{N}{m_p + N}\right)KE_{lab}, \tag{1}$$

where N is the mass of the neutral target and m_p is the mass of the ion. Equation (1) shows that the energy available for transfer into internal modes for a collision of a high m/z ion with a small neutral is very low and suggests that CID of high m/z ions should be very inefficient. However, Marzluff et al. (1994a) have shown that at least for low center-of-mass kinetic energy, the energy transfer process in collisions of large ions with small targets can be extremely efficient. In this study, a homologous series of deprotonated peptides, which all dissociate via the same well-character-ized mechanism (Marzluff et al., 1994b), was used to study the efficiency of dissociation as a function of peptide size. The results indicate that the extent of dissociation of these peptides upon SORI-CID in an FT-ICR using nitrogen collision gas increased with the size of the peptide. Rice-Ramsperger-Kassel-Marcus (RRKM) statistical dissociation rate calcula-tions (Forst, 1973; Holbrook and Robinson, 1972; Lorquet, 1994) for one of the peptides studied, gly-gly-ile, show that at the observed dissociation rate, approximately 55% of the available kinetic energy was transferred to internal energy during each collision. Trajectory calculations conducted with a molecular mechanics force field support this experimental observa-tion and indicate that for a larger system such as bradykinin (nine amino acid residues and 444 internal degrees of freedom), as much as 90% of the available kinetic energy could be transferred.

Douglas (1998) has developed an approximation for the internal energy deposited into large ions upon electron volt collisions in a linear quadru-pole, which supports the observations described earlier. Using experimen-tal dissociation cross sections (see discussion of efficiency below) measured as a function of center-of-mass collision energy, Douglas showed that electron volt collisions of bromobenzene cations with nitrogen lead to approximately 70% conversion of center-of-mass energy to internal ener-gy. Further work by Douglas aimed primarily at measuring collision cross sections for protein ions in a triple-quadrupole instrument provided further evidence for the efficiency of energy transfer (Chen et al., 1997). The collisions of large ions with small target gas atoms were shown to be most accurately modeled with a diffuse scattering model, wherein a small target atom collides with the ion at the center-of-mass collision energy, and then leaves with a thermal energy distribution; the result is that 90–95% of the collision energy is converted to internal energy of the ion during the relatively long interaction between the ion and the neutral. Experimental measurement of the dissociation threshold for the loss of heme from holomyoglobin supported this large energy transfer of more than 90% (Douglas, 1998).

A number of authors have developed an impulsive collision theory (ICT) to describe the energy-transfer process for collisions of large ions with small neutrals (Cooper *et al.*, 1993; Uggerud and Derrick, 1991). In this model, direct momentum transfer between a constituent atom of the ion and the collision partner occurs, resulting in much larger transfer of internal energy to the ion than would be predicted based only on consideration of the overall mass of the large ion (Eq. [1]). Such binary collisions between portions of a large ion and the small target have been examined by other researchers (Boyd *et al.*, 1984; Douglas, 1982; Futrell, 1986). Bradley *et al.* (1994) compared the energy transfer to large organic ions consisting of many small atoms (C, N, O, H) with that to ions of similar *m/z* consisting of large atoms (CsI clusters) and showed that the behavior was consistent with the prediction of the ICT, viz. more energy transfer to the organic ion during collisions with helium, due to the closer similarity in mass between the small atoms of the organic ion and the helium atoms. The reverse was found to be true for collisions with argon, because of a closer mass match between the Cs and I atoms of the cluster and the argon target atoms.

Turecek (1991) has shown that in principle, the center-of-mass limit on the amount of energy deposited can be overcome by invalidating the assumption made in Eq. (1), viz. that the velocity of the neutral collision partner is zero. By colliding ions with fast (kiloelectron volts) rather than thermal targets, much more energy is available for transfer to internal energy during the collision. However, the practical difficulties of crossing a fast beam of ions with a fast beam of neutrals or of colliding a fast beam of neutrals with a small stationary ion cloud, with sufficient cross section to yield measurable product signals may limit the analytical applicability of this form of CID. Note that crossed beams have been used for fundamental studies of ion–molecule reactions at very low kinetic energy (Futrell, 2000).

The number of collisions occurring during the activation period also determines the amount of energy deposited. Most analytical applications of CID use collision regions operated at pressures high enough to ensure that multiple collisions occur to increase internal energy and hence yield extensive dissociation. The disadvantage to using higher pressure is that increased scattering may result in a loss of sensitivity; however, Douglas and French (1992) have shown that in quadrupole transmission devices used to carry ions from an ESI source to the mass analyzer, collisional cooling of ions at elevated pressure ($\sim$8 millitorr) leads to improved sensitivity and resolution (by removal of kinetic energy spread in the ions). This concept has also been applied to quadrupole collision cells of tandem mass spectrometers, leading to improved product ion collection efficiency and resolution when the cells are operated at 8 millitorr compared to the 1–2 millitorr normally used (Douglas *et al.*, 1993; Thomson *et al.*, 1995).

This discussion relates to the magnitude of the energy that can be deposited; the third and fourth figures of merit relate to the distribution of energies that can be deposited and the variability of the energy distribution. The distribution of energies plays an important role in determining the appearance of CID spectra (Vekey, 1996). Ideally, the energy distribution is narrow, well defined, and variable over a wide energy range, so dissociation channels having a wide range of critical energies can be accessed. CID methods lead to energy distributions that are typically fairly broad and ill defined, at least for fast and slow CID methods. In kilo-electron-volt collisions, a distribution having a maximum at only a few electron volts, with a long tail out to higher energy (tens of electron volts), is often obtained for small ions, allowing some high critical energy channels to be observed. For electron-volt collisions, the maximum of the distribution is similar, but the high-energy tail is not present, so that only lower critical energy processes are accessed. Douglas (1998) has presented a method for estimating the internal energy distribution for multiple collision CID in a quadrupole from a measure of the dissociation cross section as a function of collision energy. For ion trap collisional activation, it has been shown that the internal energy distribution can be modeled as a Boltzmann or truncated Boltzmann distribution, depending on the relative rates of activation/deactivation and dissociation (Goeringer *et al.*, 1999). Schnier *et al.* (1999) have shown that for FT-ICR SORI, leucine enkephalin and bradykinin ions achieve effective temperatures of between 500 and 700 K, although the authors state that actual distribution of internal energy is not well characterized and may not be Boltzmann. As discussed earlier, the maximum of the internal energy distribution can be varied to some extent by varying the collision energy; CID spectra acquired at low laboratory kinetic energies (electron volts) typically show greater sensitivity to changes in energy than spectra acquired at kiloelectron-volt energy (Busch *et al.*, 1988).

The fifth figure of merit relates to whether the internal energy is present in the activated ion in electronic or vibrational modes. Statistical theories of activated ion dissociation assume that internal energy equilibration through all vibrational degrees of freedom is faster than dissociation (Forst, 1973; Holbrook and Robinson, 1972; Lorquet, 1994). Experimental evidence shows that this holds for most small organic ions; however, exceptions where dissociation from an electronic excited state is faster than relaxation of energy into vibrational modes have been reported (Shukla *et al.*, 1990). There is little evidence for or against the statistical hypothesis for large peptide and protein ions. However, it seems reasonable, at least under multiple collision conditions where energy is deposited in small increments at a variety of collision sites on the ion, that collisional activation effectively gives rise to a statistical ion energy distribution.

The sixth figure of merit relates to how fast the dissociation reactions must occur in order to be observed on the time scale of the instrument used. The kinetic window varies widely across different instrument types from a few tens to hundreds of microseconds for magnetic sector–based instruments operating at kiloelectron-volt ion kinetic energies, milliseconds in triple-quadrupole and quadrupole-TOF instruments, and from hundreds of milliseconds up to seconds in trapping instruments. The length of the kinetic window is important for large biological ions, because dissociation of these ions may be relatively slow because of their large number of degrees of freedom. The amount of energy above the critical energy for dissociation required to drive a dissociation reaction at a rate that is observable on the time scale of the instrument used is called the *kinetic shift* and may be very large (up to or even exceeding 100 eV) for large biological ions. This obviously makes the previous discussion regarding the amount of energy that can be deposited into the ions upon collisional activation very important (e.g., at kiloelectron-volt kinetic energies, dissociation rates must be driven to at least 10^4 s^{-1} in order to be observed), and hence inherently slow processes may not be observable in this regimen, due to an inaccessibly high kinetic shift.

The efficiency of the CID process relates to how much of the available precursor ion is converted to product ions. For beam-type instruments, such as sectors and triple quadrupoles, this is most conveniently represented with a Beer's law type of relationship, as shown in Eq. (2):

$$[M_p^+] = [M_p^+]_0 e^{-n\sigma l}, \tag{2}$$

where $[M_p^+]_0$ is the precursor ion flux without collision gas, $[M_p^+]$ is the precursor ion flux after the introduction of collision gas, l is the path length, n is the target number density, and σ is the total ion loss cross section. If no processes for precursor ion loss other than CID operate, then σ represents the cross section for CID. For trapping instruments, an analogous expression in terms of rate constant and reaction times can be used:

$$[M_p^+]_t = [M_p^+]_0 e^{-nkt}, \tag{3}$$

where $[M_p^+]_0$ is the precursor ion abundance before any reaction, $[M_p^+]_t$ is the precursor ion abundance after reaction time t, and k is the rate constant for all precursor ion loss processes. If no precursor ion loss processes other the CID operate, then k represents the rate constant for CID. Equation (2) shows that the only variable available to increase the efficiency of CID in beam-type instruments is *n*, the number density of the neutral collision partner (increasing the path length is possible but is usually not straightforward over a wide range). The advantages and disadvantages of operating at higher collision gas pressure were discussed earlier. At higher pressure, ion loss processes such as scattering may degrade sensitivity, although in quadrupole collision cells, the collisional cooling afforded by higher collision gas pressure has been shown to give the opposite effect. It is easier to increase the efficiency of CID in ion-trapping instruments by increasing the reaction time, *t*. Reaction times as long as seconds can be used to completely drive precursor ions to product ions. Note that operation at either higher collision gas pressure or longer activation time increases the likelihood of multiple collisions occurring and, hence, implies that the activation is either slow or very slow (slow heating), with the attendant possibility for chemistry to occur during the activation, as discussed earlier.

Summary of Commonly Used CID Conditions

The values for any individual figure of merit discussed earlier may vary widely with the type of instrument used and the design of the experiment, and the various figures are not independent but influence one another. In this section, three commonly used regimens for effecting CID of gas-phase biological ions are described in terms of the figures of merit outlined earlier. Table I summarizes the information given in this section with typical ranges of values used for the operating parameters and figures of merit for the three regimens. CID spectra for a common ion, the doubly protonated ion of the nine residue peptide bradykinin (RPPGFSPFR), are used throughout this section to illustrate the influence of the figures of merit on CID behavior.

High-Energy CID (Fast Activation)

As discussed earlier, the collision energy is only one of a number of variables that must be specified to adequately describe a CID regimen. The term "high-energy" CID is usually used to describe CID effected at kiloelectron-volt precursor ion kinetic energies, with a target gas pressure that is low enough that only single or at most a few (fewer than five) collisions can occur. Because of the high kinetic energy of the ions, the time scale for dissociation is usually on the order of a few microseconds. The vast majority of CID experiments under this regimen have been carried out using sector-based instrumentation. However, kiloelectron-volt collisions are seeing increasing use in tandem time-of-flight (TOF) instruments, as discussed later in this chapter.

TABLE I
Summary of Typical Parameters for Three Commonly Used CID Regimens

Figure of merit	"High-energy" CID (fast activation)	"Low-energy" CID (slow activation)	Trapping CID (very slow activation)
Instruments used	Magnetic/electric sectors, TOF/TOF	Tandem quadrupoles, quadrupole hybrids (e.g., QqTOF)	Quadrupole ion traps, FT-ICR traps
Collision energy	2–10 keV	1–200 eV	1–20 eV
Collision number	1–5	10–100	100 s
Activation time scale	1–10 μs	0.5–1 ms	10–100 ms
Instrument time scale (kinetic window/minimum observable reaction rate)	10–100 μs /10^6–10^4 s^{-1}	0.1–1 ms/10^4–10^3 s^{-1}	10 ms–1 s/10^2–1 s^{-1}
Distribution of internal energy	Centered at a few electron volts, high-energy tail to tens of electron volts	Centered at few eV, no high energy tail	Centered at a few electron volts, may be Boltzmann or Boltzmann-like
Variability of internal energy	Relatively invariable, scattering angle provides some energy resolved info.	Readily variable with collision energy to obtain energy resolved info.	Some variability with collision energy and number
Efficiency	<10%	5–50%	50–100%
General results	High-energy channels may be accessed together with lower energy processes, sequential dissociation observed	Lower energy processes only, isomerization of precursor may occur, sequential dissociation observed	Low-energy processes only, extensive isomerization of precursor, very slow processes can be observed, typically little sequential dissociation

Double-focusing mass spectrometers based on magnetic and electric sectors were the first instruments used for CID studies of small organic ions (Beynon *et al.*, 1973; Futrell and Tiernan, 1972; Jennings, 1968; McLafferty *et al.*, 1973; Wachs *et al.*, 1972), and were among the first used for CID of peptide ions approximately a decade later (Biemann and Papayannopoulos, 1994). Two-sector MS instruments may be used to record MS/MS spectra by using a variety of linked scans (Busch *et al.*, 1988) of the magnetic (B) and electric (E) fields, albeit with relatively poor precursor ion or product ion resolution, depending on the type of scan. A number of three-sector instruments were constructed to help overcome this limitation (Gross, 1990). Four-sector instruments, true tandem mass spectrometers that combine two double focusing instruments for high resolution of precursor and product ions, were also constructed and were commercially available from a number of manufacturers (Gross, 1990). The use of these instruments for tandem MS of biological ions was demonstrated throughout the 1980s and early 1990s, but they have now been largely superseded by tandem quadrupoles, hybrid quadrupole/TOF instruments, ion-trapping instruments, and increasingly TOF/TOF instruments.

Readily obtainable field strengths for sectors are such that these instruments must operate on ion beams accelerated to kiloelectron-volt kinetic energies; hence, most CID work with sector instruments is done with collisions in this regimen. As discussed earlier, kiloelectron-volt collisions deposit on average a few electron volts of internal energy, but the distribution can include a tail to higher internal energies, so that higher critical energy or lower rate processes (requiring large kinetic shifts) can be observed. High-energy CID spectra are relatively insensitive to changes in the kinetic energy of the precursor ion, and so obtaining energy-resolved CID spectra with high-energy CID is difficult. However, it has been shown that higher energy collisions lead to larger scattering angles, so recording MS/MS data as a function of angle (angle-resolved MS [ARMS]) yields information on dissociation as a function of internal energy (McLuckey and Cooks, 1983). It is possible to decelerate and then reaccelerate the beam to also access low-energy collisions; however, high-energy CID spectra are very reproducible on a given instrument and even between instruments of different configuration, and this is often cited as a major advantage so that operation at low energy is usually not considered to be a desirable alternative.

The transit time of a 10-keV beam of ions at m/z 1000 through a 10-cm collision cell is 2.3 ms, and 23 ms along a 1-m flight path from the collision cell to the detector. The former value establishes the maximum amount of time between activating collisions, if the cell is operated at a pressure at which multiple collisions in this time frame are likely. High-energy CID is normally performed using pressures chosen to yield a given attenuation of the ion beam, which is related to the number of collisions. Single or at most a few (5–10) collisions are typically used; beyond this, scattering of the beam leads to unacceptable losses in sensitivity. Increasing the number of collisions, of course, increases the energy deposited. At least for small ions, internal rearrangements can occur on a time scale of nanoseconds or less, hence isomerization can occur between collisions even at high kinetic energy (Holmes, 1985). The transit time through the whole instrument of some tens of microseconds establishes a lower limit on the observable rates of approximately 10^4 to 10^6 s^{-1}.

For peptides, the figures of merit described lead to CID spectra, which differ from those obtained using other commonly employed conditions primarily in that abundant dissociation of amino acid side chains is observed in addition to peptide bond cleavage. This is illustrated in Fig. 2, the MS/MS spectrum of +2 bradykinin (Downard and Biemann, 1994) obtained on a four-sector instrument using 8-keV collisions with helium at a pressure at which the primary beam intensity reduced by 70%, corresponding to between two and four collisions (Holmes, 1985). The ions labeled w_n and d_n correspond to side-chain cleavage (Biemann, 1988; Roepstorff and Fohlman, 1984). It has been suggested that these cleavages are charge remote and, hence, require higher critical energies to be accessed, implicating the high-energy tail of the internal energy distribution as the cause for these cleavages. This is supported by the absence of

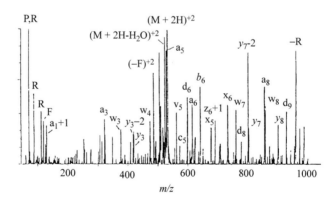

FIG. 2. Collision-induced dissociation (CID) spectrum of the $(M + 2H)^{+2}$ of bradykinin collected on a four-sector tandem mass spectrometer using 8-keV collisions with helium at a pressure that reduced the primary beam intensity by 70%, corresponding to between two and four collisions. Note the abundant side-chain cleavage ions (labeled w_n and d_n) and the immonium ions of individual amino acids (labeled with capital letters). (Reproduced with permission from Downard and Biemann, 1994.)

side-chain cleavages in CID spectra collected at electron-volt collision energy (see later discussion). These side-chain cleavages have been shown to be useful for the distinction of isomeric and isobaric amino acids in peptide-sequencing applications (Biemann and Papayannopoulos, 1994). The upper limit for CID at high energy is approximately m/z 3000, beyond which dissociation is not readily observed, probably because of the large kinetic shift required to drive dissociations at a rate of at least 10^4 s^{-1}. The efficiency of high-energy CID is typically only a few percent even for smaller molecules because of this short time scale and because of the product ion collection efficiency.

The advantages of CID conducted at kiloelectron-volt energy and low collision numbers, viz. reproducibility and access to side-chain cleavages for isomeric and isobaric ion distinction, are driving interest in an alternative instrument, the tandem TOF (TOF/TOF), for accessing this regimen. An advantage of the TOF/TOF over sector-based instruments is the well-established compatibility of TOF with the MALDI source. A number of forms of TOF/TOF have been described for CID and photodissociation of small molecules and cluster ions (Cornish and Cotter, 1993a; Hop, 1998; Jardine *et al.*, 1992). At least two groups have demonstrated the use of TOF/TOF instruments to dissociate peptide ions with kiloelectron-volt collisions (Cornish and Cotter, 1993b; Medzihradszky, 2000). Figure 3 shows a CID spectrum for a tryptic peptide dissociated using 3-keV collisions with argon (pressure unspecified) (Medzihradszky, 2000). The abundant immonium ions of individual amino acids and the w_n ions resulting from side-chain cleavage are characteristic of the high collision energy used and were helpful in establishing the peptide composition and distinguishing the isomeric leucine and isoleucine residues, respectively.

Low-Energy CID (Slow Activation)

The term "low-energy" CID is typically used to refer to CID conducted in quadrupole or other multipole collisions cells and is characterized by collision energy of up to 100 eV, target gas pressures selected to allow multiple (tens to hundreds) collisions, and a time scale on the order of a few hundred microseconds to a few milliseconds. For simplicity, throughout this discussion the term *quadrupole,* represented with the letter *q,* will be used to describe a multipole collision cell that passes all m/z ratios above a certain lower limit, with the understanding that higher multipoles such as hexapoles and octapoles are in some cases used instead of quadrupoles. The electron-volt energy regimen is used in tandem quadrupole instruments (usually referred to as *triple quadrupoles* [QqQ], where Q is a mass-resolving quadrupole) and in hybrid instruments that combine quadrupole

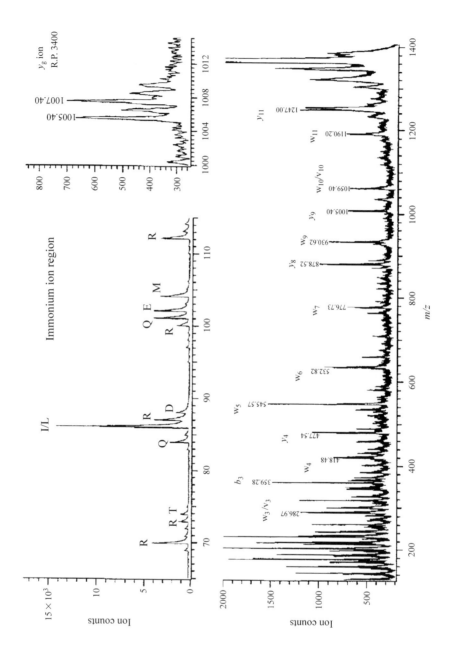

collision cells with other types of mass analyzers for precursor selection and/or product mass analysis (Yost and Boyd, 1990). A hybrid instrument that is increasing in popularity is the quadrupole-quadrupole-TOF (QqTOF) (Morris *et al.*, 1996; Shevchenko, 1997), where precursor selection is done with the first quadrupole, CID occurs in the second quadrupole, and the products are mass analyzed via TOF. The attractive mass analysis characteristics of orthogonal acceleration, reflectron TOF are driving interest in this instrument, and two commercial versions are available.

The advantage of using a radiofrequency (RF) quadrupole (or other multipole) as a collision cell is that the RF field provides a strong focusing force toward the ion optical axis, so that losses of precursor and product ions are minimized, even at relatively high collision numbers. However, to enjoy this advantage, ions must move through the quadrupole slowly enough to be influenced by the rapidly changing RF field; hence, kinetic energies less than 100 eV are typically used. For small organic ions, low-energy collisions lead to internal energy distributions that have a peak at a few electron volts of internal energy, but without the long tail to higher energy, which characterizes kiloelectron-volt CID. The observed differences between electron-volt and kiloelectron-volt dissociation of peptides is that electron-volt CID lacks the side-chain cleavage peaks often observed at higher energy. This is illustrated in Fig. 4 for the MS/MS spectrum of +2 bradykinin collected on a triple-quadrupole instrument using electron-volt collisions (between 70 and 200 eV) with argon at a pressure that yielded a target-gas thickness of 2–5 × 10^{14} atoms/cm^2 (Tang *et al.*, 1993), where target-gas thickness is equal to the product of the number density of the gas and the length of the collision cell (Thomson *et al.*, 1995). The absence of w- and d-type ions in this electron-volt kinetic energy spectrum supports the suggestion that side-chain cleavages leading to these ions are high-energy processes. Some sequential dissociation, leading to internal fragments (e.g., y_7b_5) and immonium ions (P and F), is observed.

As discussed earlier in this chapter, Douglas (1998) has estimated that collisions of large ions with small targets can be very efficient for the transfer of available kinetic energy to internal energy; therefore, provided that sufficient collisions occur, even large proteins having many degrees of

---

FIG. 3. Collision-induced dissociation (CID) spectrum for an $^{16}O/^{18}O$–labeled tryptic peptide (DLEEGIQTLMGR) collected on a tandem time-of-flight (TOF)/TOF instrument using 3-keV collisions with argon (pressure unspecified) for activation. Peak intensities in the full spectrum (lower panel) were multiplied eight times relative to the immonium ion spectrum (upper left panel). (Reproduced with permission from Medzihradszky *et al.*, 2000.)

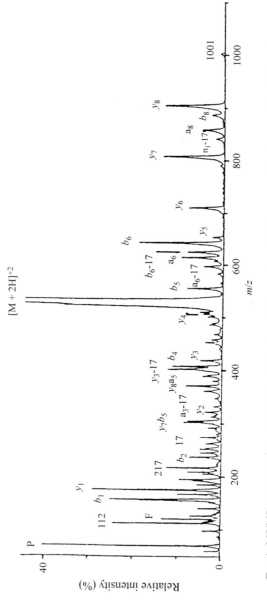

FIG. 4. MS/MS spectrum of +2 bradykinin collected on a triple-quadrupole instrument using electron volt collisions (between 70 and 200 eV) with argon at a pressure that yielded a target-gas thickness of $2–5 \times 10^{14}$ atoms/cm². Note at this lower collision energy, only b- and y-type ions were formed, and that sequential dissociation led to ammonia loss (e.g., y_{13}-17) and internal dissociation (e.g., y_7b_5). Ammonium ions were observed for proline (P) and phenylalanine (F). (Reproduced with permission from Tang *et al.*, 1993.)

freedom, and hence large kinetic shifts can be driven to dissociate at observable rates. This is illustrated in the data in Fig. 5, which shows the dissociation of the $(M + 20H)^{+20}$ ion of apomyoglobin collected on a triple-quadrupole instrument at 2-keV collision energy using collisions with argon at a pressure that yielded a target-gas thickness of 1×10^{14} atoms/ cm^2 (Smith *et al.*, 1990).

Note that this experiment is at the interface between the low-energy and high-energy regimens, highlighting the danger of describing an experiment based only on the collision energy. Because the force acting on an ion is proportional to the charge on the ion, a higher collision energy can be used for this highly charged protein ion while still enjoying the focusing advantages of the linear quadrupole discussed earlier in this chapter. At the pressure used in this study, multiple collisions were likely, so the experiment was a slow activation experiment with ample time for rearrangement before dissociation. Dissociation of this large precursor was aided by the fact that the ions were already "heated" in the electrospray interface by use of large potential differences between the lenses used to transport ions into the mass spectrometer (see the discussion of "nozzle/skimmer" dissociation below).

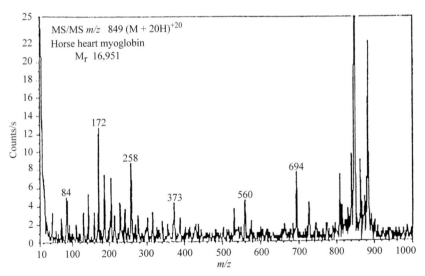

FIG. 5. Collision-induced dissociation (CID) spectrum of $(M + 20H)^{+20}$ ions of horse heart apomyoglobin collected on a triple-quadrupole instrument at 2-keV collision energy using collisions with argon at a pressure that yielded a target-gas thickness of 1×10^{14} atoms/cm^2. (Reproduced with permission from Smith *et al.*, 1990.)

CID spectra collected at electron-volt collision energy can be more sensitive to changes in the collision energy than CID at kiloelectron-volt energy, allowing the effects of internal energy on ion dissociation to be probed in this regimen. This technique is referred to as *energy-resolved MS* (ERMS) (McLuckey and Cooks, 1983). ERMS is used to generate breakdown graphs, or plots of the relative contributions of ion dissociation channels as a function of ion internal energy. Such graphs can be useful for elucidating reaction mechanisms and for distinguishing isomeric species.

At 100 eV, an *m/z* 1000 ion will travel through a 20-cm quadrupole in approximately 1.5 ms. This time scale means that extensive isomerization of the ion of interest may occur between collisions if multiple collision conditions are used. Processes with rate constants of 10^3 s^{-1} or greater can be observed in this relatively slow CID regimen. Collision gas pressures of 1–2 millitorr, leading to some tens of collisions, have traditionally been used to effect CID in a quadrupole. However, a major innovation in the use of quadrupoles for CID of larger biological ions is the demonstration that collisional damping of radial motion in the quadrupole allows higher pressures, and hence more collisions, to be used without excessive ion losses due to scattering (Douglas and French, 1992). Operation at approximately 8 millitorr has been shown to improve the efficiency of triple-quadrupole MS/MS from only a few percent to 30–50% (Thomson, 1995). Note that much of this improvement is due to the improved collection and transmission of product ions by the mass analyzing quadrupole (Q$_3$) after the collision cell due to the removal of the spatial and kinetic energy spread of the products. At higher pressure, transit times through collision cell can be very long (tens of milliseconds) unless an additional electric field in the direction of ion motion through the quadrupole is added to "pull" the ions through the gas. When such a field is added, pressures of tens to hundreds of millitorr and as high as 3 torr may be used to effect CID (Dodonov *et al.*, 1997; Javahery and Thomson, 1997; Lock and Dyer, 1999; Mansoori *et al.*, 1998). Under these conditions, dissociation reactions with rates as low as a few tens per second can be accessed.

Another form of CID conducted at electron-volt collision energy is the so-called "in-source," "cone-voltage," or "nozzle/skimmer" CID effected in the interface region of ESI sources (Bruins, 1997; Katta *et al.*, 1991; Loo *et al.*, 1998). The lens voltages that guide ions through the intermediate pressure region between the atmospheric pressure source and the high vacuum of the mass analyzer region may be manipulated to cause ions to undergo collisions with residual background gas at collision energies up to a few hundred electron volts. The time scale for this type of dissociation is on the order of a few hundred microseconds to a few milliseconds. Although this is not a true MS/MS experiment because there is no prior selection of the precursor ion, nozzle/skimmer CID can provide efficient dissociation of peptides and proteins on simple instruments. For example, the nozzle/skimmer CID spectrum of +2 bradykinin collected at 520 eV in the approximately 1–10 torr region of an ESI source interfaced to a single quadrupole is shown in Fig. 6 (bottom panel) (Katta *et al.*, 1991). The spectrum is similar to that collected via true MS/MS with a triple-quadrupole instrument (Fig. 4), in that only *b*- and *y*-type ions are observed. Nozzle/skimmer dissociation has also been demonstrated for whole protein ions. Figure 7 shows the dissociation of multiply charged carbonic anhydrase B cations in an electrospray source interfaced to an FT-ICR instrument (Senko *et al.*, 1994a). Nozzle/skimmer CID has also been used for ERMS of small peptides, with comparable results to those obtained for quadrupole ERMS (Dono *et al.*, 1997; Harrison, 1999).

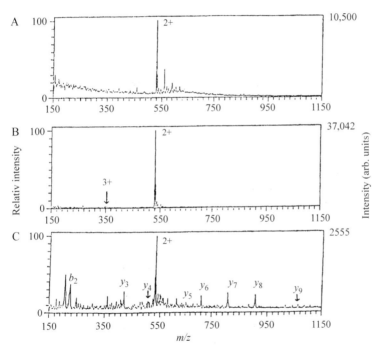

FIG. 6. Spectra of bradykinin recorded as a function of the potential difference between interface lenses of the electrospray source. (A) The potential difference was 80 V, and only the $(M + 2H)^{+2}$ ions and $(M + 2H)^{+2}$ with solvent molecules attached were observed. (B) The potential difference was increased to 160 V, so that the $(M + 2H)^{+2}$ was efficiently desolvated. (C) The potential difference was further increased to 260 V, and abundant *b*- and *y*-type product ions were observed due to nozzle/skimmer collision-induced dissociation (CID). (Reproduced with permission from Katta *et al.*, 1991.)

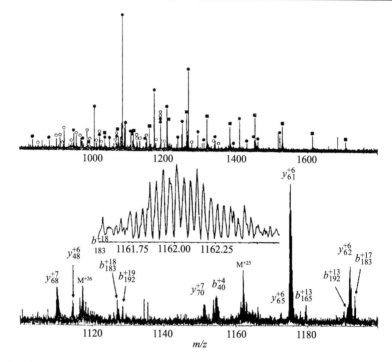

FIG. 7. Nozzle/skimmer dissociation of multiply charged carbonic anhydrase B cations in an electrospray source interfaced to an Fourier transform–ion cyclotron resonance (FT-ICR) instrument, with a potential difference of 165 V across the ion source. The inset from m/z 1161.5 to 1162.5 shows the $(M + 25H)^{+25}$ precursor ion. The squares designate protonated precursor ions, the open circles designate b-type products, and the filled circles designate y-type products. (Reproduced with permission from Senko *et al.*, 1994a.)

Ion-Trapping CID (Very Slow Activation, Slow Heating)

The two forms of ion traps used in MS, electrodynamic quadrupole ion traps and electrostatic/magnetic ICR ion traps, vary in the mechanism by which they store ions and perform mass analysis; however, their behavior with respect to CID is quite similar. CID in both instruments is characterized by many collisions (hundreds) at a few electron-volt (tens) collision energy and, as such, is qualitatively similar to the "low-energy" quadrupole CID discussed in the previous section. An important difference is that the trapping nature of these instruments means that very long time scales can be used to achieve very high dissociation efficiency (75–100%) and to access very slow dissociation channels.

Another consequence of the long time scale accessible with trapping instruments is that deactivating processes can also occur, so that a steady-state internal energy can be achieved. Details of CID in a quadrupole ion trap are discussed first in this section, followed by discussion of CID in an FT-ICR trap.

Quadrupole Ion Trap CID. Quadrupole ion traps store ions in the rapidly changing potential well generated by the application of an RF voltage to the trap electrodes. Ion motion in the field depends, *inter alia,* on the frequency and amplitude of the RF voltage and on ion m/z. For a given RF frequency and amplitude, a selected precursor m/z will have a characteristic frequency of motion in the trap, normally referred to as the *secular frequency.* Application of a supplementary voltage matching this secular frequency will excite the precursor ions, causing them to undergo energetic collisions with background gas or deliberately added collision gas (usually helium, see below), thereby increasing their internal energy. This process, commonly called *resonance excitation,* was first demonstrated by Louris *et al.* in 1987. The depth of the trapping potential well is such that the ions can only be excited to a few electron volts of kinetic energy before they are ejected from the trapping volume by the supplementary resonant voltage. For this reason, multiple collisions over a relatively long time (tens to hundreds of milliseconds) are required to build up sufficient internal energy to lead to dissociation. Clearly, extensive rearrangement of the precursor ion can occur during the activation and before dissociation. Another consequence of this long slow buildup of internal energy is that deactivating processes such as IR emission and/or cooling collisions can also occur. Goeringer and McLuckey (1996a,b) have developed a collisional energy-transfer model for resonance excitation, which shows that competition between activation and deactivation leads to a steady-state internal energy that can be characterized by a Boltzmann distribution and an internal temperature, provided the dissociation rate is low compared to the activation/deactivation rate. If dissociation is rapid enough to deplete the high-energy tail of the distribution, a truncated Boltzmann distribution results and the temperature is more correctly described as an "effective" internal temperature (Vekey, 1996). For small peptides (e.g., leucine enkephalin, 556 Da), effective temperatures of approximately 650 K can be achieved (Goeringer *et al.*, 1999). The available temperature decreases with increasing m/z as the difference between the mass of the ion and the collision partner increases. Another factor that limits the temperature to which high m/z ions can be raised is that for a given RF amplitude, higher m/z ions are stored in shallower potential wells and so must be excited with lower resonance voltage amplitudes to avoid ion losses through ejection from the trap. Current instrumentation limits the protein ion m/z on which CID can be performed to approximately 8600 Da/charge (+1 ubiquitin) (Reid *et al.*, 2001), although the upper limit for newer ion traps (e.g., the Finnigan LCQ, Bruker Esquire, or Hitachi M-8000) has not yet been evaluated. Note that this limitation is stated in terms of mass/charge. Proteins with masses greater than ubiquitin can be readily dissociated with quadrupole ion traps as multiply charged ions with m/z lower that 8600 Da/charge; for example, a study of the dissociation of $(M + 2H)^{+2}$ to $(M + 21H)^{+21}$ ions of apomyoglobin (mass 16,950 Da) has appeared (Newton *et al.*, 2001).

Shown in Fig. 8 is a CID spectrum for +2 bradykinin acquired with 200 ms of resonance excitation in an ion trap having a background pressure of 1 millitorr of helium. Note that only b- and y-type ions are formed, in similar fashion to the electron-volt collision CID performed in a quadrupole collision cell (Fig. 4) and in an ESI interface (Fig. 6). Whole-protein ions can also be dissociated via resonance excitation in a quadrupole ion trap. Figure 9 shows the CID spectrum of $(M + 8H)^{+8}$ ion of ubiquitin. The precursor ion was excited for 300 ms in the presence of 1 millitorr of helium

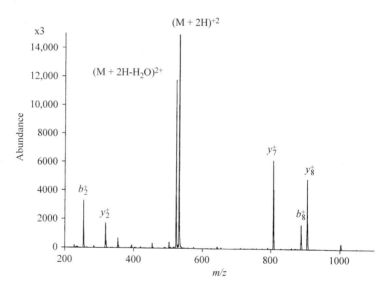

FIG. 8. Collision-induced dissociation (CID) spectrum for +2 bradykinin acquired with 200 ms of resonance excitation in a quadrupole ion trap having a background pressure of 1 millitorr of helium. Note that only b- and y-type ions are formed, in similar fashion to the electron-volt collision CID performed in a quadrupole collision cell (Fig. 4) and in an ESI interface (Fig. 6). The abundance scale was multiplied three times relative to the abundance of the remaining precursor ion.

(Reid *et al.*, 2001). Although the internal temperature accessed during ion trap CID of this large ion is relatively low, the trapping nature of the instrument allows for long reaction times and hence low reaction rates, so relatively efficient dissociation is observed. An important point about quadrupole ion trap CID is that all product ions below a certain low-mass cutoff (LMCO), determined by the RF amplitude, are not trapped. For CID of large ions, it is desirable to employ a relatively high RF amplitude to maximize the potential well depth and avoid ion losses by ejection; however, a high RF amplitude means that low m/z products are not trapped and so are not observed.

As discussed earlier in this chapter, ion trap CID is always carried out under multiple collision conditions, hence extensive rearrangement can occur during the activation period and before dissociation. The collision gas used is almost always helium, because helium bath gas is necessary to improve the resolution of ion trap mass analysis (Stafford *et al.*, 1984) and so is already present in the system at a static background pressure of 1 millitorr. The effect of adding higher mass collision gas has been studied, and it has been shown that the deposited internal energy is increased (Charles *et al.*, 1994; Gronowska *et al.*, 1990; Morand *et al.*, 1992), as predicted by Eq. (1). Several groups have reported that addition of a small fraction (5%) of, for example, argon, krypton, or xenon can improve the CID efficiency of peptide ions by allowing higher internal energy deposition (Vachet and Glish, 1996). This has the effect of increasing the extent of dissociation and accessing higher critical energy dissociation channels (Vachet and Glish, 1996). Addition of heavy gases also allows CID to be carried out at lower RF-trapping voltages, which helps to overcome the LMCO limitation discussed earlier (Doroshenko and Cotter, 1996a). Note that the upper m/z limit for CID in the ion trap given earlier (8600 Da/charge) was probably achieved at least in part because of the significant background pressure of air ($\sim 1.5 \times 10^{-4}$ torr) present in the ion trap (Reid *et al.*, 2001).

An alternative to raising the internal temperature of the ions in a quadrupole ion trap by increasing the kinetic energy of the ions via resonance excitation is to increase the kinetic energy of the neutral collision partner by heating the helium bath gas. Asano *et al.* (1999a) have shown that ions reach thermal equilibrium with the bath gas, so if the temperature of the bath gas is known, and if the rate of activation/deactivation is large relative to the rate of dissociation, then Arrhenius activation parameters can be derived from measurements of ion dissociation rate as a function of bath gas temperature. In addition, they have demonstrated the measurement of Arrhenius parameters for leucine enkephalin and

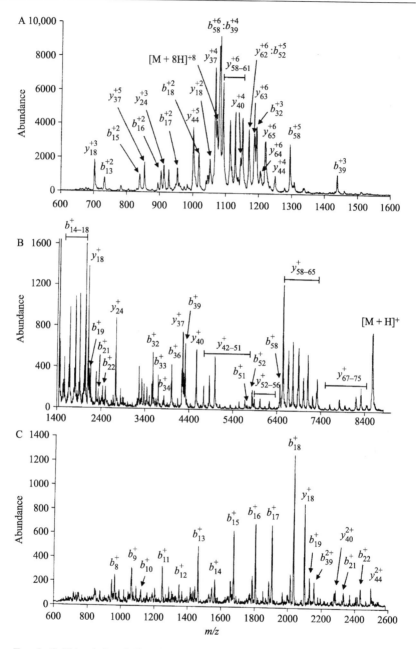

FIG. 9. Collision-induced dissociation (CID) spectrum of $(M + 8H)^{+8}$ ubiquitin ions excited in a quadrupole ion trap for 300 ms in the presence of 1 millitorr of helium. (A) The

bradykinin (Butcher *et al.*, 1999) and have used leucine enkephalin as a thermometer ion to derive the temperature achieved via resonance excitation (Goeringer *et al.*, 1999) and boundary-activated dissociation (BAD) (Asano *et al.*, 1999b). Figure 10 shows the CID spectrum obtained for +2 bradykinin at a bath gas temperature of 486 K and pressure of 1 millitorr, with a reaction time of 10 s (Butcher *et al.*, 1999). At this temperature, the facile cleavage at the N-terminal of proline dominates, with loss of a water molecule and formation of one other complementary pair also observed at low abundance.

A number of other alternatives for ion heating in quadrupole ion traps have also been demonstrated, such as direct current (DC) pulse activation

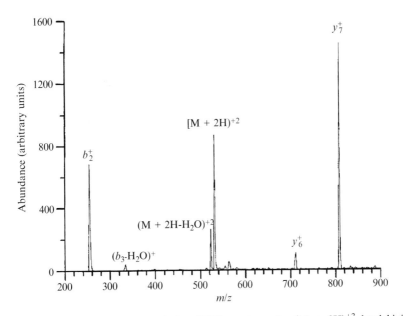

FIG. 10. Collision-induced dissociation (CID) spectrum for $(M + 2H)^{+2}$ bradykinin obtained by storing the ions in quadrupole ion trap at a helium bath gas temperature of 486 K and pressure of 1 millitorr for 10 s. (Reproduced with permission from Butcher *et al.*, 1997.)

product ions shown have charge states ranging from unity to +8. (B and C) Ion/ion proton transfer reactions with anions of perfluorodimethylcyclohexane have been used after CID to reduce the charge states of the product ions largely to +1 to simplify interpretation of the spectrum. Only *b*- and *y*-type ions are formed, and 50% sequence coverage is obtained from this single charge state. (Reproduced with permission from Reid *et al.*, 2001.)

(which yields a mixture of CID and SID products) (Lammert and Cooks, 1992); low-frequency square-wave activation (Wang *et al.*, 1996); irradiation with broadband waveforms such as filtered noise fields (FNFs) (Asano *et al.*, 1995; Volmer and Niedziella, 2000), stored waveforms generated with inverse Fourier transforms (SWIFT) (Doroshenko and Cotter, 1996b; Soni *et al.*, 1995), and random white noise (McLuckey *et al.*, 1992); BAD (Creaser and O'Neill, 1993; Glish and Vachet, 1998; Paradisi *et al.*, 1992); and off-resonance activation with a single frequency selected to be lower than the secular frequency of the precursor by approximately 5% (Qin and Chait, 1996). The last two techniques have been studied fairly extensively for their applicability to CID of peptide ions. The BAD technique involves applying a DC voltage to the ion trap electrodes to move ions close to the boundaries of stability in the trap, thereby increasing the amplitude of their oscillations and causing them to undergo energetic collisions with the bath gas. Glish and Vachet (1998) have shown that BAD can be used to dissociate peptides with the advantage that it is not necessary to tune the excitation to the exact frequency of ion oscillation to achieve maximum efficiency. However, the BAD technique suffers from relatively poor overall efficiencies due to the competition between ion ejection and dissociation. This limitation can be overcome somewhat by the combination of BAD with heavier collision gases (Glish and Vachet, 1998). Asano *et al.* (1999b) used leucine enkephalin to estimate that BAD can elevate ions to an effective internal temperature of approximately 700 K during collisions with helium, comparable to that obtained via normal resonance activation. Qin and Chait (1996) have shown that irradiating ions with a large-amplitude (21 V_{pp}) alternating current (AC) voltage at a frequency approximately 5% below the secular frequency of the ion could more efficiently dissociate large peptides than normal on-resonance excitation. They attributed the increase in efficiency to the ability to deposit larger amounts of internal energy without ion ejection from the trap.

Fourier Transform–Ion Cyclotron Resonance CID. FT-ICR MS uses the combination of a high magnetic field (from 3 to as high as 20 Tesla) with small DC potentials to trap ions. Ion motion in the magnetic field exhibits a characteristic frequency, the cyclotron orbital frequency, which is *m/z* dependent; hence, a resonance excitation method analogous to that described earlier for quadrupole ion traps can be employed in FT-ICR traps to increase ion kinetic energy and thus cause internal energy deposition via energetic collisions (Cody and Freiser, 1982). However, it has been shown that irradiating trapped ions at frequencies slightly (<1%) above or below the cyclotron frequency greatly improves the efficiency of CID in FT-ICR traps (Gauthier *et al.*, 1991), because the ions can be held at elevated kinetic energy for long periods without ejection from the trap. This technique is referred to as SORI. The mechanism for SORI essentially involves elevation of the excited ions to a steady-state kinetic energy after the first 10–20 ms of excitation due to dephasing of the ion motion with respect to the applied excitation. Ion kinetic energies are comparable to those achieved via quadrupole ion trap resonance excitation (i.e., some few electron volts and hence many multiple collisions are required to increase ion internal energy enough to cause dissociation). The maximum kinetic energy accessible depends on the strength of the trapping magnetic field. Schnier *et al.* (1999) have used Arrhenius activation parameters measured via blackbody infrared radiative dissociation (BIRD) for leucine enkephalin and bradykinin ions to show that effective temperatures of between 500 and 700 K are achieved during SORI using nitrogen as the collision gas. However, the authors are careful to state that the actual distribution of internal energy is not well characterized and may not be Boltzmann (Schnier *et al.*, 1999). Laskin and Futrell (2000) have shown that smaller organic ions, such as bromonaphthalene, achieve effective temperatures from 1300 to 4000 K during SORI (Laskin and Futrell, 2000). Preliminary attempts to use master equation modeling to develop a thermal model of SORI analogous to that developed for quadrupole ion trap CID have shown that the initial internal energy of the ions before activation and radiative relaxation during activation both play significant roles in determining the ion internal temperature (Marzluff and Beauchamp, 1996).

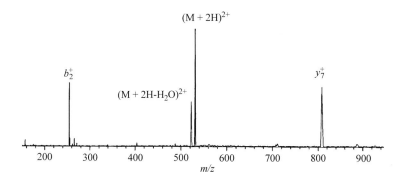

FIG. 11. Collision-induced dissociation (CID) spectrum of $(M + 2H)^{+2}$ bradykinin obtained via Fourier transform–ion cyclotron resonance (FT-ICR) sustained off-resonance irradiation (SORI) with irradiation for 500 ms at a nitrogen pressure of between 1 and 8×10^{-6} torr. Note the similarity between this spectrum and those shown in Figs. 8 and 10 for quadrupole ion trap resonance excitation and bath gas heating to 486 K, respectively. (Reproduced with permission from Schnier *et al.*, 1999.)

Figure 11 shows the CID spectrum of +2 bradykinin obtained via SORI with irradiation for 500 ms at a nitrogen pressure of between 1 and 8×10^{-6} torr (Schnier *et al.*, 1999). Note the similarity between this spectrum and those shown in Figs. 8 and 10 for quadrupole ion trap resonance excitation and bath gas heating to 486 K, respectively.

As discussed earlier in this chapter, Marzluff *et al.* (1994a) has shown that transfer of available kinetic energy to internal energy is efficient at the kinetic energies accessed via SORI. Therefore, provided that sufficient collisions occur, even large protein ions can be made to dissociate. Figure 12 shows the SORI-CID spectrum of $(M + 17H)^{+17}$ ion of apomyoglobin

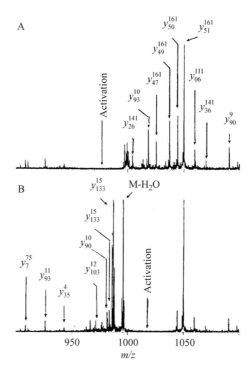

FIG. 12. Sustained off-resonance irradiation (SORI)–collision-induced dissociation (CID) spectrum of $(M + 17H)^{+17}$ apomyoglobin obtained by exciting the ions for 2 s in the presence of nitrogen. (A) The activation was shifted to the low m/z side of the precursor ion, and (B) it was shifted to the high m/z side. Product ions having m/z ratios that cause them to be on resonance with the applied activation voltage may be further dissociated or ejected from the ion cyclotron resonance (ICR) cell, so that dissociation on both sides of the precursor ion is necessary to observe all dissociation channels accessed by the precursor. (Reproduced with permission from Senko *et al.*, 1994b.)

obtained by exciting the ions for 2 s in the presence of nitrogen (pressure unspecified) (Senko *et al.*, 1994b). In analogy to quadrupole ion trap CID, the internal temperature of the ions is expected to be relatively low, but the trapping nature of the ICR instrument allows for very long excitation times, so processes with rates as low as 10/second or less can be observed.

The effects of the pressure and nature of the collision gas used in SORI-CID have not been widely studied. FT-ICR operates at very low (10^{-9} torr) pressure during mass analysis to achieve ultrahigh resolution; therefore, collision gas for CID is pulsed into the cell. Because the collision gas is not present at a constant pressure during the CID event, characterization of pressure effects is difficult. Gorshkov *et al.* (1999) have presented a model for estimating the average laboratory-frame kinetic energy, which accounts for the presence of collision gas, which can cool the ions if the pressure is too high. They have shown that there is an optimum gas pulse length and, hence, pressure peak during the CID event, which yields maximum internal energy deposition and CID efficiency. At shorter pulse times (lower pressure), an insufficient number of collisions occurs, and CID yields are reduced; at longer pulse times (higher pressure), collisional damping lowers ion kinetic energy, again limiting internal energy deposition and subsequent CID yields.

Two other methods for exciting ions for CID in an FT-ICR trap are also used, both with the same goal as SORI to increase ion kinetic energy without causing ion ejection from the trapping cell. MECA (Williams and McLafferty, 1990) relies on a number (hundreds) of low-amplitude on-resonance excitation periods, each of which increases ion kinetic energy, but not to the extent that ions are ejected. Ion kinetic energy decreases between each activation period, but internal energy does not (at least not all the way back to the starting point), and so by using several excitation periods, internal energy can be increased until ions dissociate without causing ion injection. VLE-CA also uses on-resonance activation (Boering *et al.*, 1992a,b), but with rapid 180° phase shifts of the applied excitation voltage. These phase shifts prevent the applied voltage from increasing the ion kinetic energy to the ejection point. Senko *et al.* (1994b) have evaluated the applicability of SORI, MECA, and VLE-CA (and the related resonant amplitude modulated collisional activation [RAM-CA]) to the dissociation of protein ions in an FT-ICR cell.

An important distinction between CID in both types of ion traps and the other types of CID discussed above is that because of the resonant nature of the excitation in ion traps, the product ions are not themselves subject to further activation after their formation. Product ions are formed with the same internal energy as the precursor, on a "per degree-of-freedom'' basis, and so they may still undergo further dissociation if the dissociation rate is faster than the rate of deactivation; however, they will not receive any further internal energy from the resonance excitation voltage, unless they have m/z values that place them very near the precursor ion or on-resonance with the excitation voltage in off-resonance excitation experiments such as SORI. For example, in Fig. 12, the y_{149}^{+16}, y_{147}^{+16}, y_{93}^{+10}, and y_{126}^{+14} ions are abundant products having m/z values above that of the precursor ion. These products are observed when the activation is shifted below the m/z of the precursor (Fig. 12A) but are dissociated or ejected from the trap when the activation is shifted above the m/z of the precursor (Fig. 12B). In the quadrupole ion trap, Goeringer et al. (1999) have shown that the helium bath gas actively cools product ions at a fairly high rate so that further dissociation is minimized in the quadrupole ion trap. Vachet and Glish (1996) and Doroshenko and Cotter (1996a) have reported an increase in sequential dissociation when heavier bath gases such as argon, krypton, or xenon are added to the helium during CID. The general lack of sequential dissociation observed in ion trap CID may or may not be regarded as a positive result. In terms of peptide and protein sequencing, the absence of internal fragments from the amino acid chain makes deduction of the sequence from the MS/MS dissociation easier. On the other hand, some use has been made of the immonium ions of individual amino acids that appear from sequential dissociation of the precursor to help determine the amino acid content of the peptide under study (Biemann and Papayannopoulos, 1994; Hunt et al., 1986).

Summary and Conclusions

This chapter has focused exclusively on the effects of experimental variables, described in terms of the set of figures of merit, on CID. At least an equal amount could be said about the effects of ion structure on CID behavior, even if the discussion were limited to what is known about peptides and proteins. An exhaustive summary of what is known about biological ion structural effects is beyond the scope of this chapter. The interested reader is referred to the literature on the effects of primary structure, secondary structure and conformation, and charge state on the dissociation reactions of peptides and proteins (O'Hair, 2000; Polce et al., 2000; Schlosser and Lehmann, 2000; Tsaprailis et al., 1999, 2000; Wysocki et al., 2000). The reader should also note that MS and tandem MS is seeing increasing use in the study of other important biological molecules such as nucleic acids (Hofstadler and Griffey, 2001), lipids (Murphy et al., 2001), and carbohydrates (Harvey, 2001).

CID is a widely applicable technique that can be used to obtain sequence and structural information from biologically derived ions. A wide variety of activation conditions lead to complementary information from different CID methods. CID conducted on sector or TOF/TOF instruments with kiloelectron-volt collision energy and low collision numbers leads to efficient dissociation of the amide bonds in peptides to yield sequence information, while also accessing amino acid side-chain cleavages to aid in the distinction of isomeric and isobaric amino acids. Dissociation spectra recorded in this high-energy regimen are extremely reproducible. A disadvantage of operation in this regimen is that dissociation rates must be driven at 10^4–10^6/s, which has proven to be very difficult for ions more than approximately 3000 Da. Lower energy CID in the electron-volt regimen is effected in collision quadrupoles, where time scales can be greater by at least three orders of magnitude than at high energy. This, coupled with the high efficiency of energy transfer for collisions at lower energy, allows efficient dissociation of large ions, but without the side-chain cleavages observed at higher energy. Ion-trapping instruments can use even longer dissociation times (up to several seconds) to access very slow (1/s or less) dissociation processes. This allows ion trap slow heating methods to efficiently dissociate biomolecules of increasing size, even up to very large proteins.

Acknowledgments

S. A. M. acknowledges support of his research program by the U.S. Department of Energy, Division of Chemical Sciences, Geosciences, and Biosciences, Office of Basic Sciences, U.S. Department of Energy, under Award No. DE-FG02–00ER15105 and the National Institutes of Health, under grant GM45372.

References

Aebersold, R., and Goodlet, D. R. (2001). Mass spectrometry in proteomics. *Chem. Rev.* **101,** 269–296.

Asano, K. G., Goeringer, D. E., and McLuckey, S. A. (1999a). Thermal dissociation in the quadrupole ion trap: Ions derived from leucine enkephalin. *Int. J. Mass Spectrom.* **187,** 207–219.

Asano, K. G., Butcher, D. J., Goeringer, D. E., and McLuckey, S. A. (1999b). Effective Ion internal temperatures achieved via boundary activation in the quadrupole ion trap: Protonated leucine enkephalin. *J. Mass Spectrom.* **34,** 691–698.

Asano, K. G., Goeringer, D. E., and McLuckey, S. A. (1995). Parallel monitoring for multiple targeted compounds by ion trap mass spectrometry. *Anal. Chem.* **67,** 2739.

Beynon, J. H., Cooks, R. G., Amy, J. W., Baitinger, W. E., and Ridley, T. Y. (1973). Design and performance of a mass-analyzed ion kinetic energy (MIKE) spectrometer. *Anal. Chem.* **45,** 1023A–1028A.

Biemann, K. (1988). Contributions of mass spectrometry to peptide and protein structure. *Biomed. Environ. Mass Spectrom.* **16,** 99–111.

Biemann, K., and Papayannopoulos, I. A. (1994). Amino acid sequencing of proteins. *Acc. Chem. Res.* **27,** 370–378.

Boering, K. A., Rolfe, J., and Brauman, J. I. (1992a). Low energy collision induced dissociation: Phase-shifting excitation control of ion kinetic energy in ion cyclotron resonance spectrometry. *Int. J. Mass Spectrom. Ion Processes* **117,** 357–386.

Boering, K. A., Rolfe, J., and Brauman, J. I. (1992b). Control of ion kinetic energy in ion cyclotron resonance spectrometry: Very-low-energy collision-induced dissociation. *Rapid Commun. Mass Spectrom.* **6,** 303–305.

Boyd, R. K., Kingston, E. E., Brenton, A. G., and Beynon, J. H. (1984). Angle dependence of ion kinetic energy spectra obtained by using mass spectrometers. 1. Theoretical consequences of conservation laws for collisions. *Proc. R. Soc. London Ser. A-Math. Phys. Eng. Sci.* **392,** 59–88.

Bradley, C. D., Curtis, J. M., Derrick, P. J., and Sheil, M. M. (1994). Collisional activation of large ions—energy losses and an impulsive collision theory of energy transfer. *J. Chem. Soc. Faraday Trans* **90,** 239–247.

Bruins, A. P. (1997). ESI source design and dynamic range considerations. *In* "Electrospray Mass Spectrometry: Fundamentals, Instrumentation, and Applications" (R. B. Cole, ed.). Wiley & Sons, New York.

Busch, K. L., McLuckey, S. A., and Glish, G. L. (1988). "Mass Spectrometry/Mass Spectrometry: Techniques and Applications of Tandem Mass Spectrometry." VCH Publishers, Inc., New York.

Butcher, D. J., Asano, K. G., Goeringer, D. E., and McLuckey, S. A. (1999). Thermal Dissociation of bradykinin ions. *J. Phys. Chem. A* **103,** 8664–8671.

Charles, M. J., McLuckey, S. A., and Glish, G. L. (1994). Competition between resonance ejection and ion dissociation during resonant excitation in a quadrupole ion trap. *J. Am. Soc. Mass Spectrom.* **5,** 1031–1041.

Chen, Y. L., Collings, B. A., and Douglas, D. J. (1997). Collision cross sections of myoglobin and cytochrome *c* ions with Ne, Ar, and Kr. *J. Am. Soc. Mass Spectrom.* **8,** 681–687.

Cody, R. B., and Freiser, B. S. (1982). Collision-induced dissociation in a Fourier-transform mass spectrometer. *Int. J. Mass Spectrom. Ion Processes* **41,** 199–204.

Cole, R. B. (ed.) (1997). "Electrospray Ionization Mass Spectrometry." Wiley-Interscience, New York.

Cooks, R. G. (1995). Collision-induced dissociation: Readings and commentary. *J. Mass Spectrom.* **30,** 1215–1221.

Cooks, R. G. (ed.) (1978). "Collision Spectroscopy." Plenum Press, New York.

Cooks, R. G., Beynon, J. H., Caprioli, R. M., and Lester, G. R. (1973). "Metastable Ions." Elsevier, Amsterdam.

Cornish, T. J., and Cotter, R. J. (1993a). Tandem time-of-flight mass spectrometer. *Anal. Chem.* **65,** 1043–1047.

Cornish, T. J., and Cotter, R. J. (1993b). Collision-induced dissociation in a tandem time-of-flight mass spectrometer with two single-stage reflectrons. *Org. Mass Spectrom.* **28,** 1129–1134.

Creaser, C. S., and O'Neill, K. E. (1993). Boundary-effect activated dissociation in ion trap tandem mass spectrometry. *Org. Mass Spectrom.* **28,** 564–569.

Dodonov, A., Kozlovsky, V., Loboda, A., Raznikov, V., Sulimenkov, I., Tolmachev, A., Kraft, A., and Wollnik, H. (1997). A new technique for decomposition of selected ions in molecule ion reactor coupled with ortho-time-of-flight mass spectrometry. *Rapid Commun. Mass Spectrom.* **11**, 1649–1656.

Doroshenko, V. M., and Cotter, R. J. (1996a). Pulsed gas introduction for increasing peptide CID efficiency in a MALDI/quadrupole ion trap mass spectrometer. *Anal. Chem.* **68**, 463–472.

Doroshenko, V. M., and Cotter, R. J. (1996b). Advanced stored waveform inverse Fourier transform technique for a matrix-assisted laser desorption/ionization quadrupole ion trap mass spectrometer. *Rapid Commun. Mass Spectrom.* **10**, 65–73.

Douglas, D. J. (1998). Applications of collision dynamics in quadrupole mass spectrometry. *J. Am. Soc. Mass Spectrom.* **9**, 101–113.

Douglas, D. J. (1982). Mechanism of the collision-induced dissociation of polyatomic ions studied by triple quadrupole mass spectrometry. *J. Phys. Chem.* **86**, 185–191.

Douglas, D. J., and French, J. B. (1992). Collisional focusing effects in radio frequency quadrupoles. *J. Am. Soc. Mass Spectrom.* **3**, 398–408.

Douglas, D. J., Thomson, B., Corr, J., and Hager, J. (1993). Method for increased resolution in tandem mass spectrometry. U.S Patent No. 5,248,875.

Downard, K. M., and Biemann, K. (1994). The effect of charge state and the localization of charge on the collision-induced dissociation of peptide ions. *J. Am. Soc. Mass Spectrom.* **5**, 966–975.

Forst, W. (1973). "Theory of Unimolecular Reactions." Academic Press, New York.

Futrell, J. H. (2000). Development of tandem mass spectrometry: One perspective. *Int. J. Mass Spectrom.* **200**, 495–508.

Futrell, J. H. (ed.) (1986). "Gaseous Ion Chemistry and Mass Spectrometry." Wiley, New York.

Futrell, J. H., and Tiernan, T. O. (1972). Tandem mass spectrometric studies of ion-molecule reactions. *In* "Ion-Molecule Reactions" (J. L. Franklin, ed.), Vol. 2, pp. 485–549. Plenum Press, New York.

Gauthier, J. W., Trautman, T. R., and Jacobson, D. B. (1991). Sustained off-resonance irradiation for collision-activated dissociation involving Fourier transform mass spectrometry. Collision-activated dissociation technique that emulates infrared multiphoton dissociation. *Anal. Chim. Acta* **246**, 211–225.

Glish, G. L., and Vachet, R. W. (1998). Boundary-activated dissociation of peptide ions in a quadrupole ion trap. *Anal. Chem.* **70**, 340–346.

Goeringer, D. E., Asano, K. G., and McLuckey, S. A. (1999). Ion internal temperature and ion trap collisional activation: Protonated leucine enkephalin. *Int. J. Mass Spectrom.* **183**, 275–288.

Goeringer, D. E., and McLuckey, S. A. (1996a). Kinetics of collision-induced dissociation in the Paul trap: a first-order model. *Rapid Commun. Mass Spectrom.* **10**, 328–334.

Goeringer, D. E., and McLuckey, S. A. (1996b). Evolution of ion internal energy during collisional excitation in the Paul ion trap: A stochastic approach. *J. Chem. Phys.* **104**, 2214–2221.

Gorshkov, M. V., Pasa-Tolic, L., and Smith, R. D. (1999). Pressure limited sustained off-resonance irradiation for collision-activated dissociation in Fourier transform mass spectrometry. *J. Am. Soc. Mass Spectrom.* **10**, 15–18.

Gronowska, J., Paradisi, C., Traldi, P., and Vettori, U. (1990). A study of relevant parameters in collisional-activation of ions in the ion-trap mass spectrometer. *Rapid Commun. Mass Spectrom.* **4**, 306–313.

Gross, M. L. (1990). Tandem mass spectrometry—multi-sector magnetic instruments. *Methods Enzymol.* **193**, 131–153.

Harrison, A. G. (1999). Energy-resolved mass spectrometry: A comparison of quadrupole cell and cone-voltage collision-induced dissociation. *Rapid Commun. Mass Spectrom.* **13**, 1663–1670.

Harvey, D. J. (2001). Identification of protein-bound carbohydrates by mass spectrometry. *Proteomics* **1**, 311–328.

Hillenkamp, F., and Karas, M. (2000). Matrix-assisted laser desorption/ionisation, an experience. *Int. J. Mass Spectrom.* **200**, 71–77.

Hofstadler, S. A., and Griffey, R. H. (2001). Analysis of non-covalent complexes of DNA and RNA by mass spectrometry. *Chem. Rev.* **101**, 377–390.

Holbrook, K. A., and Robinson, P. J. (1972). "Unimolecular Reactions." Wiley, Chichester.

Holmes, J. L. (1985). Assigning structures to ions in the gas phase. *Org. Mass Spectrom.* **20**, 169–183.

Hop, C. C. A. (1998). Design of an orthogonal tandem time-of-flight mass spectrometer for high-sensitivity tandem mass spectrometric experiments. *J. Mass Spectrom.* **33**, 397–398.

Hunt, D. F., Yates, J. R., Shabanowitz, J., Winston, S., and Hauer, C. R. (1986). Protein sequencing by tandem mass spectrometry. *Proc. Natl. Acad. Sci. USA* **83**, 6233–6237.

Javahery, G., and Thomson, B. A. (1997). A segmented radiofrequency-only quadrupole collision cell for measurements of ion collision cross section on a triple quadrupole mass spectrometer. *J. Am. Soc. Mass Spectrom.* **8**, 697–702.

Jardine, D. R., Morgan, J., Alderdice, D. S., and Derrick, P. J. (1992). A tandem time-of-flight mass spectrometer. *Org. Mass Spectrom.* **27**, 1077–1083.

Jennings, K. R. (1968). Collision-induced decompositions of aromatic molecular ions. *Int. J. Mass Spectrom. Ion Phys.* **1**, 227–235.

Karas, M., Buchman, D., and Hillenkamp, F. (1985). Influence of the wavelength in high-irradiance ultraviolet laser desorption mass spectrometry of organic molecules. *Anal. Chem.* **57**, 2935–2939.

Katta, V., Chowdhury, S. K., and Chait, B. T. (1991). Use of a single-quadrupole mass spectrometer for collision-induced dissociation studies of multiply charged peptide ions produced by electrospray ionization. *Anal. Chem.* **63**, 174–178.

Lammert, S. A., and Cooks, R. G. (1992). Pulsed axial activation in the ion trap: a new method for performing tandem mass spectroscopy (MS/MS). *Rapid Commun. Mass Spectrom.* **6**, 528–530.

Larsen, M. R., and Roepstorff, P. (2000). Mass spectrometric identification of proteins and characterization of their post-translational modifications in proteome analysis. *Fresenius J. Anal. Chem.* **366**, 677–690.

Laskin, J., and Futrell, J. H. (2000). Internal energy distributions resulting from sustained off-resonance excitation in Fourier transform ion cyclotron resonance mass spectrometry. II. Fragmentation of the 1-bromonaphthalene radical cation. *J. Phys. Chem. A* **104**, 5484–5494.

Lock, C. M., and Dyer, E. W. (1999). Characterisation of high pressure quadrupole collision cells possessing direct current axial fields. *Rapid Commun. Mass Spectrom.* **13**, 432–448.

Loo, J. A., Udseth, H. R., and Smith, R. D. (1988). Collisional effects on the charge distribution of ions from large molecules, formed by electrospray-ionization mass spectrometry. *Rapid Commun. Mass Spectrom.* **2**, 207–210.

Lorquet, J. C. (1994). Whither the statistical theory of mass spectra? *Mass Spectrom. Rev.* **13**, 233–257.

Louris, J. N., Cooks, R. G., Syka, J. E. P., Kelley, P. E., Stafford, G. C., and Todd, J. F. J. (1987). Instrumentation, applications, and energy deposition in quadrupole ion-trap tandem mass spectrometry. *Anal. Chem.* **59**, 1677–1685.

Mansoori, B. A., Dyer, E. W., Lock, C. M., Bateman, K., Boyd, R. K., and Thomson, B. A. (1998). Analytical performance of a high-pressure radio frequency-only quadrupole collision cell with an axial field applied by using conical rods. *J. Am. Soc. Mass Spectrom.* **9**, 775–788.

Marzluff, E. M., Campbell, S., Rodgers, M. T., and Beauchamp, J. L. (1994a). Collisional activation of large molecules is an efficient process. *J. Am. Chem. Soc.* **116**, 6947–6948.

Marzluff, E. M., Campbell, S., Rodgers, M. T., and Beauchamp, J. L. (1994b). Low-energy dissociation pathways of small deprotonated peptides in the gas phase. *J. Am. Chem. Soc.* **116**, 7787–7796.

Marzluff, E. M., and Beauchamp, J. L. (1996). Collisional activation studies of large molecules. *In* "Large Ions, Their Vaporization, Detection, and Structural Analysis" (T. Baer, C. Y. Ng, and I. Powis, eds.), pp. 115–133. John Wiley & Sons, Chichester, UK.

McLafferty, F. W. (ed.) (1983). "Tandem Mass Spectrometry." Wiley, New York.

McLafferty, F. W., Bente, R. F. I., Kornfeld, R., Tsai, S.-C., and Howe, I. (1973). Metastable ion characteristics. XXII. Collisional activation spectra of organic ions. *J. Am. Chem. Soc.* **95**, 2120–2129.

McLuckey, S. A. (1992). Principles of collisional activation in analytical mass spectrometry. *J. Am. Soc. Mass Spectrom.* **3**, 599–614.

McLuckey, S. A., and Cooks, R. G. (1983). Angle- and energy-resolved fragmentation spectra from tandem mass spectrometry. *In* "Tandem Mass Spectrometry" (F. W. McLafferty, ed.). Chapter 15 Wiley, New York.

McLuckey, S. A., and Goeringer, D. E. (1997). Slow heating methods in tandem mass spectrometry. *J. Mass Spectrom.* **32**, 461–474.

McLuckey, S. A., Goeringer, D. E., and Glish, G. L. (1992). Collisional activation with random noise in ion trap mass spectrometry. *Anal. Chem.* **64**, 1455–1460.

Medzihradszky, K. F., Campbell, J. M., Baldwin, M. A., Falick, A. M., Juhasz, P., Vestal, M. L., and Burlingame, A. L. (2000). The characteristics of peptide collision-induced dissociation using a high-performance MALDI-TOF/TOF tandem mass spectrometer. *Anal. Chem.* **72**, 552–558.

Morand, K. L., Cox, K. A., and Cooks, R. G. (1992). Efficient trapping and collision-induced dissociation of high-mass cluster ions using mixed target gases in the quadrupole ion trap. *Rapid Commun. Mass Spectrom.* **6**, 520–523.

Morris, H. R., Paxton, T., Dell, A., Langhorne, J., Berg, M., Bordoli, R. S., Hoyes, J., and Bateman, R. H. (1996). High sensitivity collisionally-activated decomposition tandem mass spectrometry on a novel quadrupole/orthogonal-acceleration time-of-flight mass spectrometer. *Rapid Commun. Mass Spectrom.* **10**, 889–896.

Murphy, R. C., Fiedler, J., and Hevko, J. (2001). Analysis of nonvolatile lipids by mass spectrometry. *Chem. Rev.* **101**, 479–526.

Newton, K. A., Chrisman, P. A., Wells, J. M., Reid, G. E., and McLuckey, S. A. (2001). Gaseous apomyoglobin ion dissociation in a quadrupole ion trap: $[M + 2H]^{2+}$-$[M + 21H]^{21+}$. *Int. J. Mass Spectrom.* **212**, 359–376.

O'Hair, R. A. J. (2000). The role of nucleophile-electrophile interactions in the unimolecular and bimolecular gas-phase ion chemistry of peptides and related systems. *J. Mass Spectrom.* **35**, 1377–1381.

Paradisi, C., Todd, J. F. J., and Vettori, U. (1992). Comparison of collisional activation by the boundary effect vs. tickle excitation in an ion trap mass spectrometer. *Org. Mass Spectrom.* **27**, 1210–1215.

Polce, M. J., Ren, D., and Wesdemiotis, C. (2000). Dissociation of the peptide bond in protonated peptides. *J. Mass Spectrom.* **35,** 1391–1398.

Qin, J., and Chait, B. T. (1996). Matrix-assisted laser desorption ion trap mass spectrometry: Efficient isolation and effective fragmentation of peptide ions. *Anal. Chem.* **68,** 2108–2112.

Reid, G. E., Wu, J., Chrisman, P. A., Wells, J. M., and McLuckey, S. A. (2001). Charge-state–dependent sequence analysis of protonated ubiquitin ions via ion trap tandem mass spectrometry. *Anal. Chem.* **73,** 3274–3281.

Roepstorff, P., and Fohlman, J. (1984). Letter to the editors. *Biomed. Mass Spectrom.* **11,** 601.

Schlosser, A., and Lehmann, W. D. (2000). Five-membered ring formation in unimolecular reactions of peptides: A key structural element controlling low-energy collision-induced dissociation of peptides. *J. Mass Spectrom.* **35,** 1382–1390.

Schnier, P. D., Jurchen, J. C., and Williams, E. R. (1999). The effective temperature of peptide ions dissociated by sustained off-resonance irradiation collisional activation in Fourier transform mass spectrometry. *J. Phys. Chem. B* **103,** 737–745.

Senko, M. W., Beu, S. C., and McLafferty, F. W. (1994a). High-resolution tandem mass spectrometry of carbonic anhydrase. *Anal. Chem.* **66,** 415–417.

Senko, M. W., Speir, J. P., and McLafferty, F. W. (1994b). Collisional activation of large multiply charged ions using Fourier transform mass spectrometry. *Anal. Chem.* **66,** 2801–2808.

Shevchenko, A., Chernushevich, I., Ens, W., Standing, K. G., Thomson, B. A., Wilm, M., and Mann, M. (1997). Rapid *de novo* peptide sequencing by a combination of nanoelectrospray, isotopic labeling and a quadrupole/time-of-flight mass spectrometer. *Rapid Commun. Mass Spectrom.* **11,** 1015–1024.

Shukla, A. K., and Futrell, J. H. (2000). Tandem mass spectrometry: Dissociation of ions by collisional activation. *J. Mass Spectrom.* **35,** 1069–1090.

Shukla, A. K., Qian, K., Anderson, S., and Futrell, J. H. (1990). Fundamentals of tandem mass spectrometry: A dynamics study of simple C-C bond cleavage in collision-activated dissociation of polyatomic ions at low energy. *J. Am. Soc. Mass Spectrom.* **1,** 6–15.

Smith, R. D., Loo, J. A., Barinaga, C. J., Edmonds, C. G., and Udseth, H. R. (1990). Collisional activation and collision-activated dissociation of large multiply charged polypeptides and proteins produced by electrospray ionization. *J. Am. Soc. Mass Spectrom.* **1,** 53–65.

Soni, M. H., Wong, P. S. H., and Cooks, R. G. (1995). Notched broad-band excitation of ions in a bench-top ion trap mass spectrometer. *Anal. Chim. Acta* **303,** 149–162.

Stafford, G. C., Kelley, P. E., Syka, J. E. P., Reynolds, W. E., and Todd, J. F. J. (1984). Recent improvements in and analytical applications of advanced ion trap technology. *Int. J. Mass Spectrom. Ion Processes* **60,** 85–98.

Tang, X.-J., Thibault, P., and Boyd, R. K. (1993). Fragmentation reactions of multiply-protonated peptides and implications for sequencing by tandem mass spectrometry with low-energy collision-induced dissociation. *Anal. Chem.* **65,** 2824–2834.

Thomson, B. A., Douglas, D. J., Corr, J. J., Hager, J. W., and Jolliffe, C. L. (1995). Improved collisionally activated dissociation efficiency and mass resolution on a triple quadrupole mass spectrometer system. *Anal. Chem.* **67,** 1696–1704.

Tsaprailis, G., Nair, H., Somogyi, A., Wysocki, V. H., Zhong, W. Q., Futrell, J. H., Summerfield, S. G., and Gaskell, S. J. (1999). Influence of secondary structure on the fragmentation of protonated peptides. *J. Am. Chem. Soc.* **121,** 5143–5154.

Tsaprailis, G., Somogyi, A., Nikolaev, E. N., and Wysocki, V. H. (2000). Refining the model for selective cleavage at acidic residues in arginine-containing protonated peptides. *Int. J. Mass Spectrom.* **196,** 467–479.

Turecek, F. (1991). Energy transfer limits in collisions with fast targets. *Rapid Commun. Mass Spectrom.* **5,** 78–80.

Turecek, F., and McLafferty, F. W. (1996). *In* "Interpretation of Mass Spectra," p. 371. University Science Books, Mill Valley, CA.

Uggerud, E., and Derrick, P. J. (1991). Theory of collisional activation of macromolecules. Impulsive collisions of organic ions. *J. Phys. Chem.* **95,** 1430–1436.

Vachet, R. W., and Glish, G. L. (1996). Effects of heavy gases on the tandem mass spectra of peptide ions in the quadrupole ion trap. *J. Am. Soc. Mass Spectrom.* **7,** 1194–1202.

Vekey, K. (1996). Internal energy effects in mass spectrometry. *J. Mass Spectrom.* **31,** 445–463.

Volmer, D. A., and Niedziella, S. (2000). Linked multiple-stage mass spectrometry experiments in an ion-trap using filtered noise fields techniques. *Rapid Commun. Mass Spectrom.* **14,** 2143–2145.

Wachs, T., Bente, R. F. I., and McLafferty, F. W. (1972). Simple modification of a commercial mass spectrometer for metastable data collection. *Int. J. Mass Spectrom. Ion Phys.* **9,** 333–341.

Wang, M., Schachterle, S., and Wells, G. (1996). Application of non-resonance excitation to ion trap tandem mass spectrometry and selected ejection chemical ionization. *J. Am. Soc. Mass Spectrom.* **7,** 668–676.

Whitehouse, C. M., Dreyer, R. N., Yamashita, M., and Fenn, J. B. (1985). Electrospray interface for liquid chromatographs and mass spectrometers. *Anal. Chem.* **57,** 675–679.

Williams, E. R., and McLafferty, F. W. (1990). High resolution and tandem Fourier-transform mass spectrometry with Californium-252 plasma desorption. *J. Am. Soc. Mass Spectrom.* **1,** 427–430.

Wysocki, V. H., Tsaprailis, G., Smith, L. L., and Breci, L. A. (2000). Mobile and localized protons: A framework for understanding peptide dissociation. *J. Mass Spectrom.* **35,** 1399–1406.

Yates, J. R. (1998). Mass spectrometry and the age of the proteome. *J. Mass Spectrom.* **33,** 1–19.

Yost, R. A., and Boyd, R. K. (1990). Tandem mass spectrometry—quadrupole and hybrid instruments. *Methods Enzymol.* **193,** 154–200.

[6] Peptide Sequencing by MALDI 193-nm Photodissociation TOF MS

By Joseph W. Morgan, Justin M. Hettick, and David H. Russell

Abstract

Ultraviolet photodissociation time-of-flight (TOF) mass spectrometry (MS) is described as a method for determination of peptide ion primary structure. Monoisotopic selection and bond-specific activation, combined with the rapidity of TOF MS analysis, render this technique invaluable to the rapidly expanding field of proteomics. Photofragment ion spectra of model peptides acquired using both post-source decay (PSD) focusing and TOF-TOF experimental methods are exhibited. Advantages of 193-nm photodissociation for *de novo* sequencing of peptide ions are discussed.

Introduction

Advances in ionization methods and instrumentation that have occurred over the past decade enable MS to play a key role in the emerging area of proteomics. For example, matrix-assisted laser desorption ionization (MALDI) (Karas *et al.*, 1987; Tanaka *et al.*, 1988) and electrospray ionization (ESI) (Whitehouse *et al.*, 1985) are now routinely used to ionize peptides and proteins (Pandey and Mann, 1985), which can then be aimed at sequencing peptides and/or proteins (Chaurand *et al.*, 1999; Cotter, 1997), determination of posttranslational modifications (Lennon and Walsh, 1999), or to study protein–protein complexes (Dikler, 2001; Young *et al.*, 2000). Although various instrument platforms can be used for the mass analysis, over the past decade great strides have been made in the development of TOF instrumentation. Advances in instrument geometry and electronics used for TOF MS allow us to interrogate large gas-phase ions with improved resolution and mass accuracy (Cotter, 1999; Mamyrin *et al.*, 1973; Wiley and McLaren, 1955). Two extremely important TOF mass spectrometric techniques to the field of proteomics are peptide mass fingerprinting (PMF)/protein database searching (Henzel *et al.*, 1993) and peptide sequencing, and both methods owe their widespread acceptance to the versatility and user friendliness of modern TOF instruments.

PMF, combined with genomic database searching to identify proteins, is now widely used, and PMF based on accurate mass measurement is now a

METHODS IN ENZYMOLOGY, VOL. 402
0076-6879/05 $35.00
DOI: 10.1016/S0076-6879(05)02006-9

routine TOF analysis (Clauser *et al.*, 1999; Edmondson and Russell, 1996; Jensen *et al.*, 1996; Russell and Edmondson, 1997). Peptide identification based on PMF/protein database searching is especially powerful when combined with peptide sequencing by tandem MS. For example, determining a partial peptide sequence or identifying a specific posttranslationally modified amino acid dramatically increases the confidence level of identification.

For proteins that are not listed in a database, peptide sequencing is essential for protein identification. Peptide sequencing using MS is based on fragmentation of the gas-phase ion. For example, fragmentation along the peptide backbone yields a series of ions separated in mass by an amount equal to the mass of the eliminated residue (Biemann, 1990a). Consequently, the ions must be formed with sufficient internal energy to undergo unimolecular dissociation (metastable ion decay or PSD), or ions formed with very little initial internal energy (i.e., by ESI) must be activated to energies above the dissociation threshold. The most commonly used ion-activation methods for peptide sequencing are collision-induced dissociation (CID) (Russell and Edmondson, 1997) and surface-induced dissociation (SID) (Cooks and Miller, 1999; Nikolaev *et al.*, 2001; Stone *et al.*, 2001); however, photodissociation has considerable potential for peptide sequencing, especially for peptide sequencing involving TOF instruments (Barbacci, 2000; Barbacci and Russell, 1999; Hettick, 2003; Hettick *et al.*, 2001a; Hunt *et al.*, 1987; Thompson *et al.*, 2004).

The physics of CID are not compatible with peptide sequencing via PSD focusing using the standard tandem TOF instrument. For instance, high-energy CID (Chaurand *et al.*, 1999; Yergey *et al.*, 2002) tends to enhance the yield of low m/z fragment ions at the expense of higher mass (e.g., more structurally/sequence informative) fragment ions. Perhaps most troublesome for tandem TOF is the physics of converting kinetic energy into internal energy. CID yields ions having a range of velocities and a broad distribution of arrival times, which limits both m/z resolution and mass measurement accuracy of the TOF measurement (Hettick *et al.*, 2001a). Another important consideration for CID of large ions is the effect of center-of-mass collision energy. In virtually all tandem TOF experiments, the mass of the ion undergoing CID is much larger than the target atom. Consequently, only a very small fraction of the laboratory energy can be partitioned as internal energy of the ion, and as the mass of the analyte ion increases, energy deposition becomes even less efficient (Shukla and Futrell, 1994; Uggerud *et al.*, 1989).

To circumvent some of the limitations of CID, several groups began investigating SID as an alternative method of ion activation (Mabud *et al.*, 1985; McCormack *et al.*, 1992). Under optimized conditions, SID yields very good fragment ion intensities; however, at higher collision energies, SID produces primarily low m/z ions, for example, immonium ions (Riederer et al., 2000; Stone et al., 2001). As the translational energy of the peptide ion is decreased, surface sticking, or "soft-landing," becomes competitive with fragmentation processes (Cooks et al., 2004; Grill et al., 2001). Also, because fragment ions in SID leave the surface with a range of velocities and angles, careful attention must be paid to instrument geometry. A comprehensive review of SID has been published by Grill et al. (2001).

Several years ago we began evaluating 193-nm photodissociation as a method to circumvent some of the limitations of CID. Radiation from an ArF excimer (193-nm, 6.43 eV/photon) laser is particularly well suited to photodissociation of peptide ions, because the amide linkage is a natural chromophore ($\lambda_{max} \sim 210$ nm), and the photon imparts sufficient energy to cleave the bond.

Relatively few tandem TOF photodissociation experiments have been reported (Gimon-Kinsel et al., 1995; Seeterlin et al., 1993; Willey et al., 1994) This is in part due to logistical issues that must be addressed in a tandem TOF photodissociation experiment. First, the timing for the trigger pulse for the excimer laser with respect to the ionization/desorption event is critical. In order to intersect the ion beam with a pulsed laser beam, both the delay generator output and the intrinsic time delay in the laser trigger must be stable to ±2 ns. (Gimon-Kinsel et al., 1995), Second, the ions should be irradiated in a region where they are spatially separated and well defined, in terms of m/z. Third, they must be given sufficient time to undergo unimolecular decay in the field-free region before entering the reflectron.

Several years ago we began a series of studies to evaluate photo-dissociation of peptide and protein ions (Gimon-Kinsel et al., 1995). The instrument consisted of two collinear TOF mass analyzers, with a photo-dissociation region positioned between the two analyzers. The instrument used for the preliminary studies was specifically designed to investigate photodissociation of MALDI-generated peptide and protein ions; however-er, the instrument did not include a reflectron to mass analyze the resultant photofragments. These studies allowed us to differentiate between frag-ment ions and neutrals produced by photodissociation of the peptide and protein ions. We also showed that photodepletion of $[M + H]^+$ ions is a single-photon process, whereas photodepletion of fragment neutrals, F_n^0, formed by photoionization of the $[M + H]^+$ is a two-photon process. Finally, based on the observation that photodepletion occurred within 1 μs of irradi-ation, it was determined that the rate constant for photodepletion was at least 10^6 s^{-1}.

Early photodissociation studies were aimed at small organic molecules, and many of these studies were carried out using Fourier transform mass spectrometers (FT-MSs). The FTMS photodissociation experiment is relatively simple; a defocused light source is passed through the cell and the photofragment ions are trapped and detected using conventional methods (Dunbar, 1989). The long trapping times (milliseconds to hundreds of seconds) and density of neutral molecules (baseline vacuum of 10^{-6} to 10^{-9} torr) is sufficient for the analyte ion and/or photofragment ions to undergo ion–molecule reactions, which can complicate the photofragment ion spectrum (Willey et al., 1994). Due to the ions being trapped in the FT-MS ion cell and bathed in a continuous radiation field, it is difficult to control the photoexcitation process (e.g., single vs. multiphoton excitation). Multiphoton excitation deposits very large amounts of energy to the molecule, resulting in extensive fragmentation. In addition, photofragment ions can also absorb a photon and yield second-generation photofragment ions. These complications do not apply to TOFMS photodissociation experiments.

Instrumentation

The results from our preliminary experiments demonstrated efficient photodissociation of protonated peptide/protein ions and motivated us to design and construct an instrument for mass analyzing the photofragment ions. The instrument used for these studies is shown in Fig. 1, and it consists of a linear TOF-1 and a reflectron that can be used as TOF-2. The effective flight-path length of TOF-1 is 0.5 m and TOF-2 is 3.4 m. Two turbo molecular pumps maintain baseline pressure of approximately 2×10^{-8} torr. It is important that such a high vacuum be maintained to eliminate the possibility of ion collisions with background gases, which contribute "background fragmentation" to the experiment.

Ions are generated in the ion source by MALDI (337 nm) and extracted by pulsing the extraction grid potential to ground (Ingendoh et al., 1994). Ions of interest are selected by the timed-ion selector and subsequently photodissociated by the excimer laser (Fig. 2). The output of the excimer laser is directed to the mass spectrometer via three reflective mirrors (front surface specific to 193 nm), and laser beam size is controlled by a mechanical iris positioned between the final mirror and the vacuum window. The laser output energy is controlled by the laser control panel, and fine control is achieved by passing the laser output through a sample cuvette filled with a methanol/water solution (Hettick et al., 2001b). Control of the laser photon density output is important, because at high laser beam energy, the photofragment ion spectrum is dominated by low m/z immonium ions,

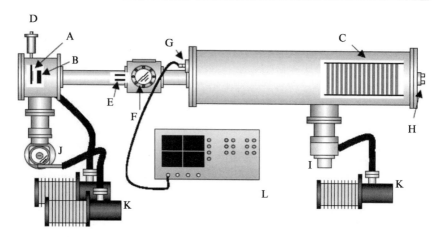

FIG. 1. Schematic of the MALDI Photofragment Tandem TOF Instrument (A) Source plate; (B) extraction grids; (C) reflectron; (D) stage manipulator; (E) timed ion selector; (F) quartz window; (G) reflected ion detector; (H) linear ion detector; (I) 170-L/s turbo molecular pump; (J) 330-L/s turbo molecular pump; (K) roughing pumps; (L) 1-GHz digital oscilloscope.

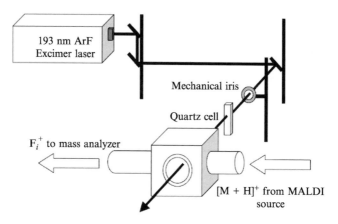

FIG. 2. Expanded view of the laser–mass spectrometer interface.

and at lower laser beam energy, the abundance of structurally informative backbone and side-chain cleavages are increased. Precautions are taken to ensure that the laser beam does not strike the surface of the vacuum chamber because this causes photoejection of electrons and neutral species that can enhance abundances of CID fragment ions.

Photofragment ion spectra are calibrated using methods commonly employed for PSD spectra (Spengler, 1997). Briefly, photofragment spectra are taken at a number of PSD reflectron ratios and then are truncated to include only that portion of the spectrum within a few microseconds of the arrival time of the parent ion. The truncated spectra are spliced together to create a complete photofragment ion spectrum.

A limitation of PSD experiments is the relatively low mass resolution of the timed-ion selector (MS^1). Currently, the timed-ion selector is limited to a 300-ns pulse. Thus, all ions that arrive at the timed-ion selector within the 300 ns window are allowed to continue toward the reflectron, and any fragment ions formed by dissociation of these ions are observed in the photofragment ion spectrum. Although the mass resolution of the timed-ion selector is low enough that $[M + H]^+$ peptide ions of m/z 1000 (20-keV translational energy) and $[M + Na]^+$ ions are not resolved, it is possible to selectively photoexcite only the $[M + H]^+$ or $[M + Na]^+$ ion (Fig. 3). For example, as the ions approach the laser-ion beam intersection, the $[M + H]^+$ and $[M + Na]^+$ ions are separated by 6.0 mm, and during a single excimer laser pulse (17 ns), the ions travel approximately 1.0 mm. Thus, even if the

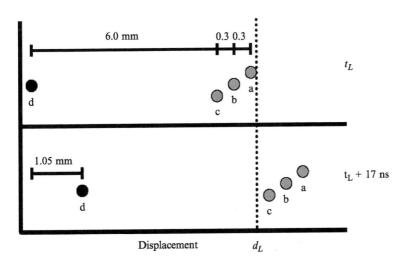

FIG. 3. Spatial separation of ions at the laser–mass spectrometer Interface Isotopes of a theoretical peptide $[M + H]^+$ of m/z 1000 (a, b, and c) are separated by approximately 0.3 mm, whereas the $[M + Na]^+$ ion (d) is approximately 6.0 mm behind. The excimer laser is fired at time t_L when the $[M + H]^+$ isotopic distribution reaches the point d_L. At time tL + 17 ns, the excimer laser beam exits the instrument. All ions have advanced approximately 1.05 mm, and only the isotopic distribution of the theoretical peptide $[M + H]^+$ ion, having traversed through point d_L, has absorbed 193-nm photons.

[M + Na]$^+$ ions are allowed to pass the TIS to the photodissociation region, they are not irradiated, and their fragment ions should not appear in the photofragment ion spectrum. It is important that fragmentation from ions such as [M + Na]$^+$ be eliminated, because they produce product ions shifted in mass (+23 Da), making spectrum analysis significantly more difficult.

Using the PSD method to analyze photofragment ions has limitations as well. Metastable ions that are formed within the first field-free drift region (TOF-1) have the same velocity as their respective parent ions. The metastable ions arising from the ion of interest are activated by the photodissociation laser, giving rise to small-neutral loss ions, which complicate the photofragment ion spectra. Our previous work also showed that metastable neutral species can be ionized by 193-nm irradiation (Gimon-Kinsel et al., 1995). The most significant drawback to the photodissociation experiment using PSD focusing is the time required for data acquisition and calibration. As with most PSD experiments, the reflectron voltage must be lowered in increments to acquire the entire fragment ion spectrum, so data acquisition must be repeated n times, where n represents the number of segments needed to stitch the spectrum. To circumvent this limitation, nonlinear reflectron fields have been proposed and used with some success (Cordero et al., 1996; Cornish and Cotter, 1993). Alternatively, ions could be activated "in-source" so that all fragment ions are accelerated to the same final kinetic energy. In this case, acquisition of the photofragment ion spectrum would have the same duty cycle as a conventional TOF mass analysis.

Our most recent photodissociation experiments used an instrument design that negates the limitations described earlier. A four-grid deceleration/acceleration cell (10 cm in length) was added, centered about the photodissociation window. The MALDI source potential for these experiments was 15 kV, and 8 kV was applied to the central element of the photodissociation cell, and the outside cell grids are held at ground potential. Ions are separated from their metastable decay products as they traverse the decelerating electric field, and consequently, only the ions of interest are irradiated at the center of the cell. A fraction of photoexcited ions dissociate within 0.5 μs as they traverse the remaining field-free region of the photodissociation cell. As ions exit the photodissociation cell, they are then accelerated to near-uniform kinetic energies and m/z analyzed by TOF-2. Because differences in final kinetic energies between parent and product ions are small, a complete photofragment ion spectrum is acquired at a single reflectron voltage. Variation in final kinetic energies between photofragment ions is dependent on parent ion mass-to-charge ratio and is accounted for in a new calibration equation:

$$TOF = a\left(\frac{m_f}{bm_f + c}\right)^{\frac{1}{2}} + t_0$$

Parameters a and c are constant for given mass spectrometer dimensions and source and cell potentials, and parameter b is inversely proportional to the m/z of the parent ion.

The $[M + H]^+$ ions for a mixture of five known peptides (HLGLAR, des-Arg^9 bradykinin, angiotensin III, bradykinin, and substance P) were sequenced by both PSD focusing and TOF-TOF photodissociation tandem MS methods. All mass spectra and photofragment ion spectra were acquired using sample preparation methods that reduce the abundance of PSD fragment ions (e.g., near-threshold laser power for MALDI) and using fructose sample preparations (Beavis et al., 1988; Castoro and Wilkins, 1993; Hettick et al., 2001a; Köster et al., 1992). CID spectra were acquired using an Applied Biosystems 4700 Proteomics Analyzer (Framingham, MA) (Medzihradsky et al., 2000). Instrument conditions used for CID included 1-kV accelerating voltage and medium gas pressure.

Mechanism of Photodissociation

Absorption of ultraviolet radiation by a gas-phase ion yields an electronically excited ion (Fig. 4), and the photoexcited ion may dissociate directly or decay by radiative and/or nonradiative channels. Although radiative decay of gas-phase ions is an important route for rigid molecules (Maier, 1982), nonradiative decay is a much higher probability process for peptide ions. Because the gas-phase ion is an isolated system (i.e., energy cannot be transferred to solvent or the local environment), the photoexcited ion may partition the electronic energy to vibrational modes, and the vibrationally excited ion can then undergo dissociation, provided sufficient energy and time are available. According to RRK theory (Steinfeld et al., 1999), the rate for unimolecular dissociation for a vibrationally activated ion having s degrees of freedom and internal energy E is calculated using the following equation:

$$k(E) = V\left(\frac{E - E_0}{E}\right)^{s-1}$$

Rates of dissociation are directly proportional to the amount of excess energy required for bond cleavage ($E - E_0$) and the frequency of the critical oscillator (ν). The process of electronic excitation of a specific oscillator via

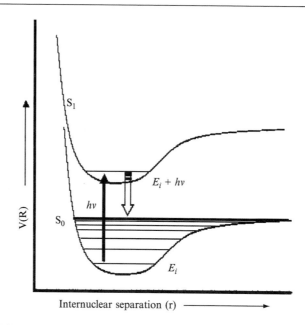

FIG. 4. Vibrational predissociation mechanism absorption of a photon of 193 nm is sufficient to induce a Franck-Condon transition to the upper electronic state. The ion may then relax via internal conversion to the lower electronic state and fragment.

single-photon absorption may increase the rate of dissociation for fragmentation pathways that are not accessible for lower energy activation methods (Tecklenburg et al., 1989). Photofragment ion spectra of peptides activated by 157-nm radiation also provide evidence that photodissociation can occur promptly at bonds that are near the chromophore (Thompson et al., 2004).

Peptide Sequencing

For peptide sequencing experiments, it is important to accurately determine the m/z values of fragment ions. Most tandem TOF sequencing experiments have average mass errors in the range of 0.1–1.0 Da (Kaufmann et al., 1996). Although the mass difference between most of the amino acids is more than 1 Da, there are several cases in which mass differences are much less. Leu and Ile have identical masses (113.084 Da) and cannot be distinguished based solely on fragmentation along the peptide backbone. Gln and Lys differ by only 0.036 Da (128.059 and 128.095 Da, respectively), so a mass measurement accuracy of ±0.01 is required to distinguish these residues. Table I lists fragment ion mass measurements

TABLE I

FRAGMENT ION MASS ACCURACY DATA FOR THE HEXAPEPTIDE HLGLAR ACQUIRED
ON THE MALDI PHOTOFRAGMENT TOF INSTRUMENT

TOF (ns)	PSD ratio	Ion identity	Meas. m/z	Calc. m/z	Error
49235.53	1.000	$[M + H]^+$	666.405	666.405	0.000
47248.32	0.870	y_5	529.395	529.346	−0.049
49195.00	0.770	$y_5\text{-}NH_3$	512.286	510.304	0.034
49110.40	0.770	$b_5 + H_2O$	510.387	492.293	−0.083
49192.09	0.740	b_5	492.275	492.293	0.018
47897.94	0.740	a_5	464.347	464.299	−0.048
47016.22	0.700	b_4	421.262	421.256	−0.006
48858.32	0.635	y_4	416.284	416.262	−0.022
47941.37	0.635	$y_4\text{-}NH_3$	399.302	399.236	−0.066
47616.95	0.635	a_4	393.294	393.261	−0.033
48375.26	0.560	y_3	359.259	359.241	−0.018
48517.89	0.530	$y_3\text{-}NH_3$	342.232	342.214	−0.018
47509.16	0.500	b_3	308.163	308.172	0.009
47372.89	0.410	b_2	251.110	251.151	0.041
47149.81	0.378	$y_2\text{-}NH_3$	229.071	229.130	0.059
48868.05	0.340	a_2	223.116	223.156	0.040
46084.94	0.240	b_1	138.073	138.067	−0.006
47695.02	0.180	R	112.082	112.087	0.005
48541.49	0.170	$A_1(H)$	110.074	110.072	−0.002
49028.12	0.130	L	86.088	86.097	0.009
				Average	0.028
				Std. Dev.	0.024

for the peptide HLGLAR acquired during a PSD focusing photodissociation experiment. The average mass measurement error of 0.028 Da is sufficient to elucidate the composition of this peptide (with the exception of differentiating between Leu and Ile) from either terminus. Both mass accuracy and resolution are higher for photodissociation TOF-TOF experiments.

The utility of photodissociation tandem TOF for peptide sequencing is illustrated by the photofragment ion spectra of bradykinin. Figure 5 contains the PSD focusing photofragment ion spectrum of bradykinin (RPPGFSPFR) taken from a five-peptide mixture. As in the previous example, immonium ions corresponding to proline, phenylalanine, and arginine are observed, and a complete set of a-type ions (except a_7) are observed. In addition, several b-type ions are observed as well, notably the $b_8 + H_2O$ ion at m/z 904. Ions corresponding to charge retention by the C-terminus are prominent in this spectrum, as evidenced by several prominent y-type ions. Because there are arginine residues at either terminus of

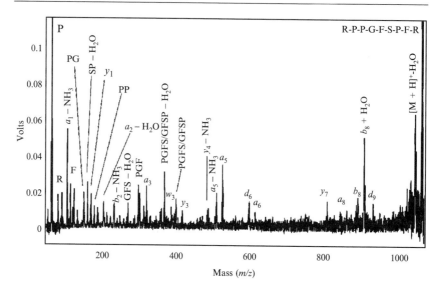

Fig. 5. Post-source decay (PSD) focusing photofragment ion spectrum of bradykinin (RPPGFSPFR) taken from the five peptide mixture.

this peptide, it is reasonable to expect significant populations of fragment ions exhibiting C-terminal charge retention from this peptide. In addition to simple backbone cleavage products, numerous side-chain cleavage products are observed in this spectrum, notably d_6, d_9, and w_3. We also observe internal fragmentation in this spectrum, which is typical for proline-containing peptides (Breci et al., 1963). Internal fragmentation is highly useful in reconstructing peptide sequence when a complete series of backbone cleavage products is not observed. For instance, using just the a-type ions observed in this photofragment ion spectrum, the sequence would be assigned "RPPGFSXXR," where XX can be either FP or PF, but knowledge of one of the internal cleavage ions such as FSP, GFSP, or PGFSP indicates the sequence must be RPPGFSPFR. Alternatively, this information could be extracted from C-terminal sequence information given by the y-type ion series. In this case, we are able to assign the sequence as XPXXXSPFR from the observed y-ions. Thus, in the case of bradykinin, the information obtained from the a- and y-type fragment ions is sufficient to unambiguously assign the complete sequence.

Figure 6 contains the TOF-TOF photofragment ion spectrum of bradykinin. Note that contributions from multiple fragmentations, such as internal cleavage ions and small neutral losses, are in much lower abundance in this spectrum. Products of prompt photodissociation, such as a-type

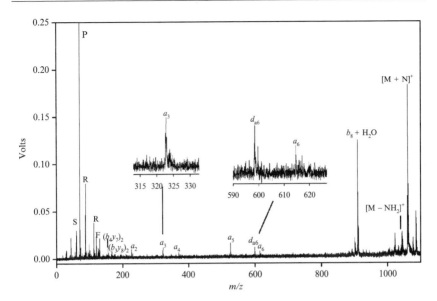

FIG. 6. Time-of-flight (TOF)-TOF photofragment ion spectrum of bradykinin (RPPGFSPFR).

sequence informative ions and d-type side-chain cleavage ions, are now the major contributors to the photofragment ion spectrum. The CID spectrum of this species is shown in Fig. 7. Note the appearance of several ion types having charge retention on either side of the dissociating bond and internal fragment ions and other unidentified species. Although much information is contained within this spectrum, its utility for *de novo* sequencing is low because of difficulties in the assignment of peaks to fragment ions.

Figure 8 contains the photofragment ion spectrum of des-Arg[9] bradykinin (RPPGFSPF) obtained from a five-peptide mixture using the PSD focusing method. The immonium ions for proline, phenylalanine, and arginine verify the presence of these three amino acids in the peptide. A number of a- and b-type fragment ions are also observed in the spectrum, which we attribute to charge retention at the N-terminus of the peptide ion. We also observe sequence ions that arise by loss of NH_3 from either the primary photofragment ions or the activated metastable ions, and these are labeled F_i-NH_3 ions. Also present are fragment ions that result from charge retention at the C-terminus of the peptide ion or at a basic amino acid side chain near the C-terminus. For example, y_1, y_2, and y_3 are observed. In addition to simple backbone cleavage reactions (e.g., a-, b-, and y-type ions), we also observe side-chain cleavage ions. For instance, v_4 and w_6 arise by side-chain cleavage products in which the charge is retained by the

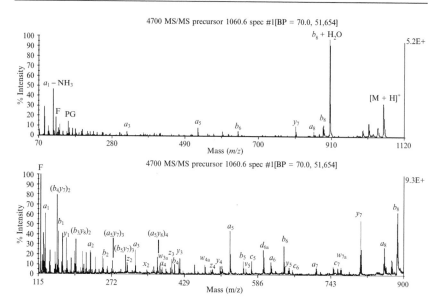

Fig. 7. Collision-induced dissociation (CID) spectrum of bradykinin (RPPGFSPFR).

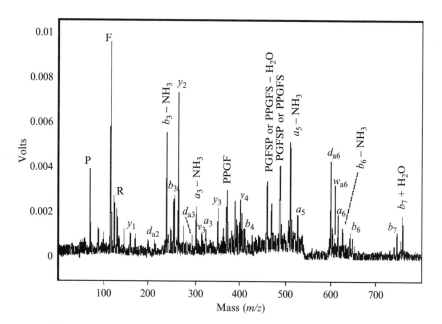

Fig. 8. Post-source decay (PSD) focusing photofragment ion spectrum of des-Arg[9] bradykinin (RPPGFSPF) taken from the five peptide mixture. (Used, with permission, from Medzihradsky *et al.*, 2000.)

C-terminus. In addition to the v- and w-side-chain cleavage ions, side-chain cleavage product ions corresponding to N-terminal charge retention, d_2, d_3, and d_6 are also observed.

The TOF-TOF photofragment ion spectrum of des-Arg[9] bradykinin is shown in Fig. 9. Photofragment ions arising from multiple dissociations and/or fragmentation of metastable ions are in much lower abundance. C-terminal charge retention products such as phenylalanine immonium ion and y-type ions are observed in lower abundance also. Analysis of these promptly formed photofragment ions provides evidence that N-terminal charge retention is dominant in the initial population of MALDI formed ions. Through comparison with results from the PSD focusing method, we postulate that b- and y-type ions are more likely to be formed from lower energy, longer-lived dissociating ions, as they appear in much lower relative abundance when the time allotted for observation of photofragment ions is decreased from 10 μs to 0.5 μs.

Figure 10 contains the CID spectrum of fibrinopeptide A (ADSGEGD-FLAEGGGVR). The spectrum consists of mostly y-type sequence ions and w-type side chain cleavage ions. Note the dominance of the y-type ions corresponding to backbone cleavage at acidic residues. These same dominant ions are observed in the PSD focusing photofragment ion spectrum

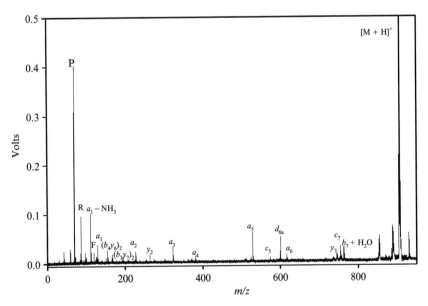

Fig. 9. Time-of-flight (TOF)-TOF Photofragment ion spectrum of des-Arg9 bradykinin.

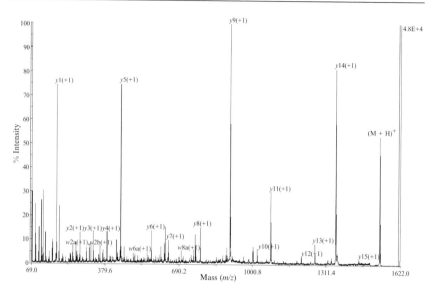

FIG. 10. Collision-induced dissociation (CID) spectrum of fibrinopeptide A (ADSGEGD-FLAEGGGVR).

shown in Fig. 11. When photoactivation is used, side-chain cleavage product w_{8a} becomes more prominent in the tandem mass spectrum. The TOF-TOF photofragment ion spectrum is shown in Fig. 12. Strong signals from y-type ions are no longer observed, rather x-type sequence ions are the most common. Side-chain cleavage products in this spectrum are observed in abundance similar to that of the other sequence informative ions. Comparison of these spectra further confirms that y-type ions are formed subsequent to internal energy reorganization, requiring a greater fragmentation time scale. Through comparison of the PSD focusing and TOF-TOF methods, we observe that photoactivation using 193-nm radiation causes both prompt fragmentations at the sites of the chromophores, as well as rotationally and vibrationally excited species, which dissociate over longer time scales.

Although the isomass residues leucine and isoleucine cannot be distinguished on the basis of m/z, these residues can be distinguished by side-chain cleavages. For example, leucine exhibits a loss of 42 Da and isoleucine loses 28 (Biemann, 1990b) Side-chain cleavage products are rarely observed by low-energy activation or in PSD spectra of MALDI-formed ions; however, side-chain cleavage reactions are frequently observed by high-energy CID and photodissociation (Barbacci and Russell, 1999; Johnson et al., 1988; Yergey et al., 2002) The TOF-TOF photofragment ion spectrum of peptide HLGLAR is shown in Fig. 13. Because the most basic

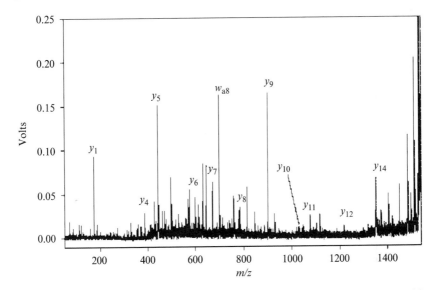

FIG. 11. Post-source decay (PSD) focusing photofragment ion spectrum of fibrinopeptide A (ADSGEGDFLAEGGGVR).

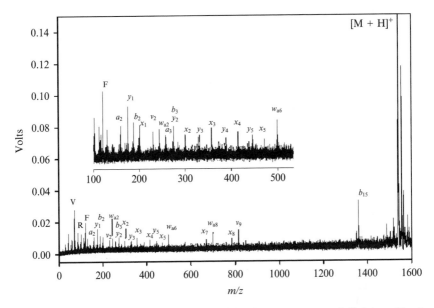

FIG. 12. Time-of-flight (TOF)-TOF photofragment ion spectrum of fibrinopeptide A (ADSGEGDFLAEGGGVR).

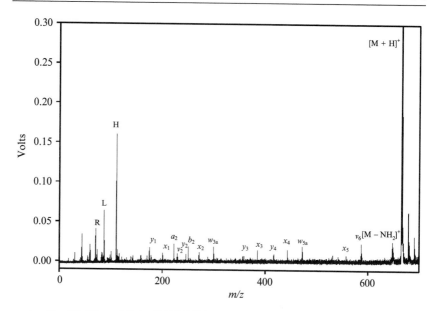

FIG. 13. Time-of-flight (TOF)-TOF photofragment ion spectrum of HLGLAR.

residue in this peptide is arginine, primarily C-terminal charge-retaining photofragment ions are observed. Side-chain cleavage ions w_{3a} and w_{5a} appear in similar abundance to other sequence informative ions, which can be used to unambiguously assign leucine to the second and fourth amino acid residues. A complete x-type ion series is observed, as well as other sequence informative ions. The TOF-TOF photofragment ion spectra of HLGLAR, fibrinopeptide A, and des-arg[9]-bradykinin serve to further illustrate the analytical utility of *de novo* peptide sequencing using 193-nm photoactivation. Peptides that are most basic at the N-terminus produce predominantly a-type ions. C-terminal charge-carrying peptides, such as peptides that result from protein digestion with trypsin, photodissociate into x-type ions. Interpretation of tandem MS data is simplified when a single complete sequence ion series dominates the mass spectrum. Further sequence information is obtained in the form of side-chain cleavage ions for both N- and C-terminal charge carriers.

To further evaluate the photodissociation tandem TOF for peptide sequencing, we analyzed a tryptic digest of the protein cytochrome *C*. A-1 μl aliquot of the tryptic digest (0.5 mg/ml) was mixed with matrix and spotted on the MALDI sample stage. It should be stressed that no sample pretreatment was performed; all buffers, salts, and all other impurities present in the digest solution are present in the MALDI sample

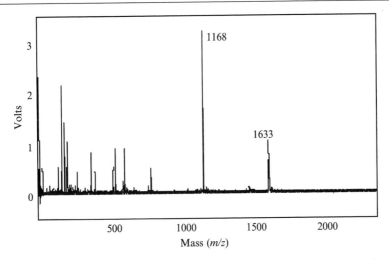

FIG. 14. Matrix-assisted laser desorption ionization (MALDI)–time-of-flight (TOF) mass spectrum of cytochrome *C* tryptic digest.

spot. Figure 14 contains the MALDI mass spectrum of this sample. Two peptides were selected, those at m/z 1633 and 1168. We previously assigned the sequence of the peptide at m/z 1633 to a heme-containing peptide (CAQCHTVEK + heme) based on accurate mass measurements of the parent ion (Edmondson *et al.*, 1996), and this assignment is consistent with the photofragment ion spectrum (Fig. 15). Note that the dominant photofragment ion is the heme at m/z 616. Both *a*- and *b*- type cleavages are observed in the spectrum, but there are too few fragment ions to completely sequence this peptide. However, confirmation of the presence of the heme group supports our earlier assignment.

Figure 16 contains the PSD focusing photofragment ion spectrum of the m/z 1168 peptide ion (TGPNLHGLFGR) of cytochrome *C*. This spectrum contains signals that correspond to the immonium ions of asparagine and phenylalanine. On the basis of *a*-, *b*-, and *y*-type fragment ions, the sequence of this peptide is assigned as TGPNXHGXFGR, where *X* is either leucine or isoleucine. This spectrum also contains a number of internal and side-chain cleavages. Of particular interest are the d_8 and w_{7a} ions, both of which correspond to leucine residues. Thus, we can assign positions 5 and 8 to leucine and the sequence can be unambiguously assigned to TGPNLHGLFGR.

The TOF-TOF photofragment ion spectrum of phosphorylated angiotensin II (DRVpYIHPF) is displayed in Fig. 17. Complete amino acid sequence information is obtained from *a*- and *y*-type ion series. The isoleucine at the fifth position is unambiguously identified using the d_{5a} ion. The identity of

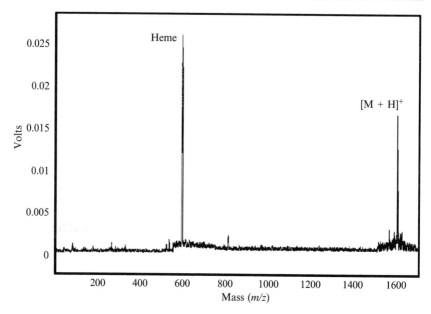

FIG. 15. Post-source decay (PSD) focusing photofragment ion spectrum of m/z 1633 tryptic peptide (CAQCHTVEK + heme).

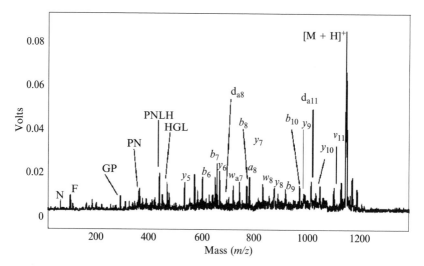

FIG. 16. Post-source decay (PSD) focusing photofragment ion spectrum of m/z 1168 tryptic peptide (TGPNLHGLFGR).

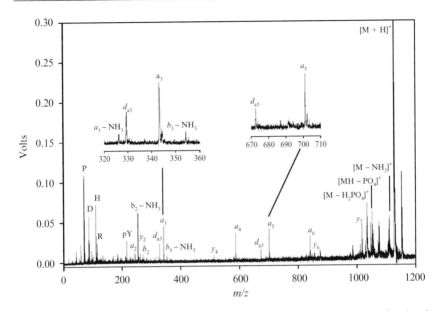

F IG . 17. Time-of-flight (TOF)-TOF photofragment ion spectrum of phosphorylated angiotensin II (DRVpYIHPF).

the modified amino acid is revealed by the phosphorylated tyrosine im-monium ion, and its position is clearly denoted by the *a*-type ion series. Photofragment ion spectra of phosphorylated peptides acquired using the PSD focusing method include ions identified as peptide fragments with loss of phosphoric acid. It was uncertain as to whether these fragment ions arise from photodissociation of the phosphorylated peptide ion or its phosphate-loss metastable ion. In the TOF-TOF photodissociation experiments, se-quence ions with phosphate loss are not observed since the phosphate-loss metastable ions are not activated.

Conclusions

Photodissociation of MALDI-generated peptide $[M + H]^+$ ions with 193-nm excimer radiation results in significant fragmentation. Current results demonstrate that this fragmentation may be used to reconstruct peptide sequences. In addition, photodissociation results in significant numbers of high-energy side-chain cleavages, which are particularly useful for differentiating between isomass residues. Our results demonstrate the ability to select peptides of interest from a digest solution and sequence them in the mass spectrometer with no prior sample treatment.

Also, photodissociation is a reliable method for probing posttranslational modification.

Acknowledgments

The authors gratefully acknowledge the support of the U.S. Department of Energy, Division of Chemical Science, and the Center for Environmental and Rural Health, Texas A&M University College of Veterinary Medicine, an NIEHS (P30-ESO9106) funded center, for supporting the photofragment project. Also, David L. McCurdy of Truman State University for assistance in acquiring some of the spectra contained herein, and Zee Yong Park for preparation of the cytochrome *C* tryptic digest.

References

Barbacci, D. C. (2000). Photodissociation of peptide ions in a matrix-assisted laser desorption ionization reflectron time-of-flight mass spectrometer. Ph.D. Dissertation, pp. 1–182. Texas A&M University.

Barbacci, D., and Russell, D. (1999). Sequence and side-chain specific photofragment (193 nm) ions from protonated substance P by matrix-assisted laser desorption ionization time-of-flight mass spectrometry. *J. Am. Soc. Mass Spectrom.* **10,** 1038–1040.

Beavis, R. C., Lindner, J., Grotemeyer, J., and Schlag, E. W. (1988). Sample-matrix effects in infrared laser neutral desorption, multiphoton-ionization mass spectrometry. *Chem. Phys. Lett.* **146,** 310–314.

Castoro, J. A., and Wilkins, C. L. (1993). Ultrahigh resolution matrix-assisted laser desorption/ionization of small proteins by Fourier-transform mass spectrometry. *Anal. Chem.* **65,** 2621–2627.

Chaurand, P., Luetzenkirchen, F., and Spengler, B. (1999). Peptide and protein identification by matrix-assisted laser desorption ionization (MALDI) and MALDI-post-source decay time-of-flight mass spectrometry. *J. Am. Soc. Mass Spectrom.* **10,** 91–103.

Clauser, K. R., Baker, P., and Burlingame, A. L. (1999). Role of accurate mass measurement (±10 ppm) in protein identification strategies employing MS or MS/MS and database searching. *Anal. Chem.* **71,** 2871–2882.

Cooks, R. G., Jo, S., and Green, J. (2004). Collisions of organic ions at surfaces. *App. Surf. Sci.* **231–232,** 13–21.

Cooks, R. G., and Miller, S. A. (1999). Fundamentals and applications of gas phase ion chemistry. *NATO ASI Ser.* **521,** 55–114.

Cordero, M. M., Cornish, T. J., and Cotter, R. J. (1996). Matrix-assisted laser desorption/ionization tandem reflection time-of-flight mass spectrometry of fullerenes. *J. Am. Soc. Mass Spectrom.* **7,** 590–597.

Cornish, T. J., and Cotter, R. J. (1993). A curved-field reflectron for improved energy focusing of product ions in time-of-flight mass spectrometry. *Rap. Comm. Mass Spectrom.* **7,** 1037–1040.

Cotter, R. J. (1999). The new time-of-flight mass spectrometry. *Anal. Chem.* **71,** 445A–451A.

Cotter, R. J. (1997). "Time-of-Flight Mass Spectrometry: Instrumentation and Applications in Biological Research." pp. 253–274. American Chemical Society, Washington, DC.

Dikler, S. (2001). Derivatization of peptides for structural studies by mass spectrometry. Ph.D. Dissertation, pp. 87–108. Texas A&M University.

Dunbar, R. C. (1989). In "Gas Phase Inorganic Chemistry" (D. H. Russell, ed.), pp. 325–352. Plenum Press, New York and London.

Edmondson, R. D., and Russell, D. H. (1996). Evaluation of matrix-assisted laser desorption ionization-time-of-flight mass measurement accuracy by using delayed extraction. J. Am. Soc. Mass Spectrom. 7, 995–1001.

Edmondson, R. D., Barbacci, D. C., and Russell, D. H. (1996). "Accurate Mass Assignments of Protein Digest Fragments, Proceedings of the 44th ASMS Conference on Mass Spectrometry and Allied Topics, Portland, Oregon, May 12–16." Elsevier, New York, NY.

Gimon-Kinsel, M. E., Kinsel, G. R., Edmondson, R. D., and Russell, D. H. (1995). Photodissociation of high molecular weight peptides and proteins in a two-stage linear time-of-flight mass spectrometer. J. Am. Soc. Mass Spectrom. 6, 578–587.

Grill, V., Shen, J., Evan, C., and Cooks, R. G. (2001). Collisions of ions with surfaces at chemically relevant energies: Instrumentation and phenomena. Rev. Sci. Instrum. 72, 3149–3179.

Henzel, W. J., Billeci, T. M., Stults, J. T., Wong, S. C., Grimley, C., and Watanabe, C. (1993). Identifying proteins from two-dimensional gels by molecular mass searching of peptide fragments in protein sequence databases. Proc. Natl. Acad. Sci. USA 90, 5011–5015.

Hettick, J. M. (2003). Optimization and utilization of MALDI 193-nm photofragment time-of-flight mass spectrometry for peptide sequencing. Ph.D. Dissertation, pp. 1–271. Texas A&M University.

Hunt, D. F., Shabanowitz, J., and Yates, J. R. (1987). Peptide sequence analysis by laser photodissociation Fourier transform mass spectrometry. J. Chem. Soc., Chem. Commun. 8, 548–550.

Ingendoh, A., Karas, M., Hillenkamp, F., and Giessmann, U. (1994). Factors affecting the resolution in matrix-assisted laser desorption-ionization mass spectrometry. Int. J. Mass Spectrom. Ion Proc. 131, 345–354.

Jensen, O. N., Podtelejnkiov, A., and Mann, M. (1996). Delayed extraction improves specificity in database searches by matrix-assisted laser desorption/ionization peptide maps. Rapid Comm. Mass Spectrom. 10, 1371–1378.

Johnson, R. S., Martin, S. A., and Biemann, K. (1988). Collision-induced fragmentation of (M + H)+ ions of peptides. Side chain specific sequence ions. Int. J. Mass Spectrom. Ion Processes. 86, 137–154.

Karas, M., Bachmann, D., Bahr, U., and Hillenkamp, F. (1987). Matrix-assisted ultraviolet laser desorption of non-volatile compounds. Int. J. Mass Spectrom. Ion Processes. 78, 53–68.

Kaufmann, R., Chaurand, P., Kirsch, D., and Spengler, B. (1996). Post-source decay and delayed extraction in matrix-assisted laser desorption/ionization-reflectron time-of-flight mass spectrometry. Are there trade-offs? Rapid Comm. Mass Spectrom. 10, 1199–1208.

Köster, C., Castoro, J. A., and Wilkins, C. L. (1992). High-resolution matrix-assisted laser desorption/ionization of biomolecules by Fourier transform mass spectrometry. J. Am. Chem. Soc. 114, 7572–7574.

Lennon, J. J., and Walsh, K. A. (1999). Locating and identifying posttranslational modifications by in-source decay during MALDI-TOF mass spectrometry. Protein Sci. 8, 2487–2493.

Mabud, M. A., Dekrey, M. J., and Cooks, R. G. (1985). Surface-induced dissociation of molecular ions. Int. J. Mass Spectrom. Ion Processes. 67, 285–294.

Maier, J. P. (1982). Open-shell organic cations: Spectroscopic studies by means of their radiative decay in the gas phase. Acc. Chem. Res. 15, 18–23.

Mamyrin, B. A., Karataev, V. I., Shmikk, D. V., and Zagulin, V. A. (1973). Mass reflectron. New nonmagnetic time-of-flight high-resolution mass spectrometer. Zh. Eksp. Teor. Fiz. 64, 82–89.

McCormack, A. L., Jones, J. L., and Wysocki, V. H. (1992). Surface-induced dissociation of multiply protonated peptides. *J. Am. Soc. Mass Spectrom.* **3**, 859–862.

Medzihradsky, K. F., Campbell, J. M., Baldwin, M. A., Falick, A. M., Juhasz, P., Vestal, M. L., and Burlingame, A. L. (2000). The characteristics of peptide collision-induced dissociation using a high-performance MALDI-TOF/TOF tandem mass spectrometer. *Anal. Chem.* **72**, 552–558.

Nikolaev, E. N., Somogyi, A., Smith, D. L., Gu, C., Wysocki, V. H., Martin, C. D., and Samuelson, G. L. (2001). Implementation of low-energy surface-induced dissociation (eV SID) and high-energy collision-induced dissociation (keV CID) in a linear sector-TOF hybrid tandem mass spectrometer. *Int. J. Mass Spectrom.* **212**, 535–551.

Riederer, D. E., Haney, L. L., Hilgenbrink, A. R., and Beck, J. R. (2000). "Investigations into the Influence of Impact Velocity and Ion Mass on the Surface Induced Dissociation Behavior of Large Ions. Proceedings of the 48th ASMS Conference on Mass Spectrometry and Allied Topics, Long Beach, California, June 11–15." Elsevier Inc., New York, NY.

Russell, D. H., and Edmondson, R. D. (1997). High-resolution mass spectrometry and accurate mass measurements with emphasis on the characterization of peptides and proteins by matrix-assisted laser desorption/ionization time-of-flight mass spectrometry. *J. Mass Spectrom.* **32**, 263–276.

Seeterlin, M. A., Vlasak, P. R., Beussman, D. J., McLane, R. D., and Enke, C. G. (1993). High efficiency photo-induced dissociation of precursor ions in a tandem time-of-flight mass spectrometer. *J. Am. Soc. Mass Spectrom.* **4**, 751–754.

Shukla, A. K., and Futrell, J. H. (1994). *In* "Experimental Mass Spectrometry" (D. H. Russell, ed.), pp. 71–112. Plenum Press, New York and London.

Spengler, B. (1997). Post-source decay analysis in matrix-assisted laser desorption/ionization mass spectrometry of biomolecules. *J. Mass Spectrom.* **32**, 1019–1036.

Steinfeld, J., Francisco, J., and Hase, W. (1999). *In* "Chemical Kinetics and Dynamics," pp. 340–343. Prentice Hall, NJ.

Stone, E., Gillig, K. J., Ruotolo, B. T., Fuhrer, K., Gonin, M., Schultz, A., and Russell, D. H. (2001). Surface-induced dissociation on a MALDI-ion mobility-orthogonal time-of-flight mass spectrometer: Sequencing peptides from an "in-solution" protein digest. *Anal. Chem.* **73**, 2233–2238.

Tanaka, K., Waki, H., Ido, Y., Akita, S., Yoshida, Y., and Yohida, T. (1988). Protein and polymer analyses up to m/z 100,000 by laser ionization time-of-flight mass spectrometry. *Rap. Comm. Mass Spectrom.* **2**, 151–153.

Tecklenburg, R. E., Miller, M. N., and Russell, D. H. (1989). Laser ion beam photodissociation studies of model amino acids and peptides. *J. Am. Chem. Soc.* **111**, 1161–1171.

Thompson, M., Cui, W., and Reilly, J. (2004). Fragmentation of singly charged peptide ions by photodissociation at lambda = 157 nm. *Angew. Chem. Int. Ed.* **43**, 4791–4794.

Uggerud, E., and Derrick, P. J. Z. (1989). Mechanism of energy transfer in collisional activation of kiloelectron-volt macromolecular ions. *Naturforsch A Phys. Sci.* **44**, 245–246.

Whitehouse, C. M., Dreyer, R. N., Yamashita, M., and Fenn, J. B. (1985). Electrospray interface for liquid chromatographs and mass spectrometers. *Anal. Chem.* **57**, 675–679.

Wiley, W. C., and McLaren, I. H. (1955). Time-of-flight mass spectrometer with improved resolution. *Rev. Sci. Instrum.* **26**, 1150–1157.

Willey, K. F., Robbins, D. L., Yeh, C. S., and Duncan, M. A. (1994). *In* "Time-of-Flight Mass Spectrometry" (R. J. Cotter, ed.), pp. 61–72. American Chemical Society, Washington, DC.

Yergey, A. L., Coorssen, J. R., Backlund, P. S., Blank, P. S., Humphrey, G. A., Zimmerberg, J., Campbell, J. M., and Vestal, M. L. (2002). *De novo* sequencing of peptides using MALDI/TOF-TOF. *J. Am. Soc. Mass Spectrom.* **13**, 784–791.

Young, M. M., Tang, N., Hempel, J. C., Oshiro, C. M., Taylor, E. W., Kuntz, I. D., Gibson, B. W., and Dollinger, G. (2000). High throughput protein fold identification by using experimental constraints derived from intramolecular cross-links and mass spectrometry. *Proc. Natl. Acad. Sci. USA* **97**, 5802–5806.

Further Reading

Biemann, K. (1990a). Peptides and proteins: Overview and strategy. *In* "Methods in Enzymology," (J. A. McClosky, ed.), pp. 351–360.

Biemann, K. (1990b). Appendix 5. *In* "Methods in Enzymology." pp. 886–887. Elsevier Inc., New York, NY.

Breci, L. A., Tabb, D. L., Yates, J. R., and Wysocki, V. H. (2003). Cleavage N-terminal to proline: Analysis of a database of peptide tandem mass spectra. *Anal. Chem.* **75**, 1963–1971.

Hettick, J. M., McCurdy, D. L., Barbacci, D. C., and Russell, D. H. (2001a). Optimization of sample preparation for peptide sequencing by MALDI-TOF photofragment mass spectrometry. *Anal. Chem.* **73**, 5378–5386.

Hettick, J. M., McCurdy, D. L., and Russell, D. H. (2001b). "Book of Abstracts, 1901P, The Pittsburgh Conference on Analytical Chemistry and Applied Spectroscopy." The Pittsburg Conference, Pittsburg, PA.

Pandey, A., and Mann, M. (2000). Proteomics to study genes and genomes. *Nature.* **405**, 837–846.

Stone, E. G., Gillig, K. J., Ruotolo, B. T., and Russell, D. H. (2001). Optimization of a matrix-assisted laser desorption ionization-ion mobility-surface-induced dissociation-orthogonal-time-of-flight mass spectrometer: Simultaneous acquisition of multiple correlated MS1 and MS2 spectra. *Int. J. Mass Spectrom.* **212**, 519–533.

[7] Peptide Sequence Analysis

By Katalin F. Medzihradszky

Abstract

In mass spectrometry (MS)–based protein studies, peptide fragmentation analysis (i.e., MS/MS experiments such as matrix-assisted laser desorption ionization [MALDI]–post-source decay [PSD] analysis, collision-induced dissociation [CID] of electrospray- and MALDI-generated ions, and electron-capture and electron-transfer dissociation analysis of multiply charged ions) provide sequence information and, thus, can be used for (i) *de novo* sequencing, (ii) protein identification, and (iii) posttranslational or other covalent modification site assignments. This chapter offers a qualitative overview on which kind of peptide fragments are formed under different MS/MS conditions. High-quality PSD and CID spectra provide illustrations of *de novo* sequencing and protein identification. The MS/MS

METHODS IN ENZYMOLOGY, VOL. 402
0076-6879/05 $35.00
DOI: 10.1016/S0076-6879(05)02007-0

behavior of some common posttranslational modifications such as acetylation, trimethylation, phosphorylation, sulfation, and O-glycosylation is also discussed.

Introduction

Within the last decade, MS has become the method of choice for protein identification. Whole cell lysates, components of protein complexes, and proteins isolated by immunoprecipitation or affinity chromatography are now being routinely studied using this technique (Aebersold and Mann, 2003; Chalmers and Gaskell, 2000; Dreger, 2003; Gygi and Aebersold, 2000; Ong et al., 2003; Roberts, 2002). The key element in this success derives from the ease of peptide sequence determination based on interpretation of peptide fragmentation spectra. Similarly, these techniques are also used for recombinant protein characterization and for structure elucidation of biologically active small peptides, such as toxins, hormones, and antibiotics (Brittain et al., 2000; Grgurina et al., 2002; Igarashi et al., 2001; Ishikawa et al., 1990; Siegel et al., 1994; Yuan et al., 1999). Peptide sequence and structural data can be obtained by collisional activation of selected singly or multiply charged precursor ions. The principles of high-energy collisional activation have been established in combination with fast atom bombardment (FAB), liquid secondary ion MS (LSIMS) ionization on four-sector instruments (Johnson et al., 1988), also with MALDI on sector-orthogonal-acceleration-time-of-flight (TOF) hybrid tandem instruments (Medzihradszky et al., 1996a), and on a TOF-TOF tandem mass spectrometer (Medzihradszky et al., 2000). Low-energy CID spectra are acquired on all the other hybrid tandem instruments, for example, quadrupole-orthogonal-acceleration-TOF (QqTOF) instruments, by triple-quadrupole mass spectrometers and ion traps regardless of the ionization method applied (Alexander et al., 1990; Hunt et al., 1986; Jonscher and Yates, 1997; Qian et al., 1995). The unimolecular decomposition of peptide ions generated by MALDI may be detected and recorded by PSD analysis in MALDI-TOF instruments equipped with reflectron (Kaufmann et al., 1994). Besides traditional CID, Fourier transform MS (FT-MS) instruments also used sustained off-resonance irradiation (SORI) (Senko et al., 1994) and infrared multiphoton dissociation (IRMPD) (Little et al., 1994) to generate MS/MS spectra of larger multiply charged molecules, and electron-capture dissociation (ECD) has been introduced for structural characterization of multiply charged ions (Axelsson et al., 1999). A novel version of this MS/MS method is hot ECD (HECD), when multiply charged polypeptides fragment upon capturing approximately 11-eV electrons (Kjeldsen et al., 2002). Another novel MS-MS technique, electron-transfer dissociation (ETD) can be performed in radiofrequency (RF) quadrupole linear ion traps (Syka *et al.*, 2004). In ETD, singly charged anions transfer an electron to multiply charged peptide ions and induce fragmentation analogous to that observed in ECD. In general, in all kinds of MS/MS experiments, the mass spectrometric detection sensitivity is high, and sequence information is obtainable in peptide mixtures, even for peptides containing unusual or covalently modified residues. Peptides yield a wide array of product ions depending on the quantity of vibrational energy they possess and the time window allowed for dissociation. The ion types formed and the abundance pattern observed are influenced by the peptide sequence, the ionization technique, the charge state, the collisional energy (if any), the method of activation, and the type of the analyzer. In this chapter, a comprehensive list of common peptide product ions is presented, and the fragmentation differences observed in different MS/MS experiments are discussed in a qualitative manner (for nomenclature, see Biemann, 1990). Examples are shown on how the MS/MS data are used for *de novo* sequence determination, protein identification, and assignment of covalent modification sites.

Peptide Fragmentation Processes

Peptides produce fragments that directly provide information on their amino acid composition: immonium (and related) ions and ions that are the products of side-chain losses from the precursor ion.

Amino acids may form immonium ions with a structure of $^+NH_2 = CH$-R, where the mass and the stability of the ion depend on the side-chain structure. Immonium ions sometimes undergo sequential fragmentation reactions yielding ion series characteristic for a particular amino acid. In high-energy CID experiments, these fragments are usually very reliable and offer a wealth of information (Table I) (Falick *et al.*, 1993). PSD spectra also feature relatively abundant immonium and related ions (Chaurand *et al.*, 1999), especially when the decomposition is enhanced by collisional activation (Stimson *et al.*, 1997). Low-energy CID experiments on singly or multiply charged precursors generated by FAB or electrospray ionization (ESI) usually yield some information on the amino acid composition of the peptide (Hunt *et al.*, 1989), whereas MALDI low-energy CID spectra acquired on QqTOF instruments (or FT-ion cyclotron resonance [ICR]) feature very few and very weak immonium ions (Baldwin *et al.*, 2001). This compositional information is usually completely lost when the CID experiment is carried out using an ion trap, in which fragments below approximately one-third of the precursor ion mass are not

TABLE I
IMMONIUM AND RELATED IONS CHARACTERISTIC OF THE 20 STANDARD AMINO ACIDS[a]

Amino acid	Immonium and related ion(s) masses		Comments
Ala	44		
Arg	129	59, 70, 73, 87, 100, 112	129, 73 usually weak
Asn	87	70	87 often weak, 70 weak
Asp	88		Usually weak
Cys	76		Usually weak
Gly	30		
Gln	101	84, 129	129 weak
Glu	102		Often weak if C-terminal
His	110	82, 121, 123, 138, 166	110 very strong
			82, 121, 123, 138 weak
Ile/Leu	86		
Lys	101	84, 112, 129	101 can be weak
Met	104	61	104 often weak
Phe	120	91	120 strong, 91 weak
Pro	70		Strong
Ser	60		
Thr	74		
Trp	159	130, 170, 171	Strong
Tyr	136	91, 107	136 strong, 107, 91 weak
Val	72		Fairly strong

detected (Jonscher and Yates, 1997). Immonium ions are not formed at all in ECD experiments (R. A. Zubarev, personal communication).

The other ion type that provides information on the amino acid composition of the peptide is formed from the precursor ion via side-chain loss. This dissociation process is very characteristic in high-energy CID spectra obtained on four-sector mass spectrometers with FAB or LSIMS ionization (Medzihradszky and Burlingame, 1994) but plays a far less significant role in other CID experiments, even in MALDI high-energy CID spectra (Medzihradszky et al., 1996a). However, some residues usually undergo this kind of fragmentation, so their presence will be indicated. Met-containing peptides may feature a loss of 47 Da (CH_3S) from the precursor ion. When this residue is oxidized, even the sequence ions containing the methionine-sulfoxide undergo an extensive fragmentation via neutral loss of 64 Da (CH_3SOH) in most MS/MS experiments (Swiderek et al., 1998). Upon MALDI ionization, due to the cleavage of the α, β-bond, Phe and Tyr may lose 91 and 107 Da, respectively (K. Vekey, personal communication). Side-chain losses and side-chain fragmentation also have been reported in ECD experiments. Basic amino acids produce the most significant fragmentation: His-containing peptides display an 82 Da loss ($-C_4H_6N_2$), whereas Arg-containing molecules may lose 101, 59, 44, and 17 Da corresponding to $C_4H_{11}N_3$, CH_5N_3, CH_4N_2, and NH_3, respectively (Cooper *et al.*, 2002). Other residues exhibiting such fragmentation in ECD experiments were Asn and Gln: -45 Da (CH_3NO), Lys: -73 Da ($C_4H_{11}N$), Met: -74 (C_3H_6S) (Cooper *et al.*, 2002), Trp: -130 Da (C_9H_8N), and -116 Da (C_8H_6N), Phe: -92 Da (C_7H_8), and -77 Da (C_6H_5); and Val/Leu: -42 Da (C_3H_6), and -43 Da (C_3H_7) (Leymarie *et al.*, 2003).

While the ions discussed above provide composition information, all the other signals in MS/MS spectra provide information on the amino acid sequence. All primary cleavages occur along the peptide backbone (for a review on peptide fragmentation pathways, see Paizs and Suhai, 2004). Most frequently the dissociation reaction occurs at the peptide bonds. When the proton (charge) is retained on the N-terminal fragment, **b** ions are formed with the structure: $H_2N\text{-}CHR_1\text{-}CO\text{-}\ldots\text{-}NH\text{-}CHR_iCO^+$ (rules for the calculation of major fragment ion masses are presented in Table II).

TABLE II
RULES FOR THE CALCULATION OF FRAGMENT ION MASSES

Fragment	Mass calculation	
	Using residue weights	From other fragments
a_i	Σresidue weights -27	$b_i - 28$
b_i	Σresidue weights $+1$	$MH^+ + 1 - y_{n-i}$
c_i	Σresidue weights $+18$	$b_i + 17$
d_i	Σresidue weights $-12 -$ side chain	$a_i - (R_i - 15)$
for Ile	Σresidue weights -55 or -41	$a_i - 28$ or -14
for Thr	Σresidue weights -43 or -41	$a_i - 16$ or -14
for Val	Σresidue weights -41	$a_i - 14$
v_i	Σresidue weights $-$side chain $+18$	$x_{i-1} + 29$
w_i	Σresidue weights $-$side chain $+17$	$x_{i-1} + 28$
for Ile	Σresidue weights $-$side chain $+31$ or $+45$	$x_{i-1} + 42$ or $+56$
for Thr	Σresidue weights $-$side chain $+31$ or $+33$	$x_{i-1} + 42$ or $+44$
for Val	Σresidue weights $-$side chain $+31$	$x_{i-1} + 42$
x_i	Σresidue weights $+45$	$y_i + 26$
y_i	Σresidue weights $+19$	$MH^+ + 1 - b_{n-i}$
Y_i	Σresidue weights $+17$	$y_i - 2$
z_i	Σresidue weights $+2$	$y_i - 17$
Internal fragments		
b-type	Σresidue weights $+1$	
a-type	Σresidue weights -27	

These sequence ions (and all the other N-terminal fragments) are numbered from the N-terminus. Normally, fragment b_2 is the first member of this series (Yalcin et al., 1995). However, the presence of N-terminal modifications, for example, acetylation leads to the formation of stable b_1 ions (Yalcin et al., 1995). If the proton (charge) is retained on the C-terminal moiety with H-transfer to that fragment, a y sequence ion is formed with the structure $^{+}NH_3\text{-}CHR_{n-i}\text{-}CO\text{-}\ldots\text{-}NH\text{-}CHR_n\text{-}COOH$. This ion series (and all other C-terminal ions) is numbered from the C-terminus. High-energy CID spectra may also display Y fragments formed by H transfer away from the C-terminal fragment, with the structure $\{NH = CR_{n-i}\text{-}CO\text{-}\ldots\text{-}NH\text{-}CHR_n\text{-}COOH\}H^{+}$. Pro residues often produce abundant Y ions (i.e., they feature a doublet separated by 2 Da). Alternative sequence ion series a and x are formed when cleavage occurs between the α-carbon and the carbonyl-group, with structures $H_2N\text{-}CHR_1\text{-}CO\text{-}\ldots\text{-}^{+}NH = CHR_i$ and $\{CO = N\text{-}CHR_{n-i}\text{-}CO\text{-};\ldots\text{-}NH\text{-}CHR_n\text{-}COOH\}H^{+}$, respectively. Alternatively, when the fragmentation occurs between the α-carbon and the amino group c and z or $z + 1$ ions are generated, with structures $H_2N\text{-}CHR_1\text{-}CO\text{-}\ldots\text{-}NH\text{-}CHR_iCO\text{-}NH_3^{+}$, $\{HC(= CR'_{n-i}R''_{n-i})\text{-}CO\text{-}\ldots\text{-}NH\text{-}CHR_n\text{-}COOH\}H^{+}$, and $\{{}^{\cdot}CHR_{n-i}\text{-}CO\text{-}\ldots\text{-}NH\text{-}CHR_n\text{-}COOH\}H^{+}$, respectively. Obviously, the only imino acid, Pro, cannot undergo this type of bond cleavage, so this residue will not yield z fragments and amino acids preceding Pro residues will not form c ions.

The product ions that are observed in any given spectrum are usually controlled by the basic groups present, such as the amino terminus itself, the ε-amino group of Lys, the imidazole-ring of His, and the guanidine side chain of Arg. This is because these groups are protonated preferentially (i.e., the charge is localized on the basic residues) and fragments containing them retain the charge and tend to dominate the spectrum. For example, tryptic peptides usually exhibit abundant C-terminal sequence ion series. However, when a tryptic peptide with C-terminal Lys contains a His residue close to its N-terminus, the ion series observed may be controlled by the latter, and in such a case, the spectrum will be dominated by N-terminal sequence ions. In general, Arg overcomes the influence of other amino acids (van Dongen et al., 1996), while His and Lys represent similar basicity in the gas phase (Gu et al., 1999).

Certain other amino acids may also promote fragmentation reactions. For example, the presence of Pro in a sequence facilitates cleavage of the peptide bond N-terminal to this residue, because of the slightly higher basicity of the imide nitrogen, yielding very abundant y fragments. Similarly, cleavage at the C-terminus of Asp residues is favored because of protonation of the peptide bond by the amino acid side chain (Yu et al., 1993). This latter effect is especially profound in MALDI-CID and PSD experiments in which abundant **y** fragments are generated via Asp-Xxx bond cleavages. It has been reported that in MALDI low-energy CID of Arg-containing peptides when using laser energies below a threshold activation level, these bonds will be cleaved exclusively (S. J. Gaskell, personal communication).

In addition, the types of ions observed in MS/MS experiments strongly depend on how the unimolecular dissociation processes are induced. In general, fragments **a**, **b**, and **y** are observed in all kinds of CID experiments and in PSD spectra. Additional backbone cleavage product ions are mostly detected only in high-energy CID experiments. Low-energy CID and PSD spectra almost never feature **x** and **z** + 1 ions, and some data suggest that the ions at m/z **y** − 17 observed in these experiments are the result of ammonia loss from amino acid side-chains (Arg, Lys, Asn, or Gln) rather than from the newly formed N-terminus. Occasionally, low-energy CID and PSD spectra may feature **c** ions, mostly when the charge is preferentially retained at the N-terminus. Interestingly, it has been reported that low-energy CID spectra acquired on a QSTAR (QqTOF) instrument always feature an abundant c_1 fragment if the second amino acid is a glutamine residue (Lee and Lee, 2004).

Electron capture leads to entirely different dissociation chemistry (Kjeldsen *et al.*, 2002). Thus, ECD and ETD spectra display almost exclusively **c** and **z** + 1(**z˙**) ions, although some **a** + 1 (**a˙**) and **y** fragments may also be detected (Axelsson *et al.*, 1999; Syka *et al.*, 2004; Zubarev *et al.*, 2000). For reasons discussed earlier in this chapter, there is usually no cleavage at the N-terminus of Pro residues in ECD and ETD experiments (Cooper *et al.*, 2002; Syka *et al.*, 2004).

The "major" sequence ions, **a**, **b**, and **y** may undergo further dissociation reactions, usually via the loss of small neutral molecules. As mentioned already, satellite ions at 17 Da lower mass are produced via ammonia loss from Arg, Lys, Asn, or Gln side chains or at a much lower extent via cleavage of the N-terminal amino group. A loss of 18 Da indicates elimination of a water molecule from the structure. Hydroxy amino acids Ser and Thr, as well as acidic residues Asp and Glu, normally undergo this type of reaction. However, it has been reported that in ion traps, peptides lacking these residues may lose water via a rearrangement reaction. Arg-containing fragments may produce satellite ions at 42 Da lower mass, most likely corresponding to the loss of NH = C = NH as a neutral moiety. In addition, any fragment that contains methionine sulfoxide may feature abundant satellite ions at 64 Da lower mass, as mentioned earlier (Swiderek *et al.*, 1998). Peptide fragments containing more than one residue capable of undergoing such dissociation reactions frequently yield series of satellite ions because of the various combinations of multiple neutral losses. In addition, in some cases, especially with multiple Arg residues, the relative intensity of sequence ions may diminish or they may completely "disappear" while satellite ions of high abundance are detected. These satellite ions can be observed in all kinds of MS/MS experiments, except in ECD and ETD.

Some satellite fragments are characteristic to high-energy CID, although **w** ions (definitions below) also have been observed in HECD experiments (Kjeldsen *et al.*, 2002, 2003) and in conventional ECD spectra of some peptides (Kjeldsen *et al.*, 2003). HECD of renin substrate also yielded two **d** ions (definition below), so far a unique observation (Kjeldsen *et al.*, 2003). Fragments **d**: $\{H_2N\text{-}CHR_1\text{-}CO\text{-}\ldots\text{-}NH\text{-}CH = CHR'_i\}H^+$, and **w**: $\{CH(= CHR'_{n-i})\text{-}CO\text{-}\ldots\text{-}NH\text{-}CHR_n\text{-}COOH\}H^+$ are formed when the fragmentation occurs between the β- and γ-carbons of the side chain of the C-terminal amino acid of an **a** + 1 ion or the N-terminal amino acid of a **z** + 1 fragment, respectively (Johnson *et al.*, 1987). These satellite fragments permit the unambiguous differentiation of isomeric amino acids Leu and Ile. The **d** and **w** ions (!) of Ile are 14 and 28 Da higher than those of the Leu residue, depending on whether the methyl or the ethyl group is retained on the β-carbon, with the lower mass product ion being dominant. Aromatic amino acids usually do not produce these fragment ions because of the strong bond between the aromatic ring and the β-carbon, but sometimes the cleavage may occur in the side chain of the adjacent amino acid. It has been reported that w-type product ions may form this way (Johnson *et al.*, 1987, 1988; Kjeldsen and Zubarev, 2003). Obviously, Pro, the only imino acid, also cannot undergo this cleavage, but it usually yields an abundant **w** ion that is formed via a different mechanism (Johnson *et al.*, 1987). High-energy CID experiments may also yield another set of C-terminal ions, the v-fragments: $^+NH_2 = CH\text{-}CO\text{-}NH\text{-}CHR_{n-1}\text{-}CO\text{-}NH\text{-}CHR_n\text{-}COOH$ (Johnson *et al.*, 1987, 1988). Pro cannot yield this product ion. In general, the presence of a basic residue in the fragment (i.e., preferential charge retention at the C-terminus) is essential for the formation of **v** and **w** ions. Similarly, the formation and further dissociation of **a** + 1 ions requires preferential charge retention at the N-terminus. This can be accomplished by the presence of a basic amino acid, or sometimes the basicity of the N-terminus itself is sufficient for **d** ion production (Medzihradszky and Burlingame, 1994). The formation of another N-terminal satellite ion, the **b** + H_2O fragment, is also dependent on the presence of a basic amino acid at the N-terminus. These ions are formed via a rearrangement reaction; "peeling off" the C-terminal amino acids one by one (Thorne and Gaskell, 1989). Usually, a one or two amino acid "loss" can be detected. These fragments are typical of all CID experiments and PSD spectra. ECD-generated fragments do not feature most of the satellite ions discussed already. However, they may display side-chain losses and "losses of some low-molecular-weight species such as H_2O, $\cdot CH_3$., $\cdot C_3H_6$, $\cdot CONH_2$" (Leymarie et al., 2003).

Peptides may undergo dissociation processes in which both termini are removed and two or more internal residues form a b-type ion. These species may also yield satellite ions losing carbon monoxide, ammonia, or water, just like the sequence ions. These product ions are called *internal ions* and are labeled with the one-letter code of the amino acids included. Abundant internal ions can be observed in all MS/MS experiments, with the exception of ECD and ETD spectra, which usually lack these ions completely. Pro residues tend to promote this type of dissociation process; complete series of internal fragments are frequently observed in CID and PSD spectra featuring Pro at their N-termini. Acidic residues Asp and Glu also appear to stabilize internal fragments.

In general, high-energy CID offers the most comprehensive information on peptide sequences. It features all the ions listed earlier in this chapter and gives reliable composition information in form of immonium ions and side-chain losses, as well as sequence and satellite ions that permit the differentiation of isomeric Ile and Leu residues. This wealth of information makes high-energy CID spectra much more complex than their low-energy counterparts or PSD data. Such detailed information may not be necessary when the purpose of the study is protein identification by database searching; and MS/MS methods that yield fewer fragments and, thus, divide the limited signal less could be advantageous for low-level sample analysis. However, high-energy CID data may prove essential for *de novo* sequence determination and structure elucidation, as well as site assignment of covalent modifications. ECD may provide complete sequence coverage for large peptides (Axelsson et al., 1999; Zubarev et al., 2000), as well as differentiate between isomeric Ile and Leu (Kjeldsen et al., 2002, 2003; Kjeldsen and Zubarev, 2003). Importantly, ECD's and ETD's unique fragmentation process also permits the localization of labile posttranslational modifications (Kelleher et al., 1999; Syka et al., 2004).

De Novo Sequence Determination

A PSD spectrum of a tryptic peptide was selected as the first example to illustrate the sequencing process. These data show that even this most fastidious method, with relatively low resolution and mass measurement accuracy for the fragment ions, may yield sufficient information for *de novo* sequence determination. The interpretation process discussed here also applies to low- and high-energy CID spectra. Obviously, in CID experiments, the fragmentation will be somewhat different, as discussed already.

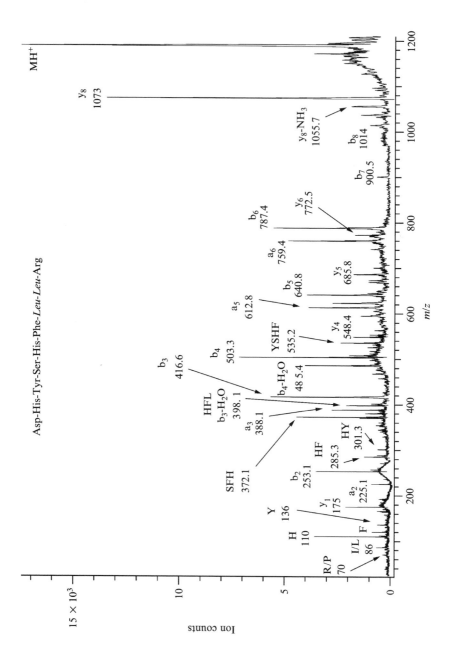

Furthermore, both the resolution and the accuracy of the mass measurement should be significantly better in spectra obtained with tandem instruments such as the QqTOF hybrid instrument or the MALDI-TOF-TOF tandem mass spectrometer.

Figure 1 shows the PSD spectrum of a tryptic peptide with MH^+ at m/z 1187.6 (1188.3 MH^+ average), isolated from an in-gel digest of the 50-kDa subunit of the DNA polymerase α of *Schizosaccharomyces pombe*. The immonium ions unambiguously identified the presence of Arg (m/z 70, 87, 112), Ile/Leu (m/z 86), His (m/z 110 and 166), Phe (m/z 120), and Tyr (m/z 136). These residue weights added together plus the water and the additional proton account for 735 Da, and the "leftover" 452 Da could correspond to 65 theoretically possible amino acid compositions. Thus, determining the amino acid composition was not possible, but identifying the terminal amino acids was straightforward. This was a tryptic peptide, featuring a y_1 ion at m/z 175, revealing a C-terminal Arg, and the corresponding b_{n-1} ion was also detected at m/z 1014. Complementary **b** and **y** ions are related, and their masses can be calculated using the $y_i + b_{n-i} = MH^+ + 1$ formula. There was a very abundant ion at m/z 1073 that corresponded to a 115-Da loss from the molecular ion. Considering the above discussed correlation, this mass difference is consistent with a b_1 mass of 116 Da, which identifies the N-terminal residue as Asp. It has been observed that cleavage at the C-terminus of Asp residues is preferred during PSD fragmentation processes, yielding abundant y-ions (Gu *et al.*, 1999). The next step is to find other sequence ions, perhaps by looking for ions that exhibit neutral losses characteristic of particular ion types. Carbonyl losses could identify b-type ions, both sequence or internal fragments. This spectrum shows some abundant ions that are good b-type candidates, such as m/z 253, 285, 398, 416, 535, 640, and 787. The last two are 147 Da apart, corresponding to a Phe residue, and the **y** ion complementary to the 640 **b** ion occurs at m/z 548. The ion at 535 obviously does not fit in the series as there is no amino acid with a 105-Da residue weight. However, there is an abundant ion at m/z 503, displaying water loss that is also characteristic of **b** ions (when containing hydroxy or acidic amino acids). The mass difference between this ion and the **b** ion at m/z

---

FIG. 1. Matrix-assisted laser desorption ionization (MALDI) post-source decay (PSD) spectrum of m/z 1187.6, a tryptic peptide of the 50-kDa subunit of DNA polymerase α from *S. pombe*. The in-gel digest was high-performance liquid chromatography (HPLC) fractionated, and fractions were manually collected and concentrated to a few microliters. The spectrum was acquired on a Voyager DE STR MALDI-TOF (Applied Biosystems) mass spectrometer in 10 steps. 4-OH-α-CN cinnamic acid was used as the matrix. The sequence was determined from these data.

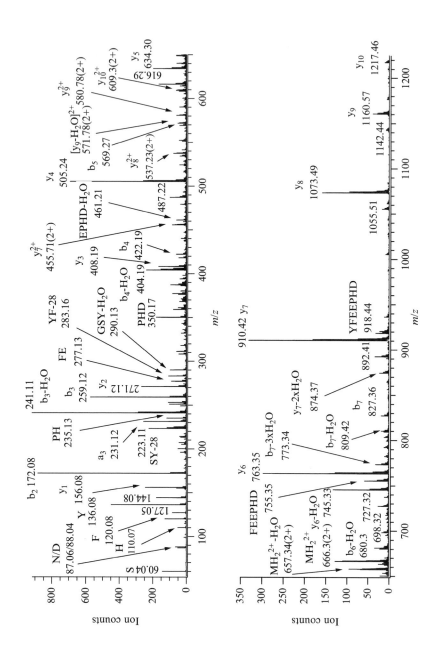

640 is 137, corresponding to a His, and the complementary **y** fragment is indeed detected at m/z 685, along with an ion at m/z 668 due to ammonia loss from the **y** ion. By continuing to "peel off" the amino acids, the next b-ion is at m/z 416 (i.e., 87 Da apart), indicating a Ser residue and explaining the water loss from the **b** ion. In addition, the corresponding y-ion is detected at m/z 772. The ion at m/z 398, the next candidate, does not fit in the series but may be formed via water loss from m/z 416. The ion at m/z 285 is 131 Da lower than the already identified **b** ion at m/z 416, which may indicate a Met residue. However, there is no other ion (i.e., immonium ion, side-chain loss, **y** ion), supporting this conclusion; thus, this ion was classified as an internal fragment (see discussion below). The adjacent residue to the Ser is a Tyr as revealed by the mass difference between the **b** ions at m/z 416 and 253. Because the N-terminal residue has already been established as Asp, and 253–116(**b₁**) = 137, the second amino acid must be a His residue. Thus, the N-terminal sequence was determined as Asp-His-Tyr-Ser-His-Phe. Considering these amino acids, plus the C-terminal Arg, the gap that has to be filled is 226 Da. Because it is known from the immonium ions that the sequence contains a Leu/Ile with a residue weight of 113, the other missing amino acid must also be either a Leu or an Ile. Indeed, there is a **y₂** ion at m/z 288, a **y₂-NH₃** ion at m/z 271, and the corresponding **b** ion at m/z 900. Thus, the sequence of this peptide is Asp-His-Tyr-Ser-His-Phe-*Leu-Leu*-Arg. Note that PSD fragmentation does not allow differentiation between the isomeric Leu/Ile residues. The b-type ions listed above that were not identified as sequence ions, such as m/z 285, 398, and 535, are internal fragments corresponding to HF, HFL, and YSHF sequences, respectively. Similarly ions at m/z 301 (HF), 372 (SHF), and 672 (HYSHF) are b-type internal fragments; only the a-type fragments formed by the carbonyl loss were not detected. The presence of basic His residues "justifies" why more abundant **b** and internal ions than C-terminal fragments were observed.

The second example (Fig. 2) is a low-energy CID spectrum acquired by LC/MS/MS analysis on a QqTOF (QSTAR) mass spectrometer in an information-dependent fashion (i.e., the precursor ion was computer-selected), and the collision energy was automatically adjusted according to its charge state and m/z value. The precursor ion was m/z 666.266(2+).

FIG. 2. Low-energy collision-induced dissociation (CID) data of m/z 666.266 (2+) acquired on a QSTAR mass spectrometer during an information-dependent LC/MS/MS analysis of a tryptic digest. The precursor ion was selected by the computer and the collision energy was also automatically adjusted. The sequence was determined from these data as Asn-Gly-Ser-Tyr-Phe-Glu-Glu-Pro-His-Asp-His.

From the low mass region, one could predict the presence of Ser, Asn, Asp, His, Phe, and Tyr immonium ions at m/z 60, 87, 88, 110, 120, and 136, respectively. The immonium ion for Asn m/z 87 also may indicate an Arg residue (see Table I), but as we will see, this particular peptide did not contain that amino acid. The next step is to determine the C-terminus. Because this peptide belonged to a tryptic digest, one would expect to see y_1 at m/z 147 or 175, indicating C-terminal Lys or Arg, respectively. This CID spectrum does not contain either. However, there is an ion at m/z 156.08 instead suggesting a C-terminal His. We may be dealing with the C-terminus of the protein or with the result of a nonspecific cleavage. Fortunately His is also a basic amino acid, preferentially retaining the charge; thus, the fragmentation pattern is similar to that displayed by "real" tryptic peptides. The abundant higher mass ions, m/z 1217.55, 1160.54, 1073.44, 910.44, 763.32, 634.25, and 505.24 represent y fragments and reveal mass differences as follows: 57, 87, 163, 147, 129, and 129 Da, corresponding to a GSYFEE sequence. Remember, we are reading the sequence backward from the y fragments! Fragments $y_7 - y_{10}$ were detected also as doubly charged ions; the instrument afforded more than sufficient resolution for charge state determination from the isotope spacing. (The potential presence of multiply charged fragment ions complicates data interpretation when working with low-resolution CID data of multiply charged ions.) The identity of the N-terminal amino acid(s) can be determined from the mass difference between the MH^+ +1 value and the mass of the highest y ion (see Table II). From this calculation, the missing **b** ion should be at m/z 115. If there were two Gly residues at the N-terminus indeed, we should observe a **b** fragment at this mass. However, if the N-terminus is an Asn residue (identical elemental composition to two Glys), we will not detect b_1 because it is not stable (Yalcin *et al.*, 1995). The first N-terminal ions, m/z 172, 144, and 127 (i.e., b_2, a_2, a_2 NH_3), indicate that the latter is the case and our working sequence by now is NGSYFEE...H. The lowest y fragment identified (m/z 505) is very abundant, suggesting that it was formed via a cleavage at the N-terminus of a Pro residue. Indeed, there is a fragment at m/z 408.19 that mass is 97 Da (Pro residue weight) lower. Thus, the mass still unaccounted for is 252 Da (y_i-y_1 = 408-156). From the immonium ions, we established that the sequence should contain an aspartic acid that would account for 115 Da and the leftover 137 Da corresponds to a second His. Thus, the options for the C-terminal sequence are PHDH or PDHH. The expected y_2 fragment for the first variation should be at m/z 271. The presence of Pro residues usually leads to the formation of abundant internal ions too. PH should be detected at m/z 235, and the carbonyl loss from this fragment would yield an ion at m/z 207. All these ions were detected in the spectrum. The other variant should yield a y_2 fragment at m/z 293 and internal fragments at m/z 213 and 185. From these three only m/z 213.09 was detected, which could also represent $\mathbf{a_3}$ H_2O. Thus, the full working sequence is NGSYFEEPHDH. To confirm this assignment, most of the remaining ions have to be assigned. As illustrated by Fig. 2, the spectrum contains an extensive $\mathbf{b}$ ion series: at m/z 172, 259, 422, 569, 698, and 827 ($\mathbf{b_2}$-$\mathbf{b_7}$), accompanied by satellite ions formed via -28 (CO) or multiple 18 (H_2O) losses. There are four residues in the sequence that may undergo the latter neutral loss. So, the $\mathbf{y}$ ions also feature satellite ions due to this process. Quite a few internal ions were also detected in this spectrum; some of them are six to seven residues long, a feature frequently observed in QSTAR CID experiments.

Protein Identification Using MS/MS Data

For protein identification, a handful of sequence ions, together with the precursor ion, may be sufficient (Nielsen *et al.*, 2002). Thus, PSD analysis and low- and high-energy CID experiments may be and have been used for this purpose (Aebersold and Mann, 2003; Chalmers and Gaskell, 2000; Dreger, 2003; Gygi and Aebersold, 2000; Ong *et al.*, 2003; Roberts, 2002). Software packages provided by the MS instrument manufacturers and by different research groups and software companies are available and can be used for data interpretation (Fenyo, 2000). Higher mass accuracy leads to more reliable protein identification even if the MS/MS analysis yields only a few ions (Nielsen *et al.*, 2002).

Figure 3 illustrates the importance of peptide sequence analysis in proteomics research. The unfractionated in-gel tryptic digest of a yeast protein yielded only a few peptide masses by MALDI TOF MS analysis. Although the yeast genome is known, a database search with these masses specifying a mass accuracy of 150 ppm (external calibration) proved inconclusive. (For all the database searches discussed in this chapter, ProteinProspector (*www.prospector.ucsf.edu*) was used.) The ion at m/z 1596.63 was subjected to MALDI CID analysis performed on an ABI 4700 TOF-TOF tandem mass spectrometer (Applied Biosystems, Framingham, MA) (Huang *et al.*, 2002; Medzihradszky *et al.*, 1996a). One of the most abundant fragments in its CID spectrum is an ion that corresponds to MH^+-64 Da, indicating that the peptide contains a Met sulfoxide. Such peptides can be identified by using the MS-Tag feature of ProteinProspector in homology mode. However, this time this search was not successful. Manual interpretation of the CID data revealed a C-terminal YAAYM(O) FK sequence. An MS-Pattern search with this sequence identified the protein unambiguously as glyceraldehyde 3-phosphate dehydrogenase. The missing N-terminal amino acids were PFITND, and the peptide was

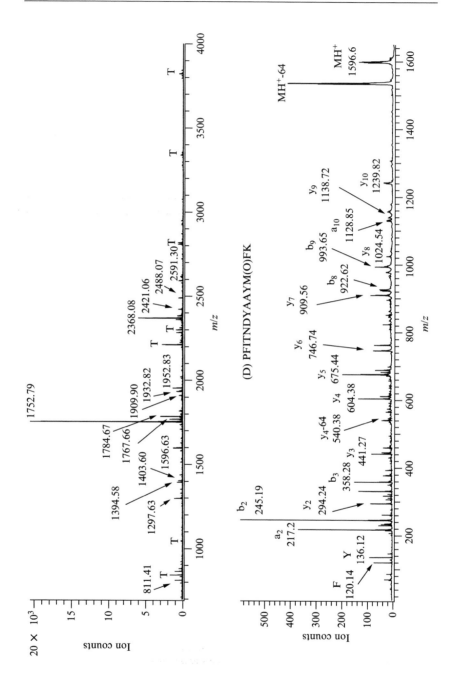

(D) PFITNDYAAYM(O)FK

formed via an Asp-Pro cleavage, which is not an expected hydrolysis site for trypsin. Only four of the peptides detected proved to correspond to predicted tryptic products for this protein: m/z 811.41, 1297.63, 1752.79, and 2591.30.

In addition to nonspecific protease cleavages and covalent modifications, CID analysis of the components in unfractionated digests may be complicated by the presence of peptides of identical or similar molecular mass. The resolution of precursor ion selection has to be considered. While the mass window for precursor ion selection is about ±5% for PSD experiments, the MALDI-TOF-TOF uses an approximately 5–10 Da wide window, triple quadrupoles, hybrid instruments, and ion traps usually operate with a 2–3 Da wide window, and four-sector instruments allow monoisotopic precursor selection. When analyzing unfractionated digests, it frequently happens that two or more different peptides are fragmented and analyzed simultaneously in a single experiment.

Figure 4 shows the electrospray low-energy CID spectrum of precursor ion at m/z 724.34 (2+). An unfractionated protein digest was analyzed using nanospray sample introduction on a QqTOF tandem mass spectrometer. The analysis yielded a few high-quality CID spectra suitable for protein identification. The theoretical tryptic digest of the protein suggested this particular spectrum might correspond to an oxidized methionine-containing peptide, ITSPLM(O)EPSSIK (MH$^+$ 1447.67), which was confirmed by the CID data. After "filtering" out the ions that were expected for this sequence, several abundant fragments remained unaccounted for and readily revealed an N-terminal sequence of PFGVALLF. This sequence belongs to the same protein, corresponding to a peptide, PFGVALLFGGVDEK formed by an Arg-Pro cleavage, giving a protonated monoisotopic molecular weight of 1448.78. The isotope peaks in the molecular ion cluster were indeed slightly higher than it would be expected for peptides in this mass range. However, isotope ratios may

---

FIG. 3. Mass spectrometry (MS)-based identification of a yeast protein. The upper panel shows the matrix-assisted laser desorption ionization (MALDI) MS spectrum of the unfractionated in-gel digest acquired on a Voyager DE STR MALDI-TOF (Applied Biosystems) mass spectrometer in reflectron mode (T indicates trypsin autolysis products). MS-Fit (*www.prospector.ucsf.edu*) search with these data was inconclusive. The lower panel shows the collision-induced dissociation (CID) spectrum of one of the peptides obtained on an ABI 4700 MALDI-TOF-TOF (Applied Biosystems) tandem instrument. The C-terminal sequence, YAAYM(O)FK, was determined by manual data interpretation. The protein was identified as glyceraldehyde 3-phosphate dehydrogenase using this sequence in an MS-Pattern search (*www.prospector.ucsf.edu*). Four additional masses, m/z 811.41, 1297.63, 1752.79, and 2591.30 corresponded to predicted tryptic peptides of this protein.

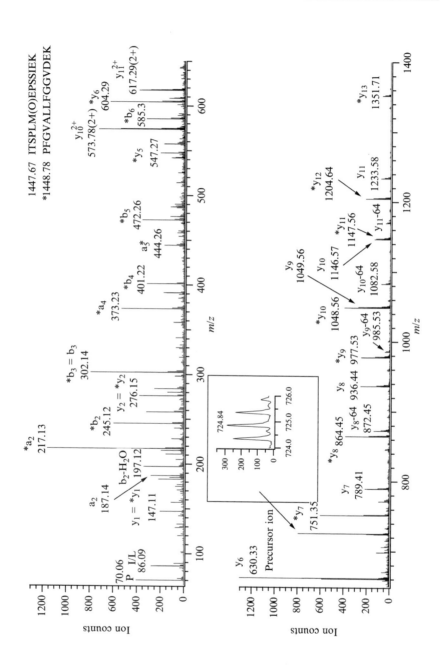

not be reliable as multiply charged peptide ions in unfractionated in-gel digests may not raise significantly above the noise level, and overlapping precursor ions may be masked by other more abundant species, possibly of different charge states. Overlapping isotope clusters do occur even in LC/MS experiments, especially when very complex mixtures are analyzed.

Figure 5 shows the MALDI low-energy CID of m/z 960.4 in a tryptic digest of a human heat shock protein. These data were also acquired on a QqTOF tandem mass spectrometer. In this case, the isotope pattern of the precursor ion clearly indicates that a mixture was analyzed. Such an isotope pattern frequently indicates the hydrolysis of Asn or Gln residues, either as a common posttranslational modification or as potential byproduct of the isolation/digestion/storage, giving a partial 1 Da mass increase. However, in this protein, the components were two unrelated peptides and the presence of both could be confirmed unambiguously.

Altering the Dissociation Processes by Chemical Derivatization

De novo sequencing of peptides and protein identification would be accomplished much faster, and at higher sensitivity if one could simplify peptide fragmentation, exclusively generating a single ion series. Of the MS/MS techniques discussed in this chapter, only ECD and ETD fulfill this description (Axelsson *et al.*, 1999; Syka *et al.*, 2004; Zubarev *et al.*, 2000). However, it has been reported that generating fixed charges in peptides by chemical derivatization (i.e., introducing positively or nega-tively charged groups at a terminal position) will lead to an altered gener-ally more simplified fragmentation pattern. First Johnson (1988) showed that peptides that featured a Lys residue converted to a 2,4,6-trimethyl-pyridinium cation exhibited a series of N-terminal or C-terminal ions depending on where the charged residue was located. A series of other derivatizations followed (Roth *et al.*, 1998), and it has been established that a fixed positive charge is beneficial for controlling the peptide fragmenta-tion in high-energy CID (Stults *et al.*, 1993; Wagner *et al.*, 1991). However, most of these derivatizations cannot be performed reproducibly at the

---

FIG. 4. Low-energy CID spectrum of m/z 724.34 (2+) from the unfractionated tryptic digest of a two-dimensional gel-purified rat protein. The data were acquired on a QqTOF mass spectrometer (QSTAR, MDS Sciex) using nanospray sample introduction. The protein was identified from other data as proteasome subunit zeta. This CID spectrum confirmed the identity of a predicted tryptic peptide and revealed the presence of another peptide with MH[+] 1448.78, from the same protein. The insert shows the precursor ion.

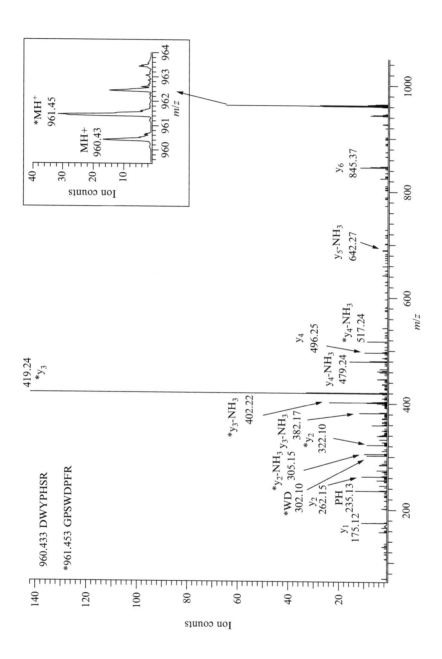

picomole level or below, may lead to side reactions, and may require purification steps. It was reported that one of the best derivatives, "C5Q," an N-terminal trimethylammonium group with a five-carbon linker (i.e., $[CH_3]_3N^+[CH_2]_5CO$ modification), could be introduced in 10 min, at the femtomole level and the derivatized peptide was readily detected by MALDI (Bartlet-Jones *et al.*, 1994). However, to generate only N-terminal fragment ions, modification of Lys and Arg side chains was recommended to reduce their basicity (Spengler *et al.*, 1997). Earlier, Burlet *et al.* (1992a) reported "that the formation of N- or C-terminal precharged derivatives is detrimental to the formation of sequence-specific product ions following low-energy collisional activation." In addition, they found that "protonation of pre-charged derivatives (yielding doubly charged ions) restores favorable fragmentation properties." This effect was attributed to the additional proton being able to move freely, occupy any site, and induce charge-directed fragmentation. This effect also explains the "y-ladder" formation of doubly charged tryptic peptides, especially with C-terminal Arg residues (Burlet *et al.*, 1992a). The same group discovered that the introduction of negatively charged residues by Cys oxidation to cysteic acid in Arg-ended tryptic peptides also promoted uniform fragmentation (i.e., comprehensive y-ion series formation even in singly charged ions). This finding was consistent with their hypothesis "that increased heterogeneity (with respect to localization of charge) of the protonated peptide precursor ion population is beneficial to the generation of high yield product ions via several charge-directed, low energy fragmentation pathways" (Burlet *et al.*, 1992b). Based on these observations, a new higher sensitivity derivatization was developed in which a sulfonic acid group was added to the N-termini of the peptides (Keough *et al.*, 1999). This derivatization with 2-sulfobenzoic acid cyclic anhydride is simple and efficient and may be performed directly on target. By donating a proton, this acidic group promotes the charge-directed fragmentation of the peptide bonds and, with a basic amino acid such as Arg at the C-terminus, yields y-ions almost exclusively. Figure 6 illustrates how this N-terminal modification alters peptide fragmentation in PSD.

FIG. 5. Low-energy CID of m/z 960.43 from an unfractionated tryptic digest of a two-dimensional gel-purified human protein. The data were acquired on a QqTOF mass spectrometer (QSTAR, MDS Sciex) with MALDI, using 2,5-dihydroxy-benzoic acid as the matrix. The insert shows the precursor ion. Its isotope distribution clearly indicated the presence of another component with a 1 Da higher molecular weight. The CID data revealed the presence of two non-related peptides of the same heat shock protein that was identified from another CID spectrum by MS-Tag search (*www.prospector.ucsf.edu*).

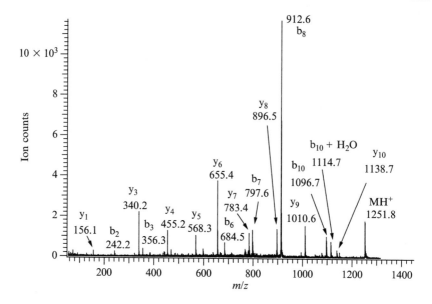

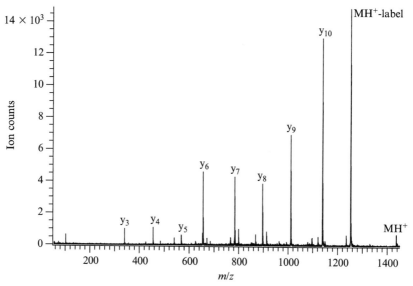

FIG. 6. Matrix-assisted laser desorption ionization (MALDI)-TOF-TOF (Applied Biosystems) post-source decay (PSD) spectra (acquired without collision gas in a single step) of cholecystokinin fragment [10]Ile-Lys-Asn-Leu-Gln-Ser-Leu-Asp-Pro-Ser-His[20]. The upper panel shows the spectrum of the peptide without modification. The lover panel shows data acquired from the N-terminally 2-sulfobenzoylated molecule.

MS/MS Analysis of Covalently Modified Peptides

MS can detect the presence of even unanticipated covalent modifications and can provide information on their structure and location. This is one of its greatest advantages over other protein analytical techniques. The presence of covalent modifications is usually detected on the peptide level in the form of masses that do not match the predicted digestion products. Obviously the covalent modifications can be the results of *in vivo* processes such as posttranslational or xenobiotic modifications, as well as the products of intentional labeling or fortuitous chemical reactions (for review, see Hunyadi-Gulyas and Medzihradszky, 2004). Common mass differences are listed, for example, at "Deltamass" at *http://www.abrf.org/index. cfm/dm.home.* However, mass measurement alone does not provide unambiguous information on the structure and the exact site of the covalent modification. To prove the structure, suspected peptide fragmentation information has to be obtained. Since for any known sequence the masses of the expected fragment ions can be calculated, the presence of covalent modifications should be detected by the observed mass shifts between the predicted and observed fragment masses. Some structures indeed are very stable upon collisional activation, and the site of the modification can be established readily from CID data. For example, acetyl groups on the N-terminus or on the ε-amino group of a Lys residue will not fall off during any MS/MS experiment. Acetylation is indicated by 42Da shifts in all the fragment masses that contain the modified residue. In addition, when the N-terminus is modified, the $\mathbf{b_1}$ ion can be detected because it is stable with acyl modification (Yalcin *et al.*, 1995). When the Lys side chain is modified, immonium ions at m/z 126.092 and 143.119—much weaker—can be detected. A good illustration of both options is shown in Fig. 7.

In addition to acetylation, the most frequently observed stable modifications are the carboxymethylation, carbamidomethylation, and acrylamide modification of Cys side chains, because these modifications are routinely used to block the −SH groups. These derivatizations result in 58-, 57-, and 71-Da mass increase of the Cys-containing fragments, respectively. Vinyl pyridine also can be used for sulfhydryl protection. This modification is also stable in that one can observe the expected mass shift and the modified immonium ion; however, because of the basic structure in it, S-ethyl-pyridyl Cys-containing peptides will also display an abundant and characteristic fragment formed via fragmentation of the protecting group: m/z 106.066 ($CH_2 = CH-C_5H_4NH^+$). The fact that covalent modifications yield abundant fragment ions can be beneficial. If the produced fragment masses are unique enough, they may serve as diagnostic markers and aid the identification of modified molecules. This approach has been

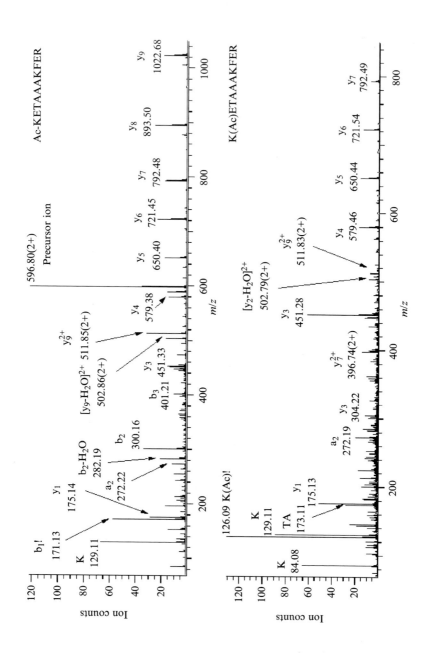

used, for example, in the localization of glycopeptides (Carr *et al.*, 1993; Huddleston *et al.*, 1993), xenobiotically modified species (Burlingame *et al.*, 1996), and ICAT-modified molecules (Baldwin *et al.*, 2001). However, if the modifying group is too successful in retaining the charge at the expense of the other parts of the peptide, it may prevent the assignment of the modification site or even obtaining just sequence confirmation (Medzihradszky *et al.*, 2002). Charged modifications may interfere with the fragmentation in other ways too. As was presented earlier in this chapter, introducing a positive or negative charge may lead to simplified but definitely altered fragmentation pattern. Trimethylation of amino groups is a known posttranslational modification contributing 42 Da to the MH^+ value. The mass difference between acetylation and trimethylation is 36 mmu. However, trimethylation differs from acetylation in other aspects, too. It also alters the fragmentation pattern. For example, one cannot observe a shifted immonium ion for a trimethylated Lys residue or b_1 fragment if the N-terminus is modified. Instead trimethyl-modified Lys residues tend to lose 59 Da, that is, $N(CH_3)_3$, as was reported for MALDI analysis (Hirota *et al.*, 2003; Zhang *et al.*, 2004) and shown for low-energy CID in this chapter (Fig. 8). In addition, in high-energy CID experiments, the presence of such a modification may lead to unexpected **d** and **w** ion formation. For example, normally a tryptic peptide with an Arg residue at its C-terminus will yield only **w** ions. However, the presence of a trimethyl-lysine close to the N-terminus will trigger **d** ion formation.

Characteristic neutral losses like the one mentioned earlier also may serve as diagnostic markers. The 64 Da loss from methionine-sulfoxide has been mentioned already. Interestingly, posttranslational modifications of Ser and Thr residues, namely, phosphorylation, *O*-glycosylation, and sulfation, also display such preferred fragmentation.

Phosphorylation has been studied extensively by MS (for reviews, see Kalume *et al.*, 2003; Mann *et al.*, 2002; McLachlin and Chait, 2001). Depending on the site of phosphorylation, the modified peptides behave very differently. Tyr-phosphorylated peptides retain the modification under most MS and MS-MS conditions. The modified Tyr displays an appropriately shifted immonium ion at m/z 216 (Steen *et al.*, 2001), and the site of modification also can be established unambiguously from the modified

---

Fig. 7. Low-energy collision-induced dissociation (CID) spectra of differently acetylated peptides acquired from the tryptic digest of an acetylated bovine RNase (P07849). The abundant b_1-ion in the upper panel clearly establishes N-terminal acetylation, while the shifted immonium ion in the lower panel indicates that not the N-terminus itself, but the ε-amino group of the N-terminal Lys residue was acetylated.

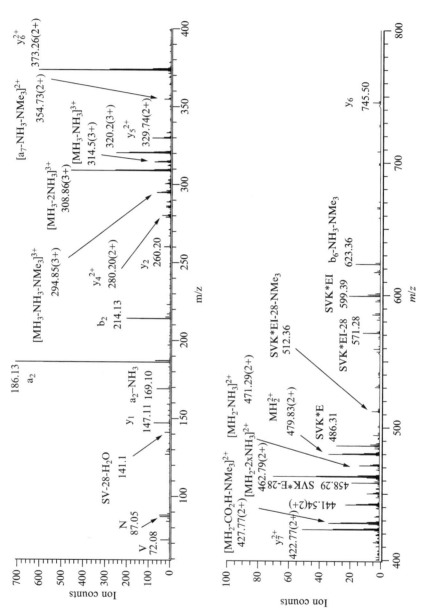

Fig. 8. Low-energy collision-induced dissociation (CID) spectrum of tryptic peptide [300]Asn-Val-Ser-Val-Lys(Me3)Glu-Ile-Lys[307] from an elongation factor 1α of *Plasmodium falciparum*.

fragment ions present; 80 Da losses have been reported only in MALDI-PSD experiments (Bonewald et al., 1999). At the other extreme, His-phosphorylated peptides undergo extensive fragmentation even under MS conditions (Medzihradszky et al., 1997). Ser- and Thr-modified peptides are stable during MS acquisition, but the favored fragmentation step upon collisional activation is the β-elimination of phosphoric acid (H_3PO_4) (i.e., 98-Da loss from the molecular ion). Based on this observation, phosphorylation sites are regularly assigned based on the presence of ions that are assumed to be the products of such β-elimination (Gronborg et al., 2002). For example, this is the way Mascot (*www.matrixscience.com*) assigns phosphorylation sites. Unfortunately, hydroxy amino acids, as well as Glu and Asp residues, readily lose water upon the fragmentation process from all kinds of sequence ions. Such a fragment derived from an unmodified ion will have the same structure (and mass) as the product of phosphoric acid β-elimination from a modified species. Most of the phosphopeptide analysis is done online, usually in a data-dependent manner, when the collision energy is set automatically. Although this approach serves quite well the protein identification studies may not be optimal for some post-translational modification assignments. Figure 9 shows the CID data of a phosphopeptide acquired from its more abundant, 3+ ion (upper panel) automatically and from the less abundant doubly charged ion, after some manual adjustments of the collision energy. The β-elimination products—both due to water or phosphoric acid losses—are very abundant in the spectrum that was taken under preset conditions suitable for the "meaningful" fragmentation of most triply charged ions. From these data—when only the β-elimination products are considered—one easily could assume that the Ser residue is modified because every single N-terminal ion detected, starting from a_3, carries the telltale double bond. However, y_4 shifted by 80 Da is the ion that reveals the real phosphorylation site. The CID spectrum of the doubly charged ion displays intact b_5 and b_6 ions, indicating that the first six residues are not modified, whereas y_3 shows the presence of modification. Interestingly, the elimination of water from b_3 still produces as abundant an ion as that derived from the phosphoric acid loss from modified b_7 and b_8.

In weaker spectra, or under less fortunate circumstances, this preferential fragmentation (i.e., the lack of modified fragments) may prevent the site assignment. However, because this fragmentation also can be triggered in the source during MS acquisition even under relatively mild conditions, it may be taken advantage of. The product of β-elimination can be selected as precursor ion and the dehydro-alanine, formed from the modified Ser or the dehydro-butyric acid, product of β-elimination from Thr readily can be identified from the 69 and 83 Da residue weights, respectively. Similarly,

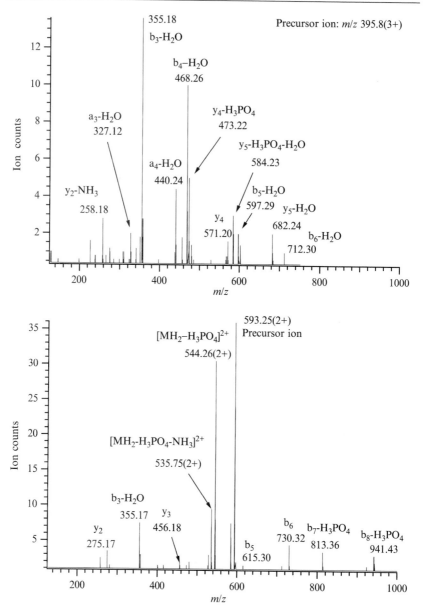

FIG. 9. Low-energy collision-induced dissociation (CID) spectra of a phosphopeptide ^{248}ESRLEDT(p)QK^{256} isolated from an endoproteinase Lys C digest of human P450 3A4. The upper panel shows CID data acquired from the triply charged ion, the lower panel shows CID data acquired from the doubly charged ion.

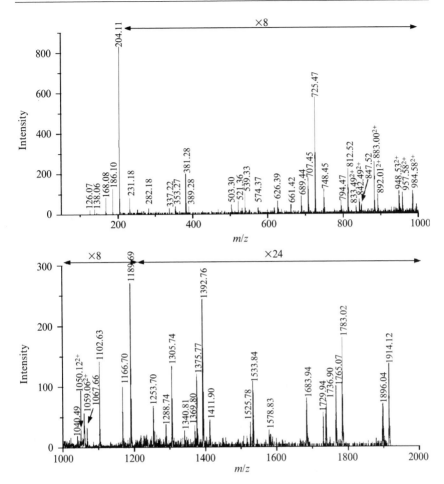

FIG. 10. Low-energy collision-induced dissociation (CID) spectrum of [303]LAPVSASVSP S(GlcNAc)AVSSANGTVL[323] from a chymotryptic digest of serum response factor. These data were acquired by online LC/MS/MS on a Q-TOF (Micromass) mass spectrometer, precursor ion at m/z 1059.01(2+). See fragment assignments in Table III. (Reprinted, with permission from Chalkley and Burlingame, 2003.)

products of the β-elimination generated in MS/MS experiments in ion traps can be subjected to subsequent MS[3] analyses. The β-elimination also can be accomplished chemically before the MS analysis (Jaffe *et al.*, 1998). Sulfated or glycosylated Ser and Thr residues are also susceptible to chemical β-elimination at about the same extent as their phosphorylated counterparts (Medzihradszky *et al.*, 2004; Wells *et al.*, 2002). However,

TABLE III

MASS ASSIGNMENTS FOR IONS DETECTED IN FIG. 9[a,b]

b_G
b

L A P V S A S V S P S A V S S A N G T V L

y
y_G

Peak	Match	Peak	Match	Peak	Match
126.07	GlcNAc Fragment	725.47	b8	1253.70	b14
138.06	GlcNAc Fragment	748.45	y8	1288.74	$y12_G$-NH_3
168.08	GlcNAc Fragment	794.47	b9-H_2O	1305.74	$y12_G$
186.10	GlcNAc Fragment	812.52	b9	1340.81	b15
204.11	GlcNAc	833.49^{2+}	y19-H_2O	1369.80	$b13_G$
231.18	y2	842.49^{2+}	y19	1375.77	$y13_G$-NH_3
282.18	b3	847.52	y9	1392.76	$y13_G$
337.27	PVSA-H_2O	883.00^{2+}	b20-H_2O	1411.90	b16
353.27	a4	892.01^{2+}	b20	1525.78	b17
381.28	b4	948.53^{2+}	MH_2^{2+}-H_2O	1553.84	y17
389.28	y4	957.58^{2+}	MH_2^{2+}	1578.83	$y15_G$
503.30	b6-$2H_2O$	984.58^{2+}	$b20_G$-H_2O	1683.94	b19
521.36	b6-H_2O	1040.49^{2+}		1729.94	y19
539.33	b6	1050.12^{2+}	$M_GH_2^{2+}$-H_2O	1736.90	$y17_G$
574.37	y6	1059.06^{2+}	$M_GH_2^{2+}$	1765.07	b20-H_2O
626.39	b7	1067.66	b12	1783.02	b20
661.42	y7	1102.63	y12	1896.04	MH^+-H_2O
689.44	b8-$2H_2O$	1166.70	b13	1914.12	MH^+
707.45	b8-H_2O	1189.69	y13		

[a] From Chalkley and Burlingame (2003).
[b] The subscript "G" indicates that the fragment listed retained the GlcNAc residue.

the favored fragmentation step in the CID fragmentation of sulfo- and O-linked glycopeptides is a rearrangement reaction that eliminates the modification from the amino acid side chain without a trace (Medzihradszky et al., 1996b, 2004). Although some of the glycopeptide fragments may retain the sugar (Alving et al., 1999; Chalkley and Burlingame, 2003; Medzihradszky et al., 1996b), the sulfate is always eliminated completely (Medzihradszky et al., 2004). Thus, while the site of glycosylation may be determined from CID data, CID-based site assignment is impossible for sulfopeptides: their fragmentation pattern is identical to that of the unmodified peptides. Figure 10 and Table III illustrate the typical low-energy CID fragmentation of peptides modified by O-linked carbohydrates. The oxonium ion of the GlcNAc residue (m/z 204) and related carbohydrate fragments dominate the low mass region. Peptide fragments that underwent the rearrangement reaction and lost the modifying sugar are usually more abundant than the glycosylated fragments. Still, there is sufficient information in the spectrum to establish Ser-313 (position 11 in the peptide shown) as the site of modification. The side chains usually do not fragment in ECD and ETD experiments. Thus, the sites of the aforementioned covalent modifications could be assigned easier from such analyses than from CID data. In addition, mass accuracy afforded by an FT-MS instrument may facilitate differentiation between isobaric phosphorylation and sulfation.

Summary

Peptide dissociation reactions that are recorded in different MS/MS experiments provide information on the amino acid sequence. These data also indicate the presence, structure, and location of covalent modifications. This chapter offers a qualitative account on the rules of dissociation processes known. At present, no computer software is available to predict peptide fragmentation patterns reliably. However, with the arrival of high-throughput proteomics, information gathered on peptide fragmentation processes will increase exponentially. As the enormous computer-generated MS/MS databases become analyzed, new patterns will most likely emerge that will aid more reliable, faster protein identification and structural characterization.

Acknowledgments

I want to thank my friends and colleagues who helped me in writing this chapter, either with providing me with samples: Ralph Davis (DNA polymerase α, in Fig. 1), Gábor Rákhely (*Sphingomonas* enzyme, in Fig. 2), Doron Greenbaum (proteasome Zeta subunit, in Fig. 4; elongation factor 1 α of *Plasmodium falciparum*, in Fig. 8), and Katherine E. Williams (human heat shock protein, in Fig. 5); or with data: David A. Maltby (Fig. 3) and Jenny Michels (Fig. 6); or with discussion and friendly advice: Al Burlingame, Roman Zubarev, and Michael A. Baldwin. Data for Fig. 7 were provided by my middle son, Mátyás Medzihradszky. My work was supported by National Institutes of Health grants NCRR RR01614, RR01296, RR014606, and RR015804 to the UCSF Mass Spectrometry Facility, Director, A. L. Burlingame.

References

Aebersold, R., and Mann, M. (2003). Mass spectrometry–based proteomics. *Nature* **422,** 198–207.
Alexander, A. J., Thibault, P., Boyd, R. K., Curtis, J. M., and Rinehart, K. L. (1990). Collision induced dissociation of peptide ions: Part 3. Comparison of results obtained using sector-quadrupole hybrids with those from tandem double-focusing instruments. *Int. J. Mass Spectrom. Ion Processes* **98,** 107–134.

Alving, K., Paulsen, H., and Peter-Katalinic, J. (1999). Characterization of O-glycosylation sites in MUC2 glycopeptides by nanoelectrospray QTOF mass spectrometry. *J. Mass Spectrom.* **34,** 395–407.

Axelsson, J., Palmblad, M., Hakansson, K., and Hakansson, P. (1999). Electron capture dissociation of substance P using a commercially available Fourier transform ion cyclotron resonance mass spectrometer. *Rapid Commun. Mass Spectrom.* **13,** 474–477.

Baldwin, M. A., Medzihradszky, K. F., Lock, C. M., Fisher, B., Settineri, C. A., and Burlingame, A. L. (2001). Matrix assisted laser desorption/ionization coupled with quadrupole/orthogonal acceleration time-of-flight mass spectrometry for protein discovery, identification and structural analysis. *Anal. Chem.* **73,** 1707–1720.

Bartlet-Jones, M., Jeffery, W. A., Hansen, H. F., and Pappin, D. J. C. (1994). Peptide ladder sequencing by mass spectrometry using a novel, volatile degradation reagent. *Rapid Commun. Mass Spectrom.* **8,** 737–742.

Biemann, K. (1990). Nomenclature for peptide fragment ions (positive-ions). *Methods Enzymol.* **193,** 886–887.

Bonewald, L. F., Bibbs, L., Khatri, A., Medzihradszky, K. F., McMurray, J. S., and Weintraub, S. (1999). Study on the synthesis and characterization of peptides containing a phosphorylated tyrosine. *J. Peptide Res.* **53,** 161–169.

Brittain, S., Mohamed, Z. A., Wang, J., Lehmann, V. K., Carmichael, W. W., and Rinehart, K. L. (2000). Isolation and characterization of microcystins from a river Nile strain of *Oscillatoria tenuis* Agardh ex Gomont. *Toxicon.* **38,** 1759–1771.

Burlet, O., Yang, C. Y., and Gaskell, S. J. (1992). Influence of cysteine to cysteic acid oxidation on the collision-activated decomposition of protonated peptides: Evidence for intraionic interactions. *J. Am. Soc. Mass Spectrom.* **3,** 337–344.

Burlet, O., Orkiszewski, R. S., Ballard, K. D., and Gaskell, S. J. (1992). Charge promotion of low-energy fragmentations of peptide ions. *Rapid Commun. Mass Spectrom.* **6,** 658–662.

Burlingame, A. L., Medzihradszky, K. F., Clauser, K. R., Hall, S. C., Maltby, D. A., and Walls, F. C. (1996). From protein primary sequence to the gamut of covalent modifications using mass spectrometry. *In* "Biochemical and Biotechnological Applications of Electrospray Ionization Mass Spectrometry" (P. Snyder, ed.), pp. 472–511. ACS Symposium Series, 619.

Carr, S. A., Huddleston, M. J., and Bean, M. F. (1993). Selective identification and differentiation of N- and O-linked oligosaccharides in glycoproteins by liquid chromatography-mass spectrometry. *Protein Sci.* **2,** 183–196.

Chalkley, R. J., and Burlingame, A. L. (2003). Identification of novel sites of O-N-acetylglucosamine modification of serum response factor using quadrupole time-of-flight mass spectrometry. *Mol. Cell. Proteomics.* **2,** 182–190.

Chalmers, M. J., and Gaskell, S. J. (2000). Advances in mass spectrometry for proteome analysis. *Curr. Opin. Biotech.* **11,** 384–390.

Chaurand, P., Luetzenkirchen, F., and Spengler, B. (1999). Peptide and protein identification by matrix-assisted laser desorption ionization (MALDI) and MALDI-post-source decay time-of-flight mass spectrometry. *J. Am. Soc. Mass Spectrom.* **10,** 91–103.

Cooper, H. J., Hudgins, R. R., Hakansson, K., and Marshall, A. G. (2002). Characterization of amino acid side chain losses in electron capture dissociation. *J. Am. Soc. Mass Spectrom.* **13,** 241–249.

Dreger, M. (2003). Proteome analysis at the level of subcellular structures. *Eur. J. Biochem.* **270,** 589–599.

Falick, A. M., Hines, W. M., Medzihradszky, K. F., Baldwin, M. A., and Gibson, B. W. (1993). Low-mass ions produced from peptides by high energy collision-induced dissociation in tandem mass spectrometry. *J. Am. Soc. Mass Spectrom.* **4,** 882–893.

Fenyo, D. (2000). Identifying the proteome: Software tools. *Curr. Op. Biotech.* **11**, 391–395.

Grgurina, I., Mariotti, F., Fogliano, V., Gallo, M., Scaloni, A., Iacobellis, N. S., Lo Cantore, P., Mannina, L., van Axel Castelli, V., Greco, M. L., and Graniti, A. (2002). A new syringopeptin produced by bean strains of *Pseudomonas syringae* pv. *syringae.* *Biochim. Biophys. Acta* **1597**, 81–89.

Gronborg, M., Kristiansen, T. Z., Stensballe, A., Andersen, J. S., Ohara, O., Mann, M., Jensen, O. N., and Pandey, A. (2002). A mass spectrometry-based proteomic approach for identification of serine/threonine-phosphorylated proteins by enrichment with phospho-specific antibodies: Identification of a novel protein, Frigg, as a protein kinase A substrate. *Mol. Cell. Proteomics* **1**, 517–527.

Gu, C. G., Somogyi, A., Wysocki, V. H., and Medzihradszky, K. F. (1999). Fragmentation of protonated oligopeptides XLDVLQ (X = L, H, K, or R) by surface-induced dissociation: Additional evidence for the "mobile proton" model. *Anal. Chim. Acta* **397**, 247–256.

Gygi, S. P., and Aebersold, R. (2000). Mass spectrometry and proteomics. *Curr. Opin. Chem. Biol.* **4**, 489–494.

Hirota, J., Satomi, Y., Yoshikawa, K., and Takao, T. (2003). Epsilon-N,N,N-trimethyllysine-specific ions in matrix–assisted laser desorption/ionization-tandem mass spectrometry. *Rapid Commun. Mass Spectrom.* **17**, 371–376.

Huang, L., Baldwin, M. A., Maltby, D., Baker, P. R., Medzihradszky, K. F., Allen, N., Rexach, M., Edmondson, R., Campbell, J., Juhasz, P., Martin, S. A., Vestal, M. L., and Burlingame, A. L. (2002). The identification of protein–protein interactions of the nuclear pore complex of *s. cerevisiae* using high throughput MALDI-TOF/TOF tandem mass spectrometry. *Mol. Cell. Proteomics.* **1**, 434–450.

Huddleston, M. J., Bean, M. F., and Carr, S. A. (1993). Collisional fragmentation of glycopeptides by electrospray ionization LC/MS and LC/MS/MS: Methods for selective detection of glycopeptides in protein digests. *Anal. Chem.* **65**, 877–884.

Hunt, D. F., Yates, J. R., III, Shabanowitz, J., Winston, S., and Hauer, C. R. (1986). Protein sequencing by tandem mass spectrometry. *Proc. Natl. Acad. Sci. USA* **83**, 6233–6237.

Hunt, D. F., Zhu, N. Z., and Shabanowitz, J. (1989). Oligopeptide sequence analysis by collision-activated dissociation of multiply charged ions. *Rapid Commun. Mass Spectrom.* **3**, 122–124.

Hunyadi-Gulyas, E., and Medzihradszky, K. F. (2004). Factors that contribute to the complexity of protein digests. *Drug Discovery Today: Targets* **3**, S3–S10.

Igarashi, Y., Kan, Y., Fujii, K., Fujita, T., Harada, K., Naoki, H., Tabata, H., Onaka, H., and Furumai, T. (2001). Goadsporin, a chemical substance which promotes secondary metabolism and morphogenesis in streptomycetes. II. Structure determination. *J. Antibiot. (Tokyo)* **54**, 1045–1053.

Ishikawa, K., Niwa, Y., Oishi, K., Aoi, S., Takeuchi, T., and Wakayama, S. (1990). Sequence determination of unknown cyclic peptide antibiotics by fast atom bombardment mass spectrometry. *Biomed. Environ. Mass Spectrom.* **19**, 395–399.

Jaffe, H., Veeranna, H., and Pant, C. (1998). Characterization of serine and threonine phosphorylation sites in beta-elimination/ethanethiol addition-modified proteins by electrospray tandem mass spectrometry and database searching. *Biochemistry* **37**, 16111–16124.

Johnson, R. S., Martin, S. A., and Biemann, K. (1988). Collision-induced fragmentation of $(M + H)^+$ ions of peptides. Side chain specific sequence ions. *Int. J. Mass Spectrom. Ion Processes* **86**, 137–154.

Johnson, R. S., Martin, S. A., Biemann, K., Stults, J. T., and Watson, J. T. (1987). Novel fragmentation process of peptides by collision-induced decomposition in a tandem mass spectrometer: Differentiation of leucine and isoleucine. *Anal. Chem.* **59,** 2621–2625.

Jonscher, K. R., and Yates, J. R., III. (1997). The quadrupole ion trap mass spectrometer—a small solution to a big challenge. *Anal. Biochem.* **244,** 1–15.

Kalume, D. E., Molina, H., and Pandey, A. (2003). Tackling the phosphoproteome: Tools and strategies. *Curr. Opin. Chem. Biol.* **7,** 64–69.

Kaufmann, R., Kirsch, D., and Spengler, B. (1994). Sequencing of peptides in a time-of-flight mass spectrometer: Evaluation of postsource decay following matrix-assisted laser desorption ionisation (MALDI). *Int. J. Mass Spectrom. Ion Processes* **131,** 355–385.

Kelleher, R. L., Zubarev, R. A., Bush, K., Furie, B., Furie, B. C., McLafferty, F. W., and Walsh, C. T. (1999). Localization of labile posttranslational modifications by electron capture dissociation: The case of gamma-carboxyglutamic acid. *Anal. Chem.* **71,** 4250–4253.

Keough, T., Youngquist, R. S., and Lacey, M. P. (1999). A method for high-sensitivity peptide sequencing using postsource decay matrix-assisted laser desorption ionization mass spectrometry. *Proc. Natl. Acad. Sci. USA* **96,** 7131–7136.

Kjeldsen, F., Haselmann, K. F., Budnik, B. A., Jensen, F., and Zubarev, R. A. (2002). Dissociative capture of hot (3–13 eV) electrons by polypeptide polycations: An efficient process accompanied by secondary fragmentation. *Chem. Phys. Lett.* **356,** 201–206.

Kjeldsen, F., Haselmann, K. F., Sorensen, E. S., and Zubarev, R. A. (2003). Distinguishing of Ile/Leu amino acid residues in the PP3 protein by (hot) electron capture dissociation in Fourier transform ion cyclotron resonance mass spectrometry. *Anal. Chem.* **75,** 1267–1274.

Kjeldsen, F., and Zubarev, R. A. (2003). Secondary losses via gamma-lactam formation in hot electron capture dissociation: A missing link to complete de novo sequencing of proteins? *J. Am. Chem. Soc.* **125,** 6628–6629.

Lee, Y. J., and Lee, Y. M. (2004). Formation of c1 fragment ions in collision-induced dissociation of glutamine-containing peptide ions: A tip for *de novo* sequencing. *Rapid. Commun. Mass Spectrom.* **18,** 2069–2076.

Leymarie, N., Costello, C. E., and O'Connor, P. B. (2003). Electron capture dissociation initiates a free radical reaction cascade. *J. Am. Chem. Soc.* **125,** 8949–8958.

Little, D. P., Speir, J. P., Senko, M. W., O'Connor, P. B., and McLafferty, F. W. (1994). Infrared multiphoton dissociation of large multiply charged ions for biomolecule sequencing. *Anal. Chem.* **66,** 2809–2815.

Mann, M., Ong, S. E., Gronborg, M., Steen, H., Jensen, O. N., and Pandey, A. (2002). Analysis of protein phosphorylation using mass spectrometry: Deciphering the phosphoproteome. *Trends Biotechnol.* **20,** 261–268.

McLachlin, D., and Chait, B. T. (2001). Analysis of phosphorylated proteins and peptides by mass spectrometry. *Curr. Opin. Chem. Biol.* **5,** 591–602.

Medzihradszky, K. F., Bateman, R. H., Green, M. R., Adams, G. W., and Burlingame, A. L. (1996a). Peptide sequence determination by matrix-assisted laser desorption ionization using a tandem double focusing magnetic/orthogonal acceleration time-of-flight mass spectrometer. *J. Am. Soc. Mass Spectrom.* **7,** 1–10.

Medzihradszky, K. F., and Burlingame, A. L. (1994). The advantages and versatility of a high energy collision-induced dissociation based strategy for the sequence and structural determination of proteins. *Methods Compan. Methods Enzymol.* **6,** 284–303.

Medzihradszky, K. F., Gillece-Castro, B. L., Hardy, M. R., Townsend, R. R., and Burlingame, A. L. (1996b). Structural elucidation of O-linked glycopeptides by high energy collision-induced dissociation. *J. Am. Soc. Mass Spectrom.* **7,** 319–328.

Medzihradszky, K. F., Phillipps, N. J., Senderowicz, L., Wang, P., and Turck, C. W. (1997). Synthesis and characterization of histidine-phosphorylated peptides. *Protein Sci.* **6,** 1405–1411.

Medzihradszky, K. F., Campbell, J. M., Baldwin, M. A., Falick, A. M., Juhasz, P., Vestal, M. A., and Burlingame, A. L. (2000). The characteristics of peptide collision-induced dissociation using a high performance MALDI-TOF/TOF tandem mass spectrometer. *Anal. Chem.* **72,** 552–558.

Medzihradszky, K. F., Ambulos, N. P., Khatri, A., Osapay, G., Remmer, H. A., Somogyi, A., and Kates, S. A. (2002). Mass spectrometry analysis for the determination of side reactions for cyclic peptides prepared from an Fmoc/tBu/DMab Protecting Group Strategy. *Lett. Pep. Sci.* **8,** 1–12.

Medzihradszky, K. F., Darula, Z., Perlson, E., Fainzilber, M., Chalkley, R. J., Ball, H., Greenbaum, D., Bogyo, M., Tyson, D. R., Bradshaw, R. A., and Burlingame, A. L. (2004). O-sulfonation of serine and threonine: Mass spectrometric detection and characterization of a new posttranslational modification in diverse proteins throughout the eukaryotes. *Mol. Cell. Proteomics* **3,** 429–440.

Nielsen, M. L., Bennett, K. L., Larsen, B., Moniatte, M., and Mann, M. (2002). Peptide end sequencing by orthogonal MALDI tandem mass spectrometry. *J. Proteome Res.* **1,** 63–71.

Ong, S. E., Foster, L. J., and Mann, M. (2003). Mass spectrometric–based approaches in quantitative proteomics. *Methods* **29,** 124–130.

Paizs, B., and Suhai, S. (2004). Fragmentation pathways of protonated peptides. *Mass Spectrom. Rev.* **24,** 508–548.

Qian, M. G., Zhang, Y., and Lubman, D. M. (1995). Collision-induced dissociation of multiply charged peptides in an ion-trap storage/reflectron time-of-flight mass spectrometer. *Rapid. Commun. Mass Spectrom.* **9,** 1275–1282.

Roberts, J. K. (2002). Proteomics and a future generation of plant molecular biologists. *Plant Mol. Biol.* **48,** 143–154.

Roth, K. D. W., Huang, Z. H., Sadagopan, N., and Watson, J. T. (1998). Charge derivatization of peptides for analysis by mass spectrometry. *Mass Spectrom. Rev.* **17,** 255–274.

Senko, M. W., Speir, J. P., and McLafferty, F. W. (1994). Collisional activation of large multiply charged ions using Fourier transform mass spectrometry. *Anal. Chem.* **66,** 2801–2808.

Siegel, M. M., Huang, J., Lin, B., Tsao, R., and Edmonds, C. G. (1994). Structures of bacitracin A and isolated congeners: sequencing of cyclic peptides with blocked linear side chains by electrospray ionization mass spectrometry. *Biol. Mass Spectrom.* **23,** 186–204.

Spengler, B., Lutzenkirchen, F., Metzger, S., Chaurand, P., Kaufmann, R., Jeffery, W., Bartlet-Jones, M., and Pappin, D. J. C. (1997). Peptide sequencing of charged derivatives by postsource decay MALDI mass spectrometry. *Int. J. Mass Spectrom. Ion Processes* **169/170,** 127–140.

Steen, H., Küster, B., Fernandez, M., Pandey, A., and Mann, M. (2001). Detection of tyrosine phosphorylated peptides by precursor ion scanning quadrupole TOF mass spectrometry in positive ion mode. *Anal. Chem.* **73,** 1440–1448.

Stimson, E., Truong, O., Richter, W. J., Waterfield, M. D., and Burlingame, A. L. (1997). Enhancement of charge remote fragmentation in protonated peptides by high-energy CID MALDI-TOF-MS using "cold" matrices. *Int. J. Mass Spectrom. Ion Processes* **169/170,** 231–240.

Stults, J. T., Lai, J., McCune, S., and Wetzel, R. (1993). Simplification of high-energy collision spectra of peptides by amino-terminal derivatization. *Anal. Chem.* **65,** 1703–1708.

Swiderek, K. M., Davis, M. T., and Lee, T. D. (1998). The identification of peptide modifications derived from gel-separated proteins using electrospray triple quadrupole and ion trap analyses. *Electrophoresis* **19,** 989–997.

Syka, J. E., Coon, J. J., Schroeder, M. J., Shabanowitz, J., and Hunt, D. F. (2004). Peptide and protein sequence analysis by electron transfer dissociation mass spectrometry. *Proc. Natl. Acad. Sci. USA* **101,** 9528–9533.

Thorne, G. C., and Gaskell, S. J. (1989). Elucidation of some fragmentations of small peptides using sequential mass spectrometry on a hybrid instrument. *Rapid Commun. Mass Spectrom.* **3,** 217–221.

van Dongen, W. D., Ruijters, H. F., Lunge, H. J., Heerma, W., and Haverkamp, J. (1996). Statistical analysis of mass spectral data obtained from singly protonated peptides under high-energy collision-induced dissociation conditions. *J. Mass Spectrom.* **31,** 1156–1162.

Wagner, D. S., Salari, A., Gage, D. A., Leykam, J., Fetter, J., Hollingworth, R., and Watson, J. T. (1991). Derivatization of peptides to enhance ionization efficiency and control fragmentation during analysis by fast atom bombardment tandem mass spectrometry. *Biol. Mass Spectrom.* **20,** 419–425.

Wells, L., Vosseller, K., Cole, R. N., Cronshaw, J. M., Matunis, M. J., and Hart, G. W. (2002). Mapping sites of O-GlcNAc modification using affinity tags for serine and threonine post-translational modifications. *Mol. Cell. Proteomics* **1,** 791–804.

Yalcin, T., Khouw, C., Csizmadia, I. G., Peterson, M. R., and Harrison, A. G. (1995). Why are B ions stable species in peptide spectra? *J. Am. Soc. Mass Spectrom.* **6,** 1165–1174.

Yu, W., Vath, J. E., Huberty, M. C., and Martin, S. A. (1993). Identification of the facile gas-phase cleavage of the Asp-Pro and Asp-Xxx peptide bonds in matrix-assisted laser desorption time-of-flight mass spectrometry. *Anal. Chem.* **65,** 3015–3023.

Yuan, M., Namikoshi, M., Otsuki, A., Rinehart, K. L., Sivonen, K., and Watanabe, M. F. (1999). Low-energy collisionally activated decomposition and structural characterization of cyclic heptapeptide microcystins by electrospray ionization mass spectrometry. *J. Mass Spectrom.* **34,** 33–43.

Zhang, K., Yau, P. M., Chandrasekhar, B., New, R., Kondrat, R., Imai, B. S., and Bradbury, M. E. (2004). Differentiation between peptides containing acetylated or tri-methylated lysines by mass spectrometry: An application for determining lysine 9 acetylation and methylation of histone H3. *Proteomics* **4,** 1–10.

Zubarev, R. A., Horn, D. M., Fridriksson, E. K., Kelleher, N. L., Kruger, N. A., Lewis, M. A., Carpenter, B. K., and McLafferty, F. W. (2000). Electron capture dissociation for structural characterization of multiply charged protein cations. *Anal. Chem.* **72,** 563–573.

[8]   Proteomics

By JOHN T. STULTS and DAVID ARNOTT

Abstract

Proteomics is the measurement of one or more protein populations or proteomes, preferably in a quantitative manner. A protein population may be the set of proteins found in an organism, in a tissue or biofluid, in a cell, or in a subcellular compartment. A population also may be the set of proteins with a common characteristic, for example, those that interact with each other in molecular complexes, those involved in the same process such as signal transduction or cell cycle control, or those that share a common posttranslational modification such as phosphorylation or glycosylation. Proteomics experiments that involve mass spectrometry are divided into five categories: (1) protein identification, (2) protein quantitation or differential analysis, (3) protein-protein interactions, (4) posttranslational modifications, and (5) structural proteomics. Each of these proteomics categories is reviewed. Examples are given for quantitative experiments involving two-dimensional gel electrophoresis, and for gel-free analysis using isotope-coded affinity tags. The impact of proteomics on biological research and on drug development is discussed. Challenges for further development in proteomics are presented, including sample preparation, sensitivity, dynamic range, and automation.

The determination of the complete genome sequences of a large number of organisms, most notably the human (Lander *et al.*, 2001; Venter *et al.*, 2001), is causing a genuine paradigm shift in biological research. The interest in genomics is stimulating a growing interest in examination of the entire set of proteins coded by a genome, the proteome. In fact, due to the surprisingly small number of genes (~25,000) now tentatively identified in the human genome, much of the responsibility for the complexity of the human organism must lie at the protein level (Fig. 1). Thus, proteome analysis (proteomics) is taking center stage in the effort to understand the function of each gene (functional genomics). Fortuitously, the necessary tools for analyzing proteins, notably mass spectrometry (MS), have been developed and enhanced to make possible the analysis of proteins on a large scale.

Proteomics has been defined in a variety of ways. It encompasses the measurements of proteins, generally in a multiplexed, large-scale, or global manner, and usually quantitatively. It is classified as research that is

METHODS IN ENZYMOLOGY, VOL. 402
0076-6879/05 $35.00
DOI: 10.1016/S0076-6879(05)02008-2

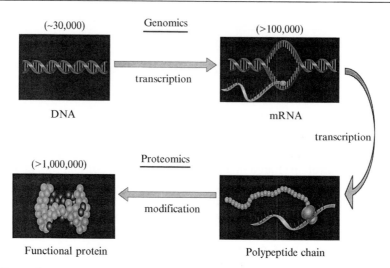

(~30,000)   Genomics   (>100,000)

transcription

DNA   mRNA

transcription

(>1,000,000)   Proteomics

modification

Functional protein   Polypeptide chain

FIG. 1. Genomics methods determine DNA sequences and the sequences and abundances of messenger RNA. Proteomics methods determine protein abundances, modifications, localization, and interactions. Analysis of the human genome sequence indicates that only about 30,000 genes exist. Alternative splicing, RNA editing, and other processes yield more than 100,000 transcripts, which are translated to protein sequences. Protein processing and modifications easily produce more than 1,000,000 different protein molecular species. The proteins are the chief functional molecules in the cell. The challenge for proteomics is to develop techniques to measure all these different proteins with high sensitivity, high dynamic range, and high throughput.

"discovery driven," rather than "hypothesis driven." That is, it seeks to produce large amounts of data that may be queried for a current or future question or to find patterns from which information may be inferred, rather than to prove or disprove a single narrowly defined hypothesis. The term *proteome* was coined by Marc Wilkins and first appeared in print in 1995 (Wilkins *et al.*, 1995). Many of the tools that have become associated with proteomics, however, date back further. Two-dimensional (2D) gel electrophoresis, the standard high-resolution method for protein separation, was developed independently in 1975 by O'Farrell (1975), Klose (1975), and Scheele (1975). Matrix-assisted laser desorption ionization (MALDI) MS was developed by Karas and Hillenkamp in the mid-1980s (Karas and Hillenkamp, 1988). Electrospray ionization (ESI) MS was developed by Fenn *et al.* (1989) at the same time. The identification of proteins by sequence database searches with peptide mass mapping was first demonstrated by Henzel *et al.* in 1989 and was first published by a number of groups in 1993 (Henzel *et al.*, 1989; James *et al.*, 1993; Mann *et al.*, 1993; Pappin *et al.*, 1993; Yates *et al.*, 1993).

The term *proteomics* has been extended in some contexts to encompass many (or all) facets of protein analysis. A number of monographs and reviews of proteomics have been written, which describe many aspects of the subject in greater detail than possible here (Aebersold and Goodlett, 2001; Aebersold and Mann, 2003; Godovac-Zimmermann and Brown, 2001; James, 2001b; Link, 1999; Pandey and Mann, 2000; Wilkins *et al.*, 1997).

Five categories of experiments may be considered to be proteomics that involve MS:

1. *Protein identification.* Sensitive and often high-throughput methods for protein identification from a sequence database are employed. Because of the large amount of sequence data available—complete genomes in some cases—proteins can be identified based on a database search instead of more traditional sequence determination. A match of tryptic (or other protease) peptide masses with those predicted from theoretical digestion of each protein in the database (peptide mass mapping) serves to identify the protein (Fig. 2). When a more complex protein mixture is analyzed or peptide mass mapping does not provide a conclusive match, tandem mass spectra for the peptides in the mixture can be used to search the database. With sufficient constraints (e.g., species or MW), a protein may be identified from a single peptide. Often the proteins are spots or bands cut from electrophoresis gels (see the section "General Methodology," later in this chapter). The use of protein identification, when combined with appropriate separations that pinpoint the subcellular location of protein, is termed *cell map proteomics* (Blackstock and Weir, 1999). Examples of subcellular or organelle proteomes are beginning to appear (Andersen *et al.*, 2002; Bell *et al.*, 2001a; Brunet *et al.*, 2003; Fox *et al.*, 2002; Garin *et al.*, 2001; Jung *et al.*, 2000; Molloy *et al.*, 2000), and the approach is gaining importance because no other method can provide protein localization with the potential for high-throughput measurements. Increasingly, proteins are identified via "shotgun proteomics" with LC-MS/MS of proteolytic digests of protein mixtures (Wolters *et al.*, 2001; Zhu *et al.*, 2004).

2. *Protein quantitation or differential analysis.* Quantitation of protein abundance is used to compare the amounts between two or more samples. These experiments seek to determine differences that may correlate with developmental state, drug treatment, disease state, or cell cycle state (Arnott *et al.*, 1998b; Epstein *et al.*, 1996; Jungblut *et al.*, 1999). The samples may be biological fluids (e.g., plasma, serum, urine), tissue, or cells grown in culture. Protein separation is commonly performed by 2D gel electrophoresis, with the intensity of an appropriate stain used for quantitation (Fig. 3). These results may be especially useful when

A

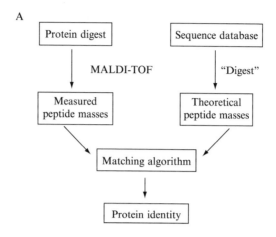

B

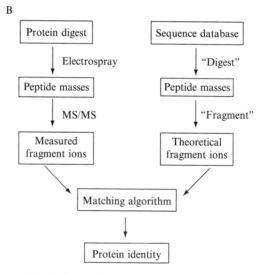

Fig. 2. Proteins are identified matching mass spectral data with protein or DNA sequence databases. (A) The masses of peptides that are generated from proteolytic digestion of a pure protein comprise a "fingerprint" of the protein. Peptide mass mapping compares measured peptide masses (from matrix-assisted laser desorption ionization [MALDI] or electrospray spectra) with masses generated by theoretical digestion of a sequence database. (B) The fragment ions of a proteolytic peptide are also a unique representation of the sequence. The fragmentation spectrum generated by tandem mass spectrometry (MS/MS) for any single peptide is sufficient to identify most proteins. Multiple peptides from the same protein provide multiple independent identifications of the protein. Because each peptide serves to identify its precursor protein, the digest of a protein mixture yields protein identifications for the mixture components.

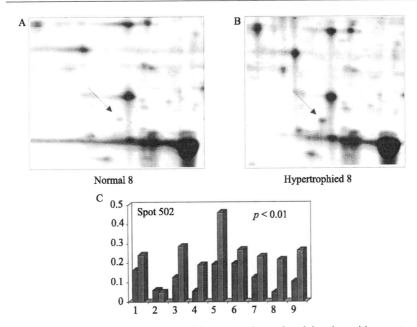

FIG. 3. Proteins may be quantitated by comparison of staining intensities on two-dimensional (2D) gels. In a comparison of normal and hypertrophied cardiomyocytes, protein spots are compared between gels. The arrow points to the same spot for a pair of gels. The integrated intensities of the spot densities are plotted in (C) for nine replicates of the experiment. The gels (A) and (B) correspond to experiment 8. Statistical analysis of the nine pairwise comparisons shows an average increase in this spot of 106%, with $p < .01$.

combined with the messenger RNA (mRNA) levels determined by microarrays (Gerhold *et al.*, 1999), serial analysis of gene expression (SAGE) (Velculescu *et al.*, 2000) or quantitative polymerase chain reaction (PCR). This area of proteomics has also been termed *expression proteomics* (James, 2001a). This area includes most of the efforts to identify biomarkers of disease (Pan *et al.*, 2005; Srinivas *et al.*, 2001; Veenstra *et al.*, 2005).

3. *Protein–protein interactions.* A protein of interest is isolated under conditions that preserve the non-covalent interactions between it and other proteins. Typically, the protein of interest (the bait protein) is expressed with an affinity tag, or an antibody that binds the wild-type protein is available. An affinity purification, often co-immunoprecipitation, is performed to isolate the protein complex. A tandem affinity tag method has been developed to produce high purity complexes (Rigaut *et al.*, 1999). The complex is analyzed directly as a mixture by LC-MS/MS or separated by one-dimensional (1D) or 2D electrophoresis before protein

identification. Similar methods may be applied to other interactions such as protein–DNA complexes (Nordhoff et al., 1999). Specific examples of this methodology include the identification of signaling pathway interactions (Husi et al., 2000; Muzio et al., 1996; Pandey et al., 2000a,b) and of components of the human spliceosome (Neubauer et al., 1998), the yeast ribosomal complex (Link et al., 1999), and the yeast nuclear pore complex (Rout MP et al., 2000). An alternative approach is the use of surface plasmon resonance measurements coupled with MS to detect protein-binding partners (Nelson et al., 2000b). More details on the methodology for protein–protein interactions are given elsewhere in this volume.

Because the interactions of molecules, intercellularly and intracellularly, form the basis of all cellular processes (e.g., signal transduction [Pawson and Nash, 2000]), the protein–protein interaction analysis aspect of proteomics may ultimately prove to have the greatest impact on basic biology. Of particular note, posttranslation modifications of proteins often control their ability to interact, and molecular genetic–based approaches to determine these interactions (the yeast two-hybrid system [Chien et al., 1991]) are not able to reveal such interactions, so protein-based methods must be used to find these interactions.

4. *Post-translational modifications.* The type and site(s) of modification are identified for a desired protein. Phosphorylation site determinations are the highest profile experiments because of the importance of phosphorylation in cellular regulatory mechanisms (Hunter, 2000). Phosphorylation site mapping can be performed on single proteins, sets of proteins such as those involved in a particular biochemical pathway (Blagoev et al., 2004), or all phosphoproteins in a cell (Ballif et al., 2004; Beausoleil et al., 2004; Ficarro et al., 2002). Methods generally rely on the detection of phospho-specific fragment ions (Annan et al., 2001), the isolation of phosphopeptides by immobilized metal ion affinity chromatography (IMAC) (Cao and Stults, 1999; Nuwaysir and Stults, 1993; Posewitz and Tempst, 1999; Zhou et al., 2000), cation exchange chromatography (Beausoleil et al., 2004), or the detection of 80-Da mass differences in spectra of phosphopeptides before and after alkaline phosphatase treatment (Larsen et al., 2001; Zhang et al., 1998). Determination of the specific amino acid(s) modified requires tandem MS of each phosphopeptide. Phosphorylation determination can be particularly challenging because of the low stoichiometry of modification at many sites and the poor ionization of some phosphopeptides in mixtures, presumably due to their acidic character. Ideally these methods can be applied at the proteome scale. However, high sequence coverage is necessary, so recovery and detection of most or all peptides from a protein digest is therefore mandatory. As a result, these analyses may require as

much as two or three orders of magnitude more material than simple protein identification; thus, modification studies continue to be performed mainly for proteins that can be isolated at high purity at picomole levels and usually on a single protein basis.

Several groups have developed methods that hold promise for proteomic-scale measurements. Han *et al.* (2001) chemically derivatize the phosphate groups to provide a site for covalent attachment of the phosphopeptide to a resin, from which the phosphopeptide is regenerated and analyzed. Oda *et al.* (2001) use conventional beta-elimination of pSer and pThr, followed by addition of a biotin label that is used for selective isolation, with subsequent mass analysis. Metal affinity chromatography, previously used to enrich for phosphopeptides in the context of single protein characterization, has been applied to proteome-wide phosphorylation site discovery. Ficcaro *et al.* (2002a) detected more than 1000 phosphopeptides and mapped 383 phosphorylation sites in yeast lysates; methyl esterification of carboxylic acids was performed to reduce nonspecific binding of acidic peptides. Each of these methods require multiple chemical reactions or sample preparation stages, which have proven challenging for low-level analyses in general. Yet, these or similar approaches will be necessary to extend phosphorylation measurements to higher sensitivity and higher throughput.

Glycosylation site and glycan structure determinations are also of great interest. The analysis is a two-step process: the identification of the glycosylated amino acid(s) and the analysis of the glycan structures at each site. Site identification for Asn-linked carbohydrates generally relies on enzymatic removal of the sugar, which leaves an aspartic acid residue as a signature for the modification site. Methods for *O*-linked carbohydrates normally require chemical release of the carbohydrate, and methods for beta-elimination followed by alkylation appear to be promising (Hanisch *et al.*, 2001; Mirgorodskaya *et al.*, 2001) for identification of the glycosylation sites. Readers are referred elsewhere for discussion of the analysis of the released oligosaccharides (Rudd *et al.*, 2001).

Analyses for the myriad other posttranslational modifications have not, in general, been developed to a state in which they can be used at the low levels or the high throughput necessary for proteomics, although progress has been made, for example, in the proteome-wide identification of sites of ubiquitination, an important event in diverse cellular processes (Peng *et al.*, 2003). For all analyses of posttranslational modifications, greater recovery of peptides at low levels is a major need and clearly a challenge for future development. The topic of posttranslational modification analysis are covered in more detail elsewhere in this volume.

5. *Structural proteomics.* Tertiary and quaternary structural (three-dimensional [3D] structure) motifs are identified for a protein, with the goal to elicit protein function and identify protein–protein interactions (Marcotte *et al.*, 1999). Although most of this field is dominated by nuclear magnetic resonance (NMR), x-ray crystallography, and structure prediction informatics, MS is playing an increasing role. Hydrogen–deuterium (H/D) exchange experiments (Engen and Smith, 2001; Miranker *et al.*, 1996; Zhang *et al.*, 1997), schemes for cross-linking (Rappsilber *et al.*, 2000; Young *et al.*, 2000), and mass analysis of intact non-covalent complexes (Sobott and Robinson, 2002) are becoming increasingly sophisticated for determining surface versus buried amino acids, for studying folding and structural dynamics, and for defining spatial relationships among protein domains. These topics are covered in considerable detail elsewhere in this volume and are not discussed here further.

Proteomics vs. Genomics

Complementary DNA (cDNA) microarrays are routinely used to measure the mRNA (also known as *message* or *transcript*) levels for thousands of genes simultaneously (Gerhold *et al.*, 1999; Shalon *et al.*, 1996). The high-throughput capability, the high sensitivity, and the ability to amplify low abundance transcripts by PCR make this method extremely valuable. Microarray results are supplemented with quantitative PCR (Heid *et al.*, 1996) for more sensitive, more precise values, when needed. The use of this technology is predicated on the belief that functional protein level is correlated with the transcript level (i.e., the protein level is regulated by transcription). Although this relationship is clearly true in many cases (Celis *et al.*, 2000), there is growing evidence (Anderson and Seilhamer, 1997; Gygi *et al.*, 1999b) that transcript levels frequently are not accurate predictors of protein levels. Protein synthesis may be regulated at the translation level, or more often, protein levels may be regulated by accumulation, translocation, and degradation processes. Furthermore, the biologically active protein is frequently produced by posttranslational modification or proteolytic processing, which may be regulated by yet other processes. Nonetheless, protein measurements have lacked the throughput and sensitivity to make proteomics a viable alternative to genomic-based measurements, even if protein measurements are the preferred source of data. Thus, it is of enormous importance that the tools for proteomics be improved to make them as reliable, fast, and sensitive as microarrays.

Why are protein measurements so much more difficult than RNA or DNA measurements? Proteins are composed of a larger set of subunits,

21 naturally occurring amino acids, which if modified, can yield a nearly unlimited variety of molecules. Estimates of the number of different protein molecules in a mammalian organism range from 1 to 20 million. This variety in primary structure gives rise to an enormous span in molecular properties, including solubility, isoelectric point, and molecular mass. The extremes in molecular properties make it nearly impossible to devise sample handling protocols and analytical strategies that are universally applicable. Modifications that affect the activity of a protein may be as subtle as the change of one or a few atoms in a molecule of more than 100,000 Da in mass. The amount of a protein in a cell may vary from a few copies per cell to several million, putting huge demands on the sensitivity and dynamic range of a measurement. Furthermore, the amount of material (e.g., diseased tissue) may be very limited in amount. Alas, there is no known amplification methodology for proteins that is analogous to PCR.

General Methodology

Sample Preparation

Regardless of the type of proteomic goal, a common set of methods is used when gel electrophoresis is used for protein separation. These are outlined in Fig. 4. Sample acquisition should be controlled as closely as possible. Tissue samples, for example, should be well matched. Differences between male and female, for example, are known to be clearly observed at the proteome level (Steiner *et al.*, 1995). If cells grown in culture are the source of the proteins, highly consistent culture conditions are required, lest artifacts such as heat shock proteins appear. Sample preparation is the most critical part of any experiment. This step may involve tissue homogenization and cell lysis. Conditions that maintain solubility of the sample components yet are compatible with the downstream methods are frequently the most difficult to achieve. The inhibition of proteases and phosphatases is crucial. Removal of artifactual or high levels of unwanted protein (e.g., BSA from cell culture media) is necessary. Additional sample preparation steps are discussed later in this chapter.

Gel Electrophoresis, Staining, Imaging, and Protein Digestion

2D electrophoresis is the high-resolution separation technique of choice for complex protein mixtures, and it is the *de facto* standard for proteomics. For less complex mixtures (e.g., from immunoprecipitation), 1D sodium dodecylsulfate (SDS)–polyacrylamide gel electrophoresis (PAGE) is frequently the method of choice because it is operationally much simpler, and

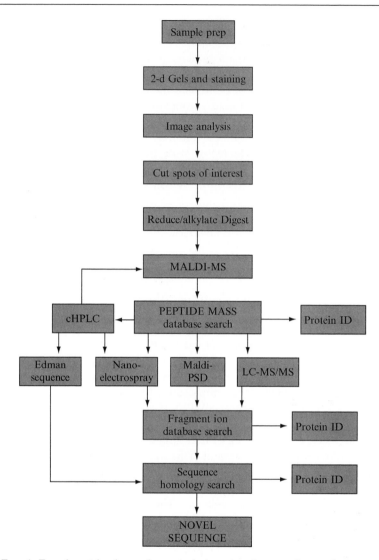

Fɪɢ. 4. Experimental scheme for a typical proteomics experiment that uses two-dimensional (2D) gel electrophoresis, with emphasis on the steps involved in protein identification. Matrix-assisted laser desorption ionization (MALDI) time-of-flight (TOF) mass spectrometry (MS) analysis is typically performed first, because of its high sensitivity and high throughput capabilities. If an unambiguous protein identification is not made with peptide mass fingerprinting, tandem MS is used to general fragment ion spectra for additional database searches.

the use of SDS in the sample buffer generally promotes greater solubility. The details of each of these separation methods is beyond the scope of this chapter, and readers are referred elsewhere (Link *et al.*, 1999). It is worth noting that for SDS-PAGE, reduction and alkylation of the proteins in the sample buffer (see Protocols at the end of the chapter) simplifies the in-gel digestion process and reduces the number of gel washing steps, which may lead to sample loss. However, reduction and alkylation of the sample before electrophoresis are usually not recommended if the sample contains an antibody, such as from an immunoprecipitation. The intact antibody appears with a molecular mass of approximately 200 kD, out of the range of most proteins, whereas the reduced heavy and light chains run at 50 and 25 kD, respectively, and their intense bands are more likely to interfere with the protein(s) of interest.

Once the separation has been completed, the gel is normally stained to locate the proteins and to provide quantitation (proportional to the staining intensity). Common stains include Coomassie blue, silver stain, reversed (imidazole) stain, and fluorescent stain. All are available in kits from a number of vendors. The authors prefer colloidal Coomassie stain and silver stain, because of their ease of use and ease of visualization. Note that the silver stain must omit glutaraldehyde, which crosslinks the proteins and prohibits subsequent proteolytic digestion. In addition, peptide signals derived from digestion of silver-stained gels are enhanced by removal of the silver ions before digestion (Gharahdaghi *et al.*, 1999). Protocols for silver staining and in-gel digestion are given at the end of the chapter. Fluorescent stains, particularly SYPRO Ruby, are becoming popular because of their high sensitivity, large dynamic range, and compatibility with proteolytic digestion (Lopez *et al.*, 2000). An alternative approach is to electroblot the protein to a polyvinylidene difluoride (PVDF) membrane with subsequent staining and digestion. This approach also permits automated Edman degradation microsequencing of the sample if the N-terminal is not blocked. In general, the blotting efficiency of proteins may be highly variable (10–90%) and unpredictable, especially for large proteins. Furthermore, a comparison of on-membrane versus in-gel digestions for sub-picomole standards has shown in-gel digestions usually to be superior for protein identification (Courchesne *et al.*, 1997).

Following staining, the gel is imaged either for documentation purposes or for use in quantitative comparison. Imaging may be done with sophisticated CCD camera or flatbed scanner systems or with inexpensive document scanners. For all but the most precise quantitation experiments, inexpensive color document scanners are usually sufficient, especially if gel documentation is the chief goal. 2D gels may be compared for spot locations and spot intensities, using software from a number of vendors.

These gel-matching programs are important because although 2D gels have become highly reproducible, they still cannot be superimposed. Despite the advances in software, the spot location and spot quantitation capabilities of available programs still necessitate subsequent manual editing of the results, which can add hours to the analysis of *each* gel.

The identity of the spot or band in a gel is determined by proteolytic digestion of the gel piece, with subsequent peptide mass mapping or tandem MS of the digest. A protocol for in-gel digestion is given at the end of this chapter. The authors prefer trypsin that is modified to prevent autodigestion of the enzyme (e.g., Promega modified trypsin). Contamination of the sample by keratin is a problem encountered in nearly every laboratory when working at the sub-picomole level. Keratin is a chief component of skin, and it is ubiquitous in dust, dandruff, skin, and materials such as wool. Steps should be taken to reduce possible sources of contamination by constant use of gloves (latex or nitrile), use of reagents of the highest purity possible, keeping all reagents and gels covered from exposure to dust, and working in a HEPA-filtered hood or ideally in a clean-room environment.

Mass Measurement

Peptide mass mapping by MALDI–time-of-flight (TOF) MS of the protein digest is usually the first mass measurement attempted (see Fig. 2). MALDI-TOF has the advantage of being exquisitely sensitive, easily automated for high-throughput measurements, and capable of excellent mass accuracy and precision. We find that a 0.2–0.5 μl aliquot of the proteolytic digest is sufficient to provide an excellent MALDI spectrum. The sample may be prepared either with the dried droplet approach, using either α-cyano-4-hydroxycinnamic acid (HCCA) or trihydroxyacetophenone (THAP), or by the fast-evaporation method (Shevchenko *et al.*, 1996). The authors prefer the fast-evaporation method or the dried droplet approach using THAP. The signal may be enhanced for the dried droplet method by desalting with a reversed-phase pipette tip (Erdjument-Bromage *et al.*, 1998; Gobom *et al.*, 1999). Internal peptide standards may be added (e.g., desArg-Bradykinin, m/z 904.4681 and ACTH-CLIP, m/z 2465.1989) may be added to achieve greater mass accuracy. In the measurement of a tryptic peptide digest, intense peaks from trypsin autolytic fragments may be used as internal standards, particularly m/z 842.5100 and 2211.1046.

The peptide masses are used to search a sequence database (see below). When the peptide mass mapping experiment does not yield an unequivocal match, fragment ion measurement by tandem MS (MS/MS) or post-source decay (PSD) is performed. Although PSD usually provides limited sequence information, it does frequently provide some ions that can be used to confirm the identity of a peptide (Gevaert and Vandekerckhove, 2000), even if the ions are only immonium ions (Fenyo *et al.*, 1998). A derivatization scheme developed by Keough *et al.* (2000) that adds a sulfonic acid group to the amino-terminus of peptides shows promise for increasing the fragmentation efficiency of peptides by PSD. The commercial availability of tandem mass spectrometers with MALDI sources, such as the Q-TOF (Shevchenko *et al.*, 2000) and TOF-TOF (Medzihradszky, 2000), also permits acquisition of more extensive fragmentation for one or more peptides to confirm the peptide mass map results.

Most often, peptide fragmentation spectra are obtained by ESI–tandem MS. Triple quadrupole, quadrupole ion trap, and quadrupole TOF are the typical mass analyzers. In one approach, the peptide mixture is desalted with a pipette-tip microcolumn and loaded into a nanospray needle (Wilm and Mann, 1996; Wilm *et al.*, 1996). Tandem spectra for each component of the mixture are acquired. The long signal that results from the low sample consumption ($\sim$1 μl/h) allows sufficient time for lengthy signal averaging to boost the signal-to-noise ratio for low-level samples. In a more frequently used method, the peptide digest is separated by reversed-phase high-performance liquid chromatography (HPLC) on-line with the mass spectrometer (Arnott *et al.*, 1998a,b; Gatlin *et al.*, 1998). The fragment ion spectra are acquired in a data-dependent manner (i.e., the presence of a peptide mass above a specified intensity threshold triggers the automated generation of the fragment ion spectrum for that precursor mass). A comparison of nanospray with LC-MS/MS showed that fragment ion spectra for more peptides in a complex mixture could be generated with greater abundance by use of LC-MS/MS rather than nanospray (Neugebauer *et al.*, 1999).

For complex peptide mixtures, LC-MS/MS has been enhanced in a number of ways. Low-level precursor masses are found in the presence of higher signals by an exclusion function (i.e., once the fragment ion spectrum for an abundant precursor is generated, that precursor is excluded from further analysis for a specified period to allow lower abundance precursors to be interrogated) (Davis *et al.*, 2001). A sample can be analyzed multiple times, each time with a different subset of the mass range for precursor ion selection, to increase the number of peptides found (Spahr *et al.*, 2001), a process termed "gas-phase fractionation." Peptide separation and identification can be enhanced further by increasing the ability of the mass spectrometer to perform MS/MS on every component. Normally the transient nature of components eluting from LC columns limits the number of co-eluting peptides that can be selected and fragmented by MS/MS. Davis and Lee (1997) showed that by temporarily stopping the flow of the HPLC column ("peak parking"), the components at the electrospray interface could be analyzed for a longer period. They found the

TABLE I

WEB ADDRESSES FOR SEQUENCE DATABASE SEARCHING

1. University of California at San Francisco mass spectrometry home page
   Searching with MS-FIT and MS-TAG for peptide masses and fragment ions
   prospector.ucsf.edu
2. Matrix Science home page
   Searching with MASCOT for peptide masses and fragment ions
   www.matrix-science.com
3. EMBL Protein Core Facility home page
   Searching with peptide masses and sequence tags
   www.narrador.embl-heidelberg.de/GroupPages/Homepage.html
4. Rockefeller University PROWL home page
   Searching with peptide masses and fragment ions
   prowl.rockefeller.edu
5. ExPASy web page
   Searching with peptide masses and amino acid composition
   www.expasy.ch/tools

eluting peptides lingered in the electrospray for several minutes, with little loss in signal, and they were able to increase the number of MS/MS spectra dramatically. This process produced little loss in chromatographic separation and provided LC-MS/MS with some of the advantages of nano-electrospray. Martin *et al.* (2000) developed a modified implementation of peak parking that uses a greatly reduced column flow rather than stopped flow.

Sequence Database Searching

The peptide masses are used to search a sequence database. A list of programs that can be used over the Internet is given in Table I. Other programs are available as part of commercial instrument software packages. In general, these programs use the protease specificity of the enzyme to predict the peptide masses for each protein in the database and find the protein(s) that match the observed masses, based on criteria such as peptide mass tolerance, protein molecular weight and pI, taxonomy, number of mismatches allowed. Many programs produce a probability score for the match, using a calculation of the probability that the match could be a random event based on the sequences in the database (Berndt *et al.*, 1999; Eriksson *et al.*, 2000; Perkins *et al.*, 1999; Zhang and Chait, 2000). The question of false-positive and false-negative matches is frequently raised. A relatively small completely sequenced genome, such as *Escherichia coli*, *Saccharomyces cerevisiae*, or *C. elegans*, increases the likelihood of a correct match. Beyond that, complete confidence in a match may necessitate the acquisition of additional sequence-based data by tandem MS.

Fragmentation spectra are matched to a sequence database by one of three types of programs. In one, the masses in the spectrum are matched to predicted fragment masses for each peptide of the same precursor mass from the theoretical digestion of each protein in the database (Clauser *et al.*, 1999; Creasy and Cottrell, 2002; Fenyo *et al.*, 1998). In a second, SEQUEST, a cross-correlation function is calculated between the measured fragment ion spectrum pattern and one generated for peptides from the sequence database (Eng *et al.*, 1994). In the third, partial interpretation of the spectrum provides a "sequence tag" that adds specificity and error tolerance (Mann and Wilm, 1994). When no match is found by these algorithms, the spectrum may need to be interpreted manually (*de novo* interpretation). A number of programs have been written for *de novo* sequence interpretation (Dancik *et al.*, 1999; Taylor and Johnson, 1997), yet most sequence interpretation continues to be done by hand.

Despite the advances in the database search programs and the probability-based scoring they provide, peptide identifications for large numbers of spectra continue to require significant amounts of manual spectrum confirmation or interpretation (Washburn *et al.*, 2001). Manual spectrum inspection is still necessary to (a) ensure the most accurate results, especially when protein identification is based on one or a few peptide spectra, (b) to account for poor signal-to-noise spectra that do not produce high-quality scores, (c) to account for peptides that produce unusual fragmentations, (d) to find peptide modifications that are not specified in the sequence database, (e) to find database errors or polymorphisms, or (f) to extract sequence when it is not found in the database. Some improvements have been made to address some of these issues. Further efforts to fully automate peptide identification and eliminate manual data checking are needed before methods can become truly high throughput.

Application: Drug-Induced Hypertrophy of Cardiomyocytes

Congestive heart failure (CHF) is the largest cause of hospitalization among patients 65 years and older. It is characterized by impaired ability of the heart to pump efficiently, and it is accompanied by the excessive, non-mitotic growth (hypertrophy) of cardiac muscle cells (myocytes). The molecular basis for CHF is not fully understood. One model system for the study of this disease is the hypertrophy of primary cardiomyocytes in culture induced by treatment with phenylephrine. A proteomics study of cardiomyocytes was undertaken to determine which proteins changed in expression level or modification in the hypertrophied cell (Arnott *et al.*, 1998b).

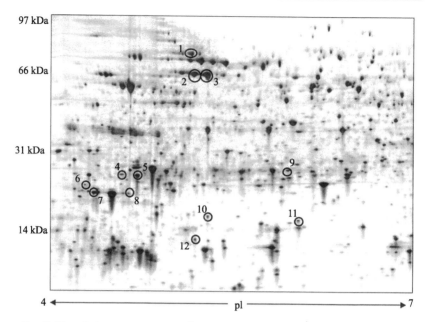

Fig. 5. The whole-cell lysate from 10^6 cultured primary rat neonatal cardiomyocytes was separated by two-dimensional gel electrophoresis and silver stained. Circled are protein spots that changed as a result of phenylephrine-induced hypertrophy, based on pairwise comparisons of nine samples (see Fig. 3). The protein identities, based on MALDI or LC-MS/MS data, are given in Table II (Arnott *et al.*, 1998b).

    Rat neonatal ventricular cardiac myocytes were grown in culture. Each culture was split, and 100 μM phenylephrine was added to the cell culture medium for half of the cells (~10^6 cells). After hypertrophy was maximal (48 h), the cells were lysed and the whole cell lysate of each sample was separated by 2D gel electrophoresis. The gels were silver stained and imaged. The image analysis software found approximately 1400 spots/gel. Pairwise comparison ($+/-$ drug for each set of cultured cells) of the protein spots for nine replicate experiments (see Fig. 3 as an example) was used to determine which spots changed in a statistically significant manner ($p < .05$). An image of one of the gels is shown in Fig. 5. Protein spots that changed are circled. The spots of interest were cut from the gel and digested in-gel with trypsin. A small aliquot of each digest was analyzed by MALDI-TOF with protein identification by our in-house database searching algorithm FRAGFIT. When no match was found, or to confirm the identity, the digests were analyzed by LC-MS/MS, with sequence matches found using the SEQUEST algorithm. The identities are given in Table II.

TABLE II
CARDIOMYOCYTE PROTEINS THAT CHANGE WITH HYPERTROPHY
(FOR CIRCLED PROTEIN SPOTS ON FIG. 5)

Spot	Protein identity	ΔExpression vs. control	Significance
1	NADH ubiquinone oxidoreductase 75 kDa subunit	−37%	$p < .01$
2	Mitochondrial matrix protein p1	−40%	$p < .01$
3	Mitochondrial matrix protein p1	−37%	$p < .01$
4	Myosin light chain 1–atrial isoform	+58%	$p = .01$
5	Myosin light chain 1–atrial isoform	+34%	$p < .01$
6	Myosin light chain 2–atrial isoform	+106%	$p < .01$
7	Myosin light chain 2–ventricular isoform	+32%	$p = .05$
8	Myosin light chain 2–ventricular isoform	+45%	$p = .02$
9	Heat shock protein 27 kDa	+66%	$p < .05$
10	XAG growth factor	+47%	$p < .01$
11	Nucleoside disphosphate kinase a	+44%	$p < .1$
12	Chaperonin cofactor a	+26%	$p = .02$

This study highlighted the need for replicate experiments and statistical analysis of the data. The differences in cells from animal to animal used for this study clearly show the need for pairwise comparison of samples from the same animal to produce useful data. Furthermore, cell culture conditions needed to be reproducibly controlled, and cells needed to be lysed immediately after removal from the incubator, lest dramatic changes in heat shock proteins be observed.

Of the proteins observed to change, five of them corresponded to isoforms of myosin light chain (MLC). In particular, increased expression of the *atrial* form of MLC in the ventricle had been previously reported to accompany cardiac hypertrophy. Changes in the ventricular form appear to be phosphorylation differences because the major spot between the two changes is also MLC-2v, but insufficient material was available to study the phosphorylation states. Note that all of the protein differences were less than twofold.

Measurement of the message level for MLC-2v by quantitative PCR (Taqman) at a number of time points (0–48 h) showed no change in the hypertrophied cells (data not shown). These data highlight an important advantage of proteomic experiments: posttranslational modification differences can be detected, yet these are transparent to message-level measurements. Measurement of the message level of a number of other genes, however, also showed some of the current shortcomings of the proteomic approach that was used. The atrial natriuretic factor transcript was elevated threefold at 48 h, yet the protein was not found in the 2D gel study because its mass (3 kD) is below the size separated with the gel system used. The multiple time points (10 points) that were easily analyzed by quantitative PCR also showed that the message was at its maximum level (12-fold overexpression) between 12 and 24 h. Finally, some low abundance, early immediate genes (e.g., c-*fos*) were highly induced (500-fold) at 0.5 h but returned to normal by the 48-h time point. These latter examples clearly show that experiments should be performed at a large number of time points to determine protein changes adequately; efficient high-throughput proteomics methods are a necessity. Similar conclusions were also reached by Nelson *et al.* (2000a) in a comparison of transcript profiling and proteomics for the analysis of prostate gene expression.

Other Methodologies

Gel-Based Methods

Although the methodology outlined here is the conventional approach to proteomics for protein quantitation and identification, useful enhancements or alternatives have been developed. To avoid the tedious gel image matching for comparison of samples, differential labeling of two samples that are mixed and analyzed simultaneously on a single gel has been demonstrated. In one method, cells that are to be studied (e.g., with drug treatment) are grown under conditions such that all proteins are metabolically labeled with ^{15}N (Lahm and Langen, 2000; Oda *et al.*, 1999; Ong *et al.*, 2002). Control (untreated) cells are grown under normal (nonlabeled) conditions. The samples are mixed, the proteins separated by 2D gel electrophoresis, digested, and analyzed by MS. The ratio of $^{14}N/^{15}N$ signals in the mass spectrometer is used to determine the relative concentrations of the protein in the two samples. The procedure works well but requires growth conditions in which the sole source of the isotope can be controlled, which can also be expensive. This method of quantitation by MS has been extended to gel-free systems (see below). An alternative to metabolic labeling is to label each sample with a different fluorescent tag and then mix the samples and run the 2D gel (Ünlü *et al.*, 1997). Although the two fluorescent labels give overlapping protein spots, only a small portion of the protein is actually labeled, and the unlabeled protein migrates to a slightly different location on the gel, complicating subsequent protein identification.

Alternatives to protein identification by in-gel or on-membrane digestion have also been demonstrated. Loo *et al.* (1999) showed direct mass analysis from isoelectric focusing gels could give accurate protein mass (and produce a "virtual 2D gel") and be useful for identification of proteins from small genomes. Eckerskorn *et al.* (1997) showed similar results from 2D gels after electroblotting, with excellent spatial resolution and detection limits comparable to silver staining. Schleuder *et al.* (1999) showed that proteins could be digested on the membrane without cutting individual spots, and the entire membrane then analyzed by MALDI-TOF MS.

Proteomics has come to be closely associated with 2D gel electrophoresis, because of the separation power and quantification provided by staining and image analysis. There are a number of reasons that 2D gels are less than optimal for proteomics studies: poor routine separations at extremes of molecular weight (MW) (<10 kD, >150 kD) or pI (<3, >10), limit of detection for staining (>1 ng), limited dynamic range for staining (<1:1000), solubility problems for some integral membrane proteins, and significant labor is involved in loading, running, staining, and imaging the gels. Although many of these limitations are being addressed, such as membrane protein separation (Santoni *et al.*, 2000) and dynamic range (Corthals *et al.*, 2000), routine 2D gels do not represent a universal separation technique for all proteins (Gygi *et al.*, 2000).

An alternative to 2D gels is to multiplex the downstream peptide detection to reduce or eliminate the need for protein separation (Fig. 6). This approach is possible because a protein can often be identified from a single peptide fragmentation spectrum. This idea has been used to identify proteins from 2D SDS-PAGE gels of complex mixtures. For example, Simpson *et al.* (2000) isolated plasma membranes from a human colon carcinoma cell

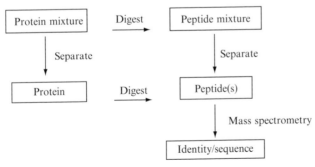

Fig. 6. Separation of a protein mixture by two-dimensional (2D) gel electrophoresis ideally produces highly resolved protein spots that contain one or a few proteins. Typical proteomics experiments use that highly purified protein for subsequent digestion and mass analysis. An alternative approach is possible because a protein can be identified (and quantitated) from a single peptide. This peptide–protein relationship allows identification of the components of a complex protein mixture, with little or no protein separation. This approach places far higher demands on the resolving power of the peptide separation.

line, separated the proteins by SDS-PAGE, cut the gel into 16 bands, and identified the proteins in each band by LC-MS/MS of the tryptic peptide mixtures. Bell *et al.* (2001b) identified 70 proteins in one lane of an SDS-PAGE gel of Golgi membrane proteins. Goodlet *et al.* (2000) showed that the high mass accuracy of Fourier transform–ion cyclotron resonance (FT-ICR) (1 ppm) allowed yeast proteins separated by SDS-PAGE to be identified solely by peptide mass for cysteine-containing peptides that were labeled with a chlorinated tag. Although protein identification in mixtures is possible by this approach, there remains a question of quantitation because most components cannot be resolved by staining.

Gel-Free Methods

Complete elimination of electrophoretic separation requires use of efficient prefractionation steps for the protein mixture or very high-resolution separation of the peptide mixture (Fig. 7). For example, a typical mammalian cell may express up to 10^4 different proteins with a 10^6 range of concentration. An average protein of 50 kD is expected to produce about 40 tryptic peptides. Thus, tryptic digestion of a whole cell lysate could produce as many as 400,000 peptides with more than 10^6 concentration dynamic range. Clearly, multiple steps of separation must be part of the analysis of such a complex mixture.

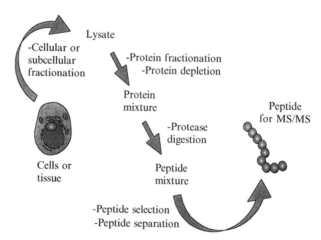

Fig. 7. At the beginning of an experiment, a single cell may be composed of up to 10,000 proteins. At the end of the experiment, only a few peptides should be present in a mass spectrum at any point in time to produce tandem mass spectra efficiently. To analyze the greatest number of components with the highest dynamic range, the mixture complexity must be reduced by some or many of the steps shown.

Link *et al.* (1999) showed the identification of more than 100 proteins from human or yeast ribosomal complexes, by use of 2D peptide HPLC (strong cation exchange coupled with reversed-phase) on-line with electrospray MS/MS. In an extension of this work with the same 2D HPLC configuration, Washburn *et al.* (2001) demonstrated identification of 1484 proteins from yeast after preseparation into just three fractions based on protein solubility. This latter work is notable in that many proteins usually difficult to observe by routine 2D gel electrophoresis (pI <4 or >11, MW <10 or >100 kD) were found, including 131 integral membrane proteins. This approach was subsequently applied to proteomic analysis of the malaria pathogen *Plasmodium falciparum*. More than 2400 proteins were catalogued with respect to four stages of the parasite's life cycle, shedding light on the identity and potential function of many previously unknown proteins (Florens *et al.*, 2002).

Gygi *et al.* (1999a) developed an important method to use the identification of proteins from individual peptides in a complex mixture to provide relative quantitation by differential labeling of samples. Their isotope-coded affinity tag (ICAT) method places a biotin-affinity tag on cysteines, and two samples can be differentially labeled by adding a deuterium label to one of the samples. The samples are subsequently mixed, the proteins digested with trypsin, and the cysteine biotin-labeled peptides extracted with avidin. The peptides are analyzed by LC-MS/MS, with the ratio of deuterated/non-deuterated peaks used for quantitation, and the tandem spectrum used for identification of the protein (Fig. 8).

As a demonstration of the ICAT methodology, two samples containing lysozyme in a 1:3 ratio were analyzed. The samples were labeled with custom synthesized thiol-reactive biotinylation reagents, one of which incorporated three deuterium atoms in place of three hydrogen atoms (Fig. 9). The samples were mixed, digested with trypsin, and the cysteine-containing tryptic peptides isolated on a monomeric avidin column, essentially as described by Gygi *et al.* (1999a). LC-MS/MS was performed on the resulting peptide mixture using an ion trap mass spectrometer. The doubly charged ion of the peptide CELAAAMK was subjected to collision-induced dissociation (CID) throughout the LC gradient, with a 5 mass unit precursor isolation window so that both heavy- and light-labeled forms of the peptide were simultaneously trapped and dissociated (Fig. 10). The identity of the peptide was verified by the presence of correct sequence-specific b and y series ions (Fig. 10B). Those fragment ions, which contain cysteine, appear as doublets separated by 3 Da and are marked with an asterisk in Fig. 10. Examination of these doublets (illustrated for two such ions in Fig. 10C and D) provides the relative abundance of the heavy- and

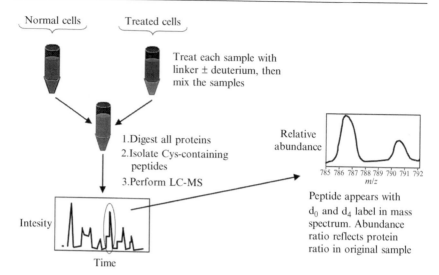

FIG. 8. The isotope-coded affinity tag (ICAT) approach to protein quantitative analysis is based on differential labeling of proteins from two samples. The protein tag links biotin to each cysteine residue, and the tags for one of the samples are labeled with deuterium (see Fig. 9). The samples are mixed, proteolytically digested, the cysteine-containing peptides are affinity isolated with avidin, and the mixture is analyzed by LC-MS. The ratio of the deuterated/non-deuterated peaks in the mass spectra provide the protein ratio in the original sample. The protein is identified by MS/MS in the same or a subsequent experiment.

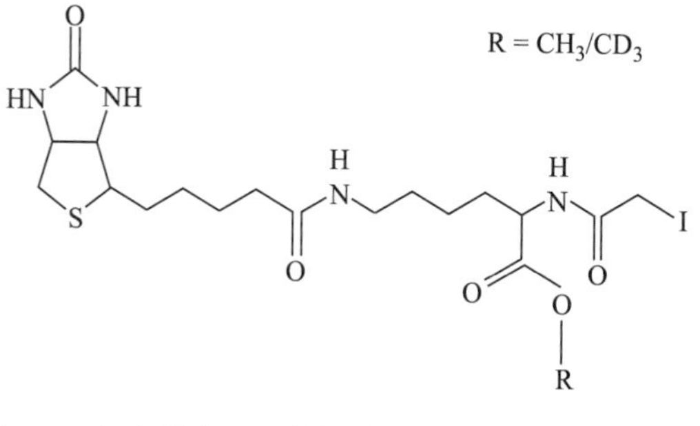

FIG. 9. The isotope-coded affinity tag (ICAT) reagent used for the experiment in Fig. 10 incorporates biotin for affinity isolation, an iodo acyl group for cysteine thiol alkylation, and a methyl ester that may be deuterated. This molecule is based on the original reagent of Gygi *et al.* (1999a).

light-labeled peptides and, therefore, the relative abundance of lysozyme in the original samples.

Griffin *et al.* (2001) extended the ICAT method by fractionation of the peptide mixture with a cation exchange step before affinity purification

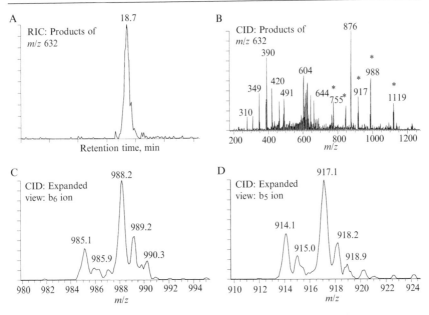

Fig. 10. Protein quantitation by the isotope-coded affinity tag (ICAT) method. The sample contains lysozyme, labeled with the ICAT reagent shown in Fig. 9 in a 1:3 (d_0:d_3) ratio, then digested with trypsin. (A) A reconstructed ion chromatogram for a tryptic peptide shows that the light and heavy ICAT versions co-elute. (B) The collision-induced dissociation (CID) fragmentation pattern serves to identify the peptide, CELAAAMK. Product ions containing Cys are marked with an asterisk. (C, D) Expanded views of two Cys-containing b-ions from (B). The relative abundance of light and heavy ICAT-labeled peptides is apparent from the fragment ions, and multiple fragment ions provide replicate measurements of the abundance ratio.

with avidin, and a MALDI-qTOF mass spectrometer was used for peptide quantitation and identification. A similar approach to the ICAT method was demonstrated by Geng et al. (2000) in which peptide amino groups were labeled by either d_0 or d_3 acetylation after tryptic digestion, the samples then mixed, separated by HPLC, and analyzed by MALDI-TOF-MS. Although this latter approach does not provide mixture simplification, the same group showed that a subset of acetyl-labeled histidine-containing peptides could be isolated with a Cu(II) affinity column (Ji et al., 2000). In another extension of the ICAT approach, Arnott et al. (2002) showed that the methodology could be used for selective targeted identification of a protein in a complex mixture. They termed this method the "Mass Western" in reference to the Western blot often used for highly sensitive, highly selective immunodetection of proteins. They used this method to show the presence of the HER2 receptor in the crude membrane preparation from a breast cancer cell line.

Because of the newness of the ICAT approaches, it is too early to predict which version of the methodology may prove superior or to determine to what extent these methodologies may complement or even replace 2D gel approaches. Early results, however, demonstrate significant promise because many of the limitations of routine 2D gels, as noted earlier in this chapter, are not found with the gel-free systems and a high degree of automation is already possible with off-the-shelf instrumentation.

Protein Arrays

An obvious approach for proteomics is to apply the array technology that has become widely used for transcript profiling in the genomics field. The most promising approach is the arraying of antibodies to which differentially labeled proteins are bound (de Wildt *et al.*, 2000; Haab and Zhou, 2004; Nielsen and Geierstanger, 2004). The arraying of proteins for detection of protein–protein interactions, protein-small molecule binding, and kinase substrate identification has also been demonstrated (MacBeath and Schreiber, 2000). Of particular note is the report of an array of 27,648 proteins produced from a human fetal brain cDNA expression library that was used to screen for antibody–antigen interactions (Holt *et al.*, 2000). Among the challenges for array-based methods are to find protein immobilization chemistries that are non-denaturing and that eliminate nonspecific binding, to find conditions for binding that are universal for all proteins on the array, and to develop high-sensitivity detection schemes to permit identification of low copy number proteins (Jenkins and Pennington, 2001; Walter *et al.*, 2000). Arrays also must prove to be more sensitive, faster, or cost effective than highly automated microtiter plate-based protein assays. A number of commercial operations are attempting to generate antibodies against large sets of proteins. When these reagents become available, and if the technical challenges for arrays are solved, protein arrays may become a dominant technique in proteomics.

Impact on Biological Research

Breakthroughs in fields of inquiry often depend on the tools that are available. For example, monoclonal antibodies are essential reagents in studies of receptor activation and signal transduction, and their use was key to the progress in this field. In a similar manner, the tools that comprise proteomics and the results derived from their use will give biologists access to information that will further increase their knowledge of the molecular mechanisms of cellular processes and dysfunction (disease). The availability of detailed information on each protein in a cell, tissue, organ, or system will help define a full biological system. These details must include *qualitative* information such as functionally important modifications, subcellular localization, intermolecular interactions, structural motifs, and signaling pathways, as well as *quantitative* information such as concentrations, binding affinities, and kinetics of synthesis, degradation, binding, and enzymatic reactions. Only when all the qualitative and quantitative aspects are fully discerned can a complex biological system be truly understood. The results of such experiments are now beginning to appear (Andersen *et al.*, 2005; Gruhler *et al.*, 2005).

The beginnings of proteome databases have been generated during the past decade, but these have been primarily small-scale compilations of 2D gel studies. Large databases of proteome information are beginning to be generated by commercial enterprises. Although a complete set of proteome data has long been a lofty goal (e.g., the *human protein index* [Anderson and Anderson, 1982]), it appeared impractical for all but microorganisms until 2001 (Anderson *et al.*, 2001). Now large, well-funded commercial endeavors are working to generate data so that the location, quantity, modification state, and intermolecular interactions will be available for most proteins in nearly every human cell type. For cell types of particular interest, this work may include data points for each stage of development or each stage of disease. These latter types of more focused proteomics experiments will continue to be done in many laboratories to provide answers to specific questions.

However, huge databases are just stockpiles of data. It is only when those data are organized efficiently and readily available, when correlations are made, when trends are observed, that understanding may begin. For it is still the development and testing of hypotheses that provide understanding. The data warehouses will make traditional hypothesis-driven research more efficient. Many ideas can be tested by conducting a database query or performing an *in silico* experiment, rather than by design and execution of a laboratory experiment.

Proteomic information will complement other growing sets of information: message-level measurements from DNA microarrays (Gerhold *et al.*, 1999) or quantitative PCR, protein location from tissue microarrays (Kononen *et al.*, 1998), immunohistochemistry, and fluorescence resonance energy transfer experiments (Mahajan *et al.*, 1998), quantities and activities from enzyme-linked immunosorbent assay (ELISA) assays, and binding information from yeast two-hybrid experiments (Chien *et al.*, 1991) and surface plasmon resonance measurements (Krone *et al.*, 1997). Indeed, it is only by the efficient linking of all these data sets (necessitating sophisticated informatics) that the greatest impact on biology will be realized. The challenge is to produce data with sufficient sensitivity, dynamic range, and throughput to actually be complementary to these other methods.

Impact on Drug Development

The time and money required to bring a drug to market, including the high failure rate of drugs during clinical trials, are the chief reasons for the high cost of pharmaceuticals and often the reason that better drugs are not available. Proteomics is beginning to have an impact on the various stages of drug development (Anderson *et al.*, 2000; Page *et al.*, 1999a; Steiner and Witzmann, 2000). The revolution in drug development of which proteomics is a part (along with combinatorial chemistry, high-throughput screening, and pharmacogenomics) is changing the approaches of the pharmaceutical industry to research and development.

Drug Target and Surrogate Marker Identification

It is the detailed knowledge of biological processes at the molecular level and how they go askew in disease that provides the targets for drug intervention. Thus, the enhancement to biological research by proteomics envisioned above will play a crucial role in the identification of drug targets and disease markers. Differential analysis between normal and diseased tissue can identify proteins that are correlated with the disease. Even when a protein is found to be correlated with a disease, however, further studies (target validation) may show that it is not pivotal to the disease process (e.g., a control point of a cellular pathway), so it is not a good target. Nonetheless, those proteins may prove highly specific and useful molecular markers for the disease (Alaiya *et al.*, 2000). The marker may subsequently be used in assays for further study of the disease, for disease diagnosis or classification, for following the efficacy of a therapeutic, or for identification of patient subpopulations that will respond to therapy.

As examples of this work, Dunn (2000) has extensively studied a form of heart disease known as *dilated cardiomyopathy,* established an extensive 2D gel database based on human heart tissue, and identified a number of disease-specific proteins. The study on cardiomyocyte described above also identified a number of heart disease–related proteins (Arnott *et al.*, 1998b).

Cancer has been the most highly studied disease by proteomics. A comparison of human breast ductal carcinoma and histologically normal tissues (without cell type separation) revealed 32 differentially expressed protein spots (Bini *et al.*, 1997). Examination of immunomagnetically purified luminal and myoepithelial cells from normal breast produced a list of proteins that will be useful for comparison with breast tumor samples (Page *et al.*, 1999b). A comparison of matched sets of human colorectal carcinoma and normal colon mucosa revealed expression differences in a number of proteins, including overexpression of the S100 family of

calcium-binding proteins (Stulik *et al.*, 1999). In the most extensive proteomic study, Celis *et al.* (1999b) studied human bladder transitional and
squamous cell carcinomas by 2D gel electrophoresis. They have established
extensive databases (Celis *et al.*, 1999b) and found markers (Celis *et al.*,
1999a; Ostergaard *et al.*, 1997, 1999) that are indicative of the degree of
disease progression and that may prove useful for early cancer detection.
The field of diagnostic biomarkers was galvanized by a 2002 report in
The Lancet that mass spectrometric profiles of low MW proteins in serum
could detect the presence of ovarian cancer at early stages of the disease
(Petricoin *et al.*, 2002). These results remain controversial and highlight the
challenges of sample collection, preparation, data analysis, and overall
study design that are inherent to clinical applications, as well as the great
need for methods that could be used to screen the general population for
diseases such as cancer (Ransohoff, 2005).

Proteomics has also been applied to the study of pathogenic microbes
in order to generate databases (Langen *et al.*, 2000), to find the molecular basis for drug efficacy or drug resistance (Cash *et al.*, 1999), or
to identify regulatory networks that may be interrupted (VanBogelen
et al., 1999).

Lead Optimization for Traditional Pharmaceuticals: Toxicology and Efficacy

Once a target is discovered and validated, lead molecules are developed. Proteomics is beginning to play a role in accelerating lead optimization, that is, the development of a molecule that ideally is efficacious, has
excellent bioavailability and pharmacokinetics, and has little or no toxicity.
A comparison of proteins that are expressed by a cell line or tissue with and
without drug treatment may identify biochemical pathways that are affected positively (efficacy) or negatively (toxicity) by the drug. In particular,
the ability to achieve accurate predictions of toxicity quickly would have a
significant impact on the speed of drug development.

As an excellent example of an efficacy study, a proposed new treatment
for hepatocellular carcinoma is a galactosyl derivative of the approved
chemotherapeutic agent 5-fluorouracil (5-FU) (Page *et al.*, 1999a). The
drug is designed to be taken up by the tumor cells and converted to
5-FU. If the activity is truly based on this conversion, then treatment of a
human hepatoma cell line with either the new drug or 5-FU should produce
a similar protein profile in the cell line. 2D gel electrophoresis of the
respective cell lysates (two drug treatments and control) showed that of
2291 spots found in common in all samples, 19 were elevated at least
fivefold, and each of the induced proteins was identical for the two drugs.

An example of a toxicity study is an examination of cyclosporin A nephrotoxicity in rats by Steiner *et al.* (1996). They examined 2D gel profiles of rat kidney proteins and observed cyclosporin-induced downregulation of the calcium-binding protein calbindin D28 that correlated with the accumulation of calcium in the tubules and with tubular toxicity. Subsequent experiments established that downregulation of calbindin also correlated with nephrotoxicity in humans treated with cyclosporin A (Aicher *et al.*, 1998). In another study of glomerular nephrotoxicity in rat, Cutler *et al.* (1999) used treatment with puromycin aminonucleoside to discern the progression of nephrotoxicity-associated proteinuria. A common liver-toxic response among a diverse set of drugs may be studied to determine whether they share a common molecular basis for the toxicity. Anderson *et al.* (1996a) studied the effects of several peroxisome proliferators on mouse liver protein abundances. They found a clear set of proteins that were induced by proxisome proliferators. Future work should be able to use these and other studies to enable proteomics to be used for toxicity testing (Chevalier *et al.*, 2000). Proteomics may be able to predict the toxicity of a drug based on the proteins with which the drug or its metabolites interact (Anderson *et al.*, 1996b). Furthermore, if mechanisms of action and metabolic pathways can be established by proteomic methods, it may also become possible to predict drug interactions by proteomic comparisons.

When a disease has multiple causes or multiple stages, such as cancer, there is frequently a subpopulation that responds to a particular drug. That subset of patients may be definable, based on the mechanism of action of the drug and the common genetic characteristics of the patients. Such correlations of drug response based on patient transcript profiling have been termed *pharmacogenomics.* Similarly, protein surrogate markers for disease or drug efficacy, found in blood or urine, could be used to predict patient populations that respond to treatment or have adverse side effects. The identification of these markers, as described earlier, has been termed *pharmacoproteomics.* The goal is to make clinical trials more efficient and, in theory, more successful, because the patient pool will be screened so that only those individuals who should respond positively to the treatment are given the drug. Further, when a drug is approved for use, tests will be used to screen patients and to tailor the treatment for each individual. At the moment, the utility of pharmacoproteomics has yet to be proven.

A methodology with multiple potential applications is tissue imaging by MS, pioneered in the laboratory of Richard Caprioli. Thin tissue sections are immobilized on target plates and overlayed with MALDI matrix. Spatially resolved mass spectra are then acquired for specific regions of the tissue section, or the laser can be rastered across the entire section. The localization of individual protein mass signatures can then be displayed as 3D images incorporating location and abundance of each protein. Pattern-recognition algorithms are applied to find sets of peaks that can be used for classification. For example, tumor tissue from patients with non–small cell lung cancer yielded markers that distinguished among major histological groups and differentiated patients with good or poor survival prognosis (Chaurand *et al.*, 2004).

Future Challenges

Despite the advances in technology and methodology, proteomics is still far from reaching the stage of productivity and utility that is necessary for it to be critical to biological research. How can this be changed? Stanley Fields (2001) suggests that "for a field so laden with razzmatazz methods, it is striking that the number one need in proteomics may be new technology." Technological progress is needed in a variety of disciplines to make proteomics live up to its promise. Areas of particular challenge are described in the following sections.

Sample Preparation

Sample preparation has been and will continue to be the crucial step in proteomics experiments, and the challenge here is in separation capabilities, recovery of material, and throughput. Many of the sample preparation steps mentioned here are familiar to a cell biologist. However, many of these methods were not developed to provide the degree of purity or the efficient recovery of material that are necessary for proteomic studies. Thus, improvements or alternatives are necessary.

Much of the truly useful information in proteomics will come directly from organisms, either model organisms or human. In either case, tissue is the starting point from which homogeneous populations of cells must be derived. Fresh tissue, taken immediately from dissection or surgery, is the best source. Frozen tissue, though generally more easily acquired, eliminates the possibility for primary cell culture or techniques that rely on live cells. Formalin-fixed paraffin-embedded tissue, found in extensive pathology tissue banks, is of no use for proteomics experiments. Tissue dissociation protocols are well established, although some studies such as the analysis of cell surface proteins necessitate dissociation without proteases. Selection of specific cell types is done by affinity selection for cell-type specific antigens using immunomagnetic beads. Important examples of cell-type–specific purification are the isolation of colorectal epithelial cells

(Reymond *et al.*, 1997), as well as the isolation of luminal and myoepithelial cells from human breast tissue (Page *et al.*, 1999b). The challenge for many cell types remains the identification of a specific antigen. An alternative approach is the use of laser-capture microdissection (LCM) of frozen tissue (Simone *et al.*, 1998). The technique can, for example, precisely cut tumor cells in the midst of surrounding stroma. Several studies have demonstrated the utility of the technique for proteomics (Banks *et al.*, 1999; Ornstein *et al.*, 2000), although it is sufficiently labor intensive (up to 16 hours to generate sufficient material for one experiment) to preclude its use as a high-throughput method.

Subcellular fractionation of cells is typically necessary to reduce the mixture complexity, to yield proteins with similar solubility properties (e.g., cytosolic proteins and membrane proteins), or to localize each protein to the area(s) of the cell where it is biologically relevant (Corthals *et al.*, 2000). Subcellular fractionation has traditionally been accomplished after cell lysis by physical methods such as density gradient centrifugation. With these methods, however, samples are often significantly contaminated with proteins from other organelles. Thus, immunomagnetic separations are also used for more highly purified subcellular fractions.

Further protein separation may still be necessary to provide adequate resolution and dynamic range (Corthals *et al.*, 2000). Depletion of known highly abundant proteins may further aid the dynamic range problem. For example, plasma proteins are typically overwhelmed by albumin and immunoglobulins. Affinity methods to remove these highly abundant proteins have been generated (Pieper *et al.*, 2003).

All of the aforementioned steps, from cell segregation to protein fractionation, rely to some extent on physical separations, but affinity methods will be crucial if truly homogeneous subpopulations of proteins are to be obtained. Highly specific protein subpopulations are necessary to increase the signal-to-noise ratio for the experiments. A high degree of homogeneity is crucial because, first, biological samples are by their nature not uniform and, second, many important biological processes occur at extremely low levels and important changes may be quite subtle. Irrelevant material (e.g., contaminating cell types or subcellular fractions) serves only to mask minor components or subtle differences. The high degree of homogeneity must be achieved with minimal losses of material. That is, all the fractionation steps must be highly resolving *and* highly efficient. Finally, all these steps must avoid contamination from the environment. Current protein analytical methods are often limited by chemical noise in the form of large amounts of keratin, found ubiquitously in dust, skin, and dandruff. Care must be taken not to introduce these or other contaminants during the sample preparation stage.

Sensitivity, Dynamic Range, and Automation

Following subfractionation, the actual protein separation, identification, and quantification will be done, whether by electrophoresis, by ICAT-type experiments, or by some other method. Protein digestion methods and peptide separations may benefit from microfluidic chips, which have the potential to reduce the sample handling, reduce surface contact, and as a result, increase sensitivity and allow for efficient automation (Ekstrom *et al.*, 2000; Figeys *et al.*, 1998; Li *et al.*, 2000).

The biggest challenge for these methods is to increase sensitivity and dynamic range. Important proteins may be present at only a few to tens of copies per cell. For a tissue sample that may only yield 10^6 cells, 10 copies/cell represents approximately 10 attomoles, assuming 100% recovery in all sample preparation steps. Although some state-of-the-art mass spectrometers are able to achieve 10-attomole detection limits for simple mixtures (Martin *et al.*, 2000), achieving this performance for all components in a complex mixture is beyond current capabilities. Finally, a high degree of sample recovery is rarely obtained with a highly fractionated sample, which makes higher sensitivity all the more important.

Sensitivity is crucial for another reason. Current protein analysis methods, despite their impressive gains, still lag in sensitivity compared to many common biological methods. Protein detection based on antibody binding (e.g., Western blot, ELISA) using fluorescent or chemiluminescent detection can be exquisitely sensitive (<1 fmol) and requires little or minimal protein purification. Metabolic labeling with ^{35}S or ^{32}P provides similar or better limits of detection by autoradiography or phosphor imaging. Proteomic methods must become competitive with these methods (or use them, as with antibody arrays). Furthermore, DNA microarrays require only a fraction of the material that protein analyses use. To be truly complementary to microarrays, proteomics must be performed, when necessary, with the same limited amounts of material.

All of the processes described here are in need of vastly improved throughput. Automation is the key to truly widespread utilization of proteomics (Harry *et al.*, 2000). And with widespread utilization will come a huge influx of data. Databases and data warehouses will need to be designed to store and make accessible in a useful format all the data that are generated. New algorithms will need to be written to correlate vast amounts of data from many experiments. Fortunately, many of the same demands are already being addressed in the DNA microarray community, and efforts are underway to establish standards for cross-platform sharing of mass spectrometric data, results of protein identification experiments, and protein–protein interactions (Orchard *et al.*, 2003).

Lest it appear that the future of proteomics is mainly "big science," it is clear that all of the automated, sensitive, high-throughput tools under development will benefit even the smallest laboratories. Although early proteomic "factories" may construct enormous robotic devices or employ scores of people, cost-effectiveness will, in the end, favor simplification. Most efficient forms of automation can be scaled up or down to fit the appropriate needs. For example, the enormous task of sequencing the human genome was completed (ahead of schedule!) not with a few, large instruments, but with rooms full of small, efficient DNA sequencing instruments that are now available to any laboratory. So it will likely be that the technological advances that make large companies successful will become readily available to smaller laboratories, who will benefit most from the speed and reliability with which important, small-scale experiments can be performed.

Conclusion

Proteomics is a new rapidly growing and evolving field that shows enormous potential to have an impact on biological research significantly. Great strides have been made in the ability to analyze proteins *en masse* with ever improving sensitivity and throughput. Nonetheless, most of the work has been intensive method development and method demonstration. Despite a large amount of effort, only recently is novel biology being discovered with proteomics. The real applications of proteomics lie ahead. Significantly more technical development is needed. The flood of data will overwhelm our capacity to generate real knowledge without further advances in informatics. It is still not clear which approach will become the standard for proteomics. Ultimately, as with all science, it will be through imaginative thinking and creative experimental design—both for data generation and data analysis—that the real impact of proteomics will be realized.

Protocols

Method for Reduction/Alkylation in Sample Buffer

To 20 μl of sample in low salt buffer
Add 20 μl of 2× reducing SDS sample buffer (3 mg DTT in 1 ml 2× sample buffer, see below).
Incubate 5 min at 85°.
Add 4 μl of iodoacetamide (4 mg IAA in 100 μl methanol).
Incubate 20 min at room temperature.

Load immediately onto Tris–glycine gel.

2× sample buffer (non-reducing), pH 8.3

0.5 *M* Tris–HCl, pH 8.3	2.0 ml
Glycerol	2.0 ml
10% w/v SDS	3.2 ml
0.1% w/v bromophenol blue	0.8 ml

Note: Basic pH of sample buffer is necessary for reduction; conventional Lammeli sample buffer is pH 6.8. Iodoacetic acid is inefficient for alkylation, presumably because of charge repulsion with the SDS-coated protein.

Method for Silver Staining Compatible with In-Gel Digestion

Fix	10% v/v acetic acid, 30% ethanol	3 × 20 min (or overnight)
Rinse	20% v/v ethanol Deionized water	20 min 10 min (ensure gel is submerged)
Sensitize	sodium thiosulfate, 0.2 g/L	1 min
Rinse	deionized water	3 × 20 s
Stain	silver nitrate, 2 g/L	30 min
Rinse	Deionized water	5–10 s
Develop	Potassium carbonate, 30 g/L; 37% formaldehyde 0.25 ml/L; sodium thiosulfate 10mg/L	10–15 s until precipitate appears; then change to fresh solution for 5–15 min until sufficiently developed
Stop	Tris base, 50 g/L; 25 ml/L gl. Acetic acid	5 min
Rinse	Deionized water	3 × 5 min

Method for In-Gel Digestion (Note: Before Each Step, Remove Solution from the Preceding Step)

Cut gel bands/spots and place each in a clear 200-μl PCR tube. Gel or bands should have been equilibrated in deionized water.

For Coomassie-Stained Gels

Destain: 5-min wash with 100 m*M* ammonium bicarbonate/50% acetonitrile. Repeat this wash step with fresh solution until gel is completely clear.

For Silver-Stained Gels

Destain: 5–10 min wash with 25 μl 30 mM potassium ferricyanide + 25 μl 100 mM sodium thiosulfate (two reagents mixed immediately before use)

For Samples Not Previously Fully Reduced and Alkylated

Reduce: 25 μl of 10 mM DTT in 100 mM ammonium bicarbonate, 1 h, 45°

Alkylate: 25 μl of 50 mM iodoacetamide in 100 mM ammonium bicarbonate, 30 min, ambient

For All Gels

Wash: 50 μl of 100 mM ammonium bicarbonate, shake for 10 min; Add 50 μl acetonitrile, shake for 10 min.

Wash: repeat preceding wash steps

Discard preceding solution; dry gel pieces in a Speed-Vac

Digestion: Reswell the gel pieces in a minimum volume of digestion buffer (0.01 μg/μl Promega modified trypsin in 50 mM ammonium bicarbonate/0.02% w/v Zwittergent-3/16 detergent). After the gel pieces have reswollen completely (~5 min, add more digest buffer if needed), add 5–10 μl of 50 mM ammonium bicarbonate to cover the pieces.

Incubate 3–15 h at 37°.

Remove the supernatant.

Extract the gel with 10 μl 50% acetonitrile/0.1% TFA, shake for 15 min, combine the extract with the supernatant.

Extract a second time with 10 μl neat acetonitrile, shake for 15 min. Combine the extract with the previous extract and supernatant.

Analyze immediately or store frozen (−70°) until analyzed.

Acknowledgments

We acknowledge the work of Kathy Stults and Bill Henzel for the protocols, Adrianne Wonnacott for ICAT protocols, Jim Marsters for synthesis of the ICAT reagent, and Mickey Williams for quantitative PCR data.

References

Aebersold, R., and Goodlett, D. R. (2001). Mass spectrometry in proteomics. *Chem. Rev.* **101,** 269–295.

Aebersold, R., and Mann, M. (2003). Mass spectrometry–based proteomics. *Nature* **422,** 198–207.

Aicher, L., Wahl, D., Arce, A., Grenet, O., and Steiner, S. (1998). New insights into cyclosporine A nephrotoxicity by proteome analysis. *Electrophoresis* **19**, 1998–2003.

Alaiya, A. A., Franzen, B., Auer, G., and Linder, S. (2000). Cancer proteomics: From identification of novel markers to creation of artificial learning models for tumor classification. *Electrophoresis* **21**, 1210–1217.

Andersen, J. S., Lam, Y. W., Leung, A. K., Ong, S. E., Lyon, C. E., Lamond, A. I., and Mann, M. (2005). Nucleolar proteome dynamics. *Nature* **433**, 77–83.

Andersen, J. S., Lyon, C. E., Fox, A. H., Leung, A. K., Lam, Y. W., Steen, H., Mann, M., and Lamond, A. I. (2002). Directed proteomic analysis of the human nucleolus. *Curr. Biol.* **12**, 1–11.

Anderson, L., and Seilhamer, J. (1997). A comparison of selected mRNA and protein abundances in human liver. *Electrophoresis* **18**, 533–537.

Anderson, N. G., and Anderson, N. L. (1982). The human protein index. *Clin. Chem.* **28**, 739–748.

Anderson, N. G., Matheson, A., and Anderson, N. L. (2001). Back to the future: The human protein index (HPI) and the agenda for post-proteomic biology. *Proteomics* **1**, 3–12.

Anderson, N. L., Esquer-Blasco, R., Richardson, F., Foxworthy, P., and Eacho, P. (1996a). The effects of peroxisome proliferators on protein abundances in mouse liver. *Toxicol. Appl. Pharmacol.* **137**, 75–89.

Anderson, N. L., Matheson, A. D., and Steiner, S. (2000). Proteomics: Applications in basic and applied biology. *Curr. Opin. Biotechnol.* **11**, 408–412.

Anderson, N. L., Taylor, J., Hofmann, J. P., Esquer-Blasco, R., Swift, S., and Anderson, N. G. (1996b). Simultaneous measurement of hundreds of liver proteins: Application in assessment of liver function. *Toxicol. Pathol.* **24**, 72–76.

Annan, R. S., Huddleston, M. J., Verma, R., Deshaies, R. J., and Carr, S. A. (2001). A multidimensional electrospray MS-based approach to phosphopeptide mapping. *Anal. Chem.* **73**, 393–404.

Arnott, D., Henzel, W., and Stults, J. T. (1998a). Rapid identification of comigrating gel-isolated proteins by ion trap-spectrometry. *Electrophoresis* **19**, 968–980.

Arnott, D., Kishiyama, A., Luis, E. A., Ludlum, S. G., Marsters, J. C., Jr., and Stults, J. T. (2002). Selective detection of membrane proteins without antibodies: A mass spectrometric version of the Western blot. *Mol. Cell Proteom.* **1**, 148–156.

Arnott, D., O'Connell, K., King, K., and Stults, J. T. (1998b). An integrated approach to proteome analysis: Identification of proteins associated with cardiac hypertrophy. *Anal. Biochem.* **258**, 1–18.

Ballif, B. A., Villen, J., Beausoleil, S. A., Schwartz, D., and Gygi, S. P. (2004). Phosphoproteomic analysis of the developing mouse brain. *Mol. Cell Proteom.* **3**, 1093–1101.

Banks, R. E., Dunn, M. J., Forbes, M. A., Stanley, A., Pappin, D., Naven, T., Gough, M., Harnden, P., and Selby, P. J. (1999). The potential use of laser capture microdissection to selectively obtain distinct populations of cells for proteomic analysis—preliminary findings. *Electrophoresis* **20**, 689–700.

Beausoleil, S. A., Jedrychowski, M., Schwartz, D., Elias, J. E., Villen, J., Li, J., Cohn, M. A., Cantley, L. C., and Gygi, S. P. (2004). Large-scale characterization of HeLa cell nuclear phosphoproteins. *Proc. Natl. Acad. Sci. USA* **101**, 12130–12135.

Bell, A. W., Ward, M. A., Blackstock, W. P., Freeman, H. N., Choudhary, J. S., Lewis, A. P., Chotai, D., Fazel, A., Gushue, J. N., Paiement, J., Palcy, S., Chevet, E., Lafreniere-Roula, M., Solari, R., Thomas, D. Y., Rowley, A., and Bergeron, J. J. (2001a). Proteomics characterization of abundant Golgi membrane proteins. *J. Biol. Chem.* **276**, 5152–5165.

Bell, A. W., Ward, M. A., Blackstock, W. P., Freeman, H. N. M., Choudhary, J. S., Lewis, A. P., Chotai, D., Fazel, A., Gushue, J. N., Paiement, J., Palcy, S., Chevet, E., Lafreniere-Roula,

M., Solari, R., Thomas, D. Y., Rowley, A., and Bergeron, J. J. M. (2001b). Proteomics characterization of abundant Golgi membrane proteins. *J. Biol. Chem.* **276,** 5152–5165.

Berndt, P., Hobohm, U., and Langen, H. (1999). Reliable automatic protein identification from matrix-assisted laser desorption/ionization mass spectrometric peptide fingerprints. *Electrophoresis* **20,** 3521–3526.

Bini, L., Magi, B., Marzocchi, B., Arcuri, F., Tripodi, S., Cintorino, M., Sanchez, J. C., Frutiger, S., Hughes, G., Pallini, V., Hochstrasser, D. F., and Tosi, P. (1997). Protein expression profiles in human breast ductal carcinoma and histologically normal tissue. *Electrophoresis* **18,** 2832–2841.

Blackstock, W., and Weir, M. (1999). Proteomics: Quantitative and physical mapping of cellular proteins. *TibTech* **17,** 121–127.

Blagoev, B., Ong, S. E., Kratchmarova, I., and Mann, M. (2004). Temporal analysis of phosphotyrosine-dependent signaling networks by quantitative proteomics. *Nat. Biotechnol.* **22,** 1139–1145.

Brunet, S., Thibault, P., Gagnon, E., Kearney, P., Bergeron, J. J., and Desjardins, M. (2003). Organelle proteomics: Looking at less to see more. *Trends Cell Biol.* **13,** 629–638.

Cao, P., and Stults, J. T. (1999). Phosphopeptide analysis by on-line immobilized metal-ion affinity chromatography-capillary electrophoresis-electrospray ionization mass spectrometry. *J. Chromatogr.* **853,** 225–235.

Cash, P., Argo, E., Ford, L., Lawrie, L., and McKenzie, H. (1999). A proteomic analysis of erythromycin resistance in *Streptococcus pneumoniae*. *Electrophoresis* **20,** 2259–2268.

Celis, J. E., Celis, P., Ostergaard, M., Basse, B., Lauridsen, J. B., Ratz, G., Rasmussen, H. H., Orntoft, T. F., Hein, B., Wolf, H., and Celis, A. (1999a). Proteomics and immunohistochemistry define some of the steps involved in the squamous differentiation of the bladder transitional epithelium: A novel strategy for identifying metaplastic lesions. *Cancer Res.* **59,** 3003–3009.

Celis, J. E., Kruhoffer, M., Gromova, I., Frederiksen, C., Ostergaard, M., Thykjaer, T., Gromov, P., Yu, J., Palsdottir, H., Magnusson, N., and Orntoft, T. F. (2000). Gene expression profiling: Monitoring transcription and translation products using DNA microarrays and proteomics. *FEBS Lett.* **480,** 2–16.

Celis, J. E., Ostergaard, M., Rasmussen, H. H., Gromov, P., Gromova, I., Varmark, H., Palsdottir, H., Magnusson, N., Andersen, I., Basse, B., Lauridsen, J. B., Ratz, G., Wolf, H., Orntoft, T. F., Celis, P., and Celis, A. (1999b). A comprehensive protein resource for the study of bladder cancer. http://biobase.dk/cgi-bin/celis *Electrophoresis* **20,** 300–309.

Chaurand, P., Schwartz, S. A., and Caprioli, R. M. (2004). Assessing protein patterns in disease using imaging mass spectrometry. *J. Proteom. Res.* **3,** 245–252.

Chevalier, S., Macdonald, N., Tonge, R., Rayner, S., Rowlinson, R., Shaw, J., Young, J., Davison, M., and Roberts, R. A. (2000). Proteomic analysis of differential protein expression in primary hepatocytes induced by EGF, tumour necrosis factor alpha or the peroxisome proliferator nafenopin. *Eur. J. Biochem.* **267,** 4624–4634.

Chien, C. T., Bartel, P. L., Sternglanz, R., and Fields, S. (1991). The two-hybrid system: A method to identify and clone genes for proteins that interact with a protein of interest. *Proc. Natl. Acad. Sci. USA* **88,** 9578–9582.

Clauser, K. R., Baker, P., and Burlingame, A. L. (1999). Role of accurate mass measurement (+10 ppm) in protein identification strategies employing MS or MS/MS and database searching. *Anal. Chem.* **71,** 2871–2882.

Corthals, G. L., Wasinger, V. C., Hochstrasser, D. F., and Sanchez, J. C. (2000). The dynamic range of protein expression: A challenge for proteomic research. *Electrophoresis* **21,** 1104–1115.

Courchesne, P., Luethy, R., and Patterson, S. (1997). Comparison of in-gel and on-membrane digestion methods at low to sub-pmol level for subsequent peptide and fragment-ion mass analysis using matrix-assisted laser-desorption/ionization mass spectrometry. *Electrophoresis* **18**, 369–381.

Creasy, D. M., and Cottrell, J. S. (2002). Error tolerant searching of uninterpreted tandem mass spectrometry data. *Proteomics* **2**, 1426–1434.

Cutler, P., Bell, D. J., Birrell, H. C., Connelly, J. C., Connor, S. C., Holmes, E., Mitchell, B. C., Monte, S. Y., Neville, B. A., Pickford, R., Polley, S., Schneider, K., and Skehel, J. M. (1999). An integrated proteomic approach to studying glomerular nephrotoxicity. *Electrophoresis* **20**, 3647–3658.

Dancik, V., Addona, T. A., Clauser, K. R., Vath, J. E., and Pevzner, P. A. (1999). *De novo* peptide sequencing via tandem mass spectrometry. *J. Comput. Biol.* **6**, 327–342.

Davis, M., and Lee, T. (1997). Variable flow liquid chromatography-tandem mass spectrometry and the comprehensive analysis of complex protein digest mixtures. *J. Am. Soc. Mass Spectrom.* **8**, 1059–1069.

Davis, M. T., Spahr, C. S., McGinley, M. D., Robinson, J. H., Bures, E. J., Beierle, J., Mort, J., Yu, W., Luethy, R., and Patterson, S. D. (2001). Towards defining the urinary proteome using liquid chromatography-tandem mass spectrometry, II. Limitations of complex mixture analyses. *Proteomics* **1**, 108–117.

de Wildt, R. M., Mundy, C. R., Gorick, B. D., and Tomlinson, I. M. (2000). Antibody arrays for high-throughput screening of antibody-antigen interactions. *Nat. Biotechnol.* **18**, 989–994.

Dunn, M. J. (2000). Studying heart disease using the proteomic approach. *Drug Discov. Today* **5**, 76–84.

Eckerskorn, C., Strupat, K., Schleuder, D., Hochstrasser, D., Sanchez, J. C., Lottspeich, F., and Hillenkamp, F. (1997). Analysis of proteins by direct scanning infrared-MALDI mass spectrometry after PD PAGE separation and electroblotting. *Anal. Chem.* **69**, 2888–2892.

Ekstrom, S., Onnerfjord, P., Nilsson, J., Bengtsson, M., Laurell, T., and Marko-Varga, G. (2000). Integrated microanalytical technology enabling rapid and automated protein identification. *Anal. Chem.* **72**, 286–293.

Eng, J. K., McCormack, A. L., and Yates, J. R. (1994). An approach to correlate tandem mass spectral data of peptides with amino acid sequences in a protein database. *J. Am. Soc. Mass Spectrom.* **5**, 976–989.

Engen, J. R., and Smith, D. L. (2001). Investigating protein structure and dynamics by hydrogen exchange MS. *Anal. Chem.* **73**, 256A–265A.

Epstein, L., Smith, D., Matsui, N., Tran, H., Sullican, C., Raineri, I., Burlingame, A., Clauser, K., Hall, S., and Andrews, L. (1996). Identification of cytokine-regulated proteins in normal and malignant cells by the combination of two-dimensional polyacrylamide gel electrophoresis, mass spectrometry, Edman degradation and immunoblotting and approaches to the analysis of their functional roles. *Electrophoresis* **17**, 1655–1670.

Erdjument-Bromage, H., Lui, M., Lacomis, L., Grewal, A., Annan, R., McNulty, D., Carr, S., and Tempst, P. (1998). Examination of micro-tip reversed-phase liquid chromatographic extraction of peptide pools for mass spectrometry analysis. *J. Chromatogr. A* **826**, 167–181.

Eriksson, J., Chait, B. T., and Fenyo, D. (2000). A statistical basis for testing the significance of mass spectrometric protein identification results. *Anal. Chem.* **72**, 999–1005.

Fenn, J. B., Mann, M., Meng, C. K., Wong, S. F., and Whitehouse, C. M. (1989). Electrospray ionization for mass spectrometry of large biomolecules. *Science* **246**, 64–67.

Fenyo, D., Qin, J., and Chait, B. T. (1998). Protein identification using mass spectrometric information. *Electrophoresis* **19**, 998–1005.

Ficarro, S. B., McCleland, M. L., Stukenberg, P. T., Burke, D. J., Ross, M. M., Shabanowitz, J., Hunt, D. F., and White, F. M. (2002). Phosphoproteome analysis by mass spectrometry and its application to *Saccharomyces cerevisiae*. *Nat. Biotechnol.* **20**, 301–305.

Fields, S. (2001). Proteomics. Proteomics in genomeland. *Science* **291,** 1221–1224.

Figeys, D., Gygi, S., McKinnon, G., and Aebersold, R. (1998). An integrated microfluidics-Tandem mass spectrometry system automated protein analysis. *Anal. Chem.* **70,** 3728–3734.

Florens, L., Washburn, M. P., Raine, J. D., Anthony, R. M., Grainger, M., Haynes, J. D., Moch, J. K., Muster, N., Sacci, J. B., Tabb, D. L., Witney, A. A., Wolters, D., Wu, Y., Gardner, M. J., Holder, A. A., SInden, R. E., Yates, J. R., and Carucci, D. J. (2002). A proteomic view of the Plasmodium falciparum life cycle. *Nature* **419,** 520–526.

Fox, A. H., Lam, Y. W., Leung, A. K., Lyon, C. E., Andersen, J., Mann, M., and Lamond, A. I. (2002). Paraspeckles: A novel nuclear domain. *Curr. Biol.* **12,** 13–25.

Garin, J., Diez, R., Kieffer, S., Dermine, J. F., Duclos, S., Gagnon, E., Sadoul, R., Rondeau, C., and Desjardins, M. (2001). The phagosome proteome: Insight into phagosome functions. *J. Cell Biol.* **152,** 165–180.

Gatlin, C., Kleemann, G., Hays, L., Link, A., and Yates, J. (1998). Protein identification at the low femtomole level from silver-stained gels using a new fritless electrospray interface for liquid chromatography-microspray and nanospray mass spectrometry. *Anal. Biochem.* **263,** 93–101.

Geng, M. H., Ji, J. Y., and Regnier, F. E. (2000). Signature-peptide approach to detecting proteins in complex mixtures. *J. Chromatogr.* **870,** 295–313.

Gerhold, D., Rushmore, T., and Caskey, C. T. (1999). DNA chips: Promising toys have become powerful tools. *Trends Biochem. Sci.* **24,** 168–173.

Gevaert, K., and Vandekerckhove, J. (2000). Protein identification methods in proteomics. *Electrophoresis* **21,** 1145–1154.

Gharahdaghi, F., Weinberg, C., Meagher, D., Imai, B., and Mische, S. (1999). Mass spectrometry identification of proteins from silver-stained polyacrylamide gel: A method for the removal of silver ions to enhance sensitivity. *Electrophoresis* **20,** 601–605.

Gobom, J., Nordhoff, E., Mirgorodskaya, E., Ekman, R., and Roepstorff, P. (1999). Sample purification and preparation technique based on nano-scale reversed-phase columns for the sensitive analysis of complex peptide mixtures by matrix-assisted laser desorption/ionization mass spectrometry. *J. Mass Spectrom.* **34,** 105–116.

Godovac-Zimmermann, J., and Brown, L. R. (2001). Perspectives for mass spectrometry and functional proteomics. *Mass Spectrom. Rev.* **20,** 1.

Goodlett, D. R., Bruce, J. E., Anderson, G. A., Rist, B., Pasa-Tolic, L., Fiehn, O., Smith, R. D., and Aebersold, R. (2000). Protein identification with a single accurate mass of a cysteine-containing peptide and constrained database searching. *Anal. Chem.* **72,** 1112–1118.

Griffin, T. J., Gygi, S. P., Rist, B., Aebersold, R., Loboda, A., Jilkine, A., Ens, W., and Standing, K. G. (2001). Quantitative proteomic analysis using a MALDI quadrupole time-of-flight mass spectrometer. *Anal. Chem.* **73,** 978–986.

Gruhler, A., Olsen, J. V., Mohammed, S., Mortensen, P., Faergeman, N. F., Mann, M., and Jensen, O. N. (2005). Quantitative phosphoproteomics applied to the yeast pheromone signaling pathway. *Mol. Cell Proteom.* **4,** 310–327.

Gygi, S. P., Corthals, G. L., Zhang, Y., Rochon, Y., and Aebersold, R. (2000). Evaluation of two-dimensional gel electrophoresis-based proteome analysis technology. *Proc. Natl. Acad. Sci. USA* **97,** 9390–9395.

Gygi, S. P., Rist, B., Gerber, S. A., Turecek, F., Gelb, M. H., and Aebersold, R. (1999a). Quantitative analysis of complex protein mixtures using isotope-coded affinity tags. *Nat. Biotechnol.* **17,** 994–999.

Gygi, S. P., Rochon, Y., Franza, B. R., and Aebersold, R. (1999b). Correlation between protein and mRNA abundance in yeast. *Mol. Cell Biol.* **19,** 1720–1730.

Haab, B. B., and Zhou, H. (2004). Multiplexed protein analysis using spotted antibody microarrays. *Methods Mol. Biol.* **264,** 33–45.

Han, D. K., Eng, J., Zhou, H. L., and Aebersold, R. (2001). Quantitative profiling of differentiation-induced microsomal proteins using isotope-coded affinity tags and mass spectrometry. *Nat. Biotechnol.* **19**, 946–951.

Hanisch, F. G., Jovanovic, M., and Peter-Katalinic, J. (2001). Glycoprotein identification and localization of O-glycosylation sites by mass spectrometric analysis of deglycosylated/alkylaminylated peptide fragments. *Anal. Biochem.* **290**, 47–59.

Harry, J. L., Wilkins, M. R., Herbert, B. R., Packer, N. H., Gooley, A. A., and Williams, K. L. (2000). Proteomics: Capacity versus utility. *Electrophoresis* **21**, 1071–1081.

Heid, C. A., Stevens, J., Livak, K. J., and Williams, P. M. (1996). Real time quantitative PCR. *Genome Resl.* **6**, 986–994.

Henzel, W. J., Aswad, D. W., and Stults, J. T. (1989). Structural Analysis of Protein Carboxyl Methyltransferase Utilizing Tandem Mass Spectrometry. *In* "Techniques in Protein Chemistry" (T. E. Hugli, ed.), pp. 127–134. Academic Press, San Diego.

Henzel, W. J., Billeci, T. M., Stults, J. T., Wong, S. C., Grimley, C., and Watanabe, C. (1993). Identifying proteins from two-dimensional gels by molecular mass searching of peptide fragments in protein sequence databases. *Proc. Natl. Acad. Sci. USA* **90**, 5011–5015.

Holt, L. J., Bussow, K., Walter, G., and Tomlinson, I. M. (2000). By-passing selection: Direct screening for antibody-antigen interactions using protein arrays. *Nucl. Acids Res.* **28**, E72.

Hunter, T. (2000). Signaling—2000 and beyond. *Cell* **100**, 113–127.

Husi, H., Ward, M. A., Choudhary, J. S., Blackstock, W. P., and Grant, S. G. (2000). Proteomic analysis of NMDA receptor-adhesion protein signaling complexes. *Nat. Neurosci.* **3**, 661–669.

James, P., Quadroni, M., Carafoli, E., and Gonnet, G. (1993). Protein identification by mass profile fingerprinting. *Biochem. Biophys. Res. Commun.* **195**, 58–64.

James, P. (2001a). Mass spectrometry and the proteome. Quo Vadis. *In* "Proteome Research: Mass Spectrometry" (P. James, ed.), pp. 259–270. Springer-Verlag, Berlin.

James, P. (2001b). "Proteome Research: Mass Spectrometry." Springer-Verlag, Berlin.

Jenkins, R. E., and Pennington, S. R. (2001). Arrays for protein expression profiling: Towards a viable alternative to two-dimensional gel electrophoresis? *Proteomics* **1**, 13–29.

Ji, J. Y., Chakraborty, A., Geng, M., Zhang, X., Amini, A., Bina, M., and Regnier, F. (2000). Strategy for qualitative and quantitative analysis in proteomics based on signature peptides. *J. Chromatogr. B* **745**, 197–210.

Jung, E., Heller, M., Sanchez, J. C., and Hochstrasser, D. F. (2000). Proteomics meets cell biology: The establishment of subcellular proteomes. *Electrophoresis* **21**, 3369–3377.

Jungblut, P. R., Zimny-Arndt, U., Zeindl-Eberhart, E., Stulik, J., Koupilova, K., Pleissner, K. P., Otto, A., Muller, E. C., Sokolowska-Kohler, W., Grabher, G., and Stoffler, G. (1999). Proteomics in human disease: Cancer, heart and infectious diseases. *Electrophoresis* **20**, 2100–2110.

Karas, M., and Hillenkamp, F. (1988). Laser desorption ionization of proteins with molecular masses exceeding 10,000 daltons. *Anal. Chem.* **60**, 2299–2301.

Keough, T., Lacey, M. P., Fieno, A. M., Grant, R. A., Sun, Y. P., Bauer, M. D., and Begley, K. B. (2000). Tandem mass spectrometry methods for definitive protein identification in proteomics research. *Electrophoresis* **21**, 2252–2265.

Klose, J. (1975). Protein mapping by combined isoelectric focusing and electrophoresis of mouse tissues. A novel approach to testing for induced point mutations in mammals. *Humangenetik* **26**, 231–243.

Kononen, J., Bubendorf, L., Kallioniemi, A., Barlund, M., Schraml, P., Leighton, S., Torhorst, J., Mihatsch, M. J., Sauter, G., and Kallioniemi, O. P. (1998). Tissue microarrays for high-throughput molecular profiling of tumor specimens. *Nat. Med.* **4**, 844–847.

Krone, J. R., Nelson, R. W., Dogruel, D., Williams, P., and Granzow, R. (1997). BIA/MS: Interfacing biomolecular interaction analysis with mass spectrometry. *Anal. Biochem.* **244,** 124–132.

Lahm, H. W., and Langen, H. (2000). Mass spectrometry: A tool for the identification of proteins separated by gels. *Electrophoresis* **21,** 2105–2114.

Lander, E. S., Linton, L. M., Birren, B., Nusbaum, C., Zody, M. C., Baldwin, J., Devon, K., Dewar, K., Doyle, M., Fitz Hugh, W., Funke, R., Gage, D., Harris, K., Heaford, A., Howland, J., Kann, L., Lehoczky, J., Le Vine, R., McEwan, P., McKernan, K., Meldrim, J., Mesirov, J. P., Miranda, C., Morris, W., Naylor, J., Raymond, C., Rosetti, M., Santos, R., Sheridan, A., Sougnez, C., Stange-Thomann, N., Stojanovic, N., Subramanian, A., Wyman, D., Rogers, J., Sulston, J., Ainscough, R., Beck, S., Bentley, D., Burton, J., Clee, C., Carter, N., Coulson, A., Deadman, R., Deloukas, P., Dunham, A., Dunham, I., Durbin, R., French, L., Grafham, D., Gregory, S., Hubbard, T., Humphray, S., Hunt, A., Jones, M., Lloyd, C., McMurray, A., Matthews, L., Mercer, S., Milne, S., Mullikin, J. C., Mungall, A., Plumb, R., Ross, M., Shownkeen, R., Sims, S., Waterston, R. H., Wilson, R. K., Hillier, L. W., McPherson, J. D., Marra, M. A., Mardis, E. R., Fulton, L. A., Chinwalla, A. T., Pepin, K. H., Gish, W. R., Chissoe, S. L., Wendl, M. C., Delehaunty, K. D., Miner, T. L., Delehaunty, A., Kramer, J. B., Cook, L. L., Fulton, R. S., Johnson, D. L., Minx, P. J., Clifton, S. W., Hawkins, T., Branscomb, E., Predki, P., Richardson, P., Wenning, S., Slezak, T., Doggett, N., Cheng, J. F., Olsen, A., Lucas, S., Elkin, C., Uberbacher, E., Frazier, M., *et al.* (2001). Initial sequencing and analysis of the human genome. *Nature* **409,** 860–921.

Langen, H., Takacs, B., Evers, S., Berndt, P., Lahm, H. W., Wipf, B., Gray, C., and Fountoulakis, M. (2000). Two-dimensional map of the proteome of *Haemophilus influenzae*. *Electrophoresis* **21,** 411–429.

Larsen, M. R., Sorensen, G. L., Fey, S. J., Larsen, P. M., and Roepstorff, P. (2001). Phosphoproteomics: Evaluation of the use of enzymatic de-phosphorylation and differential mass spectrometric peptide mass mapping for site specific phosphorylation assignment in proteins separated by gel electrophoresis. *Proteomics* **1,** 223–238.

Li, J., Wang, C., Kelly, J. F., Harrison, D. J., and Thibault, P. (2000). Rapid and sensitive separation of trace level protein digests using microfabricated devices coupled to a quadrupole–time-of-flight mass spectrometer. *Electrophoresis* **21,** 198–210.

Link, A. J. (1999). "2-D Proteome Analysis Protocols." Humana Press, Totowa, NJ.

Link, A. J., Eng, J., Schieltz, D. M., Carmack, E., Mize, G. J., Morris, D. R., Garvik, B. M., and Yates, J. R. (1999). Direct analysis of protein complexes using mass spectrometry. *Nat. Biotechnol.* **17,** 676–682.

Loo, J., Brown, J., Critchley, G., Mitchell, C., Andrews, P., and Ogorzalek Loo, R. (1999). High sensitivity mass spectrometry methods for obtaining intact molecular weights from gel-separated proteins. *Electrophoresis* **20,** 743–748.

Lopez, M. F., Berggren, K., Chernokalskaya, E., Lazarev, A., Robinson, M., and Patton, W. F. (2000). A comparison of silver stain and SYPRO Ruby Protein Gel Stain with respect to protein detection in two-dimensional gels and identification by peptide mass profiling. *Electrophoresis* **21,** 3673–3683.

Mac Beath, G., and Schreiber, S. L. (2000). Printing proteins as microarrays for high-throughput function determination. *Science* **289,** 1760–1763.

Mahajan, N. P., Linder, K., Berry, G., Gordon, G. W., Heim, R., and Herman, B. (1998). Bcl-2 and Bax interactions in mitochondria probed with green fluorescent protein and fluorescence resonance energy transfer. *Nat. Biotechnol.* **16,** 547–552.

Mann, M., Hojrup, P., and Roepstorff, P. (1993). Use of mass spectrometric molecular weight information to identify proteins in sequence databases. *Biol. Mass Spectrom.* **22,** 338–345.

Mann, M., and Wilm, M. (1994). Error-tolerant identification of peptides in sequence databases by peptide sequence tags. *Anal. Chem.* **66,** 4390–4399.

Marcotte, E. M., Pellegrini, M., Ng, H. L., Rice, D. W., Yeates, T. O., and Eisenberg, D. (1999). Detecting protein function and protein–protein interactions from genome sequences. *Science* **285,** 751–753.

Martin, S. E., Shabanowitz, J., Hunt, D. F., and Marto, J. A. (2000). Subfemtomole MS and MS/MS peptide sequence analysis using nano-HPLC micro-ESI Fourier transform ion cyclotron resonance mass. *Anal. Chem.* **72,** 4266–4274.

Medzihradszky, K. F., Campbell, J. M., Baldwin, M. A., Falick, A. M., Juhasz, P., Vestal, M. L., and Burlingame, A. L. (2000). The characteristics of peptide collision-induced dissociation using a high-performance MALDI-TOF/TOF Tandem mass spectrometer. *Anal. Chem.* **72,** 552–558.

Miranker, A., Robinson, C. V., Radford, S. E., and Dobson, C. M. (1996). Investigation of protein folding by mass spectrometry. *FASEB J.* **10,** 93–101.

Mirgorodskaya, E., Hassan, H., Clausen, H., and Roepstorff, P. (2001). Mass spectrometric determination of O-glycosylation sites using beta-elimination and partial acid hydrolysis. *Anal. Chem.* **73,** 1263–1269.

Molloy, M. P., Herbert, B. R., Slade, M. B., Rabilloud, T., Nouwens, A. S., Williams, K. L., and Gooley, A. A. (2000). Proteomic analysis of the *Escherichia coli* outer membrane. *Eur. J. Biochem.* **267,** 2871–2881.

Muzio, M., Chinnaiyan, A. M., Kischkel, F. C., O'Rourke, K., Shevchenko, A. J. N., Scaffidi, C., Bretz, J. D., Zhang, M., Gentz, R., Mann, M., Krammer, P. H., Peter, M. E., and Dixit, V. M. (1996). FLICE, a novel FADD-homologous ICE/CED-3-like protease, is recruited to the CD95 (Fas/APO-1) death-inducing signaling complex. *Cell* **85,** 817–827.

Nelson, P. S., Han, D., Rochon, Y., Corthals, G. L., Lin, B., Monson, A., Nguyen, V., Franza, B. R., Plymate, S. R., Aebersold, R., and Hood, L. (2000a). Comprehensive analyses of prostate gene expression: Convergence of expressed sequence tag databases, transcript profiling and proteomics. *Electrophoresis* **21,** 1823–1831.

Nelson, R. W., Nedelkov, D., and Tubbs, K. A. (2000b). Biomolecular interaction analysis mass spectrometry. BIA/MS can detect and characterize protiens in complex biological fluids at the low- to subfemtomole level. *Anal. Chem.* **72,** 404A–411A.

Neubauer, G., King, A., Rappsilber, J., Calvio, C., Watson, M., Ajuh, P., Sleeman, J., Lamond, A., and Mann, M. (1998). Mass spectrometry and EST-database searching allows characterization of the multi-protein spliceosome complex. *Nat. Genet.* **20,** 46–50.

Neugebauer, J., Moseley, M. A., Vissers, J. P. C., and Moyer, M. (1999). Comparison of nanospray and fully automated nanoscale capillary LC/MS/MS for protein identification. *In* "ASMS Conference on Mass Spectrometry and Allied Topics." Dallas, TX.

Nielsen, U. B., and Geierstanger, B. H. (2004). Multiplexed sandwich assays in microarray format. *J. Immunol. Methods* **290,** 107–120.

Nordhoff, E., Krogsdam, A. M., Jorgensen, H. F., Kallipolitis, B. H., Clark, B. F., Roepstorff, P., and Kristiansen, K. (1999). Rapid identification of DNA-binding proteins by mass spectrometry. *Nat. Biotechnol.* **17,** 884–888.

Nuwaysir, L. M., and Stults, J. T. (1993). Electrospray ionization mass spectrometry of phosphopeptides isolated by on-line immobilized metal-ion affinity chromatography. *J. Am. Soc. Mass Spectrom.* **4,** 662–669.

Oda, Y., Huang, K., Cross, F. R., Cowburn, D., and Chait, B. T. (1999). Accurate quantitation of protein expression and site-specific phosphorylation. *Proc. Natl. Acad. Sci. USA* **96,** 6591–6596.

Oda, Y., Nagasu, T., and Chait, B. T. (2001). Enrichment analysis of phosphorylated proteins as a tool for probing the phosphoproteome. *Nat. Biotechnol.* **19,** 379–382.

O'Farrell, P. H. (1975). High resolution two-dimensional electrophoresis of proteins. *J. Biol. Chem.* **250**, 4007–4021.

Ong, S. E., Blagoev, B., Kratchmarova, I., Kristensen, D. B., Steen, H., Pandey, A., and Mann, M. (2002). Stable isotope labeling by amino acids in cell culture, SILAC, as a simple and accurate approach to expression proteomics. *Mol. Cell Proteom.* **1**, 376–386.

Orchard, S., Hermjakob, H., and Apweiler, R. (2003). The proteomics standards initiative. *Proteomics* **3**, 1374–1376.

Ornstein, D. K., Gillespie, J. W., Paweletz, C. P., Duray, P. H., Herring, J., Vocke, C. D., Topalian, S. L., Bostwick, D. G., Linehan, W. M., Petricoin, E. F., 3rd, and Emmert-Buck, M. R. (2000). Proteomic analysis of laser capture microdissected human prostate cancer and *in vitro* prostate cell lines. *Electrophoresis* **21**, 2235–2242.

Ostergaard, M., Rasmussen, H. H., Nielsen, H. V., Vorum, H., Orntoft, T. F., Wolf, H., and Celis, J. E. (1997). Proteome profiling of bladder squamous cell carcinomas: Identification of markers that define their degree of differentiation. *Cancer Res.* **57**, 4111–4117.

Ostergaard, M., Wolf, H., Orntoft, T. F., and Celis, J. E. (1999). Psoriasin (S100A7): A putative urinary marker for the follow-up of patients with bladder squamous cell carcinomas. *Electrophoresis* **20**, 349–354.

Page, M. J., Amess, B., Rohlff, C., Stubberfield, C., and Parekh, R. (1999a). Proteomics: A major new technology for the drug discovery process. *Drug Discov. Today* **4**, 55–62.

Page, M. J., Amess, B., Townsend, R. R., Parekh, R., Herath, A., Brusten, L., Zvelebil, M. J., Stein, R. C., Waterfield, M. D., Davies, S. C., and O'Hare, M. J. (1999b). Proteomic definition of normal human luminal and myoepithelial breast cells purified from reduction mammoplasties. *Proc. Natl. Acad. Sci. USA* **96**, 12589–12594.

Pan, S., Zhang, H., Rush, J., Eng, J., Zhang, N., Patterson, D., Comb, M. J., and Aebersold, R. (2005). High throughput proteome screening for biomarker detection. *Mol. Cell Proteom.* **4**, 182–190.

Pandey, A., Fernandez, M. M., Steen, H., Blagoev, B., Nielsen, M. M., Roche, S., Mann, M., and Lodish, H. F. (2000a). Identification of a novel immunoreceptor tyrosine-based activation motif-containing molecule, STAM2, by mass spectrometry and its involvement in growth factor and cytokine receptor signaling pathways. *J. Biol. Chem.* **275**, 38633–38639.

Pandey, A., and Mann, M. (2000). Proteomics to study genes and genomes. *Nature* **405**, 837–846.

Pandey, A., Podtelejnikov, A. V., Blagoev, B., Bustelo, X. R., Mann, M., and Lodish, H. F. (2000b). Analysis of receptor signaling pathways by mass spectrometry: Identification of vav-2 as a substrate of the epidermal and platelet-derived growth factor receptors. *Proc. Natl. Acad. Sci. USA* **97**, 179–184.

Pappin, D. J. C., Hojrup, P., and Bleasby, A. J. (1993). Rapid identification of proteins by peptide-mass fingerprinting. *Curr. Biol.* **3**, 327–332.

Pawson, T., and Nash, P. (2000). Protein–protein interactions define specificity in signal transduction. *Genes Dev.* **14**, 1027–1047.

Peng, J., Schwartz, D., Elias, J. E., Thoreen, C. C., Cheng, D., Marsischky, G., Roelofs, J., Finley, D., and Gygi, S. P. (2003). A proteomics approach to understanding protein ubiquitination. *Nat. Biotechnol.* **21**, 921–926.

Perkins, D. N., Pappin, D. J. C., Creasy, D. M., and Cottrell, S. (1999). Probability-based protein identification by searching sequence databases using mass spectrometry data. *Electrophoresis* **20**, 3551–3567.

Petricoin, E. F., Ardekani, A. M., Hitt, B. A., Levine, P. J., Fusaro, V. A., Steinberg, S. M., Mills, G. B., Simone, C., Fishman, D. A., Kohn, E. C., and Liotta, L. A. (2002). Use of proteomic patterns in serum to identify ovarian cancer. *Lancet* **359**, 572–577.

Pieper, R., Su, Q., Gatlin, C. L., Huang, S. T., Anderson, N. L., and Steiner, S. (2003). Multi-component immunoaffinity subtraction chromatography: An innovative step towards a comprehensive survey of the human plasma proteome. *Proteomics* **3,** 422–432.

Posewitz, M. C., and Tempst, P. (1999). Immobilized gallium(III) affinity chromatography of phosphopeptides. *Anal. Chem.* **71,** 2883–2892.

Ransohoff, D. F. (2005). Lessons from controversy: Ovarian cancer screening and serum proteomics. *J. Natl. Cancer. Inst.* **97,** 315–319.

Rappsilber, J., Siniossoglou, S., Hurt, E. C., and Mann, M. (2000). A generic strategy to analyze the spatial organization of multi-protein complexes by cross-linking and mass spectrometry. *Anal. Chem.* **72,** 267–275.

Reymond, M. A., Sanchez, J. C., Hughes, G. J., Gunther, K., Riese, J., Tortola, S., Peinado, M. A., Kirchner, T., Hohenberger, W., Hochstrasser, D. F., and Kockerling, F. (1997). Standardized characterization of gene expression in human colorectal epithelium by two-dimensional electrophoresis. *Electrophoresis* **18,** 2842–2848.

Rigaut, G., Shevchenko, A., Rutz, B., Wilm, M., Mann, M., and Seraphin, B. (1999). A generic protein purification method for protein complex characterization and proteome exploration. *Nat. Biotechnol.* **17,** 1030–1032.

Rout, M. P., Aitchison, J. D., Suprapto, A., Hjertaas, K., Zhao, Y., and Chait, B. T. (2000). The yeast nuclear pore complex: Composition, architecture, and transport mechanism. *J. Cell Biol.* **148,** 635–652.

Rudd, P. M., Colominas, C., Royle, l., Murphy, N., Hart, E., Merry, A., Heberstreit, H., and Dwek, R. A. (2001). Glycoproteomics; High-Throughput Sequencing of Oligosaccharide Modifications to Proteins. *In* "Proteome Research: Mass Spectrometry" (P. James, ed.), pp. 207–228. Springer-Verlag, Berlin.

Santoni, V., Molloy, M., and Rabilloud, T. (2000). Membrane proteins and proteomics: Un amour impossible? *Electrophoresis* **21,** 1054–1070.

Scheele, G. A. (1975). Two-dimensional gel analysis of soluble proteins. Characterization of guinea pig exocrine pancreatic proteins. *J. Biol. Chem.* **250,** 5375–5385.

Schleuder, D., Hillenkamp, F., and Strupat, K. (1999). IR-MALDI-mass analysis of electroblotted proteins directly from the membrane: Comparison of different membranes, application to on-membrane digestion, and protein identification by database searching. *Anal. Chem.* **71,** 3238–3247.

Shalon, D., Smith, S. J., and Brown, P. O. (1996). A DNA microarray system for analyzing complex DNA samples using two-color fluorescent probe hybridization. *Genome Res.* **6,** 639–645.

Shevchenko, A., Loboda, A., Ens, W., and Standing, K. G. (2000). MALDI quadrupole time-of-flight mass spectrometry: A powerful tool for proteomic research. *Anal. Chem.* **72,** 2132–2141.

Shevchenko, A., Wilm, M., Vorm, O., and Mann, M. (1996). Mass spectrometric sequencing of proteins from silver stained polyacrylamide gels. *Anal. Chem.* **68,** 850–858.

Simone, N. L., Bonner, R. F., Gillespie, J. W., Emmert-Buck, M. R., and Liotta, L. A. (1998). Laser-capture microdissection: Opening the microscopic frontier to molecular analysis. *Trends Genet.* **14,** 272–276.

Simpson, R. J., Connolly, L. M., Eddes, J. S., Pereira, J. J., Moritz, R. L., and Reid, G. E. (2000). Proteomic analysis of the human colon carcinoma cell line (LIM 1215): Development of a membrane protein database. *Electrophoresis* **21,** 1707–1732.

Sobott, F., and Robinson, C. V. (2002). Protein complexes gain momentum. *Curr. Opin. Struct. Biol.* **12,** 729–734.

Spahr, C. S., Davis, M. T., McGinley, M. D., Robinson, J. H., Bures, E. J., Beierle, J., Mort, J., Courchesne, P. L., Chen, K., Wahl, R. C., Yu, W., Luethy, R., and Patterson, S. D. (2001).

Towards defining the urinary proteome using liquid chromatography-tandem mass spectrometry I. Profiling an unfractionated tryptic digest. *Proteomics* **1,** 93–107.

Srinivas, P. R., Srivastava, S., Hanash, S., and Wright, G. L., Jr. (2001). Proteomics in early detection of cancer. *Clin. Chem.* **47,** 1901–1911.

Steiner, S., Aicher, L., Raymackers, J., Meheus, L., Esquer-Blasco, R., Anderson, N. L., and Cordier, A. (1996). Cyclosporine A decreases the protein level of the calcium-binding protein calbindin-D 28 kDa in rat kidney. *Biochem. Pharmacol.* **51,** 253–258.

Steiner, S., Wahl, D., Varela, M. D. C., Aicher, L., and Prieto, P. (1995). Protein variability in male and female Wistar rat liver proteins. *Electrophoresis* **16,** 1969–1976.

Steiner, S., and Witzmann, F. A. (2000). Proteomics: Applications and opportunities in preclinical drug development. *Electrophoresis* **21,** 2099–2104.

Stulik, J., Koupilova, K., Osterreicher, J., Knizek, J., Macela, A., Bures, J., Jandik, P., Langr, F., Dedic, K., and Jungblut, P. R. (1999). Protein abundance alterations in matched sets of macroscopically normal colon mucosa and colorectal carcinoma. *Electrophoresis* **20,** 3638–3646.

Taylor, J., and Johnson, R. (1997). Sequence database searches in *de novo* peptide sequencing by Tandem mass spectrometry. *Rapid Commun. Mass Spectrom.* **11,** 1067–1075.

Ünlü, M., Morgan, M. E., and Minden, J. S. (1997). Difference gel electrophoresis: A single gel method for detecting changes in protein extracts. *Electrophoresis* **18,** 2071–2077.

VanBogelen, R. A., Greis, K. D., Blumenthal, R. M., Tani, T. H., and Matthews, R. G. (1999). Mapping regulatory networks in microbial cells. *Trends Microbiol.* **7,** 320–328.

Veenstra, T. D., Conrads, T. P., Hood, B. L., Avellino, A. M., Ellenbogen, R. G., and Morrison, R. S. (2005). Biomarkers: Mining the biofluid proteome. *Mol. Cell Proteom.* **4,** 409–418.

Velculescu, V. E., Vogelstein, B., and Kinzler, K. W. (2000). Analysing uncharted transcriptomes with SAGE. *Trends Genet.* **16,** 423–425.

Venter, J. C., Adams, M. D., Myers, E. W., Li, P. W., Mural, R. J., Sutton, G. G., Smith, H. O., Yandell, M., Evans, C. A., Holt, R. A., Gocayne, J. D., Amanatides, P., Ballew, R. M., Huson, D. H., Wortman, J. R., Zhang, Q., Kodira, C. D., Zheng, X. H., Chen, L., Skupski, M., Subramanian, G., Thomas, P. D., Zhang, J., Gabor Miklos, G. L., Nelson, C., Broder, S., Clark, A. G., Nadeau, J., McKusick, V. A., Zinder, N., Levine, A. J., Roberts, R. J., Simon, M., Slayman, C., Hunkapiller, M., Bolanos, R., Delcher, A., Dew, I., Fasulo, D., Flanigan, M., Florea, L., Halpern, A., Hannenhalli, S., Kravitz, S., Levy, S., Mobarry, C., Reinert, K., Remington, K., Abu-Threideh, J., Beasley, E., Biddick, K., Bonazzi, V., Brandon, R., Cargill, M., Chandramouliswaran, I., Charlab, R., Chaturvedi, K., Deng, Z., Di Francesco, V., Dunn, P., Eilbeck, K., Evangelista, C., Gabrielian, A. E., Gan, W., Ge, W., Gong, F., Gu, Z., Guan, P., Heiman, T. J., Higgins, M. E., Ji, R. R., Ke, Z., Ketchum, K. A., Lai, Z., Lei, Y., Li, Z., Li, J., Liang, Y., Lin, X., Lu, F., Merkulov, G. V., Milshina, N., Moore, H. M., Naik, A. K., Narayan, V. A., Neelam, B., Nusskern, D., Rusch, D. B., Salzberg, S., Shao, W., Shue, B., Sun, J., Wang, Z., Wang, A., Wang, X., Wang, J., Wei, M., Wides, R., Xiao, C., Yan, C., *et al.* (2001). The sequence of the human genome. *Science* **291,** 1304–1351.

Walter, G., Bussow, K., Cahill, D., Lueking, A., and Lehrach, H. (2000). Protein arrays for gene expression and molecular interaction screening. *Curr. Opin. Microbiol.* **3,** 298–302.

Washburn, M. P., Wolters, D., and Yates, J. R., 3rd. (2001). Large-scale analysis of the yeast proteome by multidimensional protein identification technology. *Nat. Biotechnol.* **19,** 242–247.

Wilkins, M., Sanchez, J., Gooley, A., Appel, R., Humphry-Smith, I., Hochstrasser, D., and Williams, K. (1995). Progress with proteome projects: Why all proteins expressed by a genome should be identified and how to do it. *Biotechnol. Genet. Eng. Rev.* **13,** 19–50.

Wilkins, M. R., Williams, R. D., Appel, R. D., and Hochstrasser, D. F. (1997). *In* "Proteome Research: New Frontiers in Functional Genomics," (M. R. Wilkins K. L. Williams R. D. Appel and D. F. Hochstrasser, eds.), p. 243. Springer-Verlag, Berlin.

Wilm, M., and Mann, M. (1996). Analytical properties of the nanoelectrospray ion source. *Anal. Chem.* **68,** 1–8.

Wilm, M., Shevchenko, A., Houthaeve, T., Breit, S., Schweigerer, L., Fotsis, T., and Mann, M. (1996). Femtomole sequencing of proteins from polyacrylamide gels by nano-electrospray mass spectrometry. *Nature* **379,** 466–469.

Wolters, D. A., Washburn, M. P., and Yates, J. R., 3rd. (2001). An automated multidimensional protein identification technology for shotgun proteomics. *Anal. Chem.* **73,** 5683–5690.

Yates, J. R., III, Speicher, S., Griffin, P. R., and Hunkapiller, T. (1993). Peptide mass maps: A highly informative approach to protein identification. *Anal. Biochem.* **214,** 397–408.

Young, M. M., Tang, N., Hempel, J. C., Oshiro, C. M., Taylor, E. W., Kuntz, I. D., Gibson, B. W., and Dollinger, G. (2000). High throughput protein fold identification by using experimental constraints derived from intramolecular cross-links and mass spectrometry. *Proc. Natl. Acad. Sci. USA* **97,** 5802–5806.

Zhang, W. Z., and Chait, B. T. (2000). Profound: An expert system for protein identification using mass spectrometric peptide mapping information. *Anal. Chem.* **72,** 2482–2489.

Zhang, X. L., Herring, C. J., Romano, P. R., Szczepanowska, J., Brzeska, H., Hinnebusch, A. G., and Qin, J. (1998). Identification of phosphorylation sites in proteins separated by polyacrylamide gel electrophoresis. *Anal. Chem.* **70,** 2050–2059.

Zhang, Z., Li, W., Logan, T. M., Li, M., and Marshall, A. G. (1997). Human recombinant [C22A] FK506-binding protein amide hydrogen exchange rates from mass spectrometry match and extend those from NMR. *Protein Sci.* **6,** 2203–2217.

Zhou, W., Merrick, B. A., Khaledi, M. G., and Tomer, K. B. (2000). Detection and sequencing of phosphopeptides affinity bound to immobilized metal ion beads by matrix-assisted laser desorption/ionization mass spectrometry. *J. Am. Soc. Mass Spectrom.* **11,** 273–282.

Zhu, W., Reich, C. I., Olsen, G. J., Giometti, C. S., and Yates, J. R., 3rd. (2004). Shotgun proteomics of *Methanococcus jannaschii* and insights into methanogenesis. *J. Proteome Res.* **3,** 538–548.

[9] Bioinformatic Methods to Exploit Mass Spectrometric Data for Proteomic Applications

By Robert J. Chalkley, Kirk C. Hansen, and Michael A. Baldwin

Abstract

The new technologies in mass spectrometric analysis of peptides and proteins necessary to accommodate proteomics-scale analyses require, in turn, concomitant development of informatics technologies suitable for very large-scale data handling and analysis. This chapter focuses on the data analysis tools available to the community for analysis of mass

METHODS IN ENZYMOLOGY, VOL. 402
Copyright 2005, Elsevier Inc. All rights reserved.

0076-6879/05 $35.00
DOI: 10.1016/S0076-6879(05)02009-4

spectrometric proteomics data. Different database searching strategies are discussed for peptide and protein identification, and approaches and tools available for comparative quantitative analysis of samples are outlined.

Introduction

Just as the study of individual genes has been transformed into the high-throughput and highly context-centered study of whole genomes, so has the study of individual proteins been enormously extended through the development of the multifaceted field of proteomics. Thus, although detailed studies of enzyme mechanisms continue to provide important insights into how biological systems function at the molecular level, larger scale studies of how proteins behave in pathways contribute new views of biological chemistry in the dynamic context of the living organism. And although critical information regarding the precise roles of phosphorylation/dephosphorylation of specific signaling molecules will continue to be important for understanding individual steps in these networks, larger scale studies of posttranslational processing across entire proteomes are being undertaken to understand the role of phosphorylation in the overall dynamic process.

The new technologies in mass spectrometric analysis of peptides and proteins necessary to accommodate proteomics-scale analyses require, in turn, concomitant development of informatics technologies suitable for very large-scale data handling and analysis. These technologies represent two primary streams of development: data handling/storage/retrieval, as represented by mass spectrometry (MS)–specific laboratory information management systems (LIMS), and data analysis, which is the subject of this chapter. Until recently, both types of bioinformatic infrastructure have been developed solely by the MS community, tied intimately and specifically to the development of the physical technologies themselves. However, information technology (IT) experts, programmers, statisticians, and computational biologists are now found in many cutting-edge MS laboratories. This type of intimate collaboration between mass spectrometrists and bioinformatics professionals will be necessary for the development of the sophisticated next-generation bioinformatic tools.

Mass Spectrometric Analysis of Peptides and Proteins

The analysis of underivatized peptides and small proteins by MS became feasible with the development of soft ionization methods, including field desorption, direct chemical ionization, fast atom bombardment (FAB), and liquid secondary ion MS. More recently, the introduction of

electrospray ionization (ESI) and matrix-assisted laser desorption ionization (MALDI) enabled the routine and highly sensitive detection of isolated
gas-phase ions of intact proteins of virtually any size (see Chapter 1).
Despite this capability, relatively little use has been made of the analysis
of intact proteins for protein identification. The mass spectrum of a protein
typically gives only a single piece of information, its molecular weight.
Furthermore, unless this measurement is made in an ultrahigh resolution
instrument such as a Fourier transform (FT) ion cyclotron resonance (ICR)
mass spectrometer, only for the smallest proteins will the isotopic components of the molecular ion be resolvable, and the measurement of molecular weight will not be sufficiently accurate to distinguish between hundreds
or even thousands of possible isobaric species. In addition, the molecular
mass of a protein will frequently differ from that predicted by the gene
sequence, which can be attributable to events such as alternative splicing,
posttranslational processing of the expressed protein or subsequent chemical modifications that can either add or remove mass, or even errors in the
databases. Thus, for an identified protein, the molecular mass (or masses
for a heterogeneously modified protein) may not be very revealing.

Analyses that start from the intact protein molecular ion have come to
be known as "top-down." Although the collision-induced dissociation
(CID) techniques that readily give sequence information for peptide ions
are generally not applicable to species as large as intact protein ions, the
fragmentation mechanisms of electron-capture dissociation (ECD) and
electron-transfer dissociation (ETD) make direct protein sequencing a
realistic possibility (Horn et al., 2000). Here, the attachment of a low
thermal energy electron to one of the positive charges in a multiply charged
ion with a recombination energy of about 6 eV gives a radical ion that will
rapidly fragment before energy equilibration occurs. Such species are
classified as *distonic ions* because the charge and radical sites are distinct
from each other. As in the well-known radical cation reactions of species
formed by electron impact, alpha-cleavage occurs not at the site of the
charge but one bond removed, giving predominant c- and z-ions for cleavage between the amide nitrogen and the α-carbon (see Fig. 1). This differs
from CID techniques that predominantly produce b- and y-type ions.

The alternative "bottom-up" approach has been the major focus for
protein identification during the last decade, based on the demonstration
that digestion of a protein with a protease such as trypsin and the analysis
of the resulting peptide mixture can provide sufficient information to
identify it in a protein database. The specificity of the protease is known,
so the molecular weights of the resulting peptides can be predicted for any
protein sequence. Even when the peptide masses are measured in an
instrument of relatively modest performance, proteins may be identified

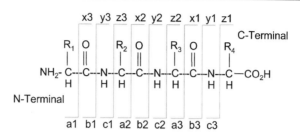

Fig. 1. Peptide fragment ion types.

from the detection of a relatively small subset of the possible peptides, using the technique referred to as *peptide mass mapping* or *mass finger-printing* (Henzel *et al.*, 1993; James *et al.*, 1993; Jensen *et al.*, 1997; Mann *et al.*, 1993; Pappin *et al.*, 1993; Yates *et al.*, 1993). As genomic databases started to grow more rapidly, it was realized that the criteria for protein identification would need to be more stringent and it became necessary to improve the accuracy of mass measurement. For routine operation, this was achieved by moving from linear to reflectron MALDI-TOF instruments with delayed extraction.

Whereas peptide mass fingerprinting is able to identify a purified protein from its digest, the technique's performance becomes unreliable if a mixture of components is present in the sample. For more confident and reliable identification, fragmentation spectra produced by tandem MS can be matched against sequence tags predicted for all proteins in a database (Clauser *et al.*, 1995, 1996, 1999; Eng *et al.*, 1994; Mann and Wilm, 1994; Medzihradszky and Burlingame, 1994; Wilm and Mann, 1996; Yates *et al.*, 1996). As originally formulated, mass tags represent a series of fragment ions that can be attributed to a coherent sequence of amino acids that is a subset of the sequence of the peptide under study. Further developments that are usually grouped with mass tag searches are actually based on comparisons between the experimentally observed fragment ions and all predicted fragments for all hypothetical peptides of the appropriate parent mass, based on known fragmentation rules. If the peptide of interest is not present in the database, *de novo* peptide sequencing can be undertaken, again based on known rules for peptide fragmentation (Biemann, 1990). This is most important when a significant proportion of peptides diverge from those predicted due to errors or incompleteness of databases, discrepancies between genomic sequence and processed or posttranslationally modified proteins, species differences, and nonspecific enzyme cleavages. Techniques have been developed to identify proteins based on sequence homologies to related proteins or equivalent proteins in different species, although these are time consuming and computationally intensive compared with straightforward searching based on sequence tags (Huang *et al.*, 2001; Taylor and Johnson, 1997).

Good mass accuracy not only increases the reliability of database searching by limiting the possible compositions of peptides for any given mass but also reduces the search times. The quality of mass spectral information that has been employed for proteomic applications has ranged from low-resolution MALDI-TOF spectra in which mass accuracy of ±1 Da was routine to FT-ICR measurements achieving better than 1 ppm accuracy. Routine operation in a DE MALDI-TOF with reflectron gives better than 50-ppm mass accuracy, and significantly better (~10 ppm) may be achieved with careful internal calibration. ESI sources are commonly used on linear quadrupole instruments or quadrupole ions traps that typically give fairly modest performance in terms of resolution and mass accuracy, although the combination of a quadrupole mass selector and quadrupole collision cell with orthogonal acceleration (QqoaTOF) gives high resolution (~10,000) and perhaps 10-ppm mass accuracy if well calibrated. FT-ICR instrumentation can routinely achieve 2-ppm mass accuracy or better, which in many cases is sufficient to specify the elemental composition of the peptide.

Sequence ion information from a MALDI-TOF instrument using post-source decay (PSD) is useful, although it is painstaking to collect and the accuracy of mass measurements is relatively low (Spengler *et al.*, 1992), so this is generally employed only when no alternative tandem MS instrument is available in a lab. Higher quality CID spectra can be obtained by MALDI-TOF/TOF, ion trap, or QqTOF geometry instruments employing either MALDI or ESI ionization.

Existing Tools for Protein Identification: What Is Available?

Several web sites offer free access to web-based database-searching programs for peptide mass fingerprinting and the identification of sequence tags, all of which provide other tools as well, such as prediction of enzyme digests and prediction of CID fragments. An alternative to Internet access is to purchase a license to have the programs resident in-house, and most mass spectrometer vendors supply software and software licenses with the purchase of an instrument. Having an in-house system provides advantages of data security and the ability to use private databases or databases configured for specific tasks. It gives the user the option of employing a more powerful computer or installing software on a computer cluster to speed up search times.

Some packages available for online searching include tools on the ExPASy proteomics server provided by the Swiss Institute of Bioinformatics (*http://us.expasy.org/tools/*), Mascot from Matrix Science (*http://www. matrixscience.com*), Peptide Search from the European Molecular Biology Laboratory (*http://www.mann.embl-heidelberg.de/GroupPages/PageLink/peptidesearchpage.html*), ProteinProspector from the University of California, San Francisco (*http://prospector.ucsf.edu*), or distributed by Applied Biosystems, Xtandem! (as part of the Global Proteome Organization machine web site at http://www.thegpm.org), and Prowl from Rockefeller University (*http://prowl.rockefeller.edu*), or related Knexus software (*http://hs1.proteome.ca/prowl/knexus.html*). Other software distributed by instrument manufacturers include Sequest from Thermo-Finnigan (San Jose, CA) and Spectrum Mill from Agilent Technologies (Palo Alto, CA), and other manufacturers produce their own software analysis packages. Most of these packages contain a number of tools, including programs for identifying proteins from a list of peptide masses (mass mapping or fingerprinting), fragment masses (mass tags), or single or multiple sections of sequence tags, which could come from *de novo* interpretation of CID spectra or from other sources.

With the availability of non-ergodic fragmentation mechanisms that can efficiently fragment small intact proteins, searching tools need to be developed to analyze these data. This type of data can be conceptually analyzed using software for analyzing peptide fragmentation such as MS-Tag (see later discussion in this chapter). However, specific software for this task is being developed, the first of which, "Prosight PTM," is designed for identifying modification sites from intact protein fragmentation (Taylor *et al.*, 2003).

The major instrument types used for acquisition of tandem MS data are ion trap instruments and QqTOF geometry instruments. Data acquired on an ion trap instrument generally have a lower information content than QqTOF data, as the lower resolution leads to lack of charge-state determination of parent and fragment ions, the lower mass accuracy requires wider mass tolerances, and the absence of the low mass region loses further information. Hence, discrimination between correct and incorrect matches is more difficult with this type of data. A popular search engine for analysis of ion trap data is Sequest, which was specifically designed for this type of data (Eng *et al.*, 1994). Several research groups have now developed tools that reinterrogate Sequest results using multiple factors about the search result to produce more reliable peptide and protein summaries (Keller *et al.*, 2002; MacCoss *et al.*, 2002; Moore *et al.*, 2002; Sadygov *et al.*, 2004).

The higher information content of data acquired on QqTOF geometry instruments allows better discrimination between correct and incorrect answers, but false assignments are still made. One has to remember that the output from the search result is a "best fit" to sequences in a database, with an attempt to apply a significance threshold. Almost all database searches result in incorrect second-best matches for particular spectra, which would pass all scoring thresholds and would be reported correct if the best matching sequence was not in the database, or if the correct peptide bears an unaccounted for modification. Thus, there are nearly always going to be incorrect peptide assignments in non-curated analyses.

Without a community standard for protein identification, a significant issue within the proteomics field is the difficulty of assessing and comparing results acquired on different instrument platforms and/or analyzed using different search algorithms. This creates a significant challenge for journal reviewers and editors with respect to assessing whether submitted results are of sufficient quality/confidence for publication. In response to this, the journal *Molecular and Cellular Proteomics* published some initial guidelines for data submission (Carr *et al.*, 2004). Although these are not expected to be rigid or in any way comprehensive, the journal's intention is to provoke discussion about the subject so some rules can be applied within the community as a whole.

Many of these problems are caused because raw MS data can currently be stored only in manufacturer-specific data formats, which are not compatible with each other. As a result, it is difficult to develop generic tools for proteomic data analysis and to accurately assess how different search engines are performing. Several groups are aiming to solve this problem. In a major initiative at the Institute for Systems Biology in Seattle, open-source tools for analysis of MS data acquired on any instrument platform are being developed, allowing direct comparison of data acquired on different instruments (*http://sashimi.sourceforge.net/*; Pedrioli *et al.*, 2004). This project includes software to convert data from any mass spectrometer into a common data format, tools for reassessing and improving reliability of search results (Keller *et al.*, 2002; Nesvizhskii *et al.*, 2003), and tools for analysis of samples labeled with isotopic quantitation reagents (Li *et al.*, 2003). The Manitoba Proteomics Center also is making available open-source software for capture and analysis of raw data acquired on many instrument platforms (*http://www.proteome.ca/opensource.html*). The Human Proteome (HUPO) initiative is also attempting to promote a common data format (Orchard *et al.*, 2004). These are all noble efforts; it remains to be seen whether any of these will be accepted on a large scale by the proteomics community.

Anatomy of a Web-Based Package: ProteinProspector

As previously stated, a large number of tools are available on the web for analysis of proteomic data. ProteinProspector is one of the more complete packages of programs for database analysis, and we describe these programs further to exemplify the types of tools available.

ProteinProspector, a suite of programs described as providing "tools for mining sequence databases in conjunction with mass proteometry experiments," was created by Karl Clauser and Peter Baker at the University of California, San Francisco (*http://www.prospector.ucsf.edu*). The software has a web-based interface and is available at *http://prospector.ucsf.edu*, as well as a mirror site at the Ludwig Institute in London. All Protein Prospector programs are written in the C++ programming language. They receive their input according to the standard common gateway interface (CGI) protocol, and initially produced output in HTML format, although XML is a recent introduction to the current version available over the web at the time of this writing (Version 4.0.5). Output of the programs can be readily parsed for downstream organization and storage in a relational database. The programs also run from the command line, which lends itself to automation.

The current web version of ProteinProspector contains the following programs:

Sequence Database Search Programs

- MS-Fit. A peptide mass fingerprinting tool that matches molecular weights of peptides observed from the enzymatic digest of a protein to the peptides predicted for each protein sequence in a database.
- MS-Tag. Identifies proteins based on comparing a list of fragment ion masses from a CID spectrum of a peptide to theoretical fragment ions for each predicted peptide of the same molecular mass in the database produced by a given enzyme cleavage specificity.
- MS-Pattern. A database-searching tool that performs text-based searches of amino acid sequence (with or without peptide mass filtering) against sequence databases.
- MS-Homology. A similar tool to MS-Pattern, it can take CID MS data as any combination of amino acid tags and masses and report peptides identical or homologous in the database. It can search data from multiple CID spectra and require a minimum number of peptides to match to the same protein in order for it to be reported. It is designed for identifying homologous proteins in the database, and because of its ability to search with multiple sequences, it is possible to identify remote homology.

• MS-Seq. An alternative to MS-Tag based on "PeptideSearch" (Mann and Wilm, 1994).

• MS-Bridge. This identifies intermolecular or intramolecular disulfide-linked peptides and can be used to identify chemical cross-linked species with user-definable structures.

Peptide/Protein Utility Programs

• MS-Digest. Performs an *in silico* enzymatic digestion of a protein sequence and reports the mass of each predicted peptide. Possible inputs include the MS-Digest index number produced by MS-Fit, MS-Tag, MS-Seq, or MS-Pattern, the accession number from a database, or a user-supplied sequence.

• MS-Product. Calculates all fragment ions resulting from fragmentation of a peptide. The fragment ion types reported are defined by specifying the instrument type used for acquisition of the data or can be selected by the user.

• MS-Isotope. Calculates stable isotope profiles of peptides (or other organic molecules) based on peak shape, resolution, and charge. It can plot two or more overlapping profiles, including peptides of different charges but similar m/z values. In addition to specifying an elemental composition or a peptide sequence, an "average mass" can be specified to give a "typical" profile for a peptide at any given mass and charge combination.

• MS-Comp. Calculates possible amino acid compositions of a peptide based on a parent or fragment m/z and partial composition from the immonium ions or other data contained in a CID spectrum. It is possible to specify various alternative ion types.

• DB-Stat. Calculates useful statistics for any specified database.

A variety of databases are available through the Internet for searching proteomic data. These range from large essentially non-curated protein sequence repositories to smaller highly annotated databases to theoretical translations of EST sequences.

The most highly curated protein database available is SwissProt (*http://www.ebi.ac.uk/swissprot/index.html*). This is a non-redundant database, in which multiple sequences for a given protein are collapsed into one entry. It contains extensive integration and links to other reference databases including information about protein domains and structure and protein families. As this database is very highly curated, new entries into the database are relatively slow to arrive. Hence, a less curated, but more rapidly expanding offshoot TrEMBL allows quicker entry of a sequence into the database. This is still annotated, but the curation is largely automated based on similarity to proteins in the SwissProt database. These two databases, along with another annotated database, PIR-PSD, have been combined into a new set of databases under the UniProt consortium (*http://www.pir.uniprot.org/*). UniProt contains additional annotation including gene ontology annotation (Consortium, 2001), which attempts to assign function, subcellular location, and processes in which a protein is involved.

The largest and most comprehensive database is the NCBI's Entrez protein database, which is formed by the combination of a number of databases including Genpept, which is a translation of the Genbank nucleic acid sequence database, an annotated collection of all publicly available nucleotide (Benson *et al.*, 2005). It is part of the International Nucleotide Sequence Database Collaboration, which also includes the DNA Data-Bank of Japan (DDBJ) and that of the European Molecular Biology Laboratory (EMBL). These three organizations exchange data regularly. Many of the proteins in Genpept are also included in other well-curated databases such as SwissProt, although SwissProt accession numbers are not always accessible through Genpept annotations pages (and vice versa). Many genes and their protein products are now represented in Entrez by "Reference Sequences." This RefSeq collection was developed to provide a comprehensive, integrated, non-redundant set of sequences, including genomic DNA, transcript (RNA), and protein products, for major research organisms. Genpept contains a large amount of redundancy, and entries have variable levels of annotation. Hence, some entries (e.g., those from SwissProt) are highly annotated, whereas other entries may be no more than hypothetical protein sequences based on gene sequence. For more detail, a review of protein database options has been published (Apweiler *et al.*, 2004).

The search programs allow effective interrogation of any FASTA-formatted database containing either protein or DNA sequences. The FASTA format was chosen primarily because of its universality, brevity, and expected ease with which database files could be shared on the same computer with other programs for sequence analysis.

When a database is first downloaded to a given search engine, most search engines run a program that analyzes the entries in the database, assigns them an index number, indexes them according to species, and may calculate protein molecular masses and theoretical pI. This permits restricting the proteins searched on the basis of species, protein molecular weight, or pI. It also allows users to add proteins to databases or create a new database containing user-defined proteins. Searches performed on these smaller databases restricted by species or mass are generally very much faster than searches performed on complete databases.

To allow users to take full advantage of the range of data available from different types of experiments, the search programs in the Protein Prospector package are structured so that the results from a search with one program can be input directly into another program. This allows a user to effectively combine information from several analyses *in silico* for maximum discriminating power in defining protein identity. The search programs are also homology tolerant, and homology matching is implemented in a general manner so it makes little difference whether a mismatched amino acid is the result of a substitution by one of the standard amino acids or a chemical, posttranslational, or xenobiotic modification. All programs based on ion fragmentation incorporate the most current knowledge of gas-phase ion fragmentation. However, MS/MS spectra from newly discovered modifications of peptides may contain previously unknown fragment ion types. Consequently, the programs have been developed so that they need only minor modifications to incorporate new fragment ion types.

Data supplied for searching in the aforementioned programs are input manually in the form of peak lists. However, the increase in robustness and reliability of tandem MS data has led to an explosion in the amount of data being acquired. With the routine acquisition of hundreds (or thousands) of fragmentation spectra in a given LC-MS analysis, the manual submission of peak lists from each spectrum is impractical. Thus, to analyze this type of data, a search engine must be able to accept as input text files that contain peak lists of all precursor ion masses and fragment ions produced and then to produce a succinct summary of the resulting identifications. Several search engines can use text files of this format as input, although the exact format of the file differs slightly between search engines. A developmental version of ProteinProspector is capable of accepting text files from formats used by other search engines such as Mascot (*http://www.matrixscience. com*). Searching data in this format is performed essentially by sequential searches of each spectrum with MS-Tag or an equivalent program. However, searching in this manner is dramatically faster than the sum of individual searches, as many of the calculations required for individual searches are in common for all searches (such as the *in silico* digestion of all proteins in the database). Thus, a search of more than 3000 MS/MS spectra against all proteins in the SwissProt database, even allowing for a few peptide modifications, is completed in about 20 min (Chalkley *et al.*, 2004b) on a single-processor desktop PC.

As well as trying to identify unmodified peptides, search engines can also be programmed to identify biologically or chemically modified peptides. These modifications can be specified as fixed; for example, it is typical to reduce and alkylate cysteine residues with, for example, iodoacetamide before protein digestion, so that all cysteines will be carbamidomethylated.

Other modifications can be variable; for example, methionines can be biologically oxidized, but more often they become oxidized during sample preparation. Therefore, peptides should be searched both as if methionines were unmodified and oxidized. Also, most proteins are acetylated on their N-terminus, and if after enzyme cleavage, a glutamine becomes an N-terminal residue, this can pyrolize to form a pyro-glutamate residue. These modifications are generally searched for as a default. Other modifications such as phosphorylation can also be searched for, although these searches do not take previous biological knowledge into consideration and thus will search for any serine, threonine, or tyrosine being modified. This greatly increases search times and leads to many more incorrect answers, so it is generally advisable not to search for such modifications as a default.

Enzyme digests typically employ trypsin, but any other enzyme can be selected. Combinations of enzyme cleavages can also be specified or data can be searched with no enzyme specificity specified for cleavage. The last of these options leads to a high number of false positives and should be used only when there is supporting evidence for such peptide products present.

For mass fingerprinting data, it is typical to specify a minimum number of peptide matches for a hit. The order of reporting results is usually by a scoring system, the most common being based on the MOWSE system (Pappin *et al.*, 1993), which is used by Mascot and MS-Fit in ProteinProspector.

When searching tandem MS data, the user must tell the search engine what type of instrumentation was used to fragment the peptides so it knows which fragment ion types to consider. For example, if fragmentation data were acquired on a MALDI-TOF/TOF instrument, additional high-energy CID fragmentation processes must be considered (Huang *et al.*, 2002). Some search engines are unable to use sequence specific rules for fragmentation, but MS-Tag in ProteinProspector uses the following rules for high-energy CID analysis:

• No side-chain fragmentation should be anticipated unless an arginine, lysine, or histidine residue is present within the ion. In general, the basic residue will not itself yield side-chain fragments: for example, C-terminal arginine does not form v- and w-ions, although it may yield a d-fragment if there is another basic amino acid in the peptide.

• No d-ions should be formed by alanine, glycine, histidine, phenylalanine, proline, tryptophan, or tyrosine.

• No w-ions should be formed by alanine, glycine, histidine, phenylalanine, tryptophan, or tyrosine.

- Because of the double substitution of the β-carbons in threonine and isoleucine, each can yield two w- and two d-fragments.
- No v-ions should be formed by glycine or proline.
- No c-fragments should occur for amino acids that are N-terminally adjacent to proline residues.
- Fragment z (i.e., y-16) cannot be formed by proline residues.
- A neutral loss of NH_3 (-17 Da) from the normal sequence ions and internal fragments is considered for species that contain arginine, asparagine, glutamine, or lysine.
- A neutral loss of H_2O (-18 Da) from the normal sequence ions and internal fragments is considered for species that contain aspartic acid, glutamic acid, serine, or threonine.
- The first members of the N-terminal ion series are not listed as such. In general b_1 is not stable unless the N-terminus is modified, the a_1-ion is equivalent to an immonium ion, and c_1 is rarely observed.

What do You do When Your Protein is Not in the Database?

Unless proteins share very high sequence identity, their tryptic peptides will not always be conserved across closely related species or even among isoforms within a single species. This, combined with the presence of sequence errors in databases, means that identification of proteins, especially from species that are not well represented in databases, may not be achieved using the conventional searching approach described earlier in this chapter. In these cases, some level of *de novo* sequencing is required. There are two levels to which *de novo* sequencing can be applied.

In one approach, a short stretch of amino acid "sequence tag" is discerned, and this, often combined with the masses of fragments before and after this amino acid tag, can be used to search a database in an error-tolerant fashion using software such as MS-Homology or PeptideSearch.

The second approach for identifying unassigned spectra is for software to attempt to *de novo* sequence the entire peptide and return a "best-fit" sequence without referencing a database. This approach relies on scoring systems to rank the quality of *de novo* sequence matches to the data. These scoring systems have to be different from those used for database searching, as there is no "look-up table" to which to compare the results. A few software packages are available for this type of analysis: Sherenga (Dancik *et al.*, 1999) (part of Spectrum Mill), Lutefisk (Taylor and Johnson, 2001) (a tool available free over the web), and commercially available PEAKS (Ma *et al.*, 2003).

Once a sequence is determined, search engines such as BLAST (Altschul and Lipman, 1990; Altschul *et al.*, 1997) and FASTA (Pearson and Lipman, 1988) can be used with sequence information obtained by *de novo* peptide sequencing of several peptides taken from a tryptic digest. However, these searches generally are performed one peptide at a time. In addition, several idiosyncratic features of fragmentation data from mass spectrometers are not adequately accommodated for in these search engines. First, sequences from MS results are almost invariably relatively short stretches of amino acid sequence (often <10 amino acids). The statistical significance of sequence matches using these search engines is length dependent (Altschul, 1991), and searches using such short sequences do not produce statistically significant matches, especially when no very close homologue exists in the databases. The fragmentation mechanism of CID generally prevents the formation of a b_1-ion; thus, identification of the N-terminal two amino acids is often not possible. Distinguishing between the pairs Ile/Leu and Gln/Lys is also generally not possible, although Ile/Leu can be differentiated using high-energy CID (Biemann, 1990; Medzihradskzy and Burlingame, 1994), and glycine and lysine can be distinguished between by accurate mass measurement of relevant sequential ion series (Clauser *et al.*, 1999). Often, a few ions are missing from an ion series in the fragmentation spectrum, leading to "gaps" in the sequence that cannot be determined.

Hence, MS-specific *de novo* sequencing and homology matching search engines have been developed that employ new scoring matrices that allow for the possible ambiguities between Ile/Leu and Gln/Lys, can cope with the mass gaps in sequence information, and can use multiple short stretches of sequence (Habermann *et al.*, 2004; Huang *et al.*, 2001; Mackey *et al.*, 2002).

Quantitation of Proteins

Although extensive resources have been applied to the development of MS as a tool for protein identification, only recently has protein quantitation been addressed seriously with respect to more global protein levels. One approach to analyzing proteins in complex mixtures is to separate proteins by two-dimensional (2D) polyacrylamide gel electrophoresis (PAGE) before mass spectrometric analysis. This technique yields a level of quantitative information based on the intensity of staining and is well suited to monitor variations in protein expression in cells or tissues under different sets of conditions. Thus, comparisons of gels enable specific proteins of interest to be identified for more detailed study by MS (Arnott *et al.*, 1998). However, many researchers have made a move away from the use of (2D) gels with more reliance on other separation methods such as multidimensional high-performance liquid chromatography (HPLC), typically strong cation exchange chromatography followed by reversed phase (Link *et al.*, 1999). There is also more emphasis on one-dimensional (1D) gels, as 2D gels are not as effective at analyzing certain classes of proteins, most notably membrane proteins. However, quantitation by staining intensity of 1D gel bands is not reliable for individual protein quantitation because bands frequently represent several unseparated proteins.

MS has been used for quantitation almost from the time of its invention, even though as a quantitative tool, it has well-known deficiencies. For example, the essential conversion of molecules to ions is generally incomplete and the relative ionization efficiencies for different compounds in different circumstances are rarely predictable. Furthermore, the actual numbers of ions released into the gas phase in ESI or MALDI may be highly dependent on the ionization method and other materials present, including other analytes, matrices, and impurities arising from purification protocols. In addition, the response of the instrument may not be the same for all ions, depending on mass, charge, and polarity.

However, it has long been appreciated that all of these deficiencies can be overcome if an internal standard can be employed with virtually identical physical and chemical properties. This is best achieved with an isotopically labeled analogue of the compound under study, which will behave identically during ionization but can be differentiated by mass. The replacement of several hydrogen atoms by deuteriums is generally the least expensive option and has been employed widely and successfully in innumerable studies. The hydrogen atoms to be replaced must not be labile under the conditions of the experiment, and there should be a sufficient number of them to separate the natural isotopic clusters for the analyte and the standard. Quantitation is then achieved by comparing the relative ion currents and relating them to the amount of the standard added.

Using a similar underlying philosophy, methods have been developed to compare protein expression under different conditions, in which one population of cells is grown in a normal expression medium and a second population is in an isotopically different environment (e.g., using ^{15}N in place of ^{14}N in the growth media). The cells are then mixed and the proteins extracted for MS. All proteins that are expressed equally should give the same peak intensities for the normal and enriched isotope, whereas upregulation or downregulation would be revealed by significant differences in the abundance ratios (Oda *et al.*, 1999). The use of isotope-depleted material has also been explored in a similar manner (Pasa-Tolic *et al.*, 1999). An adaptation of this approach has been successfully applied in which one sample to be compared is grown in media where one amino acid is isotopically labeled, and after mixing samples, peak intensities of peptides containing this amino acid are compared (Ong *et al.*, 2002).

If the samples to be compared are not produced by cell culture, which is generally the case for clinical samples or samples derived from tissue, then chemical derivatization is employed rather than metabolic labeling. Such reagents have specific reactivity for either the N-terminus of a peptide or a certain amino acid. The first widely used reagent of this type was described as an isotope-coded affinity tag (ICAT) reagent (Gygi *et al.*, 1999). This compound contains a biotin tag for affinity purification, a linker that in the light form contains no heavy isotopes and that in the heavy form has eight deuterium atoms, as well as a group that is reactive toward cysteine residues. Thus, two populations of proteins can be differentially labeled, combined, purified over a streptavidin column, digested, and the peptides analyzed by MALDI or ESI MS, giving pairs of peaks separated by 8 Da. A few studies were successful with this reagent, but unfortunately, the large reagent tag was readily fragmented in tandem MS to produce a number of fragment ions not derived from the peptide, and this had a negative impact on the quality of fragmentation spectra. In addition, the slightly less polar heavy deuterated form eluted later in RP-HPLC, making quantitation after chromatographic separation problematic. Newer reagents have switched to the use of ^{13}C, ^{15}N, or ^{18}O, which do not produce chromatographic differences. Also, cleavable linker regions have been incorporated that allow removal of the majority of the tag after affinity purification and, thus, have minimal side effects on the quality of the fragmentation spectra (Hansen *et al.*, 2003).

It was envisaged that the ability to selectively enrich only a few peptides from each protein (e.g., cysteine-containing peptides) would allow more comprehensive coverage of the proteome in a complex sample. However, this has not been the case, and the reliability of quantitation of a protein on the basis of identifying only one peptide from it is questionable. Also, there is increased interest in analyzing posttranslational modifications on these proteins. The latest commercial set of reagents for protein quantitation label the N-terminus of peptides and lysine residues (Ross *et al.*, 2004). The labeling of all peptides allows better statistics for protein quantitation and is amenable to quantitation of posttranslational modifications. In addition, this set of reagents allows the comparison of four samples in one experiment. One drawback is that labeling is not performed until after peptide digestion, and as a result, there is increased potential for introducing variability of the results due to sample handling.

Software for the analysis of these data is limited, and there is no community standard method for analyzing this type of data. Applied

Biosystems, which markets the ICAT and iTRAQ reagents, has written software called ProQuant, open-source software that is available from the Institute for Systems Biology (Li *et al.*, 2003), and the in-house developmental version of ProteinProspector has software that can perform this type of analysis (Baker *et al.*, 2004). Quantitation can be measured based on a number of factors: monoisotopic peak intensity or area or the summed intensity or peak area of the whole isotope cluster. Theoretically, the comparative peak areas of the whole isotope cluster should be most accurate, although there are potential complications with all methods: Monoisotopic peaks are often weak for high mass peptides, but using the whole isotope cluster increases the potential for problems caused by co-eluting peaks of similar mass. In our experience, any of these parameters give similar reliability and standard deviation for results. Complications arise when one or both of the isotopic pairs give peaks that barely exceed the signal to noise of the spectrum. Without accurate calculation of the background noise level, very inaccurate ratios can be calculated in these circumstances. Hence, when calculating protein quantitation ratios, it is prudent not to use peaks where one or both of a pair of isotopically differentiated peaks are close to the noise level. However, ignoring certain peptide signals will limit the dynamic range of the quantitation strategy.

The Challenge of Automation and High Throughput

Generally, the promise of high-throughput experimental genomic and proteomic sciences has developed in parallel with approaches to bioinformatic analysis of the data using semiautomated or fully automated routines. As MS-based proteomics efforts gear up to accommodate thousands of samples per day, automation of the entire process including sample preparation, MS, and informatics analysis becomes an absolute necessity (Berndt *et al.*, 1999; Krutchinsky *et al.*, 2001).

Techniques for automated gel spot identification, cutting, and digestion combined with automated MS and CID analysis by online HPLC with ESI or multiple sample deposition for off-line MALDI are rapidly establishing themselves as routine methods for large-scale protein identification. In so-called data-dependent experiments (DDEs), with HPLC coupled to a tandem mass spectrometer, repetitive recording of MS spectra is interleaved with the selection and analysis of peaks for CID analysis (Tiller *et al.*, 1998). The amount of data generated by such experiments running on a continuous basis is overwhelming for manual interpretation and automation is essential.

One issue concerning the automated transfer of mass spectrometric data to the informatics algorithms is the accuracy of so-called "peak-picking" algorithms. Such programs convert the instrument-specific and technique-specific raw mass spectra into generic lists of monoisotopic masses and intensities, usually after some spectral processing to enhance the peaks while reducing electronic and random noise. The mass value for each peak is usually defined by its centroid, determined for a certain fraction of the peak height (i.e., for the upper 50%), which is selected to avoid the inclusion of noise at the leading or trailing edge and to allow some failure to fully resolve adjacent peaks. MALDI-generated ions are assumed to be singly charged, and the challenge is to correctly identify the first peak in each isotopic cluster. For well-resolved isolated peptide ions of m/z values less than approximately 1500 with good signal-to-noise ratios, this is a relatively easy task because the first peak in each cluster (i.e., the monoisotopic peak) is the most abundant. Above m/z 2000, this is no longer true and ions in this region are frequently weaker in intensity. Therefore, the correct identification of the monoisotopic peak becomes progressively more difficult with increasing mass, particularly if it falls below a threshold selected to discriminate between actual peaks and background noise. This problem is compounded by the overlapping isotopic clusters likely to occur in the spectra of peptide mixtures. An algorithm has been described that picks monoisotopic peaks based on Poisson modeling of theoretical isotope patterns compared with the experimental data (Breen *et al.*, 2000). For ESI-generated ions, there is the further issue of identification of the charge-state (z). the spacing of the isotopic peaks on the abscissa is equal to $1/z$, the identification of which places certain demands on the resolving power of the mass analyzer. If the peak spacing can be identified correctly, conversion of the measured m/z to the molecular mass of the peptide is trivial, but for weaker peaks, or if the spectrum is noisy, then software that creates peak lists for database searching will make mistakes (Chalkley *et al.*, 2004a). Some instrument types such as ion traps do not produce spectra with distinguishable charge states in full scan spectra. A slower narrower mass range "zoom scan" would allow charge-state determination, but for high-throughput analysis of complex mixtures, many researchers decide to skip this extra scan and guess the charge states for a given spectra. For tryptic peptides, it is normal to assume a charge state of either 2+ or 3+. The free N-terminus and basic C-terminal residue are both expected to be protonated, and if there is another basic residue in the sequence, such as a histidine, this is likely to become protonated as well. Peak lists are generated as if there is a doubly and triply charged species, then whichever gives the higher, more significant match after database searching is assumed to be the correct charge. This, of course, doubles the number of spectra to be searched and does not always correctly assign the charge state, increasing the number of incorrect identifications.

A major requirement that has arisen as a result of acquisition of such large numbers of spectra is the development of scoring systems that attempt to effectively discriminate between correct and incorrect answers. It is impractical to manually verify the results from such large data sets. As a result, scoring systems for database results from search engines must be able to report the results at a measurable degree of reliability or with a statistical significance. Search engines such as "Mascot" and "Sonar" use probability-based scoring systems (Field *et al.*, 2002; *http://www. matrixscience.com*) based on the likelihood of a given number of peaks in a fragmentation spectrum matching at random. Another approach has been to apply statistical analysis of search results to create a new scoring system. The assumption is made that because the incorrect matches are random, they will form a gaussian distribution of scores that will be lower than the correct matches. Hence, from the database search results, it is possible to identify the distribution of scores for incorrect answers and assign a probability that a match is incorrect. This can be applied at both the peptide (Keller *et al.*, 2002) and the protein level (Nesvizhskii *et al.*, 2003). The latest version of ProteinProspector is using a similar approach in an attempt to reliably distinguish between correct and incorrect answers (Chalkley *et al.*, 2004b). A limitation that should also be acknowledged with statistical-based search results is that these become more reliable with the more data points (spectra) there are to model the scoring. Hence, for the analysis of small data sets, their effectiveness and reliability is diminished.

Another method that has been proposed to increase the reliability of results is to search data with multiple search engines and compare the results (Resing *et al.*, 2004). The argument goes that as most search engines use slightly different approaches and scoring if a peptide/protein is identified by multiple search engines, then it is much more likely to be correct. The developmental version of ProteinProspector allows comparison of multiple searches of a given data set, including comparison of Protein-Prospector and Mascot searches of the same data (Chalkley *et al.*, 2004b). These comparisons are also useful for the development, refinement, and benchmarking of search engines.

Although most search engines perform well, there is still room for improvement, because there is additional information within the data that is not taken into account by many search engines. For example, the probability of observing a given fragment ion is peptide sequence specific. An in-depth analysis of peptide fragmentation occurring in the N- and C-terminal to each amino acid demonstrated that certain cleavages are highly favored (e.g., cleavages N-terminal to a proline or C-terminal to an aspartic acid). The fragmentation is also affected by the charge state and number of basic residues in the sequence, with singly charged peptides being most influenced by sequence-specific fragmentation (Kapp *et al.*, 2003). Thus, incorporating this information into search engine scoring could improve the reliability of results. Other ancillary information can be exploited (e.g., by correlating actual peptide elution times from reverse-phase chromatography with predicted elution times based on hydrophobicity) (Petritis *et al.*, 2003; Resing *et al.*, 2004). Similarly, for peptides purified by cation exchange or isoelectrofocusing in the first dimension, the distribution of peptides between the various fractions can be correlated with predicted basicity or pI (Cargile *et al.*, 2004).

Through storage of previous proteomic results in a relational database, several other factors can be used to increase confidence in the results. When a protein is identified in a proteomics experiment, it is usually on the basis of only a few peptides. These peptides that are observed from a given protein are generally the same in each proteomics experiment. Therefore, a list of "favored" peptides that are observed when identifying a certain protein can be created, and this could be used as a weighting factor if the search results report identification of a protein, especially when the identification is on the basis of only one or two peptides. Similarly, rather than comparing a spectrum to a theoretical spectrum created by a search engine, one could compare the spectrum to a spectrum in which the same peptide was identified before. The Global Proteome Machine Organization is a public web site where one can search proteomic data with the search engine Xtandem!, and it stores the results from all searches performed on the site (http://www.thegpm.org). It allows users to compare their results to previous results stored on the site when identifying the same protein in terms of peptides identified, the spectrum obtained for each peptide, and other proteins other investigators have identified in association with the protein of interest.

Protein identification in simple protein mixtures using MS for genomes that are fully known has become a routine exercise. The focuses now are on providing effective methods to describe complex samples in a reliable manner, approaches to fully characterize proteins that are posttranslationally modified, and the development of techniques that can provide relative and absolute quantitation of components. These approaches will all be driven by a combination of new sample preparation and separation strategies, combined with the development of bioinformatics technologies to maximize the utilization of the data acquired.

Acknowledgments

The authors would like to thank Patricia Babbitt and Peter Baker for assistance in writing certain sections of this chapter. Financial support from the National Institutes of Health, grant No. RR01614, is gratefully acknowledged.

References

Altschul, S. F. (1991). Amino acid substitution matrices from an information theoretic perspective. *J. Mol. Biol.* **219,** 555–565.

Altschul, S. F., and Lipman, D. J. (1990). Protein database searches for multiple alignments. *Proc. Natl. Acad. Sci. USA* **87,** 5509–5513.

Altschul, S. F., Madden, T. L., Schaffer, A. A., Zhang, J., Zhang, Z., Miller, W., and Lipman, D. J. (1997). Gapped BLAST and PSI-BLAST: A new generation of protein database search programs. *Nucl. Acids Res.* **25,** 3389–3402.

Apweiler, R., Bairoch, A., and Wu, C. H. (2004). Protein sequence databases. *Curr. Opin. Chem. Biol.* **8,** 76–80.

Arnott, D., O Connell, K., King, K., and Stults, J. (1998). An integrated approach to proteome analysis: Identification of proteins associated with cardiac hypertrophy. *Anal. Biochem.* **258,** 1–18.

Baker, P. R., Chalkley, R. J., Hansen, K. C., Cutillas, P. R., Huang, L., Baldwin, M. A., and Burlingame, A. L. (2004). A novel protein quantitation algorithm within Protein Prospector. *In* "Proceedings of the 52nd Annual Conference of the American Society of Mass Spectrometry." Nashville, TN.

Benson, D. A., Karsch-Mizrachi, I., Lipman, D. J., Ostell, J., and Wheeler, D. L. (2005). GenBank. *Nucl. Acids Res.* **33,** Database Issue, D34–D38.

Berndt, P., Hobohm, U., and Langen, H. (1999). Reliable automatic protein identification from matrix-assisted laser desorption/ionization mass spectrometric peptide fingerprints. *Electrophoresis* **20,** 3521–3526.

Biemann, K. (1990). Appendix 5. Nomenclature for peptide fragment ions (positive ions). *Methods Enzymol.* **193,** 886–888.

Breen, E. J., Hopwood, F. G., Williams, K. L., and Wilkins, M. R. (2000). Automatic Poisson peak harvesting for high throughput protein identification. *Electrophoresis* **21,** 2243–2251.

Cargile, B. J., Talley, D. L., and Stephenson, J. L., Jr. (2004). Immobilized pH gradients as a first dimension in shotgun proteomics and analysis of the accuracy of pI predictability of peptides. *Electrophoresis* **25,** 936–945.

Carr, S. A., Aebersold, R., Baldwin, M. A., Burlingame, A. L., Clauser, K., and Nesvizhskii, A. I. (2004). The need for guidelines in publication of peptide and protein identification data: Working group on publication guidelines for peptide and protein identification data. *Mol. Cell Proteom.* **3,** 531–533.

Chalkley, R. J., Baker, P. R., Hansen, K. C., Medzhiradszky, K. F., Allen, N. P., Rexach, M., and Burlingame, A. L. (2005a). Comprehensive analysis of a multidimensional liquid chromatography mass spectrometry dataset acquired on a QqTOF mass spectrometer: 1. how much of the data is theoretically interpretable by search engines? *Mol. Cell Proteom.* **4**(8), 1189–1193.

Chalkley, R. J., Baker, P. R., Huang, L., Hansen, K. C., Allen, N. P., Rexach, M., and Burlingame, A. L. (2005b). Comprehensive analysis of a multidimensional liquid chromatography mass spectrometry dataset acquired on a QqTOF mass spectrometer: 2. new developments in Protein Prospector allow for reliable and comprehensive automatic analysis of large datasets. *Mol. Cell Proteom* **4**(8), 1194–1204.

Clauser, K., Baker, P., and Burlingame, A. (1999). Role of accurate mass measurement (10 ppm in protein identification strategies employing MS or MS/MS and database searching. *Anal. Chem.* **71,** 2871–2882.

Clauser, K., Hall, S., Smith, D., Webb, J., Andrews, L., Trann, H., Epstein, L., and Burlingame, A. (1995). Rapid mass spectrometric peptide sequencing and mass matching for characterisation of human melanoma proteins isolated by two-dimensional PAGE. *Proc. Natl. Acad. Sci. USA* **92**, 5072–5076.

Consortium, G. O. (2001). Creating the gene ontology resource: Design and implementation. *Genome Res.* **11**, 1425–1433.

Dancik, V., Addona, T. A., Clauser, K. R., Vath, J. E., and Pevzner, P. A. (1999). *De novo* peptide sequencing via tandem mass spectrometry. *J. Comput. Biol.* **6**, 327–342.

Eng, J. K., McCormack, A. L., and Yates, J. R. (1994). An approach to correlate tandem mass spectral data of peptides with amino acid sequences in a protein database. *J. Am. Soc. Mass Spectrom.* **5**, 976–989.

Field, H. I., Fenyo, D., and Beavis, R. C. (2002). RADARS, a bioinformatics solution that automates proteome mass spectral analysis, optimises protein identification, and archives data in a relational database. *Proteomics* **2**, 36–47.

Gygi, S. P., Rist, B., Gerber, S. A., Turecek, F., Gelb, M. H., and Aebersold, R. (1999). Quantitative analysis of complex protein mixtures using isotope-coded affinity tags. *Nat. Biotechnol.* **17**, 994–999.

Habermann, B., Oegema, J., Sunyaev, S., and Shevchenko, A. (2004). The power and the limitations of cross-species protein identification by mass spectrometry-driven sequence similarity searches. *Mol. Cell Proteom.* **3**, 238–249.

Hansen, K. C., Schmitt-Ulms, G., Chalkley, R. J., Hirsch, J., Baldwin, M. A., and Burlingame, A. L. (2003). Mass spectrometric analysis of protein mixtures at low levels using cleavable ^{13}C-isotope–coded affinity tag and multidimensional chromatography. *Mol. Cell Proteom.* **2**, 299–314.

Henzel, W. J., Billeci, T. M., Stults, J. T., Wong, S. C., Grimley, C., and Watanabe, C. (1993). Identifying proteins from two-dimensional gels by molecular mass searching of peptide fragments in protein sequence databases. *Proc. Natl. Acad. Sci. USA* **90**, 5011–5015.

Horn, D., Ge, Y., and McLafferty. (2000). Activated ion electron capture dissociation for mass spectral sequencing of larger (42 kDa) proteins. *Anal. Chem.* **72**, 4778–4784.

Huang, L., Baldwin, M. A., Maltby, D. A., Medzihradszky, K. F., Baker, P. R., Allen, N., Rexach, M., Edmondson, R. D., Campbell, J., Juhasz, P., Martin, S. A., Vestal, M. L., and Burlingame, A. L. (2002). The identification of protein–protein interactions of the nuclear pore complex of *Saccharomyces cerevisiae* using high throughput matrix-assisted laser desorption ionization time-of-flight tandem mass spectrometry. *Mol. Cell Proteom.* **1**, 434–450.

Huang, L., Jacob, R. J., Pegg, S. C., Baldwin, M. A., Wang, C. C., Burlingame, A. L., and Babbitt, P. C. (2001). Functional assignment of the 20 S proteasome from *Trypanosoma brucei* using mass spectrometry and new bioinformatics approaches. *J. Biol. Chem.* **276**, 28327–28339.

James, P., Quadroni, M., Carafoli, E., and Gonnet, G. (1993). Protein identification by mass profile fingerprinting. *Biochem. Biophys. Res. Commun.* **195**, 58–64.

Jensen, O., Podtelejnikov, A., and Mann, M. (1997). Identification of components of simple protein mixtures by high-accuracy peptide mass mapping and database searching. *Anal. Chem.* **69**, 4741–4750.

Kapp, E. A., Schutz, F., Reid, G. E., Eddes, J. S., Moritz, R. L., O'Hair, R. A., Speed, T. P., and Simpson, R. J. (2003). Mining a tandem mass spectrometry database to determine the trends and global factors influencing peptide fragmentation. *Anal. Chem.* **75**, 6251–6264.

Keller, A., Nesvizhskii, A. I., Kolker, E., and Aebersold, R. (2002). Empirical statistical model to estimate the accuracy of peptide identifications made by MS/MS and database search. *Anal. Chem.* **74**, 5383–5392.

Krutchinsky, A. N., Kalkum, M., and Chait, B. T. (2001). Automatic identification of proteins with a MALDI-quadrupole ion trap mass spectrometer. *Anal. Chem.* **73**, 5066–5077.

Li, X. J., Zhang, H., Ranish, J. A., and Aebersold, R. (2003). Automated statistical analysis of protein abundance ratios from data generated by stable-isotope dilution and tandem mass spectrometry. *Anal. Chem.* **75,** 6648–6657.

Link, A. J., Eng, J., Schieltz, D. M., Carmack, E., Mize, G. J., Morris, D. R., Garvik, B. M., and Yates, J. R., 3rd. (1999). Direct analysis of protein complexes using mass spectrometry. *Nat. Biotechnol.* **17,** 676–682.

Ma, B., Zhang, K., Hendrie, C., Liang, C., Li, M., Doherty-Kirby, A., and Lajoie, G. (2003). PEAKS: Powerful software for peptide *de novo* sequencing by tandem mass spectrometry. *Rapid Commun. Mass Spectrom.* **17,** 2337–2342.

MacCoss, M. J., Wu, C. C., and Yates, J. R., 3rd. (2002). Probability-based validation of protein identifications using a modified SEQUEST algorithm. *Anal. Chem.* **74,** 5593–5599.

Mackey, A. J., Haystead, T. A., and Pearson, W. R. (2002). Getting more from less: algorithms for rapid protein identification with multiple short peptide sequences. *Mol. Cell Proteom.* **1,** 139–147.

Mann, M., Hojrup, P., and Roepstorff, P. (1993). Use of mass spectrometric molecular weight information to identify proteins in sequence databases. *Biol. Mass Spectrom.* **22,** 338–345.

Mann, M., and Wilm, M. (1994). Error-tolerant identification of peptides in sequence databases by peptide sequence tags. *Anal. Chem.* **66,** 4390–4399.

Medzihradszky, K., and Burlingame, A. (1994). The advantages and versatility of a high-energy collision-induced dissociation-based strategy for the sequence and structural determination of proteins. *Methods Compan. Methods Enzymol.* **6,** 284–303.

Moore, R. E., Young, M. K., and Lee, T. D. (2002). Qscore: An algorithm for evaluating SEQUEST database search results. *J. Am. Soc. Mass Spectrom.* **13,** 378–386.

Nesvizhskii, A. I., Keller, A., Kolker, E., and Aebersold, R. (2003). A statistical model for identifying proteins by tandem mass spectrometry. *Anal. Chem.* **75,** 4646–4658.

Oda, Y., Huang, K., Cross, F. R., Cowburn, D., and Chait, B. T. (1999). Accurate quantitation of protein expression and site-specific phosphorylation. *Proc. Natl. Acad. Sci. USA* **96,** 6591–6596.

Ong, S. E., Blagoev, B., Kratchmarova, I., Kristensen, D. B., Steen, H., Pandey, A., and Mann, M. (2002). Stable isotope labeling by amino acids in cell culture, SILAC, as a simple and accurate approach to expression proteomics. *Mol. Cell Proteom.* **1,** 376–386.

Orchard, S., Taylor, C. F., Hermjakob, H., Weimin, Z., Julian, R. K., Jr., and Apweiler, R. (2004). Advances in the development of common interchange standards for proteomic data. *Proteomics* **4,** 2363–2365.

Pappin, D. J., Hojrup, P., and Bleasby, A. J. (1993). Rapid identification of proteins by peptide-mass fingerprinting. *Curr. Biol.* **3,** 327–332.

Pasa-Tolic, L., Jensen, P. K., Andersen, G. A., Lipton, M. S., Peden, K. K., Martinovic, S., Tolic, N., Bruce, J. E., and Smith, R. D. (1999). High throughput proteome-wide precision measurements of protein expression using mass spectrometry. *J. Am. Chem. Soc.* **121,** 7949–7950.

Pearson, W. R., and Lipman, D. J. (1988). Improved tools for biological sequence comparison. *Proc. Natl. Acad. Sci. USA* **85,** 2444–2448.

Pedrioli, P. G., Eng, J. K., Hubley, R., Vogelzang, M., Deutsch, E. W., Raught, B., Pratt, B., Nilsson, E., Angeletti, R. H., Apweiler, R., Cheung, K., Costello, C. E., Hermjakob, H., Huang, S., Julian, R. K., Kapp, E., McComb, M. E., Oliver, S. G., Omenn, G., Paton, N. W., Simpson, R., Smith, R., Taylor, C. F., Zhu, W., and Aebersold, R. (2004). A common open representation of mass spectrometry data and its application to proteomics research. *Nat. Biotechnol.* **22,** 1459–1466.

Petritis, K., Kangas, L. J., Ferguson, P. L., Anderson, G. A., Pasa-Tolic, L., Lipton, M. S., Auberry, K. J., Strittmatter, E. F., Shen, Y., Zhao, R., and Smith, R. D. (2003). Use of artificial neural networks for the accurate prediction of peptide liquid chromatography elution times in proteome analyses. *Anal. Chem.* **75,** 1039–1048.

Resing, K. A., Meyer-Arendt, K., Mendoza, A. M., Aveline-Wolf, L. D., Jonscher, K. R., Pierce, K. G., Old, W. M., Cheung, H. T., Russell, S., Wattawa, J. L., Goehle, G. R., Knight, R. D., and Ahn, N. G. (2004). Improving reproducibility and sensitivity in identifying human proteins by shotgun proteomics. *Anal. Chem.* **76,** 3556–3568.

Ross, P. L., Huang, Y. N., Marchese, J. N., Williamson, B., Parker, K., Hattan, S., Khainovski, N., Pillai, S., Dey, S., Daniels, S., Purkayastha, S., Juhasz, P., Martin, S., Bartlet-Jones, M., He, F., Jacobson, A., and Pappin, D. J. (2004). Multiplexed protein quantitation in saccharomyces cerevisiae using amine-reactive isobaric tagging reagents. *Mol. Cell Proteomics* **3,** 1154–1169.

Sadygov, R. G., Liu, H., and Yates, J. R. (2004). Statistical models for protein validation using tandem mass spectral data and protein amino acid sequence databases. *Anal. Chem.* **76,** 1664–1671.

Spengler, B., Kirsch, D., Kaufmann, R., and Jaeger, E. (1992). Peptide sequencing by matrix-assisted laser-desorption mass spectrometry. *Rapid Commun. Mass Spectrom.* **6,** 105–108.

Taylor, G. K., Kim, Y. B., Forbes, A. J., Meng, F., McCarthy, R., and Kelleher, N. L. (2003). Web and database software for identification of intact proteins using "top down" mass spectrometry. *Anal. Chem.* **75,** 4081–4086.

Taylor, J. A., and Johnson, R. S. (1997). Sequence database searches via de novo peptide sequencing by tandem mass spectrometry. *Rapid Commun. Mass Spectrom.* **11,** 1067–1075.

Taylor, J. A., and Johnson, R. S. (2001). Implementation and uses of automated de novo peptide sequencing by tandem mass spectrometry. *Anal. Chem.* **73,** 2594–2604.

Tiller, P. R., Land, A. P., Jardine, I., Murphy, D. M., Sozio, R., Ayrton, A., and Schaefer, W. H. (1998). Application of liquid chromatography-mass spectrometry(n) analyses to the characterization of novel glyburide metabolites formed in vitro. *J. Chromatogr. A* **794,** 15–25.

Wilm, M., and Mann, M. (1996). Analytical properties of the nanoelectrospray ion source. *Anal. Chem.* **68,** 1–8.

Yates, J. R., 3rd, Speicher, S., Griffin, P. R., and Hunkapiller, T. (1993). Peptide mass maps: A highly informative approach to protein identification. *Anal. Biochem.* **214,** 397–408.

Yates, J. R., Eng, J., Clauser, K. R., and Burlingame, A. L. (1996). Search of sequence databases with uninterpreted high-energy collision-induced dissociation spectra of peptides. *J. Am. Soc. Mass Spectrom.* **7,** 1089–1098.

[10] Protein Conformations, Interactions, and H/D Exchange

By Claudia S. Maier and Max L. Deinzer

Abstract

Modern mass spectrometry (MS) is well known for its exquisite sensitivity in probing the covalent structure of macromolecules, and for that reason, it has become the major tool used to identify individual proteins in proteomics studies. This use of MS is now widespread and routine. In addition to this application of MS, a handful of laboratories

METHODS IN ENZYMOLOGY, VOL. 402   0076-6879/05 $35.00
                        DOI: 10.1016/S0076-6879(05)02010-0

are developing and using a methodology by which MS can be used to probe protein conformation and dynamics. This application involves using MS to analyze amide hydrogen/deuterium (H/D) content from exchange experiments. Introduced by Linderstøm-Lang in the 1950s, H/D exchange involves using ^{2}H labeling to probe the rate at which protein backbone amide protons undergo chemical exchange with the protons of water. With the advent of highly sensitive electrospray ionization (ESI)–MS, a powerful new technique for measuring H/D exchange in proteins at unprecedented sensitivity levels also became available. Although it is still not routine, over the past decade the methodology has been developed and successfully applied to study various proteins and it has contributed to an understanding of the functional dynamics of those proteins.

Introduction

The advent of soft ESI (Fenn *et al.*, 1989) and matrix-assisted laser desorption ionization (MALDI) (Karas and Hillenkamp, 1988) techniques has accelerated the emergence of MS in the field of protein interaction studies. The coupling of these ionization methods with modern mass analyzers (e.g., ion traps, triple quadrupole, and time-of-flight [TOF] analyzers) provided breakthroughs in the biochemist's ability to more accurately determine the molecular weights of large biomolecular assemblies up to several hundred thousand daltons (Rostom *et al.*, 2000; Rostom and Robinson, 1999; Tito *et al.*, 2000; van Berkel *et al.*, 2000). Developments in microtechniques for sample preparation and handling has further advanced the role of MS for protein identification and primary structure elucidation. In particular, ESI-MS has opened up opportunities for applications of MS in the fields of analytical biochemistry and structural biology largely because of its capability of analyzing proteins from aqueous solution under nearly physiological conditions of concentration, pH, and temperature, as well as its compatibility with microseparation techniques.

Of particular interest is that new applications of biomolecular MS are emerging in the field of noncovalent interactions (Przybylski and Glocker, 1996; Smith *et al.*, 1997a,b; Smith and Zhang, 1994) and protein folding studies (Last and Robinson, 1999; Miranker *et al.*, 1993, 1996). Insight into the biological processes that occur at the molecular level and possible ways in which proteins might carry out their specific functions can be gained from a knowledge of protein folding mechanisms, their higher order structure, and the interactions of these proteins with other molecules. The question here is how can MS contribute to develop a greater understanding of the interactions in proteins that are responsible for protein stability, function, specificity, and assembly. What kind of information can be expected from MS-based approaches?

Since the pioneering work of Hvidt and Linderstrøm-Lang (1954), Hvidt and Nielsen (1966), and Linderstrøm-Lang (1958), hydrogen isotope exchange experiments have been used to gain information about the structural stability of protein conformers and an understanding of the mechanisms involved in protein mobility. Amide hydrogen isotope exchange studies provide information on protein folding and unfolding to and from the native state, on local fluctuations of the native state, and on ligand–protein interactions (Woodward, 1999). There are essentially two competing processes during which hydrogen isotope exchange takes place. One involves internal or local fluctuations from the native state, and the other involves subglobal or global unfolding. In either case, hydrogen bonds are broken and the amide hydrogens are exposed to solvent so that exchange can occur. Methods to monitor hydrogen exchange have included tritium counting (Rosenberg and Chakravarti, 1968; Rosenberg and Enberg, 1969; Rosenberg and Woodward, 1970) and hydrogen–deuterium (H/D) exchange in combination with spectroscopic methods such as infrared (Backmann et al., 1996; De Jongh et al., 1995; Zhang et al., 1995), ultraviolet (Englander et al., 1979), Raman resonance (Hildebrandt et al., 1993), and nuclear magnetic resonance (NMR) (Roder, 1989; Scholtz and Robertson, 1995). With increasing field strengths and multidimensional methods, NMR became the instrument of choice for studying hydrogen isotope exchange in proteins since the late 1970s. However, MS, particularly since the inception of ESI-MS at the beginning of the 1990s, has become increasingly popular for monitoring hydrogen isotope exchange because of some decisive advantages it has over NMR. The unique advantages of MS are that it provides the ability to (a) determine hydrogen isotope content in the micromolar as opposed to the millimolar range, (b) access proteins that aggregate at high concentrations, (c) observe the hydrogen isotope content in proteins greater than 20 kDa, (d) observe coexisting conformers simultaneously, (e) work with mixtures or nonpurified samples, (f) determine the exact hydrogen isotope content and the location by analyzing peptides, and (g) detect correlated exchange, which is fundamental for understanding the exchange and folding process. The chief advantage of NMR is that the method gives the average hydrogen isotope content at many specific sites, although MS is about to play an increasingly important role here, too. The two methods yield complementary hydrogen isotope exchange information.

MS also benefits the investigation of protein conformations in other ways. The observation that charge-state distributions displayed in ESI mass spectra of proteins are reflections of the net charges that the proteins carry in solution makes ESI-MS a powerful technique for detecting conformational changes in proteins. MS-based approaches in combination with H/D exchange experiments have proven to be highly efficient for probing the higher order structure of proteins, to study the conformational stability of proteins, and to determine the dynamics of conformational transitions. The aim of this contribution is to outline some approaches. In this chapter, emphasis is on probing the alterations in protein conformational stability and dynamics caused by environmental changes or chemical modifications. For other comprehensive reviews on H/D exchange in combination with MS, see also Smith *et al.* (1997a, 1998), Yan *et al.* (2004b), Eyles and Kaltashov (2004), and Busenlehner and Armstrong (2005).

ESI-MS for Probing Solution-Phase Protein Conformations

The ESI Method

ESI-MS has emerged as a rapidly progressing technique to study biomolecular interactions due to its unique ability to directly analyze peptides, proteins, and even noncovalent complexes from solution. In the ESI process, multiply charged protein ions or charge states are generated, transferred intact into the gas phase, then introduced into the mass spectrometer and analyzed according to their mass-to-charge (m/z) ratios. Because proteins have multiple protonation/deprotonation sites, a typical ESI mass spectrum of a protein M contains a series of ion peaks $(M + nH)^{n+}$ or $(M - nH)^{n-}$, in which each ion peak represents a population of protein molecules carrying n charges (Fig. 1). The series of ion peaks, $(M + nH)^{n+}$ or $(M - nH)^{n-}$, observed for a protein under defined conditions in an ESI mass spectrum is called the *charge-state distribution* or *charge-state envelope.*

Proteins have traditionally been analyzed by ESI-MS using aqueous solutions containing an organic modifier (e.g., 10–50% [v/v] methanol or acetonitrile) and organic acids (e.g., 1–5% formic or acetic acid). Under these conditions, ESI-MS allows mass determinations of proteins with high mass accuracy. However, because these conditions may cause denaturation of the protein, information regarding the conformation is lost. It is, therefore, essential to analyze proteins in solutions that favor the conformational states of interest. For instance, if the native conformation is of interest, then the protein should be analyzed in a solvent that favors the native conformations, for example, mildly acidic conditions (pH 5–7), no organic solvents, and low ESI source temperature.

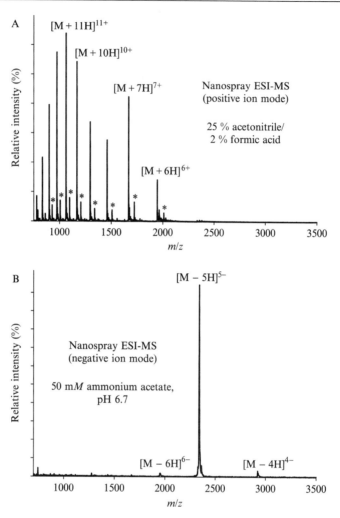

FIG. 1. Nanospray electrospray ionization (ESI) mass spectra of oxidized *Escherichia coli* thioredoxin (2 μM) (A) in 25% acetonitrile containing 2% formic acid measured in the positive ion mode and (B) in 50 mM of ammonium acetate, pH 6.7 measured in the negative ion mode. Charge-state distributions are labeled by $(M + nH)^{n+}$ and $(M - nH)^{n-}$, respectively. The charge-state distribution of an unknown impurity is marked with an asterisk. Mass spectra were recorded on an ESI time-of-flight mass spectrometer with orthogonal ion injection (LC-T, Micromass, Manchester, UK).

Charge-State Distributions

One of the first applications of ESI-MS to study protein conformational changes involved the analysis of charge-state distributions in ESI mass spectra. Proteins that are electrosprayed from solutions and retain their native folded states tend to have narrow charge-state distributions centered around molecular ions with low net charges, whereas denaturing solution conditions (i.e., nonnative physiological pH, high organic solvent concentrations, and elevated temperatures) or the reduction of intramolecular disulfide bonds give rise to charge-state distributions encompassing molecular ions carrying many more charges. The relationship between high charge states and the denatured state of the protein was first observed in 1990 for the acid-induced unfolding of cytochrome *c* (Chowdhury *et al.*, 1990). Numerous proteins have since been monitored by ESI-MS under different solution conditions, supporting the general consensus that the solution-phase conformation of a protein has dramatic effects on the charge-state distribution observed in its ESI mass spectrum. Kaltashov and coworkers have described a chemometric approach to separate the contributions of the charge-state envelope of the individual conformers to the overall charge-state envelope by using singular value decomposition for extracting the number of significantly populated conformers. Validation of this method has been carried out by monitoring acid- and alcohol-induced equilibrium states of model proteins such as chymotrypsin inhibitor 2, ubiquitin, and apomyoglobin, which obey a two-state, three-state, and four-state folding model, respectively (Mohimen *et al.*, 2003).

Still only limited information is available to fully understand the thermodynamic and kinetic processes involved in the generation of multiply charged protein molecular ions and their transfer into the gas-phase during the ESI process. Two models are discussed. In the charged residue model, the droplets generated during the ESI process shrink by solvent evaporation until only single ions remain (Dole *et al.*, 1968). In the ion evaporation model, ions are desorbed from the surface of the droplet before the ultimate droplet size is reached (Iribarne and Thompson, 1976). These two models have been refined, and at present, it is hypothesized that the direct ion emission of highly solvated molecular ions by asymmetrical droplet fission or the direct emission from the tip of a liquid cone formed under the influence of an electrical field (as in the case of nano-ESI-MS) describes best the experimental observations that (a) the charge-state distribution depends on the conformational structure present in solution and the localization of the protonation/deprotonation sites in the macromolecular ion and (b) the survival of intact noncovalent complexes in the gas phase (Przybylski and Glocker, 1996; Robinson *et al.*, 1996; Smith *et al.*,

1997b). It is believed that higher degrees of protonation as observed for proteins electrosprayed from denaturing solutions can be explained, at least in part, by the more extended or less compact conformation of the denatured proteins because of an increase in surface accessibility and changes in pK_a values at the protonation sites (Chowdhury *et al.*, 1990; Katta and Chait, 1991; Konermann and Douglas, 1997; Konermann *et al.*, 1997a,b, 1998a; Loo *et al.*, 1990, 1991; Przybylski and Glocker, 1996; Smith *et al.*, 1997a,b). However, one may speculate, on the basis of funnel-like protein folding landscapes (Dill and Chan, 1997), that the broader charge-state distributions for proteins electrosprayed from denaturing solutions may actually reflect the conformational heterogeneity and diversity associated with highly disordered unfolded conformational states. On the other hand, the narrow charge-state distributions observed for proteins electrosprayed from "native-like" solution conditions reflect the well-defined conformational properties of a highly compact folded state (Vis *et al.*, 1998).

It should be noted that the charge-state distribution of a protein depends to some degree on external parameters associated with the intrinsic properties of the analysis technique, such as ion formation, ion transmission, and ion detection. Therefore, if one attempts to quantify changes in charge-state distributions, the external factors should be carefully controlled and evaluated. However, compared to the drastic change of the appearance of the charge-state distribution in ESI mass spectra as a result of a conformational transition, these external factors influence the appearance of the charge-state distribution only to a minor extent (Przybylski and Glocker, 1996; Smith *et al.*, 1997a,b).

Comparison of ESI-MS with Other Spectroscopic Methods

While ESI-MS has emerged as a powerful tool for obtaining potential structural descriptions of proteins, it is highly desirable to correlate changes in charge-state distributions in response to conformational changes within a protein with data from other biophysical methods to confirm or deepen the understanding on which molecular events ESI-MS is capable of reporting and to be able to evaluate the ESI-MS–derived data critically. An example in which the charge-state distribution as observed in ESI-MS was successfully used for monitoring conformational changes in solution is the heat denaturation study of *E. coli* thioredoxin (TRX) (Kim *et al.*, 1999; Maier *et al.*, 1999).

For the acquisition of the temperature-dependent ESI mass spectra, a continuous flow setup was used (Fig. 2). At ambient temperature, the acquired mass spectra of TRX exhibit a narrow charge-state distribution

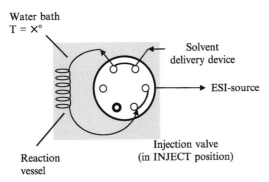

Fig. 2. The protein solution is injected into the sample loop assembled in an injection valve, immersed in a water bath maintained at the desired temperature. The sample loop served as reaction vessel. The protein solution is equilibrated at the desired temperature and directly infused via a fused silica capillary into the electrospray ionization source. Flow directions are indicated by arrows.

at higher m/z values encompassing the sixfold to ninefold protonated ions, with the $(M + 8H)^{8+}$ ion as the most abundant charge state. It is believed that these charge states represent molecular ions that were derived from the compact tightly folded conformational states of TRX. Below m/z 1200, a second series of molecular ions is observed, with low intensity representing the 10- to 15-fold protonated molecular ions, which are thought to relate to the denatured less compact conformational states of TRX (Fig. 3).

However, the charge-state distribution of TRX equilibrated at higher temperature is dramatically different. Above 40°, the charge-state distribution at lower m/z values encompassing charges 10+ to 15+ becomes more intense, whereas the intensity of the charge states centered around $(M + 8H)^{8+}$ decreases. Above 65°, the charge-state distribution is centered around the 12-fold protonated ion peak, which then dominates the appearance of the ESI mass spectrum (Fig. 3). Heat denaturation curves were deduced from the temperature-dependent charge-state distribution by plotting the average charge state $<c>$ of an ESI mass spectrum (Konermann and Douglas, 1997; Maier et $al.$, 1999) or the ion peak ratio $I_F/(I_F + I_U)$ versus the temperature (Maier et $al.$, 1999; Mirza et $al.$, 1993) (Fig. 4A).

It is important to note that at the temperatures studied, TRX showed only two distinct charge-state distributions that exhibit temperature-dependent changes in intensity without shifting of their charge-state maxima. In the case of the thermal unfolding study of $E.$ $coli$ TRX, the temperature-induced unfolding transition, derived from the ESI mass

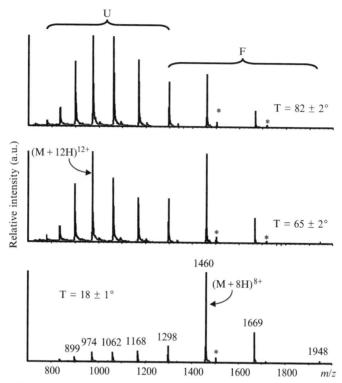

FIG. 3. Electrospray ionization (ESI) mass spectra of *Escherichia coli* thioredoxin at the temperatures (A) $82 \pm 2°$, (B) $65 \pm 2°$, and (C) $18 \pm 1°$ in 2% acetic acid. To the folded form F, ion peaks at m/z 1948, 1669, 1460, and 1298 represent charge states 6+, 7+, 8+, and 9+, respectively. Ion peaks at m/z 1168, 1062, 974, 899, 834, and 779 representing charge states 10+ to 15+ are assumed to originate from the unfolded form U. A minor unidentified impurity is marked with an asterisk.

spectra, paralleled almost exactly with the unfolding transition observed in near-ultraviolet (UV) circular dichroism (280 nm), a method that reports on changes in the environment of aromatic side chains of amino acid residues (Fig. 4B). These data together with data from H/D exchange experiments suggest (a) that the thermal unfolding of TRX under the applied conditions can be described as a two-state unfolding transition allowing the estimation of T_m, the melting temperature, and ΔH_m, the enthalpy change for unfolding (Table I) and (b) that charge-state distributions as observed by ESI-MS reflect on the intactness of the hydrogen bonding network (Maier *et al.*, 1999). These results support the emerging

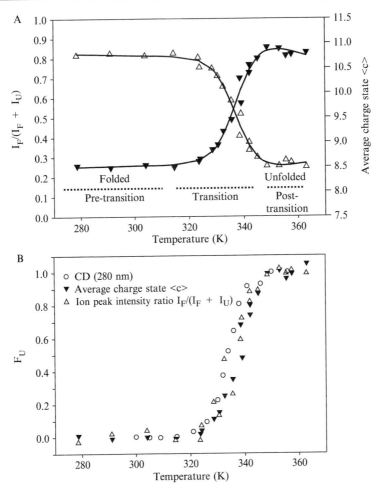

FIG. 4. (A) Heat-denaturation curves of *Escherichia coli* thioredoxin in 2 (v/v) % acetic acid deduced from the temperature-dependent charge-state distribution obtained by electrospray ionization (ESI)–mass spectrometry (MS): ($\triangle$) ion peak intensity ratio and ($\blacktriangledown$) average charge state $\langle c \rangle$ as a function of temperature. The continuous lines are theoretical curves based on the assumption that equilibrium thermal unfolding can be approximated by two-state unfolding mechanism (Creighton, 1990; Santoro and Bolen, 1988). (B) Fraction unfolded versus temperature curves; the MS-derived unfolding curves coincide with the unfolding curve derived from optical spectroscopy.

TABLE I

THERMODYNAMIC DATA FOR THE UNFOLDING TRANSITION OF *E. COLI* OXIDIZED THIOREDOXIN
DERIVED FROM NEAR-UV CIRCULAR DICHROISM SPECTROSCOPY AND THE ANALYSES OF
CHARGE-STATE DISTRIBUTIONS AS OBSERVED BY ESI-MS[a,b]

Method	T_m ($^\circ$)	ΔH_m (kcal/mol)
Near-UV circular dichroism	60	78
ESI-MS		
Ion peak intensity ratio[c]	64	74
Average charge state[d]	65	54

[a] From Maier *et al.* (1999).

[b] ESI, electrospray ionization; MS, mass spectrometry; UV, ultraviolet.

[c] The ion peak intensity ratio of an ESI mass spectrum can be deduced by calculating the ratio $I_F/(I_F + I_U)$, where I_F and I_u are obtained by summing the ion peak intensities of charge states assigned to the compact conformational state (F) and the disordered denatured state (U), respectively (Maier *et al.*, 1999; Mirza, 1993).

[d] The average charge state <c> of an ESI mass spectrum (Konermann, 1997b; Maier *et al.*, 1999) can be calculated according to Eq. (1), whereby c denotes the charge state of an ion peak and I_c its intensity (counts per second).

view that ESI-MS is a technique that is potentially capable of monitoring subtle changes in the tertiary structure.

The relative error of the thermodynamic data obtained from the analyses of charge-state distributions was estimated to be in the range of 10–20%.

$$c = \frac{\sum_c I_c c}{\sum_c I_c} \tag{1}$$

Using the same experimental approach, the global stability of oxidized *E. coli* TRX was compared to a chemically modified TRX, namely (S-(2-Cys^{32}ethyl)-glutathione)thioredoxin (GS-TRX) (Kim *et al.*, 1999). GS-TRX is a possible protein alkylation product of S-(2-chloroethyl)glutathione, a cytotoxic compound derived from glutathione and the xenobiotic 1,2-dichloroethane (Erve *et al.*, 1995). The heat denaturation curves deduced from the temperature-dependent charge-state distributions obtained by ESI-MS indicate that GS-TRX possesses a lower structural stability compared to oxidized TRX but a significantly higher stability compared to reduced TRX (Fig. 5A). These data together with data from other optical spectroscopic methods and H/D exchange experiments indicate that the glutathionyl moiety may interact with parts of TRX, which results in an increase in conformational stability (Kim *et al.*, 1999). This hypothesis has been substantiated by energy minimization studies on

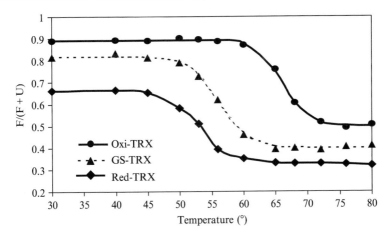

Fig. 5. (A) Comparison of the heat-denaturation curves of oxidized, reduced, and (S-(2-Cys^{32}ethyl)-glutathione)thioredoxin (GS-TRX) in 1 (v/v) % acetic acid deduced from the temperature-dependent charge-state distribution observed by electrospray ionization–mass spectrometry.

(S-(2-Cys^{32}ethyl)-glutathione)thioredoxin, which propose hydrogen bonding and electrostatic interactions between the thioredoxin core and the glutathionyl moiety (Kim et al., 2001a,b).

Charge-state analysis was also applied for determining the conformational consequences of metal ion binding to the colicin E9 endonuclease (E9 DNase) by taking advantage of the unique capability of ESI-MS of allowing simultaneous assessment of conformational heterogeneity and metal ion binding. Alterations of charge-state distributions on metal ion binding/release were correlated with spectral changes observed in far- and near-UV circular dichroism (CD) and intrinsic tryptophan fluorescence. A drastic shift to higher charges was observed for the charge-state envelopes upon collapse of the tertiary structure of Zn^{2+}-bound colicin E9 endonuclease by acid-induced release of the metal ion. Moreover, these studies revealed that the noncovalent protein–protein complexes, that is, colicin E9 endonuclease and its cognate immunity protein, Im9, dissociate in the gas phase before the metal ion complex, thereby confirming that electrostatics and not hydrophobic interactions are more important in the gas phase (van den Bremer et al., 2002).

Another example that charge-state analysis can contribute to elucidate structural properties of a protein, where other methods fail, is illustrated by studies on human recombinant macrophage colony-stimulating factor-β (rhM-CSFβ) (Zhang et al., 2001). rhM-CSFβ is a four-helical bundle cytokine and forms a compact homodimer with nine disulfide bonds. The ESI mass spectrum of rhM-CSFβ exhibits a charge-state distribution centered around the 21+ charge state. Selective reduction of two disulfide bonds ($C_{157, 159}$–$C'_{157, 159}$) near the C-terminal tail showed little change in the secondary structure as suggested by far-UV CD and fluorescence spectra, and even the three-dimensional integrity of the protein was unaltered as suggested by the unchanged biological activity. However, the ESI spectrum of this partially reduced protein showed a charge envelope centered around the 32+ charge state, suggesting additional basic sites became accessible upon removal of the two disulfide bonds. However, only Lys^{154} would appear to become available for protonation upon reduction of the disulfides, $C_{157, 159}$–$C'_{157, 159}$, as all other basic sites are in the N-terminal direction relative to the next stabilizing disulfide bond, C_{102}–C_{146}. Constructing a model for rhM-CSFβ helped rationalizing the observed charge-state alteration upon reduction. In the model, the crystal structure of rhM-CSFα was extended to aa 160, thereby providing the symmetrical C_{157}–C_{159} disulfide linkages in the dimer. The resulting structure was subjected to molecular dynamics and energy optimization by AMBER force field, and the remaining sequence to Glu^{177} was then modeled as a turn and a polyproline type II helix. H/D exchange experiments combined with computational support suggested a structure model in which disulfide bond reduction exposes additional basic sites in each subunit, which are buried by the polyproline type II helix in the intact rhM-CSFβ (Yan et al., 2002, 2004b).

Negative Ion ESI-MS Charge-State Distributions

Only a few studies have been published in which the negative ion mode ESI-MS was employed to study conformational changes in proteins. In the negative ion mode ESI-MS, proteins are measured as multiply negatively charged protein species, that is, $(M - nH)^{n-}$. Buffer systems that are commonly employed to analyze proteins in the negative ion mode include ammonium acetate/ammonia or ammonium bicarbonate solutions.

An example of the use of negative ion ESI-MS involved a study in which the influence of point mutations on the global stability of wild-type ferredoxin from the cyanobacterium *Anabaena* was elucidated (Remigy et al., 1997). Alterations within the charge-state distributions of the multiply deprotonated molecular ions were rationalized by a loss of structural stability due to disruption of stabilizing hydrogen bonds by some of the amino acid replacements.

Another example in which negative ion ESI-MS proved useful was the report on Ca^{2+}-binding in calbindin D_{28K} and two deletion mutants lacking one or both EF-hand domains (Veenstra et al., 1998). They compared changes in the charge-state distribution of multiply deprotonated molecular ions observed in the negative ion mode ESI mass spectra with changes in the near- and far-UV CD and the extrinsic fluorescence of calbindin D_{28K} and mutants upon uptake of Ca^{2+} ions. Good correlation was found between conformational changes detected by spectroscopic methods, such as near- and far-UV CD and extrinsic fluorescence, as well as changes within the charge-state distributions of the multiply deprotonated molecular ions.

However, a study by Konermann and Douglas (1998b) revealed that protein conformational changes detected by spectroscopic methods and positive ion mode ESI-MS did not necessarily result in changes of the charge-state distribution in the negative ion mode. For example, acid-induced unfolding of cytochrome c and ubiquitin, as well as the base-induced unfolding of ubiquitin, led to dramatic changes in the charge distribution observed in the positive ion mode ESI-MS, whereas only minor changes were observed in the negative ion charge-state distribution. To fully judge the potential of negative ion mode ESI-MS as a method for monitoring conformational changes of proteins in solution, we must acquire more experimental data. A better understanding of the underlying processes of negative ion ESI will also be necessary.

Protein Folding Kinetics by "Time-Resolved"-ESI-MS

Konermann and Douglas (1997a,b) introduced "time-resolved" ESI-MS to study protein folding/unfolding reactions by ESI-MS. A continuous flow mixing apparatus is directly coupled to an ESI mass spectrometer (Fig. 6). This experimental setup was first used to monitor the refolding of acid-denatured cytochrome c (Konermann and Douglas, 1997a). The analysis of the different charge-state distributions recorded during the time course of the folding experiment revealed that refolding of acid-denatured cytochrome c proceeds via a fast folding population and a slow folding subpopulation with lifetimes of 0.17 and 8.1 s, respectively. In the second study, time-resolved ESI-MS was applied to study the acid-induced denaturation of myoglobin (Konermann and Douglas, 1997b). The different charge-state distributions observed by ESI-MS in the time course of the unfolding experiment revealed the previously unknown transient formation of a folding intermediate with a lifetime of the order of 0.4 s, namely an unfolded state that was still capable of retaining the heme.

Using an online rapid mixing in conjunction with ESI-TOF MS, Wilson et al. (2005) have also published a study on the subunit disassembly and unfolding of the inducible nitric oxide synthase oxygenase domain

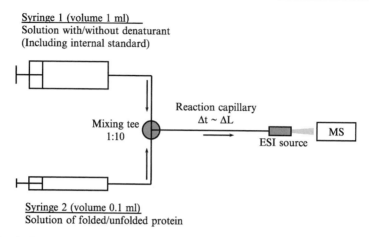

Syringe 1 (volume 1 ml)
Solution with/without denaturant
(Including internal standard)

Mixing tee
1:10

Reaction capillary
Δt ~ ΔL

ESI source

MS

Syringe 2 (volume 0.1 ml)
Solution of folded/unfolded protein

FIG. 6. Experimental apparatus used for monitoring protein folding/unfolding studies by time-resolved electrospray ionization (ESI)–mass spectrometry (MS). Two syringes are advanced simultaneously by a syringe pump. The mixing ratio is defined by the volume ratio of the syringes. The folding/unfolding reaction is initiated by 10-fold dilution in the mixing tee. Arrows indicate the flow directions of the solutions from the two syringes. The reaction time Δt is determined by the capillary length ΔL. At the end of the reaction capillary, the protein solution is electrosprayed and the multiply charged protonated gas-phase protein ions are transferred into the mass analyzer.

(iNOS$_{COD}$) following a pH jump from 7.5 to 2.8. During the denaturation process, various protein species become populated, which were distinguished by their ligand-binding behavior and the different charge-states distributions. The denaturation process can be described as follows: The first step is the disruption of the iNOS$_{COD}$ dimer, to generate heme-bound monomeric species in various degrees of unfolding. This first step is accompanied by the loss of two tetrahydrobiopterin cofactors. Subsequent heme loss generates monomeric apoproteins exhibiting various degrees of unfolding. The formation of proteins that are bound to two heme groups was also observed. Apparently, a subpopulation of holo monomers undergoes substantial unfolding while retaining contact with the heme cofactor.

Probing Protein Solution Structures of Proteins by Hydrogen/Deuterium Exchange

Hydrogen Exchange Chemistry

The chemistry of hydrogen exchange reactions has been described by Englander and Kallenbach (1984). In peptides and proteins, the exchangeable hydrogens are the polar side-chain hydrogens bound to heteroatoms

(N, O, and S), the N- and C-terminal hydrogens, and the backbone peptide amide hydrogens (Englander and Kallenbach, 1984; Woodward *et al.*, 1982). In proteins, the polar side-chain hydrogens exchange 10^3–10^6 times faster than peptide amide hydrogens at pH 7 and exchange rate constants are highly pH dependent (Englander and Kallenback, 1984; Eriksson *et al.*, 1995; Woodward *et al.*, 1982). Exchange rates for polar side chain hydrogens, as well as for N- and C-terminal hydrogens are usually too high to be readily determined. Therefore, most of the hydrogen exchange studies focus on determining the exchange characteristics of peptide amide hydrogens. Fortunately, peptide amide hydrogens represent excellent structural probes that report on surface accessibility, secondary structural bonding network, and conformational stability (Englander *et al.*, 1996).

The Influence of Solution Temperature, pH, and Composition on Amide Hydrogen Exchange

Peptide amide hydrogen exchange is almost exclusively catalyzed by OH^- down to pH 3 and below that by H_3O^+ in water-based solutions. Thus, the exchange rate constant for a freely exposed peptide amide hydrogen $k_{ex,NH}$ can be approximated as

$$k_{ex,NH} = k_{OH}[OH^-] + k_H[H^+] + k_0, \qquad (2)$$

where k_{OH}, k_H, and k_0 are the rate constants for base-catalyzed, acid-catalyzed exchange, and direct exchange with water, respectively (Englander and Kallenbach, 1984; Eriksson *et al.*, 1995; Woodward *et al.*, 1982). Amide H/D exchange studies using poly-DL-alanine as a random coil-like model peptide provided reference rate constants for $k_{OH} = 1.12 \times 10^{10}$ and $k_H = 41.7$ $M^{-1}min^{-1}$, respectively at $20°$ and low salt concentrations (Bai *et al.*, 1993; Molday *et al.*, 1972). Direct exchange with water is commonly considered as insignificant in most of the exchange studies ($k_0 = 0.03\,min^{-1}$) (Bai, 1993 *et al.*; Molday *et al.*, 1972; Woodward *et al.*, 1982).

There is a pH dependence of exchange rate constants of polar side-chain hydrogens and exposed peptide amide hydrogens (Creighton, 1993) (Fig. 7). The $log(k_{ex,NH})$ versus pH plot has a pronounced minimum at pH 2.5–3.0, indicating that exchange rates for base and acid catalysis are equal. The extreme pH sensitivity of peptide amide hydrogen exchange is apparent by an approximate 10-fold increase of the exchange rate for each unit change in pH.

Hydrogen exchange rates are temperature dependent. The rate constants follow the Arrhenius equation, $lnk = lnA - E_a/(RT)$. Thus, to predict exchange rate constants at different temperatures the following equation can be used:

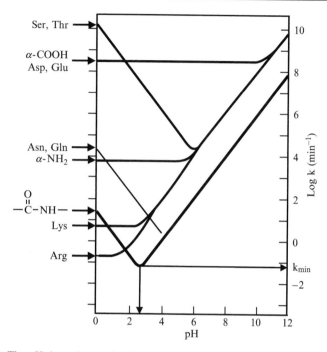

Fɪɢ. 7. The pH dependence of exchange-rate constants of different types of hydrogens occurring in peptides and proteins.

$$k_{ex,T2} = k_{ex,T1} \exp(-E_a\{1/T_2 - 1/T_1\}/R) \qquad (3)$$

whereby $k_{ex,T2}$ and $k_{ex,T1}$ are the exchange rate constants at temperatures T_2 and T_1 (in Kelvin), R is the gas constant 8.314 $Jmol^{-1}K^{-1}$ and the activation energies E_a for the acid catalyzed, the base-catalyzed, and water-catalyzed exchange are 14 kcal/mol, 17 kcal/mol, 19 kcal/mol, respectively (1 kcal = 4.184 kJ) (Bai et al., 1993). This corresponds to an approximate threefold increase of the exchange rate with every 10-K increase in temperature.

Temperature and pH adjustments are used in exchange rate studies of peptide amide hydrogens to (a) probe fluctuations of conformational states, (b) monitor conformational alterations as a consequence of ligand binding or environmental changes, and (c) monitor the cooperativity of unfolding transitions. At pH 7 and 25°, half-lives of exposed peptide amide hydrogen are in the range of 0.05–0.01 s, whereas at 0° and pH 2.7 the half-lives are 1–2 hr. The latter represent conditions used for quenching exchange reactions to preserve the isotope labels during analysis.

In addition to the effects of temperature and pH, solvent composition has a decisive effect on amide hydrogen isotope exchange rates (Englander et al., 1985). Miscible organic solvents used in reverse-phase HPLC (RP-HPLC) generally slow down the exchange rate, largely because of a decrease in the equilibrium constant K_w for water or hydroxide ion activity. The pH_{min} shifts to progressively higher values with increasing organic solvent fraction. A change in solvent composition can be used to reduce back-exchange of the isotopic label when working outside the pH_{min} limits, for example, during HPLC analysis.

Side-Chain Effects on Amide Hydrogen Exchange

Peptide amide hydrogen exchange rates are influenced by neighboring side chains through inductive and steric blocking effects (Bai et al., 1993; Molday et al., 1972). Polar side chains withdraw electrons, thereby increasing the acidity of the peptide hydrogen group. Thus, the exchange rate by base catalysis is increased and the exchange rate by acid catalysis decreased. Steric blocking, and hence, retardation of exchange, is caused by bulky side chains, such as β- and γ-branching of side chains and by aromatic residues. The additivity of the inductive and steric blocking effects on peptide hydrogen exchange has been demonstrated and quantified by Bai et al. (1993) and Molday et al. (1972) through amide H/D exchange kinetics in dipeptide models of all 20 naturally occurring amino acids. These values are commonly used to calculate the exchange rate constant k_{rc} for a peptide amide hydrogen imbedded in a particular amino acid sequence assuming random coil-like behavior. The ratio $k_{rc}/k_{obs} = P$, where P is the protection factor and k_{obs} is the experimentally determined rate constant for the peptide amide hydrogen in question. The protection factor enables an assessment of the retardation factor of the amide hydrogen exchange caused by secondary and tertiary structural hydrogen bonding and/or solvent inaccessibility (Englander and Kallenbach, 1984; Jeng et al., 1990; Wand et al., 1986).

Hydrogen Exchange Mechanism: Ex1 and Ex2

Since its introduction in the early 1950s (Hvidt and Linderstrøm-Lang, 1954; Hvidt and Nielsen, 1966), hydrogen exchange has been a powerful method for studying protein structure and conformational stability. Labile hydrogens in the side chains (e.g., O-H, S-H, and N-H), N- and C-terminal hydrogens, and the backbone amide hydrogens of proteins are exchangeable. However, information regarding structure and conformational stability can best be extracted from the exchange behavior of the backbone amide hydrogens, because these build up the secondary structural bonding network. The rate at which a peptide amide hydrogen exchanges with solvent provides information about the structural stability and surface accessibility of that specific peptide amide hydrogen. Protection against exchange results predominantly from hydrogen bonding, but solvent exclusion of amide hydrogens buried in the core of a protein contributes as well. Labile protons on the surface of the protein that are not involved in hydrogen bonding exchange more readily (Englander and Kallenbach, 1984; Englander *et al.*, 1996; Woodward, 1994; Woodward *et al.*, 1982). A dual-pathway model for hydrogen exchange of peptide amide hydrogens is most often considered (Li and Woodward, 1999; Woodward *et al.*, 1982). The first pathway describes exchange directly from the folded state via local transient opening reactions or "breathing" motions. It is assumed that peptide amide hydrogens located at or near the surface of the protein or in close proximity to solvent channels exchange via this pathway (Eq. [4], Fig. 8). In the second mechanism, exchange is preceded by transient reversible unfolding reactions that are accompanied by the breaking of hydrogen bonds (Fig. 9A) (Englander and Kallenbach, 1984; Englander *et al.*, 1996; Linderstrøm-Lang, 1958; Woodward, 1994; Woodward *et al.*, 1982). Unfolding of proteins can occur via (a) global or whole-molecule transitions, (b) subglobal transitions, or (c) local fluctuations (Chamberlain and Marqusee, 1997; Englander *et al.*, 1996). Which of the two exchange mechanisms dominates depends on the experimental conditions (i.e., the pH, temperature, or chemical denaturants present).

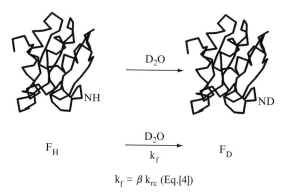

$$k_f = \beta \, k_{rc} \ (Eq.[4])$$

FIG. 8. Pictorial presentation of the underlying principle of peptide amide hydrogen exchange from folded proteins. The rate constant for exchange at any backbone amide is given by $k_{obs} = \beta \, k_{rc}$, where β is the probability that D_2O and OD^- are present at the back amide and k_{rc} is the intrinsic chemical rate constant for isotopic exchange at amide hydrogens in random coil-like peptides.

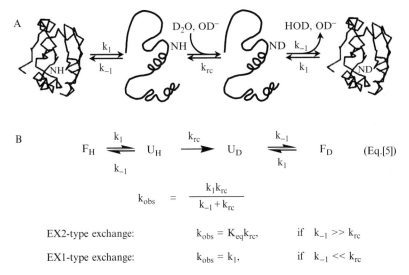

$$k_{obs} = \frac{k_1 k_{rc}}{k_{-1} + k_{rc}}$$

EX2-type exchange: $k_{obs} = K_{eq}k_{rc}$,   if $k_{-1} \gg k_{rc}$

EX1-type exchange: $k_{obs} = k_1$,   if $k_{-1} \ll k_{rc}$

FIG. 9. (A) Pictorial presentation of the underlying principles of peptide amide hydrogen exchange from folded proteins via partial, subglobal, or global unfolding. (B) The Linderstrøm-Lang model (Eq. [5]), which connects observed exchange rates with structural unfolding reactions in proteins. Rate constants k_1 and k_{-1} describe the unfolding and refolding reaction of a protein, and k_{rc} is the intrinsic chemical rate constant for isotopic exchange at amide hydrogens in random coil-like peptides.

The Linderstrøm-Lang model connects the second mechanism with the observed exchange rates (Eq. [5], Fig. 9B).

The observed exchange reaction will be second order, if the refolding rate of the transient opening reaction is much greater than the rate with which the intrinsic chemical exchange occurs (i.e., $k_{-1} \gg k_{rc}$). Thus, the experimentally observed rate constant is given by $k_{obs} = k_{rc} (k_1/k_{-1})$. This type of exchange is termed the *EX2-type mechanism,* and exchange via this mechanism is observed for most proteins under neutral pH and in the absence of denaturants. In this mechanism, refolding (i.e., large k_{-1}) is more likely than the exchange reaction and the exchange is random or uncorrelated. Exchange occurs via the so-called *EX1-type mechanism,* if the rate of the chemical reaction is much greater than refolding (i.e., $k_{rc} \gg k_{-1}$). Thus, the observed exchange rate is directly related to the protein unfolding rate (i.e., the observed exchange rate constant is given by $k_{obs} = k_1$) (Clarke and Itzhaki, 1998; Englander *et al.*, 1996; Linderstrøm-Lang, 1958; Woodward *et al.*, 1982), and correlated motion can be detected by MS (Arrington and Robertson, 2000; Arrington *et al.*, 1999; Deng and Smith, 1998a,b; Maier *et al.*, 1999; Yi and Baker, 1996; Zhang *et al.*, 1997a).

Hydrogen Isotope Exchange Monitored by Mass Spectrometry

Typically, H/D exchange experiments are monitored by NMR techniques because deuterons in proton NMR experiments do not give a signal. The exchange behavior of individual amide protons as a function of exchange periods is deduced from the intensity of resolved NH-resonances in one-dimensional (1D) NMR spectra or NH-related cross peaks in two-dimensional (2D) NMR spectra. Thus, deuterium exchange experiments in which the average proton occupancies is monitored in a site-specific manner yield exchange rates of individual backbone amides in proteins but averaged over all molecules (Roder, 1989; Scholtz and Robertson, 1995).

The advantages of using MS to monitor H/D exchange reactions have become recognized. MS-based approaches use the fact that the difference in mass between hydrogen and deuterium is 1 Da, and, thus, changes in the number of deuteriums incorporated in a molecule can be directly related to mass shifts. The application of MS to monitor H/D exchange in peptides and proteins was first introduced by Katta and Chait (1991, 1993) to study the native and denatured states of ubiquitin. Since then, the approach has been widely used to (a) study the high-order proteins structures in different environments (Katta and Chait, 1991, 1993; Wang *et al.*, 1997; Zhang and Smith, 1993; Zhang *et al.*, 1996, 1997b), (b) assess the consequences of point mutations (Guy *et al.*, 1996a,b; Nettleton *et al.*, 1998) and ligand binding (Nemirovskiy *et al.*, 1999), (c) probe the solution dynamics of proteins and peptides (Arrington and Robertson, 2000; Arrington *et al.*, 1999; Maier *et al.*, 1999), (d) obtain structural descriptions for partially folded states (Chung *et al.*, 1997; Maier *et al.*, 1997; Robinson *et al.*, 1994), and (e) examine protein folding (Miranker *et al.*, 1993) and unfolding pathways (Chung *et al.*, 1997; Deng and Smith, 1998a,b; Deng *et al.*, 1999b; Maier *et al.*, 1999; Yi and Baker, 1996).

In principal, two experimental approaches are used. In the first approach, the deuterium exchange-in reaction is followed by dilution of the protonated protein in deuterating solvents. This approach is beneficial for proteins that are difficult to prepare in a fully deuterated form, but it suffers from the fact that back-exchange during the ESI-MS analysis may occur. Therefore, this methodology is more suitable for experiments in which comparisons between different proteins or conformations are made. In this context, it is important to note that if deuterium exchange-in experiments are directly monitored by ESI-MS, multiply deuterated ions $(M+nD)^{n+}$ will be observed. Thus, deuterium incorporation can be monitored directly as mass increase and expressed as the percentage of mass increase relative to the maximum number of exchangeable hydrogens (Fig. 10A).

In the second approach, the exchange is initiated from a fully deuterated protein by dilution in protonating solvent. Under these conditions, deuterium exchange-out kinetics are monitored. This process has the obvious advantage that the exchange rates observed are directly related to the exchange behavior of the backbone amide deuterons because exchange of the side-change deuterons is several orders of magnitude faster than exchange from the backbone amides. Fortunately, the exchange behavior of the backbone amides is of particular interest for structural studies, because the backbone amides provide the framework for the hydrogen bonding network that build up the tertiary structure. Under these circumstances, if deuterium exchange-out experiments are directly monitored by ESI-MS, multiply protonated $(M + nH)^{n+}$ ions are observed and the mass decrease as exchange-out proceeds can be best expressed as the number of deuteriums remaining after a certain exchange-out period. The deuteriums remaining can be determined by subtracting the observed mass shift (i.e., the deuterium loss) from the number of deuteriums observed in the fully deuterated protein (Fig. 10B).

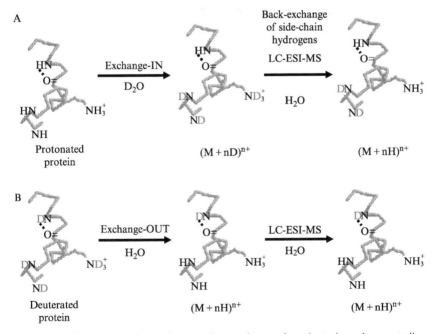

Fig. 10. The two experimental approaches used to perform isotopic exchange studies: (A) the exchange-in and (B) the exchange-out approach. (See color insert.)

Monitoring and analyzing D/H exchange-out is particularly powerful for studying protein interfaces. The protein–protein interfaces are indicated by regions that retain more deuterons in the complex compared with control experiments in which only the individual, non-complexed protein is present. Monitoring exchange-out was used, for example, by Mandell *et al.* (1998), for the identification of the kinase inhibitor PKI(5–24) and ATP-binding sites in the cyclic AMP–dependent protein kinase. The same authors have also published the identification of the thrombin–thrombomodulin interfaces by monitoring the kinetics of amide deuterium exchange-out (Mandell *et al.*, 2001).

The influence of a disulfide bond on the thermodynamic stability of a small single-domain protein can be seen in deuterium exchange-in experiments, with oxidized and reduced *Escherichia coli* thioredoxin (TRX) using ESI-MS to measure the mass increase (Fig. 11). Oxidized *E. coli* TRX possesses 173 exchangeable hydrogens; 102 are backbone amide hydrogens, 68 are side-chain hydrogens, and 3 are terminal hydrogens. The reduced protein has a 175 exchangeable sites because of two additional thiol protons. For oxidized TRX, a molecular mass increase of approximately 71 ± 5 Da ($40 \pm 3\%$) is observed after 60-min incubation in 1% acetic acid-d_1 (v/v) at room temperature (Fig. 11), indicating that approximately 102 ± 5 of the hydrogens did not yet exchange, and these are protected against exchange and, presumably, are part of the stable protein core. By contrast, in reduced TRX, approximately 82 ± 5 ($47 \pm 3\%$) protons have undergone exchange, leaving approximately 93 ± 5 hydrogens on the protein. A comparison of the global hydrogen exchange profiles over a 2-h period demonstrated that reduced TRX is less protected against deuterium exchange-in over the entire time course sampled, thereby confirming that reduced TRX is thermodynamically less stable than oxidized TRX (Kim *et al.*, 1999).

Zhu *et al.* (2003a) introduced a methodology to quantify protein–ligand interaction by MS, ligand titration, and H/D exchange, termed *PLIMSTEX*. This approach can determine conformational change, binding stoichiometry, and affinity in protein–ligand interactions including those that involve small molecules, metal ions, and peptides. The method yields plots of the mass difference between the deuterated and nondeuterated protein (deuterium uptake) versus the total ligand concentration. These titration curves are then fit to a 1:n (*n* is the number of binding sites) protein:ligand sequential binding model. This method was applied to determine the Ca^{2+} interactions in calmodulin (Zhu *et al.*, 2003b) and to study the interaction of the intestinal fatty acid–binding protein (IFABP) with a fatty acid carboxylate (Zhu *et al.*, 2004).

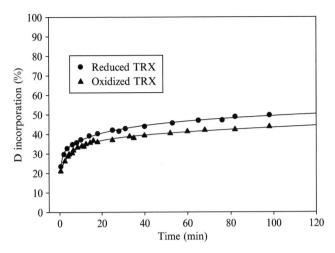

Fig. 11. H/D exchange-in kinetic profiles of reduced and oxidized *Escherichia coli* thioredoxin (TRX) at 22° in 1% acetic acid-d_1 (v/v).

Determination of H/D-Exchange Mechanisms: EX1 and EX2 Types

Proteins are not rigid structures but highly dynamic systems that undergo constant fluctuations in which hydrogen bonds are transiently opened and re-formed. The extent of these fluctuations is critically dependent on the environment that can stabilize or destabilize the native conformational state. Thus, backbone H/D exchange in proteins under solution-phase conditions depends on two processes: (a) the unfolding or opening event during which the backbone amides are transiently exposed to the solvent and (b) the intrinsic chemical exchange of the exposed amide hydrogen with deuterating solvent. The two rate-limiting mechanisms (i.e., EX2 and EX1 types for isotopic exchange reactions in proteins) can be distinguished (Fig. 9B) (Clarke and Itzhaki, 1998; Englander *et al.*, 1996; Linderstrøm-Lang, 1958; Woodward *et al.*, 1982). When a large excess of the hydrogen isotopically labeled water is present, the overall exchange reaction is essentially irreversible.

The conformational dynamics of TRX during thermal unfolding have been studied by solution-phase H/D exchange-in as a function of time and temperature (Maier *et al.*, 1999) (Fig. 12). To study the dynamic processes of TRX by H/D exchange-in during heat denaturation, mild acidic solution conditions were chosen to slightly destabilize the native state. H/D exchange was initiated by diluting the protein solution into a deuterating solution, which consisted of 2% acetic-acid-d_1 (v/v). A capillary assembled

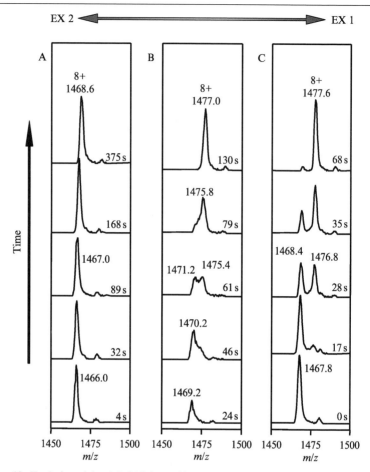

FIG. 12. Evolution of the eightfold charged ion peak of *Escherichia coli* thioredoxin during the online H/D exchange-in experiments at 40° (A), 60 ± 2° (B), and 80° (C) in 1% acetic acid-d₁ (v/v). The time points given refer to incubation periods (time).

in a conventional injection valve was used as a reaction vessel (Fig. 13). The solvent delivery line, the reaction vessel, and the injection valve were immersed in a water bath equilibrated at the desired temperature. Thermal denaturation and H/D exchange-in reactions were performed in the reaction vessel that was directly connected by a fused silica capillary to the ESI source. This infusion capillary was immersed in an ice bath to quench H/D exchange-in and initiate refolding of thermally denatured TRX.

Exchange according to a pseudo–first-order mechanism (EX1-type mechanism) occurs if interconversion between the native closed con-

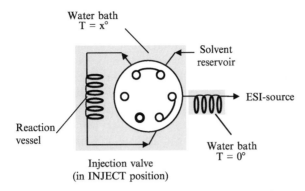

FIG. 13. Experimental setup used for the H/D exchange-in experiments monitored online by electrospray ionization–mass spectrometry.

formational states (F) and the unfolded open states is slow with respect to the intrinsic hydrogen exchange rate. In this case, the rate-limiting step is determined by k_{-1} and the experimentally observed exchange rate constant k_{ex} is directly related to the unfolding rate constant k_1 (i.e., $k_{ex} = k_1$). These are normal denaturing conditions (e.g., high temperature or in the presence of chaotropic reagents) under which the native state is only marginally stable. Because the chemical exchange rate is highly pH dependent, alkaline pH favors exchange according to an EX1 type of mechanism (Clarke and Itzhaki, 1998; Englander *et al.*, 1996; Linderstrøm-Lang, 1958; Woodward *et al.*, 1982). Mass spectrometric approaches are uniquely able to detect the cooperativity of exchange of multiple sites because the correlated exchange of several sites results in a mass difference, which results in a bimodal ion peak pattern (Arrington *et al.*, 1999; Arrington and Robertson, 2000; Deng and Smith, 1998a,b; Maier *et al.*, 1999; Yi and Baker, 1996). Such a scenario is observed during thermal denaturation coupled with deuterium exchange-in experiments of oxidized TRX above 65° (Fig. 12). After a short incubation period, a bimodal ion peak pattern was observed for each charge state detected. The ion peak at lower m/z values represents the population of protein molecules that had not undergone cooperative unfolding, and their core amide protons were still protected against exchange-in, whereas the ion peak at higher m/z values arises from the deuterated population of molecules that had undergone cooperative unfolding. Hence, these molecules had spent enough time in the unfolded state to allow their core amide protons to exchange. The number of amides that participate in the cooperative unfolding event can be extracted from the m/z difference of these two ion peaks under consideration of the charge state. In the case of oxidized TRX, 68 ± 6 deuterium underwent correlated exchange (Fig. 12). If exchange occurs exclusively according to an EX1 mechanism, the intensity of the ion peak at lower m/z values decreases during exchange-in, whereas the intensity of the ion peak at higher m/z values increases, in which case no mass shift is observed. The rate of conversion allows an estimation of the unfolding rate constant k_1. For oxidized TRX in 2% acetic acid-d_1 (v/v) at 80°, approximately 68 ± 6 backbone amides underwent cooperative unfolding with a rate constant k_1 of approximately 2 ± 0.2 min (Maier et al., 1999).

Under conditions that favor the compact native state, exchange occurs via a second-order mechanism (EX2-type mechanism) that is characterized by a fast interconversion between native and unfolded open states. Thus, the overall exchange rate of the backbone amide hydrogen depends on the intrinsic exchange rate and the average fraction of time during which the backbone amide hydrogen is transiently in the open state and exposed to bulk solvent (i.e., $k_{obs} = k_{rc}(k_1/k_{-1})$) (Clarke and Itzhaki, 1998; Englander et al., 1996; Linderstrøm-Lang, 1958; Woodward et al., 1982). H/D-exchange behavior of the EX2 type has been observed for oxidized TRX when exchange experiments are performed below 40°. Under these conditions, the ESI mass spectra showed ion peaks that shifted gradually to higher m/z values with increasing exchange-in time because each amide will exchange to a varying degree depending on its individual exchange rate (Maier et al., 1999) (Fig. 12).

Another example that demonstrates the beauty of using MS-based H/D-exchange approaches to investigate conformational dynamics is the study on E. coli heat shock transcription factor σ^{32} by Rist et al. (2003). Using an HPLC-MS setup, they monitored the incorporation of deuterons into full-length σ^{32}. In combination with immobilized pepsin for digestion, these authors localized slow- and fast-exchanging regions within the entire sequence of σ^{32}. These studies revealed that σ^{32} adopts a highly flexible structure at 37°, as indicated by a rapid exchange of about 220 of the 294 amide hydrogens. However, at 42°, a slow-correlated exchange of 30 additional amide hydrogens was observed. The domain that exhibited correlated exchange was part of a helix-loop-helix motif within domain sigma 2, which is responsible for the recognition of the -10 region in heat shock promoters. The correlated exchange is related to reversible unfolding with a half-life of about 30 min due to a temperature-dependent decrease in stabilization energy. These data support a role for σ^{32} as a thermosensor by undergoing conformational alterations in response to heat shock (Rist et al., 2003).

Finally, it is important to point out that mass spectrometric approaches are uniquely able to distinguish between EX1 and EX2 exchange regimens because the distribution of deuteriums can be obtained from the mass spectrum. NMR-based approaches are not able to directly distinguish between these two exchange mechanisms because NMR methods monitor the average proton occupancy at individual sites (Arrington and Robertson, 2000; Arrington et al., 1999 ; Miranker et al, 1993; Yi and Baker, 1996). A number of authors (Arrington and Robertson, 2000; Arrington et al., 1999; Yi and Baker, 1996) have emphasized the complementary use of both techniques (i.e., NMR and MS) to investigate the native-state H/D-exchange experiments of small single-domain proteins. In the latter study, the third domain of turkey ovomucoid was subjected to native-state H/D-exchange experiments. NMR studies suggested that 9 of 14 of the slowest exchanging peptide hydrogens possess very similar unfolding rate constants and free energies, suggesting that the exchange of these 9 backbone amide hydrogens participates in a cooperative unfolding event. However, ESI-MS data and mass peak simulations revealed that only three to five (not nine) backbone amide hydrogens exchange cooperatively. In this case, the combined use of NMR and MS to study native-state amide hydrogen exchange resulted in a very detailed picture of the thermodynamic and kinetic properties of the very slowest exchanging backbone amide hydrogens of the turkey ovomucoid domain. This kind of knowledge is important for understanding the nature of conformational fluctuations and cooperativity of folding/unfolding events in the modern view of funnel-like energy landscapes of proteins (Arrington and Robertson, 2000; Arrington et al., 1999; Dill and Chan, 1997).

Peak Width Analysis

The peak width of a molecular ion in an ESI mass spectrum of a protein arises from the contribution of the natural isotope, primarily that of carbon-13 abundance to the individual charge state and the instrumental resolving power. If mass spectrometric measurements are done on small proteins, the ion peak width of a certain charge state at half height compares well with the natural peak width simulated (Chapman, 1996). For example, the peak width at half height of approximately 1.2 amu observed for the eightfold protonated molecular ion for oxidized TRX compares reasonably well with the theoretically expected peak width at half height of approximately 1 amu for the eightfold protonated ion peak of oxidized TRX, which has the empirical formula $C_{528}H_{836}N_{132}O_{159}S_3$ (Fig. 14A).

In experiments in which MS is used to monitor H/D exchange, the analysis of the ion peak width may reveal unique information regarding the

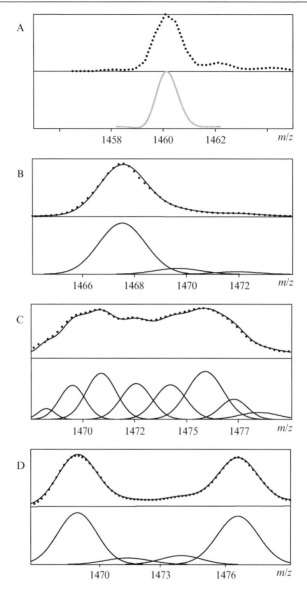

Fig. 14. Peak width analyses of the eightfold charged ion peak of *Escherichia coli* thioredoxin observed during the exchange-in experiments under different exchange conditions. (A) Comparison of the eightfold charged ion peak (black dotted line) with the corresponding calculated isotopic distribution for the eightfold protonated ion peak (gray continuous line) of oxidized TRX ($C_{528}H_{836}N_{132}O_{159}S_3$). (B) Comparison of the eightfold charged ion peak (dotted line) with a Gaussian deconvolution (continuous line) of the eightfold charged ion peak after 90 ± 4 s if exchange occurs according to an EX2-type

conformational diversity of the protein and the exchange kinetics. A consequence of H/D-exchange experiments is a broadening of the ion peak caused by the convolution of the natural isotope distribution with an assumed random distribution of deuterium throughout the protein (Chung *et al.*, 1997). For example, during H/D experiments under EX2 conditions, the peak width at half height of the eightfold protonated molecular ion of oxidized *E. coli* TRX was approximately 2.5 amu throughout the time course of the exchange-in experiments, which suggests that a relatively small distribution of conformational states is present and that peak broadening is primarily caused by random, uncorrelated deuterium incorporation via local unfolding events (Chung *et al.*, 1997; Maier *et al.*, 1999) (Fig. 14B).

In contrast, when the exchange experiment was performed at elevated temperature, 60°, the appearance of the ion peak changed dramatically after only short incubation periods. There was significant broadening of the observed ion peaks. For instance, for the eightfold protonated ion peak of oxidized *E. coli,* TRX showed a width at half height of approximately 8 amu. This type of broadening signifies that under conditions favoring the transition zone between the EX1 and EX2 mechanism, a wide range of conformational states with differing extents of protection against exchange-in exist (Maier *et al.*, 1999; Yi and Baker, 1996) (Fig. 14C).

As exchange-in experiments are performed under conditions favoring the EX1-type mechanism, two distinct ion peaks evolve for each charge state, indicating that the compact folded and unfolded states are simultaneously populated. For example, the peak widths at half height for the eightfold protonated ion peaks representing the folded and unfolded population are approximately 2.5 amu, which can be interpreted on the basis that the folded and the unfolded conformational states are represented as distinct populations of molecules. Gaussian deconvolution of the ion pattern reveals two minor additional populations (<10 %) exhibiting deuterium incorporation that might be expected for partially unfolded states. In addition, during the course of the exchange-in at elevated temperature, a slight shift to higher m/z values and a concomitant narrowing of the ion

---

mechanism. The extracted peak width at half maximum height (~2.5 Da) compares well with a Gaussian approximation of the ion peak under consideration of the natural isotope distribution and a random deuterium exchange model. (C) Gaussian deconvolution (continuous line) of the eightfold charged ion peak after 60 ± 4 s (dotted line) if exchange occurs between EX2-and EX1-type mechanism. Gaussian deconvolution was used with a variable mean and coefficients but a fixed standard deviation (derived from the approximate average width at half maximum height from the eightfold charged ion peak of the protonated and fully deuterated protein). (D) Gaussian deconvolution (continuous line) of the eightfold charged ion peak pattern after 28 ± 4 s (dotted line) if exchange approaches the EX1-regime.

peak at higher m/z value were observed, which indicates that the exchange mechanism was not exclusively EX1, but a combination involving both the EX1 and the EX2 type of mechanism (Maier *et al.*, 1999; Yi and Baker, 1996) (Fig. 14D). Yi and Baker (1996) and Arrington *et al.* (1999) have described an algorithm that used NMR-derived exchange rate constants to simulate accurate binomial mass spectral data under consideration of a different contribution of the two rate-limiting exchange mechanisms (i.e., EX1 and EX2), various degrees of cooperativity, and the natural isotope distribution.

Regional Structural and Dynamic Information from Medium Spatial Resolution Data

In addition to probing the global conformational stability and dynamics of a protein, more detailed information with a medium spatial resolution can be obtained by digesting the deuterium-labeled protein with pepsin under conditions of slow exchange (i.e., pH ~2.5 and 0°) and then analyzing the resulting peptic fragments by liquid chromatography online–coupled MS. Both fast-atom bombardment (FAB) and ESI-MS have been used. The deuterium levels of the individual peptides can be extracted from the observed mass shifts between the unlabeled and labeled peptides. This approach has been used in numerous applications to characterize the higher order structure and dynamics of proteins (Engen *et al.*, 1997; Johnson, 1996; Johnson and Walsh, 1994; Maier *et al.*, 1997; Resing and Ahn, 1998; Resing *et al.*, 1999; Zhang and Smith, 1993, 1996; Zhang *et al.*, 1996) or to study the protein conformational influence on ligand binding (Wang *et al.*, 1997, 1998a,b, 1999). The protein proteolysis approach is especially attractive because of its universality, but it is relatively time consuming and suffers from increased probability of back-exchange of hydrogen isotopes during chromatographic separation and the limited proteolysis often observed for protease-resistant proteins. Procedures to adjust for deuterium loss during the analysis have been published (Zhang and Smith, 1993). Other authors have described HPLC setups that combine online pepsin digestion and LC-ESI-MS analysis (Wang *et al.*, 2002; Wang and Smith, 2003).

An example in which conformational and solvent accessibility changes were studied by H/D exchange, peptic proteolysis, and MS is the study of human retinoid X receptor homodimer upon 9-cis retinoic acid binding (Yan *et al.*, 2004a). The RXRα LBD is a 54.4-kDa non-covalently linked homodimeric protein. Time course labeling studies in the presence and absence of its ligand 9-cis retinoic acid were performed. Deuterium exchange levels with medium spatial resolution were obtained by peptic proteolysis followed by mass analysis. The deuterium exchange profiles of peptic peptides generated from the apo- and holo-RXRα LBD homodimer were compared. It became apparent that approximately half of the regions that show significant exchange protection are regions that have direct contact surfaces with the ligand, as indicated by the X-ray structure: The amino termini of helixes 3 and 9, the two β-sheets, helix 8, the H8-H9 loop, and the C-terminus of helix 11. Unexpectedly, protection was also observed in peptides derived from helixes 7, 10, 11, and the H7-H8 and H10-H11 loops, regions that are not directly in contact with bound 9-cis-RA, and these changes were not apparent in the X-ray structure of the holo protein (Yan *et al.*, 2004a). Similarly, the ability of MS-based H/D exchange, in combination with proteolysis to determine both the location of binding sites and the effects of binding on distal locations, was also reported for different states of actin (Chik and Schriemer, 2003).

Continuous Vs. Pulse Labeling

Essentially, two experimental H/D-exchange approaches are used: continuous exchange and pulse labeling. Continuous exchange involves exposing the protein to the deuterated medium during the incubation time. The labeled protein is thus a measure of all unfolding events that took place, provided the exchange rate is much faster than the unfolding event. In pulse labeling, the deuterated protein is a snapshot of the unfolded protein that existed during the pulse. Many unfolding and refolding events may have taken place during the incubation, but these are not recorded because no deuterium is available for exchange between the pulses (Deng *et al.*, 1999b).

Proteins can undergo H/D exchange either from their folded states (Eq. [4]) in which case the exchange involves amide hydrogens at the surface of the protein that are not hydrogen bonded, or they can exchange after unfolding (Eq. [5]), which means that hydrogen bonds were broken in order for the exchange to take place. A number of authors (Deng and Smith, 1998a,b; Deng *et al.*, 1999b) have shown that the combination of continuous versus pulse labeling in D_2O can help to reveal when both mechanisms are operative in different regions of the protein. The approach is particularly useful when labeling under denaturing conditions occurs to give bimodal isotope distributions. Thus, for rabbit muscle aldolase in 3 M urea, when exchange in a given region of the folded state was slow, a bimodal isotope distribution displayed in the mass spectra of peptic peptides derived from the intact protein was observed in both experiments, but the relative intensity of the protonated distribution by pulse labeling diminished more slowly with time. When exchange was rapid, poorly resolved isotope distributions observed under continuous exchange conditions in the examined peptides of certain regions of the protein were nicely resolved into bimodal distributions in the pulsed labeling experiment.

In Source Collisionally Activated Dissociation of Proteins

There are two basic ESI dissociation strategies: one takes place in the ESI source, between the end of the capillary and the skimmer cone (Light-Wahl et al., 1993; Loo et al., 1993; Senko et al., 1994), and the other takes place after the ESI source. In the latter, ions are selected for CID and the products are then analyzed by standard MS/MS strategies. This procedure is commonly used for sequencing peptides. In the former, collision-activated dissociation (CAD) of protein ions is induced in the ESI interface region by a large voltage difference between the capillary and the skimmer cone. This method is a useful alternative to the usual procedure of digesting the protein by proteolytic enzymes—in the case of H/D exchange experiments by pepsin—and introducing the digest into the mass spectrometer by infusion or HPLC. Fragmentation by CAD eliminates the sample workup time and reduces the amount of sample required. Solution-phase H/D exchange and CAD in the nozzle-skimmer region of the electrospray source in combination with ESI-FT-ICR MS was used to probe the conformational stability of cellular retinoic acid-binding protein 1 (Eyles et al., 2000). A nonstructured 22-residue N-terminal His tag underwent complete amide isotope exchange, while the peptide fragments from regions of stable β-sheets and α-helixes were found to be well protected against exchange.

High-Resolution Mass Spectrometry (FT-ICR-MS) for H/D Exchange Studies

The combination of electrospray ionization with Fourier transform ion cyclotron resonance (FT-ICR) MS allows for the analysis at unit mass resolution of proteins whose masses exceed 40 kDa (McLafferty et al., 1998a). By double depletion of ^{13}C and ^{15}N isotopes to less than 0.01%, the mass range at which unit mass resolution is observable can be extended to nearly 100 kDa (Marshall et al., 1997). Unambiguous assignment of charge states is a key advantage to high-resolution FT-ICR-MS. Studies of gas-phase conformational states and changes between states and protein folding by the H/D exchange methodology depend on the accuracy of the charge-state assignments (McLafferty et al., 1998b). When low-resolution MS is used, the determination of charge states may require HPLC separation of the peptides. Such separation schemes take time and increase the probability for back-exchange of the isotope (Zhang and Smith, 1993).

Identification of the peptic peptides or other proteolytic fragments is necessary for successful studies on protein conformations by H/D exchange, and the identification of these peptides can readily be achieved in most cases by accurate mass determinations alone. Low-resolution MS may require MS/MS experiments, which take time.

Tandem Mass Spectrometry and Site-Specific Protein Amide
 Hydrogen Exchange

The b_n-Ions

In comparison to NMR methods, MS offers certain advantages for probing conformational structures. The one advantage enjoyed by NMR was the ability to determine deuterium levels at the amide nitrogen of individual peptide linkages. Amide H/D content can also be measured by tandem MS at the individual amino acid residue level; however, the results must be interpreted with caution because of the potential for randomization during the collision process. Deuterium present on amide nitrogens will be evident in the shift to higher masses of the isotopic clusters of sequence ions, such as the b_n- and y_n-ions, when compared to the non-deuterated peptides. From these data, the H/D ratios at individual amide nitrogens can be determined.

For these determinations to succeed, a sequential set of ions must be observed. Although earlier studies used several ion series, such as the b_n, y_n, and z_n (Anderegg *et al.*, 1994), subsequent studies rely mostly on the b_n-ions (Deng *et al.*, 1999a). A potential problem in using y_n-ions is that they are believed to be formed by the transfer of a hydrogen from the N-terminal side of the cleavage site (Kenhy *et al.*, 1992). This transfer of hydrogen or deuterium during analysis obviously would lead to errors if not accounted for. Thus, analysis of the b_n-ions is preferable because no transfer of hydrogens is involved during fragmentation to the acylium ions. However, the collision process required for fragmentation in collisional-induced dissociation (CID) tandem MS (i.e., MS/MS) can cause intramolecular scrambling of the amide hydrogens. For example, Demmers *et al.* (2002) reported gas-phase deuterium scrambling in peptide ions of transmembrane peptides, in which the near-terminal amino acids that anchor the peptide at the lipid–water interface were systematically varied. Using nanoelectrospray and CID on a quadrupole TOF instrument, the extent of intramolecular hydrogen scrambling was influenced by the amino acid sequence of the peptide, the nature of the charge carrier, and, therefore, most likely also by the gas-phase structure of the peptide ion. Protonated b_n- and y_n-ions were evaluated, as well as sodiated a- and y-type fragment ions. Although the observed scrambling seems to be mostly independent of the peptide fragment ion type, scrambling seems to be reduced by using alkali metal cationization instead of protonation in the ionization process (Demmers et al., 2002).

Randomization occurs by the "mobile proton" (i.e., by the added proton that forms the MH^+). All exchangeable hydrogens are subject to interchange (Johnson et al., 1995). Extensive H/D scrambling was observed in sustained off-resonance irradiation (SORI) CID in ESI-FT ICR-MS analysis (McLafferty et al., 1998b), although α-helical regions were found to remain intact by salt bridge stabilization. The degree of hydrogen scrambling depends to a large extent on experimental conditions and the nature of the peptides being analyzed. High internal energy of the peptide ions in the gas phase results in decreased H/D exchange rates (Cheng and Fenselau, 1992; Hoerner et al., 2004; Kaltashov et al., 1997). Slow heating involved in SORI activation leads to extensive intramolecular hydrogen scrambling, whereas rapid heating should reduce scrambling (Eyles et al., 2000; Hoerner et al., 2004). There is evidence that electron-capture dissociation (ECD) to fragment proteins into a series of c_n- and z_n-ions causes much less H/D scrambling (Zubarev et al., 1999), but fragmentation by ECD alone appears to yield sequence information only for the terminal regions of the peptide (Horn et al., 2001). To obtain a more complete ion series, or to analyze larger proteins, a combination of ECD and collisional activation by a background gas (CAD) or blackbody infrared irradiation (IR) is necessary (Horn et al., 2000). How this combination of ECD and CAD or IR will affect hydrogen scrambling is not yet known. Scrambling appears to be much less prevalent for short helical peptides (Anderegg et al., 1994; Waring et al., 1998) or short peptides generally (Deng et al., 1999a).

In addition to hydrogen scrambling during the collision process, there are some other limitations in using MS for determining the amide H/D ratios. Experience has shown that CID MS/MS will not always yield complete sequence information for peptides. Usually, intermediate-size peptide fragment ions are observed, but the smaller ones (dipeptides, tripeptides, etc.) are sometimes missing. As the size of the peptides increases, the completeness of the sequence usually diminishes, and peptides of 30 or more residues almost never yield complete sequence ions. Some information can in principle be obtained for these ions from the reverse direction sequences and possibly from other ion series, but again, the potential for hydrogen scrambling in other ions must be kept in mind. The total deuterium content in the stretch of amino acid residues representing the gaps in the sequence, which often involves no more than two or three residues, will, of course, be recorded and the average number of deuteriums per residue can be determined.

The charge state of peptide ions selected for collisional activation also is relatively important. Higher charge states of the selected peptide ions (i.e., +3 and +4) can be used, but it is often more difficult to interpret the results because of the uncertainty in the charge states in the fragment ions. In addition, errors in mass measurements are multiplied by a factor equal to the charge. High-resolution FT-ICR instruments overcome these problems and are particularly advantageous for studying higher charged ions (Wu *et al.*, 1995).

Correlation of H/D Ratios and NMR-Derived Exchange Rates

Despite the potential problems, recent pulse-labeling studies with cytochrome *c* suggest that it is, in fact, possible to obtain reliable H/D exchange information on backbone amide groups of peptides by tandem MS/MS (Deng *et al.*, 1999a). These new results confirm the results of some limited continuous labeling studies performed earlier with cytochrome *c* (Maier *et al.*, 1995). If it is assumed that randomization of exchangeable amide hydrogen by the "mobile proton" is statistical (Johnson *et al.*, 1995), in other words no preference for any given amide site, once in the gas phase, then it is reasonable that the fractional deuterium content measured at a particular amide site has some relationship to what it was before collisional activation. The studies by Deng *et al.* have shown that the level of deuteration of amide groups in the b-ion series and even the y-ion series following exchange-in from D_2O for 10 s can be qualitatively corroborated with H/D exchange rate constants as determined by NMR (Deng *et al.*, 1999a). Fast exchange rates usually result in high levels of deuterium, whereas slow exchange rates result in low deuterium levels at the amide groups. These authors have concluded that intramolecular migration of deuteriums for the b_n-ions is negligible. The y-ions, on the other had, were found to be less reliable, as some discrepancies were observed, suggesting H/D transfer during CID fragmentation.

Studies with oxidized and reduced *E. coli* thioredoxin for which NMR exchange rate data are available show good qualitative correlations between the hydrogen rates of exchange and the amount of deuterium label at the amide sites in the b_n-ions (Fig. 15), but this was not the case for the y_n-ions (Kim *et al.*, 2001a) (Fig. 16). Negative values and values greater than 1 for site-specific amide deuterium levels are likely due to hydrogen scrambling. This problem is most severe for the y_n-ions. To determine accurate deuterium levels on the b_n-ions, it is critical to obtain the centroids

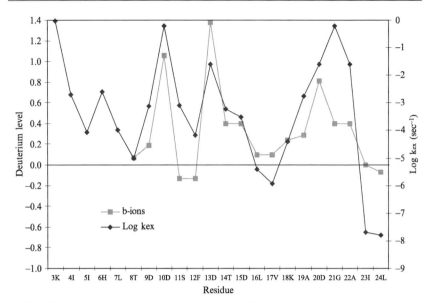

FIG. 15. Comparison of the deuterium levels at amide sites in the peptide 1–24 of oxidized thioredoxin based on the analysis of b_n-ions and H/D exchange rate constants from nuclear magnetic resonance data. (From Jeng and Dyson, 1995.)

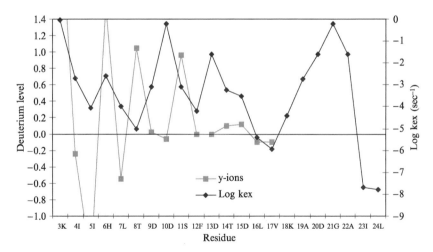

FIG. 16. Comparison of the deuterium levels at amide sites in the peptide 1–24 of oxidized thioredoxin based on the analysis of y_n-ions and H/D exchange rate constants from nuclear magnetic resonance data. (From Jeng and Dyson, 1995.)

of the isotopic peak clusters. Algorithms have been published to calculate these values (Zhang and Marshall, 1998), and adjustments for deuterium loss during the analysis can also be made (Zhang and Smith, 1993).

Conformations and Site-Specific Hydrogen Bonds

Chemical modifications of proteins cause changes that may alter the conformations, the folding and unfolding profiles, and the protein's stability toward denaturation under different pH levels and temperature conditions (Englander *et al.*, 1985; Eyles *et al.*, 1994). A C35A *E. coli* thioredoxin mutant shows little change in the proton NMR resonances relative to the wild-type protein, although significant changes are observed in the resonances of I75 and P76, which are within van der Waals' contact of C35 in the wild type. A more significant effect from the modification of the protein is the large change in pK_a of D26 due to shielding of the region from solvent by a covalently attached peptide at C32 in the mutant (Jeng *et al.*, 1998). Alkylation of *E. coli* thioredoxin at Cys-32 by *S*-(2-chloroethyl) glutathione or *S*-(2-chloroethyl)cysteine yields protein structures that show a reduction in the melting temperature of the protein and an increase in the deuterium incorporated into the overall structure relative to oxidized thioredoxin (Kim *et al.*, 2002). But the effects are not strictly related to the fact that a disulfide linkage is lost through alkylation, because reduced thioredoxin incorporates still more deuteriums and has even a lower melting temperature than the alkylated proteins.

The ethylglutathionyl thioredoxin structure when subjected to energy minimization using AMBER force-field showed that the Arg-73 side chain is rotated and forms a salt bridge with the carboxylate group of the glutathione γ-Glu residue (Fig. 17). A second salt bridge was predicted between the carboxy terminus of the glutathione Gly residue and the ε-amino side chain of the Lys-90 residue. In addition, in the model the glutathione moiety forms three hydrogen bonding interactions with amide nitrogens and carbonyl oxygens of the protein backbone. The amino termini of the γ-Glu, the γ-carbonyl oxygen of γ-Glu, and the carbonyl oxygen of the Cys residue of the glutathionyl moiety form hydrogen bonds to the following moieties of the protein backbone, the carbonyl oxygen of Arg-73, the amide nitrogen of Ile-75, and the amide nitrogen of Ala-93, respectively. From these interactions, it was predicted that the amido nitrogen of Ile-75 and Ala-93 would be protected from solvent by the glutathionyl moiety and, therefore, will show a reduction in the rates of H/D exchange. The H/D exchange experiments confirmed the predictions and showed, indeed, low levels of deuterium incorporation at the backbone nitrogens of theses residues (Kim *et al.* 2001b).

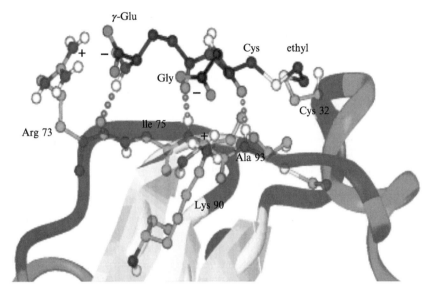

FIG. 17. Part of the structural model of the ethylglutathionyl thioredoxin conjugate highlighting the interactions of ethylglutathionyl moiety with the thioredoxin backbone. Hydrogen bonding interactions are depicted as purple dots. The positively charged side chains of Arg-73 and Lys-93 are shown to form salt bridges to the negatively charged carboxylate groups of the glutathionyl γ-Glu and Gly residues. Carbons of the glutathione chain are shown as black spheres, while those in the thioredoxin protein are shown in green. Oxygen atoms are in red, sulfurs in yellow, and nitrogens in blue. The protein backbone is shown as blue ribbon, α-helical regions are colored in red and β-strands in yellow. (See color insert.)

Deuterium content following H/D exchange on amides near the active site region of oxidized and reduced thioredoxin generally show excellent correlation with NMR hydrogen isotope exchange rate constants (Table II). In striking contrast is the deuterium content for Ile-75 and to some degree for Ala-93 for the $(S$-$(2$-$Cys^{32})$ethyl)glutathione)-thioredoxin (GS-TRX), both of which show low levels of deuterium. $(S$-$(2$-(Cys^{32})ethyl)-cysteinyl)-thioredoxin (Cys-TRX) does not show low incorporation of deuterium at Ile-75 and Ala-93, and there is little difference in labeling for this region of Cys-TRX protein compared to Oxi- and Red-TRX.

Conformations in the Gas Phase

Most H/D exchange studies for probing protein conformations have been conducted in solution. Increasing evidence indicates that proteins retain some of their conformational properties during ionization and analysis by MS (Hoaglund-Hyzer *et al.*, 1999). Accordingly, there has been greater emphasis in studying proteins in the gas phase because the absence of water is expected to lead to results that correlate better with theoretical predictions. Comparisons of H/D exchange data between these two phases should provide information on the role of water in protein structures, particularly where conformational changes take place such as in the folding pathways. In the absence of water, a multitude of transient conformers result from strong intramolecular interactions that precede the final native conformation in a funnel-like fashion (Chan and Dill, 1998). FT-ICR-MS is ideally suited for probing the gas-phase conformational preferences of proteins because the ions can be trapped and then exposed to D_2O for varying time periods before their excitation and detection (McLafferty *et al.*, 1998b; Suckau *et al.*, 1993; Wood *et al.*, 1995).

To a first approximation, the number of exchangeable hydrogens of a given protein conformation is independent of charge state (Hoaglund-Hyzer *et al.*, 1999; Suckau *et al.*, 1993), despite that collision cross sections (Chen *et al.*, 1997) and ion mobility studies show that the diffuseness of proteins is highly dependent on the charge state (Clemmer *et al.*, 1995; Shelimov *et al.*, 1997; Valentine and Clemmer, 1997). Thus, if the number of exchangeable protons changes significantly with the charge state, different conformations are suspected of being present and the coexistence of different conformational states in the gas phase can be detected by FT-ICR-MS (McLafferty *et al.*, 1998b; Suckau *et al.*, 1993; Wood *et al.*, 1995). Under these conditions, cytochrome *c* was found to exist in multiple conformational states that were stable for hours, whereas in solution, H/D exchange involves nearly all of the exchangeable hydrogens, indicating rapid equilibration between conformers. By MS/MS, it has been shown that terminal α-helices that are most stable in solution are not favored in the gas phase. Conversion between states can be achieved by infrared laser heating, which resembles thermal unfolding in solution, by high-velocity collisions, which completely denature proteins, or by charge stripping with a proton-affinity agent to mimic solution pH increases and promote fewer open states (Wood *et al.*, 1995).

Despite ion mobility and collision cross-section studies that show strong correlation with the charge state and reflect open conformations, it is surprising that the rates of H/D exchange are at best not affected (Suckau *et al.*, 1993; Valentine and Clemmer, 1997). It is expected that a systematic increase in the exposed amide sites would lead to an increase in the numbers of amide hydrogens exchanged when exposed to D_2O in the gas phase. In fact, an increase in the number of charges actually results in less H/D exchange. A number of explanations have been offered such as local chemical environments and energetically favorable hydrogen bonding that protect amide sites even in diffuse proteins (Hoaglund-Hyzer *et al.*, 1999; Valentine and Clemmer, 1997). Many peptide ions and even energetic metastable ions retain their secondary structures on the microsecond time scale, which serves to protect the hydrogens involved in hydrogen bonding for the duration of the mass spectrometric experiment (Kaltashov and Fenselau, 1997; Li *et al.*, 1998). It is still more difficult to rationalize the linear decrease in the numbers of exchangeable amide hydrogens with increasing charge (McLafferty *et al.*, 1998b). A possible explanation is that the extra charges appearing on the side chains become solvated and, therefore, do not affect exchange (F. W. McLafferty, personal communication, 2000).

Acknowledgments

This work was supported in part by the National Institute of Environmental Health Sciences (ES00040 and ES00210) and the National Science Foundation (BIR-921 4371).

References

Anderegg, R. J., Wagner, D. S., and Stevenson, C. L. (1994). The mass spectrometry of helical unfolding in peptides. *J. Am. Soc. Mass Spectrom.* **5**, 425–433.

Arrington, C. B., and Robertson, A. D. (2000). Microsecond to minute dynamics revealed by EX1-type hydrogen exchange at nearly every backbone hydrogen bond in a native protein. *J. Mol. Biol.* **296**, 1307–1317.

Arrington, C. B., Teesch, L. M., and Robertson, A. D. (1999). Defining protein ensembles with native-state NH exchange: Kinetics of interconversion and cooperative units from combined NMR and MS analysis. *J. Mol. Biol.* **285**, 1265–1275.

Backmann, J., Schultz, C., Fabian, H., Hahn, U., Saenger, W., and Naumann, D. (1996). Thermally induced hydrogen exchange processes in small proteins as seen by FTIR spectroscopy. *Proteins* **24**, 379–387.

Bai, Y., Milne, J. S., Mayne, L., and Englander, S. W. (1993). Primary structure effects on peptide group hydrogen exchange. *Proteins* **17**, 75–86.

Busenlehner, L. S., and Armstrong, R. N. (2005). Insights into enzyme structure and dynamics elucidated by amide H/D exchange mass spectrometry. *Arch. Biochem. Biophys.* **433**, 34–46.

Chamberlain, A. K., and Marqusee, S. (1997). Touring the landscapes: Partially folded proteins examined by hydrogen exchange. *Structure* **5**, 859–863.

Chan, H. S., and Dill, K. A. (1998). Protein folding in the landscape perspective: Chevron plots and non-Arrhenius kinetics. *Proteins* **30**, 2–33.

Chapman, J. R. (1996). Protein and peptide analysis by mass spectrometry. *In* "Methods in Molecular Biology," vol. 61, pp. 320–321. Humana Press, Totowa, NJ.

Chen, Y.-L., Collings, B. A., and Douglas, D. J. (1997). Collision cross sections of myoglobin and cytochrome *c* ions with Ne, Ar, and Kr. *J. Am. Soc. Mass Spectrom.* **8**, 681–687.

Cheng, X., and Fenselau, C. (1992). Hydrogen/deuterium exchange of mass-selected peptide ions with ND_3 in a tandem sector mass spectrometer. *Int. J. Mass Spectrom. Ion Processes.* **122**, 109–119.

Chik, J. K., and Schriemer, D. C. (2003). Hydrogen/deuterium exchange mass spectrometry of actin in various biochemical contexts. *J. Mol. Biol.* **334,** 373–385.

Chowdhury, S. K., Katta, V., and Chait, B. T. (1990). Probing conformational changes in proteins by mass spectrometry. *J. Am. Chem. Soc.* **112,** 9012–9013.

Chung, E. V., Nettleton, E. J., Morgan, C. J., Gross, M., Miranker, A., Radford, S. E., Dobson, C. M., and Robinson, C. V. (1997). Hydrogen exchange properties of proteins in native and denatured states monitored by mass spectrometry and NMR. *Protein Sci.* **6,** 1316–1324.

Clarke, J., and Itzhaki, L. S. (1998). Hydrogen exchange and protein folding. *Curr. Opin. Struct. Biol.* **8,** 112–118.

Clemmer, D.E, Hudgins, R. R., and Jarrold, M. F. (1995). Naked protein conformations: Cytochrome c in the gas phase. *J. Am. Chem. Soc.* **117,** 10141–10142.

Creighton, T. E. (1990). Protein folding. *Biochem. J.* **270,** 1–16.

Creighton, T. E. (1993). "Proteins: Structures and Molecular Properties." Freeman, New York.

De Jongh, H. H., Goormaghtigh, E., and Ruysschaert, J. M. (1995). Tertiary stability of native and methionine-80 modified cytochrome c detected by proton-deuterium exchange using on-line Fourier transform infrared spectroscopy. *Biochemistry* **34,** 172–179.

Demmers, J. A., Rijkers, D. T., Haverkamp, J., Killian, J. A., and Heck, A. J. (2002). Factors affecting gas-phase deuterium scrambling in peptide ions and their implications for protein structure determination. *J. Am. Chem. Soc.* **124,** 11191–11198.

Deng, Y., and Smith, D. L. (1998a). Hydrogen exchange demonstrates three domains in aldolase unfold sequentially. *J. Mol. Biol.* **294,** 247–258.

Deng, Y., and Smith, D. L. (1998b). Identification of unfolding domains in large proteins by their unfolding rates. *Biochemistry* **37,** 6256–6262.

Deng, Y., Pan, H., and Smith, D. L. (1999a). Selective isotope labeling demonstrates that hydrogen exchange at individual peptide amide linkages can be determined by collision-induced dissociation mass spectrometry. *J. Am. Chem. Soc.* **121,** 1966–1967.

Deng, Y., Zhang, Z., and Smith, D. L. (1999b). Comparison of continuous and pulsed labeling amide hydrogen exchange/mass spectrometry for studies of protein dynamics. *J. Am. Soc. Mass. Spectrom.* **10,** 675–684.

Dill, K. A., and Chan, H. S. (1997). From Leventhal to pathways to funnels. *Nat. Struct. Biol.* **4,** 10–19.

Dole, M., Mack, L. L., Hines, R. L., Mobley, R. C., Fergusson, L. P., and Alice, M. B. (1968). Molecular beams of macroions. *J. Chem. Phys.* **49,** 2240–2249.

Engen, J. R., Smithgall, T. E., Gmeiner, W. H., and Smith, D. L. (1997). Identification and localization of slow, natural, cooperative unfolding in the hematopoietic cell kinase SH3 domain by amide hydrogen exchange and mass spectrometry. *Biochemistry* **36,** 14384–14391.

Englander, J. J., Calhoun, D. B., and Englander, S. W. (1979). Measurement and calibration of peptide group hydrogen-deuterium exchange by ultraviolet spectrophotometry. *Anal. Biochem.* **92,** 517–524.

Englander, J. J., Rogero, J. R., and Englander, S. W. (1985). Protein hydrogen exchange studied by the fragment separation method. *Anal. Biochem.* **147,** 234–244.

Englander, S. W., and Kallenbach, N. (1984). Hydrogen exchange and structural dynamics of proteins and nucleic acids. *Quart. Rev. Biophys.* **16,** 521–655.

Englander, S. W., Sosnick, T. R., Englander, J. J., and Mayne, L. (1996). Mechanisms and uses of hydrogen exchange. *Curr. Opin. Struct. Biol.* **6,** 18–23.

Eriksson, M. A. L., Haerd, T., and Nilsson, L. (1995). On the pH dependance of amide proton exchange rates in proteins. *Biophys. J.* **69,** 329–339.

Erve, J. C., Barofsky, E., Barofsky, D. F., Deinzer, M. L., and Reed, D. J. (1995). Alkylation of *Escherichia coli* Thioredoxin by *S*-(2-chloroethyl)glutathione and identification of the adduct on the active site cysteine-32 by mass spectrometry. *Chem. Res. Toxicol.* **8,** 934–941.

Eyles, S. J., and Kaltashov, I. A. (2004). Methods to study protein dynamics and folding by mass spectrometry. *Methods* **34,** 88–99.

Eyles, S. J., Radford, S. E., Robinson, C. V., and Dobson, C. M. (1994). Kinetic consequences of the removal of a disulfide bridge on the folding of hen lysozyme. *Biochemistry* **33,** 13038–13048.

Eyles, S. J., Speir, J. P., Kruppa, G. H., Gierasch, L. M., and Kaltashov, I. A. (2000). Protein conformational stability probed by Fourier transform cyclotron resonance mass spectrometry. *J. Am. Chem. Soc.* **122,** 495–500.

Fenn, J. B., Mann, M., Meng, C. K., Wong, S. F., and Whitehouse, C. M. (1989). Electrospray ionization for mass spectrometry of large biomolecules. *Science* **246,** 64–71.

Guy, P., Jaquinod, M., Remigy, H., Andrieu, J. P., Gagnon, J., Bersch, B., Dolla, A., Blanchard, L., Guerlesquin, F., and Forest, E. (1996a). New conformational properties induced by the replacement of Tyr-64 in *Desulfovibrio vulgaris Hildenborough* ferricytochrome *c*553 using isotopic exchanges monitored by mass spectrometry. *FEBS Lett.* **395,** 53–57.

Guy, P., Remigy, H., Jaquinod, M., Bersch, B., Blanchard, L., Dolla, A., and Forest, E. (1996b). Study of the new stability properties induced by amino acid replacement of tyrosine 64 in cytochrome c553 from *Desulfovibrio vulgaris Hildenborough* using electrospray ionization mass spectrometry. *Biochem. Biophys. Res. Commun.* **218,** 97–103.

Hildebrandt, P., Vanhecke, F., Heibel, G., and Mauk, A. G. (1993). Structural changes in cytochrome *c* upon hydrogen-deuterium exchange. *Biochemistry* **32,** 14158–14164.

Hoaglund-Hyzer, C. S., Counterman, A. E., and Clemmer, D. E. (1999). Anhydrous protein ions. *Chem. Rev.* **99,** 3037–3079.

Hoerner, J. K., Xiao, H., Dobo, A., and Kaltashov, I. A. (2004). Is there hydrogen scrambling in the gas phase? Energetic and structural determinants of proton mobility within protein ions. *J. Am. Chem. Soc.* **126,** 7709–7717.

Horn, D. M., Breuker, K., Frank, A. J., and McLafferty, F. W. (2001). Kinetic intermediates in the folding of gaseous protein ions characterized by electron capture dissociation mass spectrometry. *J. Am. Chem. Soc.* **123,** 9792–9799.

Horn, D. M., Ge, Y., and McLafferty, F. W. (2000). Activated ion electron capture dissociation for mass spectral sequencing of larger (42 kDa) proteins. *Anal. Chem.* **72,** 4778–4784.

Hvidt, A., and Linderstrøm-Lang, K. (1954). Exchange of hydrogen atoms in insulin with deuterium atoms in aqueous solutions. *Biochim. Biophys. Acta* **14,** 574–575.

Hvidt, A., and Nielsen, S. O. (1966). Hydrogen exchange in proteins. *Adv. Protein Chem.* **21,** 287–386.

Iribarne, V., and Thompson, B. A. (1976). On the evaporation of small ions from charged droplets. *J. Chem. Phys.* **64,** 2287–2294.

Jeng, M. F., and Dyson, H. J. (1995). Comparison of the hydrogen-exchange behavior of reduced and oxidized *Escherichia coli* thioredoxin. *Biochemistry* **34,** 611–619.

Jeng, M. F., Englander, S. W., Eloeve, G. A., Wand, A. J., and Roder, H. (1990). Structural description of acid-denatured cytochrome *c* by hydrogen exchange and 2D NMR. *Biochemistry* **29,** 10433–10437.

Jeng, M. F., Reymond, M. T., Tennant, L. L., Holmgren, A., and Dyson, H. J. (1998). NMR characterization of a single-cysteine mutant of *Escherichia coli* thioredoxin and a covalent thioredoxin-peptide complex. *Eur. J. Biochem.* **257,** 299–308.

Johnson, R. S. (1996). Mass spectrometric measurement of changes in protein hydrogen exchange rates that result from point mutations. *J. Am. Soc. Mass Spectrom.* **7**, 515–521.

Johnson, R. S., Krylov, D., and Walsh, K. A. (1995). Proton mobility within electrosprayed peptide ions. *J. Mass Spectrom.* **30**, 386–387.

Johnson, R. S., and Walsh, K. A. (1994). Mass spectrometric measurement of protein amide hydrogen exchange rates of apo- and holo-myoglobin. *Protein Sci.* **3**, 2411–2418.

Kaltashov, I. A., Doroshenko, V. M., and Cotter, R. J. (1997). Gas phase hydrogen/deuterium exchange reactions of peptide ions in a quadrupole ion trap mass spectrometer. *Proteins* **28**, 53–58.

Kaltashov, I. A., and Fenselau, C. (1997). Stability of secondary structural elements in a solvent-free environment: The alpha helix. *Proteins* **27**, 165–170.

Karas, M., and Hillenkamp, F. (1988). Laser desorption ionization of proteins with molecular masses exceeding 10,000 daltons. *Anal. Chem.* **60**, 2299–2301.

Katta, V., and Chait, B. T. (1991). Conformational changes in proteins probed by hydrogen-exchange electrospray ionization mass spectrometry. *Rapid Commun. Mass Spectrom.* **5**, 214–217.

Katta, V., and Chait, B. T. (1993). Hydrogen/deuterium exchange electrospray ionization mass spectrometry: A method for probing protein conformational changes in solution. *J. Am. Chem. Soc.* **115**, 6317–6321.

Kenny, P. T., Nomoto, K., and Orlando, R. (1992). Fragmentation studies of peptides: The formation of y ions. *Rapid Commun. Mass Spectrom.* **6**, 95.

Kim, M.-Y., Maier, C. S., and Deinzer, M. L. (1999). Noncovalent intramolecular interaction in glutathionylated *E. coli* thioredoxin monitored by electrospray ionization mass spectrometry. "Proceedings of the 47th Conference on Mass Spectrometry and Allied Topics."

Kim, M. Y., Maier, C. S., Reed, D. J., and Deinzer, M. L. (2001a). Site-specific amide hydrogen/deuterium exchange in *E. coli* thioredoxins measured by electrospray ionization mass spectrometry. *J. Am. Chem. Soc.* **123**, 9860–9866.

Kim, M. Y., Maier, C. S., Ho, S., Deinzer, and M.L. (2001b). Intramolecular interactions in chemically modified *Escherichia coli* thioredoxin monitored by hydrogen/deuterium exchange and electrospray ionization mass spectrometry. *Biochemistry* **40**, 14413–14421.

Kim, M. Y., Maier, C. S., Reed, D. J., and Deinzer, M. L. (2002). Conformational changes in chemically modified *Escherichia coli* thioredoxin monitored by H/D exchange and electrospray ionization mass spectrometry. *Protein Sci.* **11**, 1320–1329.

Konermann, L., Collings, B. A., and Douglas, D. J. (1997a). Cytochrome *c* folding kinetics studied by time-resolved electrospray ionization mass spectrometry. *Biochemistry* **36**, 5554–5559.

Konermann, L., and Douglas, D. J. (1997). Acid-induced unfolding of cytochrome c at different methanol concentrations: Electrospray ionization mass spectrometry specifically monitors changes in the tertiary structure. *Biochemistry* **36**, 12296–12302.

Konermann, L., and Douglas, D. J. (1998a). Equilibrium unfolding of proteins monitored by electrospray ionization mass spectrometry: Distinguishing two-state from multi-state transitions. *Rapid Commun. Mass Spectrom.* **12**, 435–442.

Konermann, L., and Douglas, D. J. (1998b). Unfolding of proteins monitored by electrospray ionization mass spectrometry: A comparison of positive and negative ion modes. *J. Am. Soc. Mass Spectrom.* **9**, 1248–1254.

Konermann, L., Rosell, F. I., Mauk, A. G., and Douglas, D. J. (1997b). Acid-induced denaturation of myoglobin studied by time-resolved electrospray ionization mass spectrometry. *Biochemistry* **36**, 6448–6454.

Last, A. M., and Robinson, C. V. (1999). Protein folding and interactions revealed by mass spectrometry. *Curr. Opin. Chem. Biol.* **3**, 564–570.

Li, A., Fenselau, C., and Kaltashov, I. A. (1998). Stability of secondary structural elements in a solvent-free environment. II: The beta-pleated sheets. *Proteins* Suppl., 22–27.

Li, R., and Woodward, C. (1999). The hydrogen exchange core and protein folding. *Protein Sci.* **8**, 1571–1590.

Light-Wahl, K. J., Loo, J. A., Edmonds, C. G., Smith, R. D., Witkowska, H. E., Shackleton, C. H. L., and Wu, C.-S. (1993). Collisionally activated dissociation and tandem mass spectrometry of intact hemoglobin b-chain variant proteins with electrospray ionization. *Biol. Mass Spectrom.* **22**, 112–120.

Linderstrøm-Lang, K. U. (1958). Deuterium exchange and protein structure. *In* "Symposium on Protein Structure" (A. Neuberger, ed.), pp. 23–34. Methuen, London.

Loo, J. A., Edmonds, C. G., Udseth, H. R., and Smith, R. D. (1990). Effect of reducing disulfide-containing proteins on electrospray ionization mass spectra. *Anal. Chem.* **62**, 693–698.

Loo, J. A., Edmonds, C. G., and Smith, R. D. (1993). Tandem mass spectrometry of very large molecule 2. Dissociation of multiply charged proline-containing proteins from electrospray ionization. *Anal. Chem.* **65**, 425–438.

Loo, J. A., Ogorzalek, R. R., Udseth, H. R., Edmonds, C. G., and Smith, R. D. (1991). Solvent-induced conformational changes of polypeptides probed by electrospray-ionization mass spectrometry. *Rapid Commun. Mass Spectrom.* **5**, 101–105.

Maier, C. S., Kim, O.-H., and Deinzer, M. L. (1997). Conformational properties of the A-state of cytochrome c studied by hydrogen/deuterium exchange and electrospray mass spectrometry. *Anal. Biochem.* **252**, 127–135.

Maier, C. S., Kim, O.-H., and Deinzer, M. L. (1995). Probing high order structure of proteins by H/D exchange/ESI-MS. "Proceedings of the 43th Conference on Mass Spectrometry and Allied Topics" p. 301.

Maier, C. S., Schimerlik, M. I., and Deinzer, M. L. (1999). Thermal denaturation of *Escherichia coli* thioredoxin studied by hydrogen/deuterium exchange and electrospray ionization mass spectrometry: Monitoring a two-state protein unfolding transition. *Biochemistry* **38**, 1136–1143.

Mandell, J. G., Baerga-Ortiz, A., Akashi, S., Takio, K., and Komives, E. A. (2001). Solvent accessibility of the thrombin-thrombomodulin interface. *J. Mol. Biol.* **306**, 575–589.

Mandell, J. G., Falick, A. M., and Komives, E. A. (1998). Identification of protein–protein interfaces by decreased amide proton solvent accessibility. *Proc. Natl. Acad. Sci. USA* **95**, 14705–14710.

Marshall, A. G., Senko, M. W., Li, W., Li, M., Dillon, S., Guan, S., and Logan, T. M. (1997). Protein molecular weight to 1Da by ^{13}C, ^{15}N double-depletion and FT-ICR mass spectrometry. *J. Am. Chem. Soc.* **119**, 433–434.

McLafferty, F. W., Kelleher, N. L., Begley, T. P., Fridriksson, E. K., Zubarev, R. A., and Horn, D. M. (1998a). Two-dimensional mass spectrometry of biomolecules at the subfemtomole level. *Curr. Opin. Chem. Biol.* **2**, 571–578.

McLafferty, F. W. G., Haupts, U., Wood, T. D., and Kelleher, N. L. (1998b). Gaseous conformational structures of cytochrome c. *J. Am. Chem. Soc.* **120**, 4732–4740.

Miranker, A., Robinson, C. V., Radford, S. E., and Dobson, C. M. (1996). Investigation of protein folding by mass spectrometry. *FASEB J.* **10**, 93–101.

Miranker, A., Robinson, C. V., Radford, S. E., Aplin, R. T., and Dobson, C. M. (1993). Detection of transient protein folding populations by mass spectrometry. *Science* **262**, 896–899.

Mirza, U. A., Cohen, S. L., and Chait, B. T. (1993). Heat-induced conformational changes in proteins studied by electrospray ionization mass spectrometry. *Anal. Chem.* **65**, 1–6.

Mohimen, A., Dobo, A., Hoerner, J. K., and Kaltashov, I. A. (2003). A chemometric approach to detection and characterization of multiple protein conformers in solution using electrospray ionization mass spectrometry. *Anal. Chem.* **75,** 4139–4147.

Molday, R. S., Englander, S. W., and Kallen, R. G. (1972). Primary structure effects on peptide group hydrogen exchange. *Biochemistry* **11,** 150–158.

Nemirovskiy, O., Giblin, D. E., and Gross, M. L. (1999). Electrospray ionization mass spectrometry and hydrogen/deuterium exchange for probing the interaction of calmodulin with calcium. *J. Am. Soc. Mass Spectrom.* **10,** 711–718.

Nettleton, E. J., Sunde, M., Lai, Z., Kelly, J. W., Dobson, C. M., and Robinson, C. V. (1998). Protein subunit interactions and structural integrity of amyloidogenic transthyretins: Evidence from electrospray mass spectrometry. *J. Mol. Biol.* **281,** 553–564.

Przybylski, M., and Glocker, M. O. (1996). Electrospray mass spectrometry of biomolecular complexes with noncovalent interactions—new analytical perspectives for supramolecular chemistry and molecular recognition processes. *Angew. Chem. Int. Ed. Engl.* **35,** 806–826.

Remigy, H., Jaquinod, M., Petillot, Y., Gagnon, J., Cheng, H., Xia, B., Markley, J. L., Hurley, J. K., Tollin, G., and Forest, E. (1997). Probing the influence of mutations on the stability of a ferredoxin by mass spectrometry. *J. Protein Chem.* **16,** 527–532.

Resing, K. A., and Ahn, N. G. (1998). Deuterium exchange mass spectrometry as a probe of protein kinase activation. Analysis of wild-type and constitutively active mutants of MAP kinase kinase-1. *Biochemistry* **37,** 463–475.

Resing, K. A., Hoofnagle, A. N., and Ahn, N. G. (1999). Modeling deuterium exchange behavior of ERK2 using pepsin mapping to probe secondary structure. *J. Am. Soc. Mass Spectrom.* **10,** 685–702.

Rist, W., Jorgensen, T. J., Roepstorff, P., Bukau, B., and Mayer, M. P. (2003). Mapping temperature-induced conformational changes in the *Escherichia coli* heat shock transcription factor sigma 32 by amide hydrogen exchange. *J. Biol. Chem.* **278,** 51415–51421.

Robinson, C. V., Chung, E. W., Kragelund, B. B., Knudsen, J., Aplin, R. T., Poulsen, F. M., and Dobson, C. M. (1996). Probing the nature of noncovalent interactions by mass spectrometry. A study of protein-CoA ligand binding and assembly. *J. Am. Chem. Soc.* **118,** 8646–8653.

Robinson, C. V., Groß, M., Eyles, S. J., Ewbank, J. J., Mayhew, M., Hartl, F. U., Dobson, C. M., and Radford, S. E. (1994). Conformation of GroEL-bound a-lactalbumin. *Nature* **372,** 646–372.

Roder, H. (1989). Structural characterization of protein folding intermediates by proton magnetic resonance and hydrogen exchange. *Methods Enzymol.* **176,** 446–473.

Rosenberg, A., and Chakravarti, K. (1968). Studies of hydrogen exchange in proteins: I. The exchange kinetics of bovine carbonic anhydrase. *J. Biol. Chem.* **243,** 5193–5201.

Rosenberg, A., and Enberg, J. (1969). Studies of hydrogen exchange in proteins: II. The reversible thermal unfolding of chymotrypsinogen A as studied by exchange kinetics. *J. Biol. Chem.* **244,** 6153–6159.

Rosenberg, A., and Woodward, C. K. (1970). Studies of hydrogen exchange in proteins: III. The effects of the chymotrypsinogen-a-chymotrypsin conversion on hydrogen exchange kinetics. *J. Biol. Chem.* **245,** 4677–4683.

Rostom, A. A., Fucini, P., Benjamin, D. R., Juenemann, R., Nierhaus, K. H., Hartl, F. U., Dobson, C. M., and Robinson, C. V. (2000). Detection and selective dissociation of intact ribosomes in a mass spectrometer. *Proc. Natl. Acad. Sci. USA* **97,** 5185–5190.

Rostom, A. A., and Robinson, C. V. (1999). Detection of the intact GroEl chaperonin assembly by mass spectrometry. *J. Am. Chem. Soc.* **121,** 4718–4719.

Santoro, M. M., and Bolen, D. W. (1988). Unfolding free energy changes by the linear extrapolation method. 1. Unfolding of phenylmethanesulfonyl a-chymotrypsin using different denaturants. *Biochemistry* **27,** 8063–8068.

Scholtz, J. M., and Robertson, A. D. (1995). Hydrogen exchange techniques. *In* "Methods in Molecular Biology" (B. A. Shirely, ed.), Vol. 40, "Protein Stability and Folding: Theory and Practice," pp. 291–311. Humana Press, Totowa, NJ.

Senko, M. W., Speir, J. P., and McLafferty, F. W. (1994). Collisional activation of large multiply charged ions using Fourier transform mass spectrometry. *Anal. Chem.* **66,** 2801–2808.

Shelimov, K. B., Clemmer, D. E., Hudgins, R. R., and Jarrold, M. F. (1997). Protein structure in vacuo: Gas-phase conformations of BPTI and cytochrome *c*. *J. Am. Chem. Soc.* **119,** 2240–2248.

Smith, D. L. (1998). Local structure and dynamics in proteins characterized by hydrogen exchange and mass spectrometry. *Biochemistry (Moscow)* **63,** 285–293.

Smith, D. L., Deng, Y., and Zhang, Z. (1997a). Probing the non-covalent structure of proteins by amide hydrogen exchange and mass spectrometry. *J. Mass Spectrom.* **32,** 135–146.

Smith, D. L., and Zhang, Z. (1994). Probing noncovalent structural features of proteins by mass spectrometry. *Mass Spectrom. Rev.* **13,** 411–429.

Smith, R. D., Bruce, J. E., Wu, Q., and Lei, Q. P. (1997b). New mass spectrometric methods for the study of noncovalent associations of biopolymers. *Chem. Soc. Rev.* **26,** 191–202.

Suckau, D., Shi, Y., Beu, S. C., Senko, M. W., Quinn, J. P., Wampler, F. M. D., and McLafferty, F. W. (1993). Coexisting stable conformations of gaseous protein ions. *Proc. Natl. Acad. Sci. USA* **90,** 790–793.

Tito, M. A., Tars, K., Valegrad, K., Hajdu, J., and Robinson, C. V. (2000). Electrospray time-of-flight mass spectrometry of intact MS2 virus capsid. *J. Am. Chem. Soc.* **112,** 3550–3551.

Valentine, S. J., and Clemmer, D. E. (1997). H/D exchange levels of shape-resolved *cytochrome c* conformers in the gas phase. *J. Am. Chem. Soc.* **119,** 3558–3566.

van Berkel, W. J., van den Heuvel, R. H., Versluis, C., and Heck, A. J. R. (2000). Detection of intact megaDalton protein assemblies of vanillyl-alcohol oxidase by mass spectrometry. *Protein Sci.* **9,** 435–439.

van den Bremer, E. T., Jiskoot, W., James, R., Moore, G. R., Kleanthous, C., Heck, A. J., and Maier, C. S. (2002). Probing metal ion binding and conformational properties of the colicin E9 endonuclease by electrospray ionization time-of-flight mass spectrometry. *Protein Sci.* **11,** 1738–1752.

Veenstra, T. D., Johnson, K. L., Tomlinson, A. J., Kumar, R., and Naylor, S. (1998). Correlation of fluorescence and circular dichroism spectroscopy with electrospray ionization mass spectrometry in the determination of tertiary conformational changes in calcium-binding proteins. *Rapid Commun. Mass Spectrom.* **12,** 613–619.

Vis, H., Heinemann, U., Dobson, C. M., and Robinson, C. V. (1998). Detection of a monomeric intermediate associated with dimerization of protein Hu by mass spectrometry. *J. Am. Chem. Soc.* **120,** 6427–6428.

Wand, A. J., Roder, H., and Englander, S. W. (1986). Two-dimensional ^{1}H NMR studies of cytochrome *c*: Hydrogen exchange in the N-terminal helix. *Biochemistry* **25,** 1107–1114.

Wang, F., Blanchard, J. S., and Tang, X. (1997). Hydrogen exchange/electrospray ionization mass spectrometry studies of substrate and inhibitor binding and conformational changes of *Escherichia coli* dihydrodipicolinate reductase. *Biochemistry* **36,** 3755–3759.

Wang, F., Li, W., Emmett, M. R., Hendrickson, C. L., Marshall, A. G., Zhang, Y. L., Wu, L., and Zhang, Z. Y. (1998a). Conformational and dynamic changes of Yersinia protein tyrosine phosphatase induced by ligand binding and active site mutation and revealed by H/D exchange and electrospray ionization Fourier transform ion cyclotron resonance mass spectrometry. *Biochemistry* **37,** 15289–15299.

Wang, F., Li, W., Emmett, M. R., Marshall, A. G., Corson, D., and Sykes, B. D. (1999). Fourier transform ion cyclotron resonance mass spectrometric detection of small Ca(2+)–induced conformational changes in the regulatory domain of human cardiac troponin C. *J. Am. Soc. Mass Spectrom.* **10,** 703–710.

Wang, F., Scapin, G., Blanchard, J. S., and Angeletti, R. H. (1998b). Substrate binding and conformational changes of *Clostridium glutamicum* diaminopimelate dehydrogenase revealed by hydrogen/deuterium exchange and electrospray mass spectrometry. *Protein Sci.* **7,** 293–289.

Wang, L., Pan, H., and Smith, D. L. (2002). Hydrogen exchange-mass spectrometry: Optimization of digestion conditions. *Mol. Cell Proteom.* **1,** 132–138.

Wang, L., and Smith, D. L. (2003). Downsizing improves sensitivity 100-fold for hydrogen exchange-mass spectrometry. *Anal. Biochem.* **314,** 46–53.

Waring, A. J., Mobley, P. W., and Gordon, L. M. (1998). Conformational mapping of a viral fusion peptide in structure-promoting solvents using circular dichroism and electrospray mass spectrometry. *Proteins* Suppl., 38–49.

Wilson, D. J., Rafferty, S. P., and Konermann, L. (2005). Kinetic unfolding mechanism of the inducible nitric oxide synthase oxygenase domain determined by time-resolved electrospray mass spectrometry. *Biochemistry* **44,** 2276–2283.

Wood, T. D., Chorush, R. A., Wampler, F. M., 3rd, Little, D. P., O'Connor, P. B., and McLafferty, F. W. (1995). Gas-phase folding and unfolding of cytochrome *c* cations. *Proc. Natl. Acad. Sci. USA* **92,** 2451–2454.

Woodward, C. (1999). Advances in protein hydrogen exchange by mass spectrometry. *J. Am. Soc. Mass Spectrom.* **10,** 672–674.

Woodward, C., Simon, I., and Tuechsen, E. (1982). Hydrogen exchange and the dynamic structure of proteins. *Mol. Cell Biochem.* **48,** 135–160.

Woodward, C. K. (1994). Hydrogen exchange rates and protein folding. *Curr. Opin. Struct. Biol.* **4,** 112–116.

Wu, Q., Van Orden, S., Cheng, X., Bakhtiar, R., and Smith, R. D. (1995). Characterization of cytochrome c variants with high-resolution FTICR mass spectrometry: Correlation of fragmentation and structure. *Anal. Chem.* **67,** 2498–2509.

Yan, X., Broderick, D., Leid, M. E., Schimerlik, M. I., and Deinzer, M. L. (2004a). Dynamics and ligand-induced solvent accessibility changes in human retinoid X receptor homodimer determined by hydrogen deuterium exchange and mass spectrometry. *Biochemistry* **43,** 909–917.

Yan, X., Watson, J., Ho, P. S., and Deinzer, M. L. (2004b). Mass spectrometric approaches using electrospray ionization charge states and hydrogen-deuterium exchange for determining protein structures and their conformational changes. *Mol. Cell Proteom.* **3,** 10–23.

Yan, X., Zhang, H., Watson, J., Schimerlik, M. I., and Deinzer, M. L. (2002). Hydrogen/deuterium exchange and mass spectrometric analysis of a protein containing multiple disulfide bonds: Solution structure of recombinant macrophage colony stimulating factor-beta (rhM-CSFbeta). *Protein Sci.* **11,** 2113–2124.

Yi, Q., and Baker, D. (1996). Direct evidence for a two-state protein unfolding transition from hydrogen-deuterium exchange, mass spectrometry, and NMR. *Protein Sci.* **5,** 1060–1066.

Zhang, Y. H., Yan, X., Maier, C. S., Schimerlik, M. I., and Deinzer, M. L. (2001). Structural comparison of recombinant human macrophage colony stimulating factor beta and a partially reduced derivative using hydrogen deuterium exchange and electrospray ionization mass spectrometry. *Protein Sci.* **10**, 2336–2345.

Zhang, Y.-P., Lewis, R. N. A. H., Henry, G. D., Sykes, B. D., Hodges, R. S., and McElhaney, R. N. (1995). Peptide models of helical hydrophobic transmembrane segments of membrane proteins: 1. Studies of the conformation, intrabilayer orientation, and amide hydrogen exchangeability of Ac-K_2-$(LA)_{12}$-K_2-amide. *Biochemistry* **34**, 2348–2361.

Zhang, Z., Li, W., Li, M., Logan, T. M., Guan, S., and Marshall, A. G. (1997a). Higher-order structure and dynamics of FK506-binding protein probed by backbone amide hydrogen/deuterium exchange and electrospray fourier transform ion cyclotron resonance mass spectrometry. *In* "Techniques in Protein Chemistry VIII" (D. R. Marshak, ed.), pp. 703–713. Academic Press, San Diego, CA.

Zhang, Z., Li, W., Logan, T. M., Li, M., and Marshall, A. G. (1997b). Human recombinant [C22A] FK506-binding protein amide hydrogen exchange rates from mass spectrometry match amd extend those from NMR. *Protein Sci.* **6**, 2203–2217.

Zhang, Z., and Marshall, A. G. (1998). A universal algorithm for fast and automated charge state deconvolution of electrospray mass-to-charge ratio spectra. *J. Am. Soc. Mass Spectrom.* **9**, 225–233.

Zhang, Z., Post, C. B., and Smith, D. L. (1996). Amide hydrogen exchange determined by mass spectrometry: Application to rabbit muscle aldolase. *Biochemistry* **35**, 779–791.

Zhang, Z., and Smith, D. L. (1993). Determination of amide hydrogen exchange by mass spectrometry: A new tool for protein structure elucidation. *Protein Sci.* **2**, 522–531.

Zhang, Z., and Smith, D. L. (1996). Thermal-induced unfolding domains in aldolase identified by hydrogen exchange and mass spectrometry. *Protein Sci.* **5**, 1282–1289.

Zhu, M. M., Rempel, D. L., Du, Z., and Gross, M. L. (2003a). Quantification of protein-ligand interactions by mass spectrometry, titration, and H/D exchange: PLIMSTEX. *J. Am. Chem. Soc.* **125**, 5252–5253.

Zhu, M. M., Rempel, D. L., and Gross, M. L. (2004). Modeling data from titration, amide H/D exchange, and mass spectrometry to obtain protein-ligand binding constants. *J. Am. Soc. Mass Spectrom.* **15**, 388–397.

Zhu, M. M., Rempel, D. L., Zhao, J., Giblin, D. E., and Gross, M. L. (2003b). Probing Ca2+–induced conformational changes in porcine calmodulin by H/D exchange and ESI-MS: Effect of cations and ionic strength. *Biochemistry* **42**, 15388–15397.

Zubarev, R. A., Kruger, N. A., Fridriksson, E. K., Lewis, M. A., Horn, D. M., Carpenter, B. K., and McLafferty, F. W. (1999). Electron capture dissociation of gaseous multiply-charged proteins is favored at disulfide bonds and other sites of high hydrogen atom affinity. *J. Am. Chem. Soc.* **121**, 2857–2862.

[11]   Ligand–Metal Ion Binding to Proteins: Investigation by ESI Mass Spectrometry

By NOELLE POTIER, HÉLÈNE ROGNIAUX, GUILLAUME CHEVREUX, and ALAIN VAN DORSSELAER

Abstract

The objective of this chapter is to show the general mass spectrometry (MS)–based strategies that can be used to retrieve information regarding protein–metal and protein–ligand noncovalent complexes. Indeed, when using carefully controlled conditions in the atmospheric pressure–vacuum interface of the mass spectrometer, and when sample preparation is optimized, it is possible to preserve large specific multiprotein–metal–ligand noncovalent complexes during MS analysis. Examples describing the possibilities of electrospray ionization MS (ESI-MS) are shown. For instance, it can be used to probe cooperativity in the binding of a ligand or a metal to a protein or may constitute a new methodology for a more rational approach for drug discovery and for human genome annotation. Thanks to its ability to directly give information on stoichiometry or dynamics of the interactions formed in solution, MS offers new possibilities to tackle more and more various applications.

Introduction

Noncovalent binding of small molecules to proteins is a recognition phenomenon deeply involved in most cellular processes. Interactions with cofactors, substrates, and metal ions, as well as with other proteins, are important both in the structural integrity and in the activity of many proteins or enzymes. In the biomedical field, numerous therapeutic strategies exploit the noncovalent recognition of drug candidates by the target enzymes.

Existence of these complexes has been commonly investigated using established techniques such as gel filtration chromatography, ultracentrifuge, calorimetry, scattering methods (X-ray diffraction, circular dichroism), fluorescence-quenching, ultraviolet (UV) spectroscopy, or nuclear magnetic resonance spectroscopy (NMRS). Each of these techniques presents both strengths and weaknesses. Encountered difficulties include large sample requirement, long time experiments, lack of specificity of the technique for a given complex, poor mass resolution, and absence of any spectroscopic property of the ligand or the protein–ligand complex.

METHODS IN ENZYMOLOGY, VOL. 402
0076-6879/05 $35.00
DOI: 10.1016/S0076-6879(05)02011-2

Some of these difficulties could be alleviated with the introduction of ESI MS (Fenn *et al.*, 1989; Meng *et al.*, 1988). In fact, it appeared quite rapidly that this ionization method could allow the preservation of some noncovalent interactions during desorption in the gas phase: Intact non-covalent complexes preexisting in solution could then be characterized by ESI-MS completely desolvated in the mass spectrometer.

First examples of noncovalent complexes characterized by ESI were reported for synthetic molecules. The term *supramolecular chemistry* was coined by chemists to name these noncovalent synthetic assemblies. Accordingly, *supramolecular mass spectrometry* emerged to designate this related novel MS (Lehn, 1995) in which instrumentation and analytical conditions were optimized to preserve noncovalent interactions, which was absolutely unusual in classic MS. Similar applications were soon developed in the biological area. In 1991, two groups simultaneously demonstrated that by carefully controlling several experimental parameters, specific protein–ligand noncovalent interactions could survive the ESI-MS analysis (Ganem *et al.*, 1991; Katta *et al.*, 1991). Since these initial reports, the use of ESI-MS to characterize increasingly heavy or fragile noncovalent complexes has been widely illustrated. The growing number of publications reported on the topic reflects the potential of applying this technique to solve various biological problems (Heck *et al.*, 2004; Last *et al.*, 1999; Loo, 1997, 2000; Pramanik *et al.*, 1998; Przybylski, 1995; Przybylski *et al.*, 1996; Smith *et al.*, 1997).

When the formation of a noncovalent complex between a protein and any metal ion or ligand is investigated, many questions arise:

- Which molecules are interacting specifically within the complex?
- What is the stoichiometry of the complex? Is this stoichiometry unique or heterogeneous?
- How tight are the interactions?
- Are the interactions specific or artifacts (simple aggregation)?
- Where are the interactions located?

Obviously, MS will not provide a high-resolution three-dimensional (3D) image of the complex formed in solution, as other techniques might do. However, most of the aforementioned issues can be addressed by appropriate ESI-MS experiments. The objective of this chapter is to describe the general MS-based strategies that can be used to retrieve information regarding the noncovalent complex formation, its stoichiometry, and its stability.

Examples cited in the following sections were selected from work performed in our laboratory on commercially available instruments; they cover a variety of applications of supramolecular MS in biology.

Important Experimental Parameters

Buffer

The buffers commonly used in biochemistry are very unfavorable for MS. Indeed, the presence of nonvolatile salts (even at trace level) has a large impact on the quality of the spectra and can even prevent ion detection. On the other hand, usual solvents in ESI-MS (organic and acidic media) do not preserve the protein's folding, which is necessary to keep specific noncovalent interactions. For this reason, the protein has to be dialyzed against a buffer compatible with both the complex stability and the MS. Because of their reasonably high volatility, ammonium salts appeared to be good candidates, and their use for the characterization of noncovalent complexes by MS has been widely illustrated (Lemaire *et al.*, 2001). No or few ammonium acetate adducts are observed in the mass spectra, resulting in narrow multiply charged peaks with a good mass accuracy.

Accessible m/z *Range of the Mass Analyzer*

When proteins are analyzed under non-denaturing conditions (aqueous solution and controlled pH of 6–9), the number of effective charges is greatly decreased, in comparison to that detected for the individual species under denaturing conditions (strongly acidic and organic solutions, e.g., H_2O/CH_3CN: 1/1, 1% HCOOH). Ions are then often detected at high mass-to-charge (m/z) ratios.

Many commercially available ESI instruments are coupled to quadrupole mass analyzers, with a fairly limited measurable m/z range (<4000 m/z). This, in many cases, constitutes a major technical limitation to achieve successful supramolecular MS experiments. However, the development of ESI instruments coupled to time-of-flight (TOF) mass analyzers has overcome this limitation, because this was already convincingly illustrated by several groups (Ayed *et al.*, 1998; Fitzgerald *et al.*, 1996; Rostom *et al.*, 1999). The coupling of continuous flow ES is achieved using an orthogonal TOF. In theory, the m/z range is unlimited, and it makes no doubt that these advents will contribute to boost application of supramolecular MS in many laboratories.

Interface Parameters (Vc, P, T)

Interface conditions are optimized to obtain optimum sensitivity and spectrum quality while preventing dissociation of the complex. Energy transferred to ions must then be controlled in order: (i) to provide a sufficient ion desolvation, (ii) to ensure a good transmission of the ion beam, and (iii) to preserve the specific noncovalent interactions. The following variables affecting the ion desolvation and transmission appear to be critical: the source temperature T, the cone voltage Vc, and the interfacial pressure P. A stability diagram can be drawn for each system (Fig. 1).

In some cases, it might be difficult to completely remove all solvent molecules from the protein complex without affecting the complex stability. In this case, the resulting mass spectrum shows variable amount of solvent molecules, still attached to the protein complex leading to lower sensitivity, broader ion peaks, and less accurate molecular mass measurement.

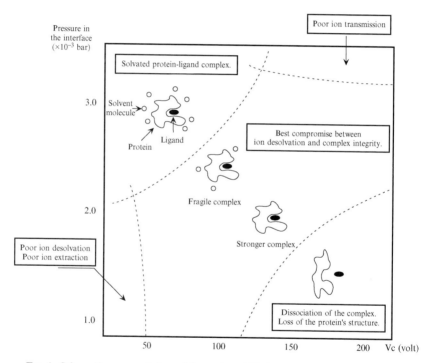

FIG. 1. Schematic representation of the tuning of Vc (accelerating voltage of ions) and P (interfacial pressure) in order to both preserve the protein–ligand noncovalent interaction and detect the signal with a reasonable quality (good desolvation and transmission of ions). The best compromise between intact complex detection and ion desolvation/transmission is obtained for the (Vc, P) couples that are in the center region (dashed lines). The source temperature is not represented in this diagram, but this parameter also has to be controlled. The source temperature is set to a value that optimizes ion desolvation while avoiding thermal dissociation of the complex.

Introduction System

All experiments were performed using classic probe at a flow rate of 4 μl/min. When a small amount of material is available, a nanospray probe can be used, yielding typical flow rates of 50 nl/min as shown by Robinson (Benesch *et al.*, 2003; Fandrich *et al.*, 2000; Keetch *et al.*, 2003).

Direct Information Provided by an ESI-MS Analysis

Determination of the Complex Stoichiometry

ESI-MS was shown to be a rapid and sensitive tool to determine protein–ligand stoichiometry. Determining the number of ligands that are involved in a biologically active complex is a unique advantage of this technique. The stoichiometry of the complex (n) can be easily obtained from the direct molecular mass measurement of the complex, *even if multiple stoichiometries coexist in solution*:

$$\text{measured mass} = M_P + \mathbf{n}\, M_L,$$

where M_P and M_L are the molecular masses of the protein and the ligand, respectively.

Protein–Metal Interactions. Metal ions are essential for the activity of many metalloenzymes. Counting the number of metal ions that are involved in a biologically relevant complex is an important issue and is not always straightforward with usual techniques (CD, spectroscopic absorption, NMRS). The potential of MS for the determination of the protein–metal stoichiometry is very promising and now well illustrated (Brewer *et al.*, 2000; Fekkes *et al.*, 1999; Gehrig *et al.*, 2000; Hill *et al.*, 2000; Hu *et al.*, 1994; Pinkse *et al.*, 2004; Veenstra *et al.*, 1998a,b).

As an example, ESI-MS has been used to investigate metal binding stoichiometry of a matrix metalloproteinase, the stromelysin 3 (ST3; molecular weight [MW] = 20,097 Da). ST3 is a zinc-dependent extracellular enzyme that requires zinc and calcium for activity (Veenstra *et al.*, 1998a). Under denaturing conditions (H_2O/CH_3CN: 1/1, 1% HCOOH), a molecular mass of 20,097 $\pm$ 1 Da was measured for ST3, which agrees with the mass expected from the primary structure. As shown in Fig. 2, metal attachment can be preserved by using non-denaturing conditions: In 25 mM AcONH$_4$, a unique series of multiply charged ion was observed, which could be attributed to multiple protonated states (8+ to 10+) of a species of mass 20,258 $\pm$ 1 Da. The mass increment of 161 Da allowed us to directly establish the binding stoichiometry as being stromelysin/Zn/Ca:1/2/1, which agrees with previous data obtained by other methods (accession number:

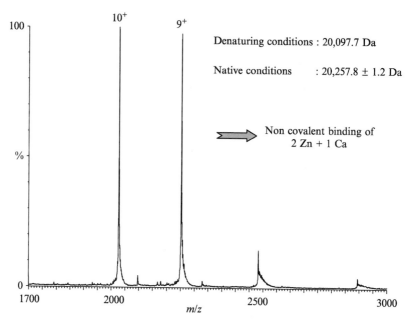

Fig. 2. Determination of the metal-binding stoichiometry of stromelysin 3 by electrospray ionization (ESI)–mass spectrometry (MS). The protein was dialyzed against 25 mM of AcONH$_4$ pH 6.5 and the cone voltage set at Vc = 80 V. A comparison of the measured molecular mass and the calculated mass as expected from the sequence directly gives the binding stoichiometry as being ST3/Zn/Ca = 1/2/1.

P24347). This complex was shown to be resistant to gas-phase collision, and mild interface conditions were not necessary for the observation of the intact complex.

Additional experiments might help if any ambiguity in the binding stoichiometry persists. For instance, intermediate stoichiometries can be revealed by coupling MS to time-coarse dialyses or to the use of chelating agents (EDTA) in solution. Tandem MS experiments can also provide complementary information on the stoichiometry, if a gradual dissociation of the complex can be achieved. Most often, however, metal ions dissociate simultaneously in the gas phase and no intermediate products are detected.

Competitive binding to another metal ion monitored by MS can also be used to ascertain the stoichiometry of the complex. For example, comparing the mass measured under denaturing conditions (8889 ± 1 Da) to that measured under non-denaturing conditions (9016 ± 1 Da) for the C-terminal part of the protein p44 reveals that this protein likely binds two zinc ions (mass increment of 127 Da). This result is in good agreement with the zinc finger motif expected in this region of p44, particularly rich in histidine and cysteine residues, and it could also be confirmed by the successive loss of the two zinc ions observed by MS after the controlled addition of EDTA in solution (data not shown). This interpretation could be reinforced by further competitive binding experiments against several metal ions; in fact, it was found that dialysis against cadmium ions resulted in the replacement of the initially bound Zn^{2+} ions by two Cd^{2+} ions. This result perfectly agrees with the known similarity in the coordination behavior of these two ions (Fribourg *et al.*, 2000) (Fig. 3).

Protein–Ligand Interactions. As it was previously shown for protein–metal interactions, ESI-MS can also provide information concerning the binding of substrate molecules to a protein. Comparison of the molecular masses of a species measured under denaturing and under non-denaturing conditions again indicates whether the species is or is not interacting with another molecule.

As an example, interactions between an enzyme, pig lens aldose reductase (AR, M = 35780 Da), its cofactor (NADPH, M = 742 Da), and an inhibitor (the tolrestat, M = 357 Da) have been studied by ESI-MS. AR is known to require the presence of NADPH to catalyze the reduction of its substrates (aldehydes or aldoses) to their corresponding alcohol (Jaquinod *et al.*, 1993). It follows an ordered addition of substrates, with NADPH binding first, and an ordered release of products, with $NADP^+$ being released last. Figure 4 shows the mass spectra obtained for a preparation of AR apoenzyme in presence of one (Fig. 4B and D) or both ligands (Fig. 4C). The measured molecular masses ($36,512 \pm 5$ Da and $36,877 \pm 3$ Da) directly show that AR forms binary and ternary complexes, respectively, with one molecule of cofactor and one molecule of inhibitor.

Rigorous control experiments were performed to establish that the noncovalent complexes detected by ESI-MS were really the result of in-solution interactions rather than artefactual nonspecific associations occurring during the ESI process. Indeed, modifications of solution conditions, as well as complex components, would provide good support for specificity if they produce substantial change in the mass spectrum. In the case of AR, no binary or ternary complexes were observed anymore by ESI-MS after modification of the pH or after addition of organic solvent in solution, both modifications being known to destroy many specific interactions in solution. The fact that the inhibitor binds to AR only if the cofactor is already bound to its site also provides good support of specificity (Fig. 4D). Crystal structure confirmed this result because it clearly showed that the inhibitor site was incompletely formed in the absence of the cofactor, so that the tolrestat could not bind with a high enough affinity (Potier *et al.*, 1997; Urzhumtsev *et al.*, 1997).

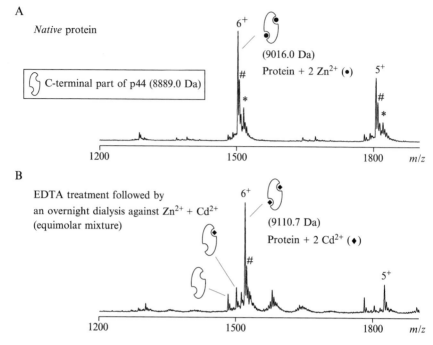

FIG. 3. Metal ion content of the C-terminal part of p44 (residues 321–395, M = 8889.0 Da). The p44(321–395) protein is diluted to 10 μM in 10 mM ammonium acetate (pH 7.0). (A) Electrospray ionization (ESI)–mass spectrometry (MS) analysis of *native* p44(321–395) protein. (B) ESI-MS analysis of the protein after an EDTA treatment followed by overnight dialysis (at 4°) against an equimolar mixture of Zn^{2+} and Cd^{2+} ions; both metal ions are in a large molar excess over the total protein binding sites. Comparison of the expected mass for *native* p44(321–395) to that measured in non-denaturing MS conditions (A) (M = 9016.0 Da) shows a mass increment of 127 Da. This mass increment might arise from the noncovalent binding of two Zn^{2+} ions (2 × 63.5 Da). A similar comparison reveals that after the EDTA treatment and the subsequent competitive addition of Zn^{2+} and Cd^{2+} ions, the protein has bound two Cd^{2+} ions (mass increment of 210 Da) (B). Peaks labeled with a sharp (#) correspond to sodium adducts (+22 Da); those labeled with an asterisk (*) correspond to 2-mercaptoethanol adducts (+76 Da).

Determination of the Complex Stability

Principle. Ion intensities (peak heights or peak area) of protein–ligand complexes observed in the ESI mass spectrum are compared to assess their binding constants in solution. Several studies have been reported that show that quantitative or semiquantitative information derived from ESI mass spectra permit the investigation of solution-phase thermodynamics of these systems.

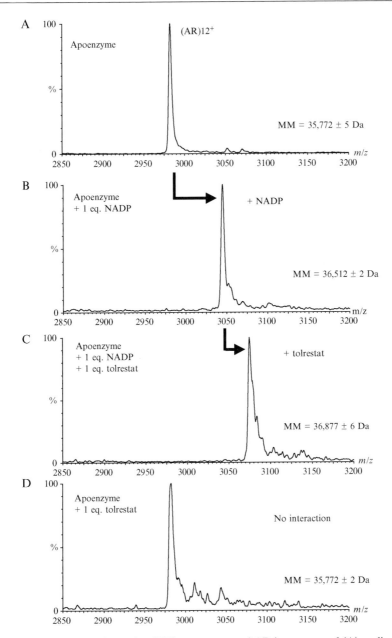

Fig. 4. Electrospray ionization (ESI) mass spectra of AR in presence of (A) no ligand, (B) 1 eq of NADP, (C) 1 eq of NADP + 1 eq of tolrestat, (D) 0 eq of NADP + 1 eq of tolrestat. The binding stoichiometries are directly deduced form the measured molecular masses. The inhibitor binds to AR only when NADP is bound to its site. Spectra recorded at Vc = 50 V.

In semiquantitative approaches, a competitive binding experiment is performed between several ligands in solution, where the total protein concentration is less than the total ligand concentration. Under these competitive conditions, abundance of a protein–ligand complex depends on the value of its binding constant relative to those of other complexes. Performing these competitive binding experiments with a ligand of known binding affinity, it is possible to yield absolute values for the binding constant of the targeted complex (Cheng et al., 1995; Jorgensen et al., 1998; Loo et al., 1997). The use of one protein–ligand complex with a known binding constant value as internal reference allows to normalize the ES efficiency of the protein–ligand complex of interest (Kempen et al., 2000).

A direct quantitation of all complexed and dissociated species detected on the ESI mass spectrum during a titrimetric experiment may also be used in certain cases. The ion abundance of both the bound protein and the unbound protein is then used to draw Scatchard-type binding plots from which absolute binding constants can be deduced without requiring any "reference" ligand (Ayed et al., 1998; Gabelica et al., 2003; Greig et al., 1995; Lim et al., 1995). In this case, one makes the assumption that ES efficiencies for the bound and unbound complex are similar.

Limitations in the Determination of Binding Constants by ESI-MS. In all the studies cited, the results obtained from ESI-MS experiments were consistent with the binding constants measured by other techniques in solution. This means that for these particular systems, the mass spectra were faithfully reflecting the distribution of the species in solution.

However, this conclusion should not be thoughtlessly extended to any protein–ligand system. As pointed out by several authors, relative abundance of certain complexes may be dramatically affected by the desorption process, because of the partial loss of hydrophobic interactions in the gas phase (Li et al., 1994; Robinson et al., 1996; Rogniaux et al., 1999; Wu et al., 1997). When it happens, such a distortion evidently makes the mass spectrum improper to investigate distribution of the species in the condensed phase.

Additional difficulties appear when one attempts to compare largely different species, because of possible differences in ionization efficiency or in ion transmission of these species. Thus, only closely related species are usually compared, for instance, complexes of a protein with a series of ligands. Note that this very often gives rise to a further analytical difficulty, namely the ability to separate close *m/z* ions.

Possible distortion arising from ion desorption or ion transfer to the mass analyzer constitutes the major risk of a false quantitative interpretation of ESI data (this aspect is further developed and discussed in the second part of this chapter). One then easily figures that quantitation from ESI mass spectrum should be used only with great care. General applicability of the method is far from being established, and it is crucial to investigate its validity for each new system under study.

Nonetheless, sensitivity and rapidity of ESI-MS experiments still make the technique uniquely attractive to characterize ligand binding to proteins, as this was beautifully illustrated by the group of R. D. Smith with the "BACMS" approach.

The "BACMS" Approach. The concepts of BACMS (Bio-Affinity Characterization Mass Spectrometry) approach were introduced by Smith and coworkers (Bruce *et al.*, 1995; Cheng *et al.*, 1995). BACMS exploits the high-resolution and MS^n capability of Fourier transform (FT) ion cyclotron resonance (ICR) MS to identify tightly bound inhibitors of a target enzyme in a complex mixture. Possibilities afforded by this method were illustrated by Gao *et al.* (1996), with the screening of a 290-compound combinatorial library of carbonic anhydrase II inhibitors. Wigger *et al.* (2002) succeeded in identifying the preferred ligands of the HckSrc homology 2 domain protein among a library of 324 compounds in a single FT-ICR MS experiment.

A solution containing the affinity target and the ligand library prepared under competitive binding conditions is ionized by ESI. Complexes of interest are trapped and selectively accumulated in the FT-ICR. Tandem MS experiments are then performed on the isolated complexes: the resulting mass spectrum allows the dissociated ligands to be unequivocally identified based on their different molecular weights; moreover, relative ion intensities are measured to derive the relative binding affinities of the ligands. Obviously, such experiments require ionization of the ligands in the gas phase.

Determination of the Cooperativity in Ligand Binding to Multimeric Enzymes

Note that in this paragraph, E designates a multisites enzyme, L is the ligand, and K_n is the thermodynamic constant describing the binding of the nth ligand molecule to the enzyme (Eq. [1]).

$$EL_{n-1} + L \leftrightarrows EL_n \quad K_n = [EL_n]/[EL_{n-1}] \, [L] \tag{1}$$

Definition of Cooperative Binding. When a multimeric enzyme is composed of identical and *independent* (i.e., noninteracting) subunits, the ligand molecules are expected to be distributed *statistically* on all its binding sites. Binding constants associated to the successive binding steps (K_n) then verify Eq. (2) (s is the total number of binding sites).

$$K_{n+1}/K_n = n(s-n)/(n+1) \ (s-n+1) \tag{2}$$

When a *significant deviation* from the statistical binding is observed, one speaks of a *cooperative binding*. If the $(n+1)$th ligand binds more readily than the nth ligand, the binding occurs with a *positive cooperativity;* the K_{n+1}/K_n ratio is then higher than it would be for a statistical binding. On the contrary, if the binding of the nth ligand hinders the subsequent binding of the $(n+1)$th ligand, the binding occurs with a *negative cooperativity* and the K_{n+1}/K_n ratio is lower than it would be for a statistical binding.

Application of Supramolecular MS to Cooperativity Studies

PRINCIPLE. A common approach to investigate cooperativity in binding of a ligand to a multimeric receptor consists of measuring the variations of the K_n as the sites are progressively filled by the ligand: K_{n+1}/K_n ratios are then compared to the values predicted for a statistical binding (Eq. [2]).

Because of its ability to distinguish species of different molecular masses, ESI-MS has the potential to give a direct view of all distinct enzymatic species ($EL_{n, \ n=[0 \ . \ . \ . \ 4]}$) that are in equilibrium as increasing amount of ligand is added in solution.

A semiquantitative interpretation of the ESI data permits to yield relative abundances of all enzymatic species and to subsequently deduce the K_{n+1}/K_n ratios, because $K_{n+1}/K_n = [EL_{n-1}] \times [EL_{n+1}]/[EL_n]^2$.

EXAMPLE. Figure 5 shows illustrative ESI mass spectra recorded during addition of the oxidized cofactor NAD^+ to tetrameric Sturgeon muscle glyceraldehyde-3-phosphate dehydrogenase (GPDH) (Fig. 5A) or to tetrameric baker's yeast alcohol dehydrogenase (ADH) (Fig. 5B).

When analyzed under non-denaturing conditions in the absence of the cofactor, both enzymes display four main charge states ($z = 26$ to $z = 30$) that match the apo form of the tetramer.

As shown in Fig. 5, addition of NAD^+ in solution gives rise to other species, which masses correspond to the attachment of one to four molecules of NAD^+ ($M = 663.4$ Da) to the tetrameric enzyme. These species are detected in the same charge states as the initial unbound tetramer (i.e., $z = 26$ to $z = 30$).

Experimental K_{n+1}/K_n values deduced from the semiquantitative interpretation of the MS data for these two systems are given in Table I. Comparison to the values expected for a statistical binding shows that for the case of baker's yeast ADH, NAD^+ binding occurs without significant cooperative effect, whereas for the case of Sturgeon muscle GPDH, a slight positive deviation to the statistical binding is observed from site 1 to site 2 and from site 3 to site 4 (with K_2/K_1 and K_4/K_3 ratios about four times higher than that expected for a statistical binding).

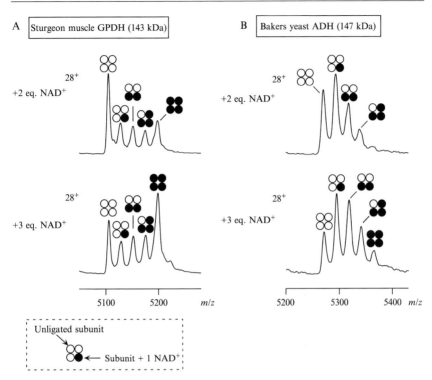

FIG. 5. Typical electrospray ionization (ESI) mass spectra (here presented for the 28-fold protonated state of the tetramer) recorded during addition of NAD^+ to two tetrameric enzymes. (A) Sturgeon muscle GPDH (M = 142,890 Da, the tetrameric protein is diluted to 20 μM in 15 mM ammonium bicarbonate, pH 8.9). (B) Baker's yeast ADH (M = 147,005 Da, the tetrameric protein is diluted to 10 μM in 10 mM ammonium acetate, pH 7.0). Abundance of each tetrameric species ($EL_{n,\ n=[0...4]}$) relative to the others is obtained by measuring the ion intensity (peak height) of all the charge states displayed by the species in the ESI mass spectrum; abundance is then normalized to 1. For example, in the lower mass spectrum of (B), the following ion intensities are measured for the EL_n species: 0.159, 0.293, 0.277, 0.180, and 0.091 for EL_0, EL_1, EL_2, EL_3, and EL_4, respectively. From the above intensities, the following K_{n+1}/K_n ratios can be deduced ($K_{n+1}/K_n = [EL_{n+1}] \times [EL_{n+1}]/[EL_n]^2$, see Eq. [1] in the text): $K_2/K_1 = 0.512$, $K_3/K_2 = 0.689$ and $K_4/K_3 = 0.775$.

VALIDITY OF THE APPROACH. As mentioned previously for quantitative or semiquantitative treatment of ESI-MS data, an essential prerequisite is that relative peak intensities observed in the mass spectra are reliable to investigate the solution-phase thermodynamics (i.e., the equilibrium concentrations of the species). Great care in the data acquisition and in the interpretation must be taken, because it is well known that the

TABLE I
COOPERATIVITY STUDIES BY MS: BINDING OF NAD$^+$ TO STURGEON MUSCLE GPDH
AND TO BAKER'S YEAST ADH^a

	K_2/K_1	K_3/K_2	K_4/K_3
Sturgeon muscle GPDH	$\mu = 1.77$	$\mu = 0.94$	$\mu = 1.50$
38 mass spectra	$\delta = 28\%$	$\delta = 20\%$	$\delta = 28\%$
Baker's yeast ADH	$\mu = 0.56$	$\mu = 0.67$	$\mu = 0.79$
10 mass spectra	$\delta = 12\%$	$\delta = 17\%$	$\delta = 18\%$

^{a} Ten to about forty ESI mass spectra were recorded at various enzymatic and NAD$^+$ concentrations for the two tetrameric systems. The K_2/K_1, K_3/K_2, and K_4/K_3 values were derived from the relative peak intensities as described in the legend of Fig. 4. Of the K_2/K_1, K_3/K_2, and K_4/K_3 values obtained from all the measurements, 95% are inside the interval $\mu' \pm \delta\%$, where μ is the median and δ is the maximum deviation from the median (Rogniaux *et al.*, 2001).

solution-phase image might be distorted during the ESI mass analysis, especially during ion-desorption and ion-transfer processes.

The following points support the validity of this approach for the cooperativity study described in the previous paragraphs:

1. The EL_n species have nearly the same mass ($\Delta M \approx 664/145{,}000$) and they all have the same charge states. They are, thus, detected at very close m/z ratios, so there will not be a strong discrimination effect of the different species due to focalization of the ion beam through the interface region.

2. 3D changes that may accompany the binding of NAD$^+$ to the enzymatic subunits do not change dramatically the surface of the protein, which is exposed to the solvent; thus, EL_n species likely display comparable solvation energies relative to the free enzyme E. Together with their close mass and charge, the assumption that their response factors to the ESI process will be similar is reasonable.

3. Experiments performed at different accelerating voltages (Vc) show that distribution of EL_n species is not at all modified by gas-phase collisions of increasing energies. This result permits to rule out the possibility that some species may be dissociated by gas-phase collisions and be abnormally represented on the mass spectrum.

The major advantage of MS to study cooperativity, in comparison to established methods in the field (UV spectroscopy, fluorescence-quenching measurement, capillary electrophoresis, etc.) is to enable the view of all individual species in solution. Furthermore, rapidity and sensitivity of MS make it attractive for such studies (Rogniaux *et al.*, 2001).

Application to the Characterization of Orphan Proteins. An important question concerning the function of orphan proteins (i.e., proteins for which function and natural ligand are unknown) is whether their activity is mediated through ligand binding or not. Indeed, some orphan receptors have been shown to be constitutively active (such as CAR, ERR, and ROR), but this constitutive activity may in fact be only apparent. In this context, supramolecular MS seems to emerge as a new technique for addressing this question at the two following levels:

- To control the homogeneity of the expressed protein in terms of interactions with small molecules present in the expression system or purification environment.
- To identify a potential ligand whose activity could then be verified by an appropriate biological assay.

The ability of supramolecular MS to determine existence and stoichiometry of a protein–ligand complex can be used to control the homogeneity of new expressed proteins in terms of interaction with small molecules.

Indeed, some studies have shown that fortuitous ligands can be captured by the protein in the expression host with the proper stoichiometry (Bourguet *et al.*, 2000; de Urquiza *et al.*, 2000; Elviri *et al.*, 2001). These molecules act as "fillers" and are auto-selected by the protein in the available chemical library constituted by the host cell media. They may not be the physiological ligands, but they prove to be essential to stabilize the active conformation of the receptors. The characterization of such ligands is then essential because if a fortuitous ligand is present, but the binding is not quantitative, the resulting chemical and conformational heterogeneity will hamper production and crystallization of the protein because its ligation cannot be controlled. In turn, the knowledge of the 3D structure of the ligand-binding pocket is a good starting point for the design of new high-affinity ligands. These ligands can then be used for the functional characterization of the orphan receptor.

If not expected and not displaying any spectroscopic properties, the detection of such small ligands bound to the protein may be very challenging. In the case of the Ultra Spiracle Protein (USP), the insect ortholog of the vertebrate retinoid X nuclear receptor (RXR), ESI-MS is the only technique that has been able to show the presence of an unexpected ligand molecule non-covalently attached to the receptor and to assign a molecular mass of 745 Da for this molecule (Potier *et al.*, 2003). Combining ESI-MS under denaturing and native conditions, solvent extraction, and data from X-ray diffraction, it has been possible to demonstrate that the ligand captured inside the ligand-binding pocket of the USP was a phosphatidyl-ethanolamine.

As the number and diversity of both clone products and host expression systems increases each day, it becomes more and more important to have in hand a technique that enables the rapid and detailed characterization of the material derived from gene expression before performing biochemical or structural studies.

Screening of Ligands Using Mass Spectrometry

The ability of ES-MS to show whether a ligand is or is not bound to a protein is now more often used in the pharmaceutical field for screening potential drugs (Cancilla *et al.*, 2000; Geoghegan *et al.*, 2005; McCammon *et al.*, 2002; Wright *et al.*, 2000). Indeed, supramolecular MS might constitute a new methodology for a more reasoned and rational approach for drug discovery and for human genome annotation (Fig. 6). By characterizing the interaction (positive or negative) between the target protein and a molecule (potential ligand or drug), MS will be helpful to do the following:

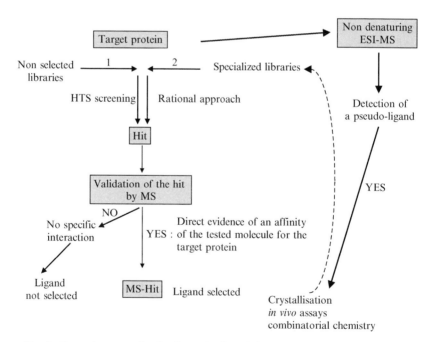

FIG. 6. General strategy for the determination of the function of orphan proteins using supramolecular mass spectrometry.

- Validate a hit obtained from HTS tests: an interaction between the tested molecule and the target protein will be directly observed.
- Validate a target protein evidenced by differential proteomic. Indeed, the observed variation of the protein's expression rate is not necessarily related to an affinity of the protein for the tested drug.
- Identify the physiological ligand in case of orphan proteins.

Figure 7 displays a typical screening experiment using supramolecular MS. In such experiments, library compounds are incubated with the target protein one by one and analyzed by ESI-TOF under gentle interface conditions. Those ligands displaying an affinity for the target protein will be selected, such an affinity being evidenced by a corresponding mass shift on the mass spectrum.

In our case, the target protein is observed as a doublet because of an additional N-terminal amino acid (Fig. 7A). Incubation conditions and ESI parameter settings are controlled using a reference ligand (ν), known to bind to the active site of the protein (Fig. 7B). After addition of 2 eq of ligand, the signal corresponding to the apo protein has completely disappeared and a new species displaying a Δm^* corresponding to the mass of the reference ligand is observed. Referring to our general strategy shown in Fig. 6, a compound will be retained as soon as a signal corresponding to a protein–ligand complex is detected (Fig. 7C). However, of major importance is the necessity to discriminate specific against nonspecific binders (both protein–ligand complexes displaying the same molecular mass). This can be easily performed through competition experiments between the potential ligand and the reference ligand (Fig. 7D). Two situations can occur: (i) the measured molecular mass of the resulting complex indicates that the reference ligand was able to displace the tested compound (Fig. 7D left, compound σ) and (ii) the measured molecular mass of the resulting complex indicates that the reference ligand was not able to replace the tested compound (Fig. 7D right, compound λ). In the first case, the compound σ is, thus, shown to be specific and will be definitively selected as a MS hit; in the second case, both ligands (the reference one and compound λ) are bound to the protein, suggesting a nonspecific binding for the tested molecule because the protein is known to possess a single binding site. The lack of specificity of compound λ could have been already suspected from the multi-addition effect observed in Fig. 7C.

Drug-screening methods using MS have been reviewed by Siegel (2002). Supramolecular MS constitutes an attractive middle-throughput analytical technique that can be used as a direct detector of affinity between a target protein and a potential ligand with possible capability of automation (Zhang et al., 2003a,b).

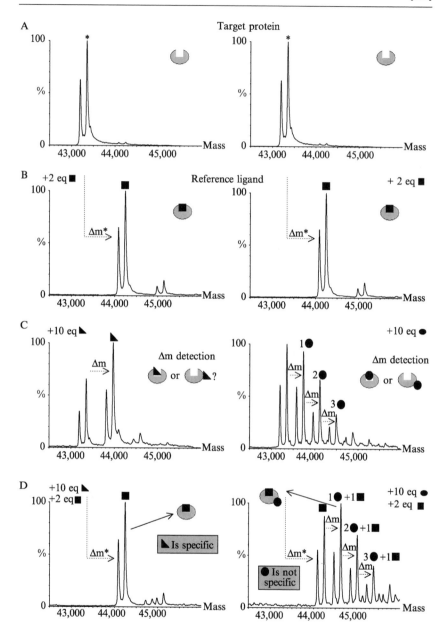

FIG. 7. Typical drug screening experiment using supramolecular mass spectrometry (MS). Electrospray ionization (ESI) mass spectrum of the target protein diluted to 10 μM in 50 mM ammonium acetate (A) in absence of ligands; (B) in presence of 2 eq of a reference ligand (ν); (C) in presence of 10 eq of a compound stemming from a library (left: compound σ; right:

Information Deduced from the Gas-Phase Behavior

During the ESI process, progressive desolvation of the droplets produces the noncovalent complexes as dried species in vacuum. On most conventional instruments, it is possible to control the kinetic energy of gas-phase ions, either in the first vacuum stage of the mass spectrometer, where the pressure is still of the order of 1 mbar, or in a collision-gas cell (tandem MS). Increasing the internal energy transferred to ions through collisions with gaseous molecules provokes disruption of the weakest interactions (i.e., the noncovalent interactions. Stability of noncovalent complexes can, thus, be measured by their resistance to energetic collisions as a function of either the accelerating voltage Vc or the collision energy (tandem MS).

Gas-Phase Stability Measurements

Progressive dissociation of a noncovalent protein–ligand complex can be monitored by the appearance of ions of the unbound species in the mass spectrum by increasing of the accelerating voltage Vc. An example is given in Fig. 8 for an AR–inhibitor complex: Going from Vc = 65 V to Vc = 85 V leads to detection of ions corresponding to the unbound enzyme (ions for the free inhibitor also appear in the low m/z range but are not shown in this figure).

A widely used criteria to evaluate gas-phase stability of a given noncovalent complex is the accelerating voltage needed to achieve 50% of dissociation in the mass spectrometer (Vc_{50}). This value is deduced from dissociation curves as those drawn on the right part of Fig. 8, when abundance of the complex is equal to that of the dissociated species. Here, again, abundance of the complexed species relative to the dissociated species is measured from ion intensity displayed in the ESI mass spectrum (i.e., peak heights summed on all the charge states).

Interpretation of Dissociation Experiments

Relationship between Gas-Phase and Solution-Phase Stability. Several authors tried to find a correlation between the gas-phase dissociation energy (i.e., the Vc_{50}) and the solution-phase binding strength (i.e., the binding constants) for a series of protein–ligand noncovalent complexes.

compound γ). (D) Competition experiments between the library compound and the reference compound. The cone voltage (Vc) was optimized in each case to prevent gas-phase dissociation of the protein–ligand complexes. Replacement of the protein–ligand complex by the protein–reference complex shows that the tested compound was effectively bound in the active site of the target protein. Peaks labeled with an asterisk (*) correspond to the target protein modified by an additional N-terminal amino acid.

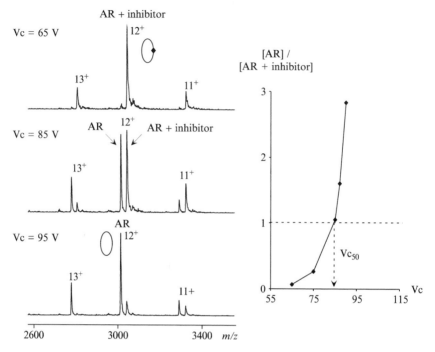

FIG. 8. Gas-phase stability study of an enzyme–inhibitor complex. The enzyme is human aldose reductase (AR) complexed with its natural cofactor $NADP^+$ ($M_{AR\text{-}NADP} = 36,880$ Da); the inhibitor is a carboxylic acid of mass 345 Da. AR ($10~\mu M$ in 10 mM ammonium acetate) is shortly incubated with 1 eq of its cofactor and 1 eq of the inhibitor. The mixture is then continuously infused into the mass spectrometer. Mass spectra were recorded for increasing Vc voltages (the interfacial pressure P is equal to 2 mbar) (left side). The enzyme–inhibitor complex gradually dissociates when Vc increases; the extent of dissociation at a given Vc voltage is evaluated by measuring the intensity (peak heights averaged on the three charge states) of the dissociated species (D) relative to that of the non-dissociated enzyme–inhibitor complex (C). The experimental D/C ratios are then plotted versus Vc to deduce the so-called Vc_{50}, the accelerating voltage of ions leading to 50% of the complex dissociation (right side).

For some systems, a good correlation was reported, with the order of stability in the gas phase following that in the solution phase (Hunter *et al.*, 1997; Potier *et al.*, 1997). However, for other systems, striking discrepancies were found between solution- and gas-phase behavior (Li *et al.*, 1994; Wu *et al.*, 1997). For a while, reasons for the observed disagreement remained unclear, even though the possible influence of hydrophobic interactions—which would be partly lost in the gas phase—was mentioned as early as 1994 (Li *et al.*, 1994). However, as noted by Joseph A. Loo (1997), "It has yet to be conclusively demonstrated that a gas-phase dissociation energy can be used to predict or even to reflect the solution-phase binding strength."

Formation of hydrophobic interactions in proteins is driven by water molecules surrounding nonpolar groups. These nonpolar groups tend to aggregate to minimize contacts with polar solvent molecules. This results in formation of hydrophobic interactions. Progressive disappearance of water as species are dried in the mass spectrometer removes the driving force for hydrophobic interactions. It is, therefore, natural to assume that hydrophobic interactions do not contribute significantly to stability of the complex in the gas phase and do not intervene in the dissociation energy as measured by the Vc_{50}. This was unequivocally verified in our group through a parallel study of several enzyme–inhibitor complexes by X-ray crystallography and ESI-MS (Rogniaux *et al.*, 1999).

Comparison of Crystallographic and Mass Spectrometric Data for a Series of Enzyme–Inhibitor Complexes. Gas-phase stability of noncovalent complexes of AR with a series of synthetic inhibitors was evaluated by ESI-MS, as described earlier in this chapter. Among all inhibitors, four compounds representative of the different families of AR inhibitors were extensively studied by both ESI-MS and X-ray crystallography. The aim of this parallel study was to validate or invalidate the loss of hydrophobic interactions during the ion-desorption process in the mass spectrometer.

X-ray diffraction data allowed us to calculate the contribution of non-hydrophobic interactions (i.e., in the present case, sum of electrostatic and H-bond interactions) in the binding energy of each inhibitor to AR, while solution-phase measurements gave the overall binding energy (expressed by the IC_{50} of each inhibitor, i.e., the concentration leading to 50% inhibition of the enzymatic activity).

Table II reports the IC_{50} values obtained for the four inhibitors hereby studied, the corresponding electrostatic and H-bond energies calculated from the crystallographic data ($E_{electrostatic} + E_{H\text{-}bond}$) and the Vc_{50} measured by gas-phase collisions in the interfacial region of our ESI mass spectrometer. It is striking that for these four inhibitors, the Vc_{50} values do not follow the order of the IC_{50} values, whereas a clear correlation is observed between calculated $E_{electrostatic} + E_{H\text{-}bond}$ binding energies and the Vc_{50} values.

This example demonstrates that hydrophobic contacts poorly contribute to gas-phase stability of noncovalent complexes.

From Solution Phase to the Gas Phase: Possible Origins of Distortion

From the preceding paragraphs, it is obvious that MS will provide an important tool for characterization of supramolecular complexes. Information as important as the stoichiometry of the complex, its stability in

TABLE II
GAS-PHASE STABILITY MEASUREMENTS: AR–INHIBITOR COMPLEXES[a]

	Vc_{50} (volt)	IC_{50} (nanomole)	$E_{electrostatic} + E_{H\text{-}bond}$ (Kcal/mol)
AminoSNM	70 ± 1	52	-24
Imirestat	76 ± 1	10	-43
LIPHA3071	84 ± 1	9	-91
IDD384	103 ± 1	108	-94

[a] The gas-phase stability experimentally measured by mass spectrometry (Vc_{50}), and the binding energies measured in solution (IC_{50}) are listed for four inhibitors of AR. For these inhibitors, it was possible to calculate the energies of the electrostatic and H-bond interactions from the crystal structure of the complex: The sum of these energies ($E_{electrostatic} + E_{H\text{-}bond}$) is given in the third column. Clearly, $E_{electrostatic} + E_{H\text{-}bond}$ leads to the same classification of the four inhibitors than the Vc_{50}. On the contrary, it is obvious that the Vc_{50} and the IC_{50} do not lead to the same classification of the inhibitors (Rogniaux et al., 1999).

solution, or the nature of the interaction involved in the complex formation can be deduced from MS experiments.

However, as we pointed out repeatedly in this chapter, the technique is not yet "routine" and still requires constant care. Inappropriate or uncontrolled experimental conditions, as well as the intrinsic nature of certain complexes, might be responsible for artifacts occurring in the MS analysis. If understanding of these artifacts has indeed continuously improved in the last few years, it remains essential to systematically perform controlled experiments to establish the validity of the MS results and to answer the following question: Do the data from ESI mass spectra reflect the associations existing in solution?

Ionization Efficiency and Ion Transmission

Species are expected to display different response factors in ESI-MS analysis depending on their charge states and solvation energies.

Solvation energy was found to play a key role in the ESI response factor of small alkaline cations (Leize et al., 1996). In this case, huge differences in solvation energies from one cation to the other resulted in largely unequal ion intensities detected for these ions by ESI-MS. To a lesser extent, solvation energy is also expected to play an important role in the response factor of larger systems, especially peptides and proteins. The resulting mass spectrum might then misrepresent the real abundance of the species in solution.

Focalization of the ion beam through the interface region may similarly discriminate some species relative to others. Significant discrimination effects between larger versus smaller m/z ratios were commonly observed by our group and others, depending on the tuning of the lenses. Comparative relative abundances between species are then not likely to be quantitative when their m/z ratios are largely different.

Observation of Artifact Associations

The simple observation of ions corresponding to a protein–ligand noncovalent complex does not constitute sufficient evidence of *specific* structurally interactions existing in solution (Aplin *et al.*, 1994; Cunniff *et al.*, 1995; Robinson *et al.*, 1996; Smith *et al.*, 1993). Control experiments are, thus, necessary to differentiate *specific* from *nonspecific* (simple aggregation) interactions that might be formed during the ESI process.

If the observed complex results from specific interactions in solution, it must be sensitive to modification of the experimental conditions affecting its stability. Modifying the conditions in solution should then produce a substantial change in the mass spectrum. For example, changing the nature of the ligand, denaturing the protein (by changing the pH, the solvent, the temperature, etc. of the solution), or diluting the sample should be reflected by the appropriate change in the intact complex abundance versus dissociated complex abundance on the ESI mass spectrum.

Influence of the Nature of the Interactions Involved in the Complex

From results of our group and other groups, a relationship seems to emerge between the stability of noncovalent complexes in the ESI process and the nature of the interactions involved in the complex formation.

In the case of leucine-zipper (Li *et al.*, 1993, 1994) or of complexes between acyl-coenzyme A (acyl-CoA) binding protein and acyl-CoA derivatives (Robinson *et al.*, 1996), ESI data did not reflect the relative abundance of the noncovalent interactions in solution. The reason suggested by the authors was that interactions in these systems were mostly hydrophobic and were partly lost in the gas phase.

On the contrary, other complexes where electrostatic interactions were predominant displayed a unusual stability in the gas phase. For example, protein–DNA and protein–RNA noncovalent complexes appeared to be very resistant to gas-phase collisions (Cheng *et al.*, 1996; Feng, 1995; Potier *et al.*, 1998). For certain systems like the AR-NADP system, it was shown that dissociation of the noncovalent complex could not be achieved even though the collision energy produced the breakage of a covalent bond (Potier *et al.*, 1997).

Going from solution to a solvent-less environment, it is obvious that the strength of noncovalent interactions will be affected: hydrophobic effects are likely to be weakened because their driving force (water) has disappeared, but electrostatic interactions are reinforced because the dielectric constant in vacuum is 80 times lower than in an aqueous medium. Therefore, success in studying a noncovalent complex by ESI-MS will crucially depend on the nature of the interactions.

The loss of water molecules during the desorption process may thus prevent the study of certain systems by MS. This has long been considered a major drawback of this technique and even gave rise to skepticism about the future of the technique. However, it appeared that one can take advantage of this drawback and exploit it to measure the contribution of non-hydrophobic relative to hydrophobic interactions (Robinson *et al.*, 1996; Rogniaux *et al.*, 1999). As proposed by J. A. Loo (1997), why not use ESI-MS to assess "the type of bonding interaction that keeps complexes together?" This indeed may be of special interest in drug discovery, because specificity of ligand binding to a protein is mostly determined by non-hydrophobic oriented interactions.

Conclusion

When using carefully controlled conditions in the atmospheric pressure–vacuum interface, and when sample preparation is optimized, it is possible to preserve large specific multiprotein– metal–ligand noncovalent complexes during MS analysis. In fact, the mass measurement of complexes of more than 1 million Da is only limited by the sample preparation quality (in particular, the removal of nonvolatile salts and detergents). MS analysis can then yield precious information on the stoichiometry. In addition, in some cases, information on the relative contribution of hydrophobic and electrostatic or hydrogen bond interactions can be determined through gas-phase stability experiments. It is important to keep in mind that gas-phase stability will be greater when the dominant interactions are ionic than for hydrophobically driven binding. Higher analyzer resolution will also be useful to assess more information from mass spectra. Obviously, ESI-MS will not provide direct structural data as NMRS or X-ray do, but the amount of material required is usually less than 1 nmol for a series of experiments.

Acknowledgments

We are deeply indebted to Brian Green (Micromass, Manchester, UK) for numerous helpful discussions on the optimization of the ESI interface in the study of fragile noncovalent complexes. Guillaume Chevreux thanks Sanofi-Aventis for a grant. We also acknowledge Michel Robin (Sanofi-Aventis) for fruitful discussions on drug-screening methodology.

References

Aplin, R., Robinson, C. V., Schofield, C. J., and Westwood, N. J. (1994). Does the observation of noncovalent complexes between biomolecules by electrospray ionization mass spectrometry necessarily reflect solution interactions? *J. Chem. Soc. Chem. Commun.* **241,** 5–2417.

Ayed, A., Krutchinsky, A. N., Ens, W., Standing, K. G., and Duckworth, H. W. (1998). Quantitative evaluation of protein–protein and ligand–protein equilibria of a large allosteric enzyme by electrospray ionization time-of-flight mass spectrometry. *Rapid Commun. Mass Spectrom.* **12,** 339–344.

Benesch, J. L., Sobott, F., and Robinson, C. V. (2003). Thermal dissociation of multimeric protein complexes by using nanoelectrospray mass spectrometry. *Anal. Chem.* **75,** 2208–2214.

Brewer, D., and Lajoie, G. (2000). Evaluation of the metal binding properties of the histidine-rich antimicrobial peptides histatin 3 and 5 by electrospray ionization mass spectrometry. *Rapid Commun. Mass Spectrom.* **14,** 1736–1745.

Bourguet, W., Andry, V., Iltis, C., Klaholz, B., Potier, N., Van Dorsselaer, A., Chambon, P., Gronemeyer, H., and Moras, D. (2000). Heterodimeric complex of RAR and RXR nuclear receptor ligand binding domains: Purification, crystallization and preliminary X-ray diffraction analysis. *Protein Expr. Pur.* **19,** 284–288.

Bruce, J. E., Anderson, G. A., Chen, R., Cheng, X., Gale, D. C., Hofstadler, S. A., Schwartz, B. L., and Smith, R. D. (1995). Bio-affinity characterization mass spectrometry. *Rapid Commun. Mass Spectrom.* **9,** 644–650.

Cancilla, M. T., Leavell, M. D., Chow, J., and Leary, J. A. (2000). Mass spectrometry and immobilized enzymes for the screening of inhibitor libraries. *Proc. Natl. Acad. Sci. USA* **97,** 12008–12013.

Cheng, X., Chen, R., Bruce, J. E., Schwartz, B. L., Anderson, G. A., Hofstadler, S. A., Gale, D. C., and Smith, R. D. (1995). Using electrospray ionization FTICR mass spectrometry to study competitive binding of inhibitors to carbonic anhydrase. *J. Am. Chem. Soc.* **117,** 8859–8860.

Cheng, X., Harms, A. C., Goudreau, P. N., Terwilliger, T. S., and Smith, R. D. (1996). Direct measurement of oligonucleotide binding stoichiometry of gene V protein by mass spectrometry. *Proc. Natl. Acad. Sci. USA* **93,** 7022–7027.

Cunniff, J. B., and Vouros, P. (1995). False positives and the detection of cyclodextrin inclusion complexes by electrospray mass spectrometry. *J. Am. Soc. Mass Spectrom.* **6,** 437–447.

de Urquiza, A. M., Liu, S., Sjöberg, M., Zetterström, R. H., Griffiths, W., Sjövall, J., and Perlmann, T. (2000). Docosahaenoic acid, a ligand for the retinoid X receptor in mouse brain. *Science* **290,** 2140–214.

Elviri, L., Zagnoni, I., Careri, M., Cavazzini, D., and Rossi, G. L. (2001). Non-covalent binding of endogenous ligands to recombinant cellular retinol-binding proteins studied by mass spectrometric techniques. *Rapid Commun. Mass Spectrom.* **15,** 2186–92.

Fandrich, M., Tito, M. A., Leroux, M. R., Rostom, A. A., Hartl, F. U., Dobson, C. M., and Robinson, C. V. (2000). Observation of the noncovalent assembly and disassembly pathways of the chaperone complex MtGimC by mass spectrometry. *Proc. Natl. Acad. Sci. USA* **97,** 14151–14155.

Fekkes, P., de Wit, J. G., Boorsma, A., Friesen, R. H., and Driessen, A. J. (1999). Zinc stabilizes the SecB binding site of SecA. *Biochem.* **38,** 5111–5116.

Feng, R. (1995). Unusually strong binding of a noncovalent gas-phase spermine–peptide complex and its dramatic temperature dependence. "43rd ASMS Conference on Mass Spectrometry and Allied Topics Atlanta, Georgia, May 21–26," p. 1264.

Fenn, J. B., Mann, M., Meng, C. K., Wong, S. F., and Whitehouse, C. M. (1989). Electrospray ionization for mass spectrometry of large biomolecules. *Science* **246**, 64–71.

Fitzgerald, M. C., Chernushevich, I., Standing, K. G., Whitman, C. P., and Ken, S. B. (1996). Probing the oligomeric structure of an enzyme by electrospray ionization time-of-flight mass spectrometry. *Proc. Natl. Acad. Sci. USA* **93**, 6851–6856.

Fribourg, S., Kellenberger, E., Rogniaux, H., Poterszman., A., Van Dorsselaer, A., Thierry, J. C., Egly, J. M., Moras, D., and Kieffer, B. (2000). Structural characterization of the cysteine-rich domain of TFIIH p44 subunit. *J. Biol. Chem.* **275**, 31963–31971.

Gabelica, V., Galic, N., Rosu, F., Houssier, C., and De Pauw, E. (2003). Influence of response factors on determining equilibrium association constants of non-covalent complexes by electrospray ionization mass spectrometry. *J. Mass Spectrom.* **38**, 491–501.

Ganem, B., Li, Y. T., and Henion, J. D. (1991). Detection of noncovalent receptor–ligand complexes by mass spectrometry. *J. Am. Chem. Soc.* **113**, 6294–6296.

Gao, J., Cheng, X., Chen, R., Sigal, G. B., Bruce, J. E., Schwartz, B. L. A., Hofstadler, S., Anderson, G. A., Smith, G. M., and Whitesides, G. M. (1996). Screening derivatized peptide libraries for tight binding inhibitors to carbonic anhydrase II by electrospray ionization-mass spectrometry. *J. Med. Chem.* **39**, 1949–1955.

Gehrig, P. M., You, C., Dallinger, R., Gruber, C., Brouwer, M., Kagi, J. H., and Hunziker, P. E. (2000). Electrospray ionization mass spectrometry of zinc, cadmium, and copper metallothioneins: Evidence for metal-binding cooperativity. *Protein Sci.* **9**, 395–402.

Greig, M., Gaus, H., Cummins, L. L., Sasmor, H., and Griffey, R. H. (1995). Measurement of macromolecular binding using electrospray mass spectrometry determination of dissociation constants for oligonucleotide: Serum albumin complexes. *J. Am. Chem. Soc.* **117**, 10765–10766.

Geoghegan, K. F., and Kelly, M. A. (2005). Biochemical applications of mass spectrometry in pharmaceutical drug discovery. *Mass Spectrom. Rev.* **24**, 347–366.

Heck, A. J., and Van Den Heuvel, R. H. (2004). Investigation of intact protein complexes by mass spectrometry. *Mass Spectrom. Rev.* **23**, 368–389.

Hill, T. J., Laffite, D., Wallace, J. I., Cooper, H. J., Tsvetkov, P. O., and Derrick, P. J. (2000). Calmodulin-peptide interactions: Apocalmodulin binding to the myosin light chain kinase target-site. *Biochemistry* **39**, 7284–7290.

Hu, P., Ye, O. Z., and Loo, J. A. (1994). Calcium stoichiometry determination for calcium binding proteins by electrospray ionization mass spectrometry. *Anal. Chem.* **66**, 4190–4194.

Hunter, C. L., Mauk, A. G., and Douglas, D. J. (1997). Dissociation of heme from myoglobin and cytochrome *b*5: Comparison of behavior in solution and the gas phase. *Biochemistry* **36**, 1018–1025.

Jaquinod, M., Potier, N., Klarskov, K., Reymann, J. M., Sorokine, O., Kieffer, S., Barth, P., Andriantomanga, V., Biellmann, J. F., and Van Dorsselaer, A. (1993). Sequence of pig lens aldose reductase and electrospray mass spectrometry of non-covalent and covalent complexes. *Eur. J.* **218**, 893–903.

Jorgensen, T. J. D., Roepstorff, P., and Heck, A. J. R. (1998). Direct determination of solution binding constants for noncovalent complexes between bacterial cell wall peptide analogues and vancomycin group antibiotics by electrospray ionization mass spectrometry. *Anal. Chem.* **70**, 4427–4432.

Katta, V., and Chait, B. T. (1991). Observation of the heme-globin complex in native myoglobin by electrospray ionization mass spectrometry. *J. Am. Chem. Soc.* **113**, 8534–8535.

Keetch, C. A., Hernandez, H., Sterling, A., Baumert, M., Allen, M. H., and Robinson, C. V. (2003). Use of a microchip device coupled with mass spectrometry for ligand screening of a multi-protein target. *Anal Chem.* **75**, 4937–4941.

Kempen, E. C., and Brodbelt, J. S. (2000). A method for the determination of binding constants by electrospray ionization mass spectrometry. *Anal. Chem.* **72,** 5411–5416.

Last, A. M., and Robinson, C. V. (1999). Protein folding and interactions revealed by mass spectrometry. *Curr. Opin. Chem. Biol.* **3,** 564–570.

Lehn, J. M. (1995). "Supramolecular Chemistry." VCH Verlagsgesellschaft MbH, Weinheim, Germany.

Leize, E., Jaffrezic, A., and Van Dorsselaer, A. (1996). Correlation between solvation energies and electrospray mass spectrometric response factors study by electrospray mass spectrometry of supramolecular complexes in thermodynamic equilibrium in solution. *J. Mass Spectrom.* **31,** 537–544.

Lemaire, D., Marie, G., Serani, L., and Laprevote, O. (2001). Stabilization of gas-phase noncovalent macromolecular complexes in electrospray mass spectrometry using aqueous triethylammonium bicarbonate buffer. *Anal. Chem.* **73,** 1699–1706.

Li, Y. T., Hiesh, Y. L., Henion, J. D., Senko, M. W., McLafferty, F. W., and Ganem, B. (1993). Mass spectrometric studies on noncovalent dimers of leucine zipper peptides. *J. Am. Chem. Soc.* **115,** 8409–8413.

Li, Y.-T., Hsieh, Y.-L., Henion, J. D., Ocain, T. D., Schiehser, G. A., and Ganem, B. (1994). Analysis of the energetics of gas-phase immunophilin–ligand complexes by ion spray mass spectrometry. *J. Am. Chem. Soc.* **116,** 7487–7493.

Lim, H.-K., Hsieh, Y. L., Ganem, B., and Henion, J. D. (1995). Recognition of cell-wall peptides ligands by vancomycin group antibiotics studies using ion spray mass spectrometry. *J. Mass Spectrom.* **30,** 708.

Loo, J. A. (1997). Studying noncovalent protein complexes by electrospray ionization mass spectrometry. *Mass Spectrom. Rev.* **16,** 1–23.

Loo, J. A., Hu, P., McConnell, P., Mueller, W. T., Sawyer, T. K., and Thanabal, V. (1997). A study of Src SH2 domain protein-phosphopeptide binding interactions by electrospray ionization mass spectrometry. *J. Am. Soc. Mass Spectrom.* **8,** 234–243.

Loo, J. A. (2000). Electrospray ionization mass spectrometry: A technology for studying noncovalent macromolecular complexes. *Int. J. Mass Spectrom.* **200,** 175–186.

McCammon, M. G., Scott, D. J., Keetch, C. A., Greene, L. H., Purkey, H. E., Petrassi, H. M., Kelly, J. W., and Robinson, C. V. (2002). Screening transthyretin amyloid fibril inhibitors: Characterization of novel multiprotein, multiligand complexes by mass spectrometry. *Structure* **10,** 851–863.

Meng, C. K., Mann, M., and Fenn, J. B. (1988). Of proteins or protons—A beams a beam for a that. *Z. Phys. D.* **10,** 31.

Pinkse, M. W., Heck., A. J., Rumpel, K., and Pullen, F. (2004). Probing noncovalent protein–ligand interactions of the cGMP-dependent protein kinase using electrospray ionization time of flight mass spectrometry. *J. Am. Soc. Mass Spectrom.* **15,** 1392–1399.

Pramanik, B. N., Bartner, P. L., Mirza, U. A., Liu, Y. H., and Ganguly, A. K. (1998). Electrospray ionization mass spectrometry for the study of non-covalent complexes: An emerging technology. *J. Mass Spectrom.* **33,** 911–920.

Potier, N., Barth, P., Tritsch, D., Biellmann, J. F., and Van, Dorsselaer.A. (1997). Study of non-covalent enzyme-inhibitor complexes of aldose reductase by electrospray mass spectrometry. *Eur. J. Biochem.* **243,** 274–282.

Potier, N., Donald, L. J., Chernushevich, I., Ayed, A., Ens, W., Arrowsmith, C. H., Standing, K. G., and Duckworth, H. W. (1998). Study of a noncovalent trp repressor: DNA operator complex by electrospray ionization time-of-flight mass spectrometry. *Protein Sci.* **7,** 1388–1395.

Potier, N., Billas, I. M., Steinmetz, A., Schaeffer, C., Van Dorsselaer, A., Moras, D., and Renaud, J. P. (2003). Using nondenaturing mass spectrometry to detect fortuitous ligands in orphan nuclear receptors. *Protein Sci.* **12,** 725–733.

Przybylski, M., and Glocker, M. O. (1996). Electrospray mass spectrometry of biomacromolecular complexes with noncovalent interactions—new analytical perspectives for supramolecular chemistry and molecular recognition processes. *Angew. Chem. Int. Ed. Engl.* **35,** 807–826.

Robinson, C. V., Chung, E. W., Kragelund, B.B, Knudsen, J., Aplin, R. T., Poulsen, F. M., and Dobson, C. M. (1996). Probing the nature of noncovalent interactions by mass spectrometry. A study of protein CoA ligand binding and assembly. *J. Am. Chem. Soc.* **118,** 8646–8653.

Rogniaux, H., Van Dorsselaer, A., Barth, P., Biellmann, J.-F., Barbanton, J., van Zandt, M., Chevrier, B., Howard, E., Mitschler, A., Potier, N., Urzhumtseva, L., Moras, D., and Podjarny, A. D. (1999). Binding of aldose reductase inhibitors: correlation of crystallographic and mass spectrometric studies. *J. Am. Soc. Mass Spectrom.* **10,** 635–647.

Rogniaux, H., Sanglier, S., Strupat, K., Azza, S., Roitel, O., Ball, V., Tritsch, D., Branlant, G., and Van Dorsselaer, A. (2001). Mass spectrometry as a novel approach to probe cooperativity in multimeric enzymatic systems. *Anal. Biochem.* **291,** 48–61.

Rostom, A. A., and Robinson, C. V. (1999). Disassembly of intact multiprotein complexes in the gas phase. *Curr. Opin. Struct. Biol.* **9,** 135–141.

Siegel, M. M. (2002). Early discovery drug screening using mass spectrometry. *Curr. Top. Med. Chem.* **2,** 13–33.

Smith, R. D., and Light-Wahl, K. (1993). The observation of noncovalent interactions in solution by electrospray ionization mass spectrometry: Promise, pitfalls and prognosis. *J. Biol. Mass Spectrom.* **22,** 493–501.

Smith, R. D., Bruce, J. E., Wu, Q., and Lei, P. (1997). New mass spectrometric methods for the study of noncovalent associations of biopolymers. *Chem. Soc. Rev.* **26,** 191.

Urzhumtsev, A., Tête-Favier, F., Mitschler, A., Barbanton, J., Barth, P., Urzhumtseva, L., Biellmann, J.-F., Podjarny, A. D., and Moras, D. (1997). A "specificity" pocket inferred from the crystal structures of the complexes of aldose reductase with the pharmaceutically important inhibitors tolrestat and sorbinil. *Structure* **5,** 601–612.

Van den Heuvel, R. H., and Heck, A. J. (2004). Native protein mass spectrometry: From intact oligomers to functional machineries. *Curr. Opin. Biol.* **8,** 519–526.

Veenstra, T. D., Johnson, K. L., Tomlinson, A. J., Craig, T. A., Kumar, R., and Naylor, S. (1998a). Zinc-induced conformational changes in the DNA-binding domain of the vitamin D receptor determined by electrospray ionization mass spectrometry. *J. Am. Soc. Mass Spectrom.* **9,** 8–14.

Veenstra, T. D., Benson, L. M., Craig, T. A., Tomlinson, A. J., Kumar, R., and Naylor., S. (1998b). Metal mediated sterol receptor-DNA complex association and dissociation determined by electrospray ionization mass spectrometry. *Nat. Biotechnol.* **16,** 262–266.

Wigger, M., Eyler, J. R., Benner, S. A., Li, W., and Marshall, A. G. (2002). Fourier transform-ion cyclotron resonance mass spectrometric resolution, identification, and screening of non-covalent complexes of Hck Src homology 2 domain receptor and ligands from a 324-member peptide combinatorial library. *J. Am. Soc. Mass Spectrom.* **13,** 1162–1169.

Wright, P. A., Rostom, A. A., Robinson, C. V., and Schofield, C. J. (2000). Mass spectrometry reveals elastase inhibitors from the reactive centre loop of alpha1-antitrypsin. *Bioorg. Med. Chem. Lett.* **10,** 1219–1221.

Wu, Q., Gao, J., Joseph-McCarthy, D., Sigal, G. B., Bruce, J. E., Whitesides, G. M., and Smith, R. D. (1997). Carbonic anhydrase-inhibitor binding: from solution to the gas phase. *J. Am. Chem. Soc.* **119,** 1157–1158.

Further Reading

Zhang, S., Van Pelt, C. K., and Wilson, D. B. (2003). Quantitative determination of noncovalent binding interactions using automated nanoelectrospray mass spectrometry. *Anal. Chem.* **75,** 3010–3018.
Zhang, S., Van Pelt, C. K., and Henion, J. D. (2003). Automated chip-based nanoelectrospray-mass spectrometry for rapid identification of proteins separated by two-dimensional gel electrophoresis. *Electrophoresis* **24,** 3620–3632.

[12] Site-Specific Hydrogen Exchange of Proteins: Insights into the Structures of Amyloidogenic Intermediates

By Zhong-ping Yao, Paula Tito, and Carol V. Robinson

Abstract

We describe the use of nano-electrospray ionization (nano-ESI) mass spectrometry (MS) for monitoring hydrogen exchange. Using this approach, we have compared the fluctuations in structure of the wild-type human lysozyme with those of the Asp67His and Ile56Thr variants, the two amyloidogenic forms of the protein. The results revealed that a significant region of the structure was transiently unfolded in both variants compared with the wild-type protein. Using peptic digestion, we located the region of the protein involved in the unfolding reaction to the β-domain and adjacent C-helix. This unfolding reaction is proposed to facilitate the initial stages of the fibril formation process. Also by this approach, we discovered that binding of an antibody fragment to the proteins prevents the unfolding events. These observations, therefore, not only highlight the use of MS to monitor and locate regions of enhanced hydrogen exchange kinetics, even in proteins that are prone to aggregation, but also demonstrate the use of such an approach to discover potential therapies.

Introduction

Studies on the structure and dynamics of proteins play an important role in understanding mechanisms of diseases, particularly the amyloidoses. These diseases are now known to involve partial unfolding of proteins that are then susceptible to aggregation and formation of amyloid fibrils and plaques. Deposition of these insoluble species in various organs and tissues

METHODS IN ENZYMOLOGY, VOL. 402
Copyright 2005, Elsevier Inc. All rights reserved.

0076-6879/05 $35.00
DOI: 10.1016/S0076-6879(05)02012-4

underlies a group of more than 20 human disorders including Alzheimer's disease and the transmissible spongiform encephalopathies (Selkoe, 2003). Hydrogen exchange has proven to be a powerful tool in probing protein folding and dynamics (Englander and Krishna, 2001; Krishna et al., 2004). Nuclear magnetic resonance (NMR) is established for the measurement of hydrogen exchange in proteins, but in the last decade, MS has emerged as a complementary approach (Eyles and Kaltashov, 2004; Hoofnagle et al., 2003; Lanman and Prevelige, 2004; Last and Robinson, 1999). Here, we show the application of hydrogen exchange methods in combination with MS to probe the partially folded states implicated in the conversion of normally soluble protein into amyloid fibrils.

The mechanism of hydrogen exchange of proteins involves reversible unfolding and isotope exchange at exposed regions (Hoofnagle et al., 2003; Hvidt and Nielsen, 1966). The process is represented by the following equation:

$$F \underset{k_c}{\overset{k_o}{\rightleftarrows}} U \overset{k_{ex}}{\rightarrow} U_{ex}$$

where F, U, and U_{ex} refer to folded, unfolded, and unfolded exchanged forms of the protein, and k_o, k_c, and k_{ex} represent the opening, closing, and intrinsic exchange rates, respectively. The observed exchange rate can be expressed as $k_{obs} = k_o \times k_{ex}/(k_{ex} + k_c)$. For proteins in their native states, the refolding rate is usually much faster than the intrinsic exchange rate (i.e., $k_c \gg k_{ex}$), and the observed exchange rate can, thus, be interpreted as $k_{obs} = k_o \times k_{ex}/k_c$. This process, known as the EX2 exchange mechanism, occurs when random fluctuations of the native state of the protein expose regions of the backbone for exchange with solvent. A broad single peak is observed in the mass spectrum of the exchanging protein, corresponding to an ensemble of partially exchanged protein molecules (Chung et al., 1997). If the closing rate is much slower than the exchange rate (i.e., $k_c \ll k_{ex}$), for example, under high pH conditions when the intrinsic rate of hydrogen exchange is fast (Bai et al., 1993), and when proteins are destabilized, the equation is simplified as $k_{obs} = k_o$. In this so-called EX1 exchange regime, any exposed region induced by unfolding will be completely exchanged. The resulting mass spectrum will, therefore, contain ostensibly two peaks, corresponding to the protected and exchanged states.

To locate regions of the protein that have undergone hydrogen exchange, we subject proteins to proteolysis before mass spectrometric analysis (Zhang and Smith, 1993). The key to success in this analysis is to maintain the deuterium contents of peptides until their release from solution into the gas phase. High-performance liquid chromatography (HPLC) separation followed by ESI analysis (Zhang and Smith, 1993) and matrix-assisted laser desorption/ionization (MALDI) (Mandell *et al.*, 1998) are two separation and analysis methods that have been effectively used in conjunction with hydrogen exchange MS. In both methods, significant efforts are made to minimize back-exchange that takes place during HPLC separation or matrix mixing and sample loading. In this chapter, we describe a simple method for monitoring hydrogen exchange in which the peptic digestion products are analyzed directly using nano-ESI MS without separation. Because no chromatographic separation or mixing with matrix is required, the method is rapid and minimum back-exchange is encountered. Moreover, because each peptide fragment is compared within the same spectrum, very subtle differences in hydrogen exchange, typically observed for partially folded forms of proteins, can be addressed.

We demonstrate this approach with the amyloidogenic forms of human lysozyme Asp67His and Ile56Thr, which are known to form amyloid fibrils *in vivo* (Booth *et al.*, 1997). Using hydrogen exchange, we show that both variants can populate partially unfolded forms in solution more readily than the wild-type protein. The region that is susceptible to unfolding was identified in the Asp67His variant (Canet *et al.*, 2002). We also investigate a therapeutic strategy involving an antibody fragment designed to combat these unfolding events through binding to the variant proteins. This antibody is a fragment of the "heavy chain" camel antibody with high specificity for native human lysozyme (Dumoulin *et al.*, 2002).

Methods

Protein Expression and Purification

The ^{15}N labeled wild-type, D67H and I56T variants of human lysozyme were expressed in *Aspergillus niger* and purified as described (Canet *et al.*, 1999; Spencer *et al.*, 1999). Guanidinium hydrochloride (GuHCl) and pepsin were obtained from Sigma. Deuterium oxide (D$_2$O, 99.9%) was from Fluorochem. Acetic acid-d$_4$ (DLM-12, 99.5%) was from Cambridge Isotope Laboratories. Deuterated GuHCl (GuDCl) was prepared by three cycles of dilution of GuHCl in D$_2$O followed by lyophilization.

Optimization of Peptic Digestion

Before hydrogen exchange is carried out, conditions for the proteolysis reaction can be determined with unlabeled proteins. These conditions typically include the ratio of pepsin to protein, pH, temperature, and incubation time. Digestions are usually carried out at pH 2.5–3.0 and

0° to reduce back-exchange because the intrinsic rate for hydrogen exchange is low under these conditions (Bai *et al.*, 1993). The enzyme ideally suited to coupling with hydrogen exchange is pepsin because it has been shown to cleave protein backbone at low pH levels and at low temperature (Zhang and Smith, 1993). Relatively high ratios of pepsin (typically 2:1 molar ratio of pepsin to protein) are used to generate the required amount of cleavage within the shortest time. Despite the fact that human lysozyme is highly constrained with four disulfide bridges and extensive helical structure, by using a high ratio of pepsin to protein and incubation on ice for only 1 min, a total of 19 peptides were formed (Fig. 1), giving rise to a sequence coverage of 100%. Autolysis of pepsin under these conditions was not observed. This combination of high enzyme concentration, low temperature, and short digestion time was found to yield extensive cleavage and keep back-exchange to a minimum.

Strategies for the Assignment of Peptic Fragments

The most time-consuming and difficult aspect of this work is the assignment of the digest fragments to the sequence of the protein. Although pepsin is effective under the regime required for hydrogen exchange, its nonspecific cleavage renders the assignment process extremely difficult. The number of possibilities for a given mass within an estimated error increases as a function of the length of peptides and the number of disulfide linkages. As human lysozyme contains four disulfide bonds, the number of possible assignments for a given mass is extremely large.

For fragments below 600 Da, these could be assigned solely on the basis of their mass. For those fragments between 600 and 3000 Da, the mass accuracy was not sufficiently high to delineate all possible combinations, and these peptides were analyzed using a tandem MS (MS/MS) approach. For peptides above mass 3000, many possibilities exist within 0.3 Da of the measured mass, and MS/MS would not normally provide adequate sequence information for disulfide-bonded peptides. These peptides were, therefore, isolated by HPLC and hydrolyzed with a specific protease, in this case trypsin, and the digestion products were analyzed by MS (Fig. 1A). We validated our assignments by carrying out digestion of ^{15}N-labeled human lysozyme under identical conditions to those of the variant proteins. An example of this approach is given in Fig. 1B. The nitrogen content calculated for the proposed sequence can be used to delineate various amino acid compositions (Canet *et al.*, 2002; Miranker *et al.*, 1996). A summary of the peptides observed reproducibly for human lysozyme under these experimental conditions is given in Fig. 1C.

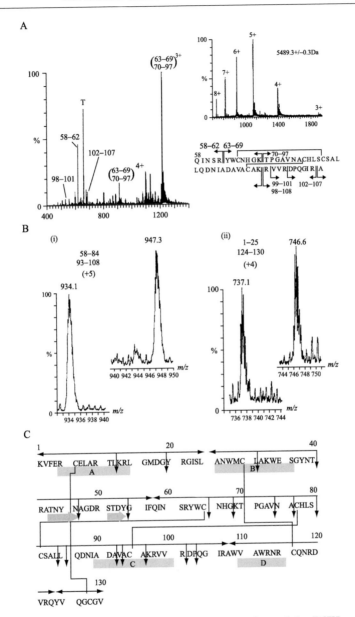

FIG. 1. Strategies for assigning the peptic digestion products of the D67H variant of human lysozyme. (A) Tryptic digestion of the 5489-Da peptic peptide followed by mass spectrometry (MS) analysis enabled us to assign it as residues (58–84)-(85–108). (B) The nitrogen content obtained from comparison of peptic peptides with digestion of ^{15}N-labeled protein under identical conditions confirms sequences of (58–84)-(93–108) and (1–25)-(124–130). (C) Summary of the cleavage sites formed by peptic digestion denoted with vertical arrows. Two β-sheets and four α-helixes A, B, C, and D are also indicated in the diagram.

Hydrogen Exchange

The deuterated ^{15}N wild-type lysozyme and the two deuterated ^{14}N mutational variants D67H and I56T were prepared by unfolding in GuDCl followed by refolding in deuterated buffer. The deuterated ^{15}N wild type was mixed with either deuterated [14]N D67H or I56T in equal proportions for the pulse-labeling hydrogen exchange experiments. The pulse-labeling experiments were carried out on a Bio-Logic QFM5 mixer connected to a circulating water bath set to 37°. The protein solution was diluted 1:15 with 100 mM ammonia/formic acid buffer in H_2O at pH 8.2. The exchange was allowed to proceed for various times between 0.4 and 600 s before acetic acid (7 volume equivalents, final pH 3.5) was added to quench exchange and samples stored on ice before analysis by MS. For sequence-specific analysis, the D67H variant protein was pulse-labeled manually at pH 10.0 and 37° for 3 s so that an equal proportion of exchanged and unexchanged peaks can be detected in the mass spectra. The sample was then subjected to peptic digestion and nano-ESI analysis. For analysis in the presence of the camellid antibody fragment (cAb-HuL6), D67H or I56T, and an equimolar amount of protonated cAb-HuL6 was mixed immediately before the experiment and exchange was carried out at pH 8.0 (100 mM ammonia/formic acid in H_2O) and 37° for various lengths of time and then quenched by decreasing the temperature and pH as described earlier.

Mass Spectrometry

Mass spectra of intact proteins and the antibody complex were recorded on an LCT mass spectrometer, and peptic peptides and hydrogen exchange measurements were carried out on a Q-TOF mass spectrometer (Micromass, UK). Samples were introduced into the spectrometers via gold-coated nano-electrospray needles prepared in the laboratory. The instrument was calibrated using cesium iodide, and MS/MS was carried out using argon as the collision gas. The instrumental settings are standard settings for analysis of proteins and peptides by nano-ESI. Typically, the first six scans were accumulated and used for data analysis. The samples of the antibody lysozyme complexes, unless stated otherwise, were electrosprayed at the base pressure of the Platform mass spectrometer with a cone voltage of 150 V such that the complex was dissociated in the gas phase to yield free lysozyme variant for comparison with the unbound protein.

Simulations of the isotope patterns were made for peptide fragments by using isotope-modeling in the Masslynx software (Micromass, UK). For EX1 exchange, two populations were calculated. One corresponds to the natural abundance isotopes and the other to the natural abundance isotope plus the number of deuterons calculated from the average mass difference between the two peaks. Addition of the two isotope distributions was then carried out using Sigmaplot version 4.01 (SPSS). For EX2 exchange, only one isotope distribution was calculated, which corresponded to the isotopic abundance with the average number of deuterons calculated from the m/z value of the peak.

Results

The Two Mutational Variants Populate Partially Folded States

Spectra obtained for a solution of deuterated wild-type and D67H human lysozyme proteins mixed in equimolar amounts, exchanged simultaneously, and quenched after various times are shown in Fig. 2. By using ^{15}N-labeled wild-type protein, whose molecular weight is distinctly larger than that of the D67H variant, the two proteins are sufficiently distinct to allow complete separation in the mass spectrum. Thus, any small differences that could be encountered through slight changes in the solution and mass spectrometric conditions are avoided. The wild-type protein is visible as a single peak whose mass decreases with the exchange time. Similar behavior is observed for the wild type at all the pH values studied (Canet et al., 2002). The results demonstrate that hydrogen exchange of wild-type human lysozyme obeys an EX2 mechanism, in which the refolding rate is much higher than the intrinsic exchange rate.

The data for the D67H variant at pH 8.2, in contrast to those of the wild-type protein, show a clear bimodal distribution of masses as the exchange process takes place (Canet et al., 2002; Dumoulin et al., 2005) (Fig. 2). This pattern is typical of an EX1 type of hydrogen exchange mechanism, in which the closing rate of a fluctuation is much lower than the intrinsic exchange rate. Similar behavior was observed for the D67H variant at pH 10.0. At the lower pH values of 5.0 and 6.1, however, only a single broad peak can be observed, indicating that the intrinsic exchange rate at these lower pH values is lower than the closing rate, and the exchange obeys EX2 kinetics (data not shown). EX1 hydrogen exchange is usually associated with destabilization of the protein structure. In the present case, the mass difference between the exchanged and unexchanged peaks is too small to correspond to complete global unfolding. The clear bimodal distribution observed at pH values between 8.2 and 10.0 thus indicates a cooperative unfolding of a significant region of the protein.

To determine whether this structural property was common to both variants, we investigated the I56T mutation under identical solution and MS conditions as those used to study the D67H variant. The results show that despite the fact that the two mutations are in different regions of the

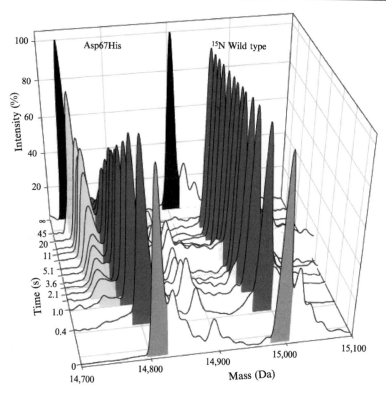

FIG. 2. Nano-electrospray mass spectra recorded for an approximately equimolar mixture of ^{15}N-labeled wild-type lysozyme and the D67H variant following exposure to hydrogen exchange conditions at pH 8.2 and 37°. The spectra represent mass-transformed data from the +9 charge state. The peaks colored light gray were observed in spectra of the D67H variant but not in the wild-type protein. The peaks colored dark gray arise from the gradual loss of deuterium during the course of the exchange reaction. The peaks in the spectra of control samples recorded for time zero and time infinity are colored gray and black, respectively.

molecule, the partially unfolded forms from both variant proteins possess the same characteristics (Dumoulin *et al.*, 2005).

Localization of the Unfolding Domain in the D67H Variant

To locate the regions of the structure of the D67H variant involved in the cooperative unfolding processes, we exposed a sample of the protein to hydrogen exchange conditions to generate two populations of approximately equal intensity. This sample was then subjected to peptic digestion and mass spectrometric analysis. Comparison with ^{15}N-labeled protein was also carried out to verify the peptide assignment (Fig. 1B). The initial digestion products correspond to cleavage at Leu-25, Phe-57, Ala-108, and Gln-123. These cleavage sites are remote from the interchain disulfides and are not in regions containing secondary structure. These three distinct regions of the protein yield 100% sequence coverage (Fig. 1C). Interestingly, of the regions containing helical and β-sheet structure, only the C-helix shows extensive proteolytic cleavage at residues 91, 92, and 96, suggesting that under these conditions, the structure in helixes A, B, D, and the β-sheet region are less susceptible to peptic digestion than that in helix C.

The spectra for the predominant peptic peptides after exchange are shown in Fig. 3. For peptides where two populations are clearly visible, the experimental isotopic distributions have been compared with simulations on the basis of an EX1 exchange mechanism and the measured number of protected sites. The bimodal isotope distribution corresponding to peptide 58–108, which includes the C helix (residues 89–100) as well as residues 58–64 from the β-domain, closely matches a simulation in which 10 labile hydrogens from this region are involved in the EX1 unfolding event. However, the isotope distribution in the shorter fragment from this region (58–84)-(93–108) (peptides 58–84 and 93–108 linked by a disulfide bond) is best described by a simulation for an EX1 exchange process involving eight labile hydrogens. Thus, removal of the residues 85–92, which includes four residues from the C helix, involves a reduction of two in the number of protected hydrogens involved in the EX1 unfolding event. Previous hydrogen exchange studies of human lysozyme have shown that only residues 85 and 92 are highly protected in this region of the sequence (Redfield and Dobson, 1990). The present results of the MS analysis, therefore, indicate that these two residues, located in the C helix and the loop preceding it, are involved in the cooperative unfolding transition detected by hydrogen exchange. The peptide fragment (26–57)-(109–123), which includes the residues forming the B and D helixes (26–34 and 109–115, respectively) of the α-domain and residues 41–57 of the β-domain, also shows a distinct bimodal distribution, corresponding to two hydrogen exchange populations involving 10 backbone amides (data not shown). The isotope distribution in an additional peptide (26–40)-(109–123), which differs by the absence of residues 41–57 in the β-domain, is also bimodal and correlates well with a simulation involving five backbone amide hydrogens. For a yet shorter peptide fragment derived from this region, (26–31)-(109–123), in which residues 32–40 are removed, the isotope distribution shows no evidence of EX1 exchange (Fig. 3). The distribution is well described by a single hydrogen exchange population having an average of four protected hydrogens that exchange via an EX2 mechanism. Thus, comparison of the exchange properties of a series of peptides generated from the sequence (26–57)-(109–123) indicates that the amide hydrogens undergoing EX1 exchange

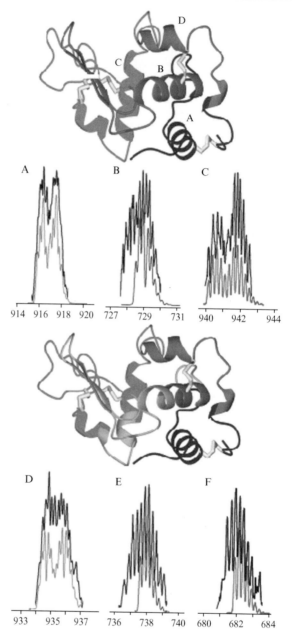

FIG. 3. Nano-electrospray mass spectra of the partially exchanged peptide fragments of the D67H variant. The multiply charged peptides are assigned as follows: (A) $(58-108)^{+6}$; (B) $([1-25]-[124-130])^{+4}$; (C) $([26-40]-[109-123])^{+5}$; (D) $([58-84]-[93-108])^{+5}$; (E) $([26-31]-[109-123])^{+4}$; and

in this region are located within the sequence 31–57, corresponding to the four C-terminal residues of the B helix and encompassing the antiparallel β-sheet of the β-domain. In contrast, the peptide fragments containing the sequence 26–40, which includes the residues forming helix B, are consistent with a single hydrogen exchange population with an average of eight protected amides. Overall, the MS analysis of these peptide fragments indicates that the regions involved in the cooperative unfolding process are all located within the sequence 31–104 (Canet *et al.*, 2002).

Binding to an Antibody Fragment Inhibits Partial Unfolding of Lysozyme Variants

To investigate the effects of binding a soluble antibody fragment to the lysozyme variants, the 1:1 complex was formed before hydrogen exchange was initiated. The mass of the antibody fragment: D67H complex measured under non-dissociating conditions confirms the 1:1 binding stoichiometry, but given the broadness of the peaks, it is not possible to observe subtle differences in hydrogen exchange within this complex (Fig. 4A). We, therefore, dissociated the complex in the gas phase and measured the hydrogen exchange protection of the liberated lysozyme (Fig. 4B). In the absence of the antibody, and under the conditions used for this experiment, two equal populations of the folded and partially folded species are observed for both variants after approximately 15 s of exchange. By contrast, in the presence of the antibody fragment, the unfolding events that occur in both variants are effectively inhibited throughout the time scale of the hydrogen exchange reaction (Dumoulin *et al.*, 2003, 2005).

Conclusions

By using hydrogen exchange in combination with analysis by MS, we have been able to compare the fluctuations in structure of the D67H and I56T variant proteins of human lysozyme with those of the wild-type

---

(F) ($[1–19]$-$[124–130])^{+4}$. Before digestion, the protein was exposed to hydrogen exchange at pH 10 for 3 s. The location of the principal peptic digestion products is shown within the native structure of lysozyme in the upper panel. Similarly, the smaller peptide fragments, derived from three regions (D), (E), and (F), are indicated in the lysozyme structure in the lower panel. The lysozyme structure was generated from the X-ray structure (1REX) and produced using MOLMOL. The peaks in the spectra are compared with simulations of the isotope patterns color-coded for each peptide fragment. The peptide fragments (A), (C), and (D) fit well to an exchange model in which the total number of protected sites undergo exchange exclusively via an EX1 mechanism to give two distinct hydrogen exchange populations. The peptide fragments (B), (E), and (F) show a single hydrogen exchange population that undergoes exchange via an EX2 mechanism. (Reprinted, with permission, from Canet *et al.* [2002].) (See color insert.)

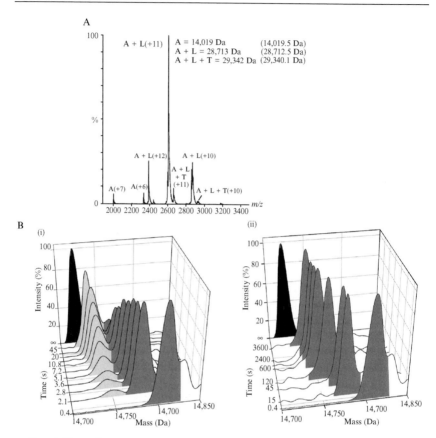

Fig. 4. Nano-electrospray mass spectra of the D67H lysozyme: cAb-HuL6 complex and hydrogen exchange of D67H lysozyme in the absence (i) and presence (ii) of an equimolar amount of the antibody fragment. (A) The spectrum of the 1:1 complex formed from addition of equimolar antibody and D67H lysozyme recorded under conditions designed to maintain noncovalent interactions confirms the 1:1 stoichiometry. *A*, antibody; A + L, antibody + lysozyme; A + L + T, antibody + lysozyme + substrate (tri-nag) complex. Theoretical masses are indicated in parentheses. (B) Time course hydrogen exchange experiments reveal that binding to the antibody fragment inhibits partial unfolding of D67H lysozyme. The peaks colored black represent the control sample that has undergone complete exchange. The peaks colored gray arise from the gradual loss of deuterium throughout the time course of the experiment. Peaks colored light gray are observed in the spectra of the D67H variant only in the absence of the antibody fragment. (Reprinted, with permission, from Dumoulin *et al.* [2003]).

protein. The results reveal that a substantial portion of the structure is transiently unfolded in both variants compared with the wild-type protein. Using proteolysis and several approaches to assign the resultant disulfide-bonded peptides, we assigned the regions that were unfolding transiently to the β-domain residues and the adjacent C helix. Moreover, we used this observation to discover the mode of action of an antibody fragment that arrests fibril formation *in vitro*. Binding the antibody fragment to both variants was found to prevent the initial unfolding events. More generally, the results demonstrate that the powerful approach of monitoring hydrogen exchange by using MS can be applied even to unfavorable proteins, such as those that are prone to aggregation, heavily disulfide bonded, and present in a binary complex.

References

Bai, Y., Milne, J. S., Mayne, L., and Englander, S. W. (1993). Primary structure effects on peptide group hydrogen exchange. *Proteins Struc. Func. Genet.* **17,** 75–86.

Booth, D., Sunde, M., Bellotti, V., Robinson, C. V., Hutchinson, W. L., Fraser, P. E., Hawkins, P. N., Dobson, C. M., Radford, S. E., Blake, C. C. F., and Pepys, M. B. (1997). Instability, unfolding and aggregation of human lysozyme variants underlying amyloid fibrillogenesis. *Nature* **385,** 787–793.

Canet, D., Last, A. M., Tito, P., Sunde, M., Spencer, A., Archer, D. B., Redfield, C., Robinson, C. V., and Dobson, C. M. (2002). Local cooperativity in the unfolding of an amyloidogenic variant of human lysozyme. *Nat. Struct. Biol.* **9,** 308–315.

Canet, D., Sunde, M., Last, A. M., Miranker, A., Spencer, A., Robinson, C. V., and Dobson, C. M. (1999). Mechanistic studies of the folding of human lysozyme and the origin of amyloidogenic behavior in its disease related variants. *Biochemistry* **38,** 6419–6427.

Chung, E. W., Nettleton, E. J., Morgan, C. J., Grob, M., Miranker, A., Radford, S. E., Dobson, C. M., and Robinson, C. V. (1997). Hydrogen exchange properties of proteins in native and denatured states monitored by mass spectrometry and NMR. *Protein Sci.* **6,** 1316–1324.

Dumoulin, M., Canet, D., Last, A. M., Pardon, E., Archer, D. B., Muyldermans, S., Wyns, L., Matagne, A., Robinson, C. V., Redfield, C., and Dobson, C. M. (2005). Reduced global cooperativity is a common feature underlying the amyloidogenicity of pathogenic lysozyme mutations. *J. Mol. Biol.* **346,** 773–788.

Dumoulin, M., Conrath, K., Van Meirhaeghe, A., Meersman, F., Heremans, K., Frenken, L. G., Muyldermans, S., Wyns, L., and Matagne, A. (2002). Single-domain antibody fragments with high conformational stability. *Protein Sci.* **11,** 500–515.

Dumoulin, M., Last, A. M., Desmyter, A., Decanniere, K., Canet, D., Larsson, G., Spencer, A., Archer, D. B., Sasse, J., Muyldermans, S., Wyns, L., Redfield, C., Matagne, A., Robinson, C. V., and Dobson, C. M. (2003). A cameloid antibody fragment inhibits the formation of amyloid fibrils by human lysozyme. *Nature* **424,** 783–788.

Englander, S. W., and Krishna, M. M. (2001). Hydrogen exchange. *Nat. Struct. Biol.* **8,** 741–742.

Eyles, S. J., and Kaltashov, I. A. (2004). Methods to study protein dynamics and folding by mass spectrometry. *Methods* **34,** 88–99.

Hoofnagle, A. N., Resing, K. A., and Ahn, N. G. (2003). Protein analysis by hydrogen exchange mass spectrometry. *Annu. Rev. Biophys. Biomol. Struct.* **32,** 1–25.

Hvidt, A., and Nielsen, S. O. (1966). Hydrogen exchange in proteins. *Adv. Protein Chem.* **21,** 287–386.

Krishna, M. M., Hoang, L., Lin, Y., and Englander, S. W. (2004). Hydrogen exchange methods to study protein folding. *Methods* **34,** 51–64.

Lanman, J., and Prevelige, P. E., Jr. (2004). High-sensitivity mass spectrometry for imaging subunit interactions: Hydrogen/deuterium exchange. *Curr. Opin. Struct. Biol.* **14,** 181–188.

Last, A. M., and Robinson, C. V. (1999). Protein folding and interactions revealed by mass spectrometry. *Curr. Opin. Chem. Biol.* **3,** 564–570.

Mandell, J. G., Falick, A. M., and Komives, E. A. (1998). Measurement of amide hydrogen exchange by MALDI-TOF mass spectrometry. *Anal. Chem.* **70,** 3987–3995.

Miranker, A., Kruppa, G. H., Robinson, C. V., Aplin, R. T., and Dobson, C. M. (1996). Isotope-labeling strategy for the assignment of protein fragments generated for mass spectrometry. *J. Am. Chem. Soc.* **118,** 7402–7403.

Redfield, C., and Dobson, C. M. (1990). ^{1}H NMR studies of human lysozyme: Spectral assignment and comparison with hen lysozyme. *Biochemistry* **29,** 7201–7214.

Selkoe, D. J. (2003). Folding proteins in fatal ways. *Nature* **426,** 900–904.

Spencer, A., Morozov-Roche, L. A., Noppe, W., MacKenzie, D. A., Jeenes, D. J., Joniau, M., Dobson, C. M., and Archer, D. B. (1999). Expression, purification, and characterization of the recombinant calcium-binding equine lysozyme secreted by the filamentous fungus *Aspergillus niger:* Comparisons with the production of hen and human lysozymes. *Protein Expr. Purif.* **16,** 171–180.

Zhang, Z., and Smith, D. L. (1993). Determination of amide hydrogen exchange by mass spectrometry: A new tool for protein structure elucidation. *Protein Sci.* **2,** 522–531.

[13] Quantitating Isotopic Molecular Labels with Accelerator Mass Spectrometry

By John S. Vogel and Adam H. Love

Abstract

Accelerator mass spectrometry (AMS) traces isotopically labeled biochemicals and provides significant new directions for understanding molecular kinetics and dynamics in biological systems. AMS traces low-abundance radioisotopes for high specificity but detects them with MS for high sensitivity. AMS reduces radiation exposure doses to levels safe for use in human volunteers of all ages. Total radiation exposures are equivalent to those obtained in very short airplane flights, a commonly accepted radiation risk. Waste products seldom reach the Nuclear Regulatory Commission (NRC) definition of radioactive waste material for ^{14}C and [3]H. Attomoles of labeled compounds are quantified in milligram-sized samples, such as 20 μl of blood. AMS is available from several facilities that offer services and new spectrometers that are affordable. Detailed examples of designing AMS studies are provided, and the methods of analyzing AMS data are outlined.

METHODS IN ENZYMOLOGY, VOL. 402   0076-6879/05 $35.00
                                                   DOI: 10.1016/S0076-6879(05)02013-6

Introduction

Isotopic labels remain the standard for tracing chemically exact analogues of biochemicals, whether they are drugs, toxins, nutrients, or macromolecules, through biological systems. Isotopes are one or more equivalent atomic forms of a chemical element that contain that element's signature number of nuclear protons and atomic electrons, but that may have varying numbers of nuclear neutrons that give each isotope of an element a distinctive mass. The number of neutrons in an isotope nucleus may render it unstable against several forms of radioactive decay. These radioisotopes may have lifetimes of billions of years or fractions of a second. Some elements form stable nuclei from only one configuration of protons and neutrons and have a single isotopic mass, such as aluminum (atomic weight = 27 g/mol) or indium (113 g/mol). The next element beyond indium, tin, has 10 stable isotopic forms from mass 112 g/mol with a natural terrestrial abundance of 0.97% to mass 124 g/mol at a natural abundance of 5.79%. Most elements have a dominant atomic mass, such as carbon-12 (^{12}C) (98.89% abundance), with one or more low abundance stable isotopes like carbon-13 (^{13}C) (1.11%). Hydrogen is 99.985% mass 1 and 0.015% mass 2 (deuterium). Both radioisotopes and stable isotopes are used as labels that give a molecule a distinctive molecular signature in an isolated biological component. The altered molecular mass generally has no effect on the molecule's chemical interactions, although an entire field of enzyme kinetic research does depend on effects resulting from the careful placement of an isotope in a particularly reactive molecular location (Cleland, 2003, and references therein).

The ultimate sensitivity for quantifying an isotopic label depends on three factors: the rarity of the isotope, the precision and sensitivity of the isotopic detection technique, and the number of isotope labels in the traced compound. The number of isotope labels in a molecule has linear significance at most; the number may not exceed the natural amount of the element in the compound without changing its chemical properties. Isotope abundance and detection provide a more complex space from which to choose an experimental design. The more rare an isotope, the more distinctive it will be as a molecular label, assuming equivalent efficiency and precision of detection. The rarest stable isotope, ^{3}He, has a natural occurrence of 1.4 ppm but is useless as a label because of its noble chemistry. The decay of radioisotopes ensures that they have very low natural abundance, unless there are continuous production mechanisms, such as the creation of ^{14}C in the upper atmosphere. The natural concentration of ^{14}C in the living biosphere remains approximately a part per trillion (ppt) ($1:10^{12}$), compared to the part-per-100 abundance of ^{13}C. Doubling the concentration of both isotopes in 1 mg of carbon requires 900 nmol of ^{13}C label or 98 amol of ^{14}C. Equivalent detection of the two isotopes gives a 10 billion increase in sensitivity for compounds labeled with ^{14}C rather than ^{13}C. Routine laboratory measurements of these isotopes are not equivalent, however, and fail to exploit the inherent sensitivity of low abundance radioisotope labels. Isotope quantitation falls into two broad technologies: detection of decay products from radioactive isotopes or isotope ratio MS (IRMS) for stable isotopes. AMS, a type of high-energy IRMS, now quantifies certain long-lived radioisotopes for biochemical research at efficiencies similar to those for stable isotopes but with much higher sensitivity.

AMS arose in the late 1970s from two distinct research threads with a common goal: an improvement in radiocarbon dating that would make efficient use of datable material and that would extend the routine and maximum reach of radiocarbon dating. Attempts to use IRMS for ^{14}C were hampered by difficult isobar backgrounds (Aitken, 1978), but nuclear physicists were able to use negatively charged ions and high-energy collisions to remove these interferences (Bennett et al., 1977; Nelson et al., 1977). AMS is routinely used in geochronology and archaeology, but biological applications began appearing in 1990 (Turteltaub et al., 1990) and were available only from a few laboratories. AMS is now more accessible for biochemical quantitation from several industrial concerns, at a National Institutes of Health (NIH) Research Resource at Lawrence Livermore National Laboratory, or through the development of smaller affordable spectrometers. This chapter provides a tutorial for biochemical tracing of ^{14}C- and ^{3}H-labeled compounds using AMS.

Methods

Decay Counting Efficiency

The contrast between radioisotope quantitation by decay counting and by MS is indicated by the definition of "activity," A, in which the decays per time (e.g., dpm) are related to the total number of that isotope present through the mean isotopic life (τ = half-life divided by the natural log of 2) in Eq. (1). The mean life of ^{14}C is 8270 yr (4.35×10^9 min), and that of tritium is 17.8 yr (9.36×10^6 min).

$$A = \frac{\partial N}{\partial t} = \frac{N}{t_{1/2}/\ln(2)} = \frac{N}{\tau} \tag{1}$$

The maximum efficiency (ε) of detecting a fraction of the ^{14}C depends directly on the length of time used to detect the decays, no matter which decay counting method is used:

$$\partial t = \frac{\partial N}{N} \times \tau = \epsilon \times \tau \qquad (2)$$

Thus, 1% of the ^{14}C in a sample can be counted in, at best, 83 yr. A sample decaying at 1 dpm (0.45 picocurie [pCi], 17 millibecquerel [mBq]) requires a week of counting to provide 10,000 counts for 1% statistical precision, momentarily ignoring the usual instrumental background that prevents quantitating 1 dpm with most decay counters. The sample contains 7.2 fmol of ^{14}C:

$$N = \frac{\partial N}{\partial t} \times \tau = \frac{1}{min} \times 4.35 \times 10^9 \ min \times \frac{mol}{6.02 \times 10^{23}} = 7.22 \ fmol \qquad (3)$$

AMS has isotope-counting efficiencies greater than 1% in short measurements of a few minutes (Vogel et al., 1989). New spectrometers count natural ^{14}C (ppt) at 300 counts/s or more (Brown et al., 2000). The above 7.2 fmol ^{14}C in 1 mg of carbon is 87 ppt, which provides 10,000 ^{14}C counts in less than 1s by AMS counting. Sensitivity is only slightly less striking for shorter lived isotopes, such as tritium. One dpm of tritium arises from 15.5 amol (10^{-18} mol) of the isotope. In 1 μl of water, this amount creates an isotope ratio of 0.14 ppt ($^3H/H$), from which AMS counts 10,000 counts in 1000s (17 min).

Replacing decay counting by MS yields a sensitivity gain of 1000 to 1 billion for radioisotope quantitation. This gain can be used to reduce ^{14}C and ^3H use below operational definitions of radioactive, isolate much smaller and more specific biochemical fractions for measurement, trace compounds at physiological rather than pharmacological levels in host animals or in all human populations (including children), obtain higher sample throughput for studies requiring many samples, or some combination of all of these goals.

Accelerator Mass Spectrometry

AMS produces high detection efficiency using techniques outlined in Fig. 1, which is drawn to diagram ^{14}C analysis but applies to a number of isotopes without fundamental changes. Atoms from about 1 mg of carbon are ionized with an added electron and accelerated to a moderate energy before a mass analysis at 14 atomic mass units (amu). About 1% of the ion beam from the source passes this analysis, although the true ^{14}C represents only 1 ppt, emphasizing the early difficulty in detecting ^{14}C with simple mass spectrometers. The vast majority of these ions are molecular hydrides of the lighter carbon isotopes: $^{12}CH_2$, ^{12}CD, and ^{13}CH. Lithium enters the

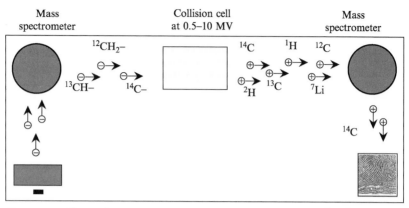

Mass spectrometer   Collision cell at 0.5–10 MV   Mass spectrometer

$^{12}CH_2-$ ^{14}C ^{1}H ^{12}C

$^{13}CH-$ $^{14}C-$ ^{2}H ^{13}C ^{7}Li

^{14}C

Ion source with sample   Ion detector

FIG. 1. Schematic of the basic ideas behind accelerator mass spectrometry (AMS): Tandem MS (MS/MS) with a charge-changing collision cell at million-volt potentials between the spectrometer elements. Each ion is "fingerprinted" by how it loses energy in the final detector.

prepared samples from the glass reaction vessels used in reducing all carbon samples to a graphitic form (Loyd *et al.*, 1991) and forms a mass 14 molecule, $^{7}Li_2$. Nitrogen, which is 99.63% the mass 14 isotope, does not form a negative ion and does not appear in the initial ion beam. This discovery was one of the fundamental steps in making AMS possible for ^{14}C detection (Middleton, 1974). As in tandem collision MS, molecules are broken by collisions in a thin foil or gas cell. Macromolecules dissociate at kiloelectron-volt (keV) energies, but an acceleration through several million volts is necessary to destroy hydrides in a single collision. Multiple collisions at sub-megavolt potentials have been used to destroy these molecules, ushering in a series of smaller and more affordable spectrometers (Hughey *et al.*, 1997; Suter *et al.*, 1997). Molecular dissociation produces positive ions that accelerate away from the high positive potential back to ground potential. Multiple magnetic and electrostatic elements follow the accelerations to select the desired high-energy ^{14}C ions from the molecular debris composed of H, D, ^{7}Li, ^{12}C, and ^{13}C. This selection is effective to parts per billion or better and reduces the filtered ion stream to hundreds of particles per second. Scattered background ions are rare in the high-energy spectrometer because ion energies are too high for atomic scattering and the nuclear scattering probabilities are very low. Transport through the high-energy spectrometer ends at a detector that uniquely identifies the ion by quantifying how it loses energy.

The specificity of AMS lies in the two processes that require high energies provided by inclusion of an accelerator in the tandem mass spectrometer: destruction of molecular isobars in energetic collisions and identification of each high-energy ion through quantifying the rate of energy loss. These two processes are present in AMS detection of all isotopes and are straightforward for ^{3}H and ^{14}C, because their nuclear isobars (isotopes of other elements having the same mass as the counted ion) do not form negative ions. Other elements often require "tricks" of chemistry or physics to reduce the counting rate of the detector to acceptable levels from an interfering isobar. For example, ^{41}Ca is initially accelerated as ^{41}CaF$_3$ because ^{41}KF$_3$ is not stable, and a thick foil is placed before the final detector in quantifying ^{10}Be because the higher atomic number of ^{10}B forces it to lose energy faster than ^{10}Be, which passes through.

Table I lists long-lived isotopes of elements that are routinely measured at the LLNL Center for AMS. Most AMS facilities have ^{14}C capabilities, and many quantify other isotopes. Chemical isolation of the chosen element is often a limiting technology in applying AMS to a given isotope. Suppression of nuclear isobars in the chemical preparation is fundamental to the use of AMS for many elements, because high count rates in the ion detector degrade its ability to identify individual ions. Methods for detection of the isotopes in Table I are published in the AMS literature with the last AMS conference proceedings as a suitable resource (Nakamura et al., 2004).

TABLE I
LONG-LIVED RADIOISOTOPES MEASURED AT CAMS/LLNL

Element	Isotope	Half-life (yr)	Mean life (yr)	Mean life (min)
Hydrogen	^{3}H	12.33	17.79	9.36e6
Beryllium	^{7}Be	0.15	0.22	1.14e5
	^{10}Be	1.51 M	2.18 M	1.15e12
Carbon	^{14}C	5730	8270	4.35e9
Aluminum	^{26}Al	740 k	1.07 M	5.61e11
Chlorine	^{36}Cl	301 k	434 k	2.28e11
Calcium	^{41}Ca	103 k	149 k	7.82e10
Nickel	^{59}Ni	108 k	156 k	8.20e10
	^{63}Ni	100.1	144	7.60e7
Technetium	^{99}Tc	210 k	303 k	1.59e11
Iodine	^{129}I	15.7 M	22.6 M	1.19e13
Uranium	Various	Various	Various	Various
Plutonium	Various	Various	Various	Various

Units

Radiocarbon occupies a special place among isotopes, because it is produced in the atmosphere and is incorporated in all living systems, whether plant, animal, microbe, etc. The "distance" of a living creature along the food chain from the atmospheric source is reflected in the concentration of ^{14}C that it contains. Fresh deepsea fish or soil microbes, for example, have lower ^{14}C levels than a new blade of grass. The popular application of ^{14}C is for radiocarbon dating that depends on the steady radioactive decay of the isotope within the carbon of a previously living creature or plant. Atmospheric ^{14}C is variable over many centuries, because of production influences of the solar cycle and the earth's magnetic field. Testing nuclear weapons in the 1950s and 1960s almost doubled the amount of ^{14}C in the atmosphere, but that excess has been absorbed into the oceans with an atmospheric mean lifetime of 15 years. The atmosphere is now only a few percentages above its "natural" level. The ^{14}C dating community and many AMS facilities report ^{14}C concentration in a unit that reflects this expected natural concentration of atmospheric ^{14}C: Modern. The unit is convenient in biochemical AMS to give an immediate feel for the level of labeled compound in a living host. It is also a convenient way to report ^{14}C concentrations that can be robustly converted to one of several other units desired for analysis. "Modern" represents a ^{14}C concentration, an amount of ^{14}C within a specific amount of carbon. Contrast this with a measure of decay products (e.g., dpm, microcurie) that represents an amount of ^{14}C that gives rise to the radioactivity of a sample independently of the sample size, as defined in Eq. (1). Table II lists several equivalent numerical values of Modern, with 13.56 dpm/g carbon being the fundamental definition (Stuiver and Polach, 1977). AMS samples often contain about 1 mg of total carbon, so some units are expressed per-milligram of carbon.

TABLE II
EQUIVALENT CONCENTRATIONS OF ^{14}C AND ^{3}H IN VARIOUS UNITS

^{14}C Value	Unit	^{3}H Value	Unit
1.0	Modern (Mod)	1.0	Tritium unit (TU)
1.176×10^{-12}	atom/atom C	1.0×10^{-18}	atom/atom H
1.176	pmol/mol C	1	amol/mol H
13.56	dpm/g C	64.38	μ dpm/mg H
226.0	μ Bq/mg C	1.073	μ Bq/mg H
6.108	fCi/mg C	29.00	aCi/mg H
97.89	amol/mg C	0.992	zmol/mg H

Tritium naturally arises from radioactive decay products in the earth and from atmospheric sources similar to ^{14}C, but at much lower concentrations. The unit of natural ^{3}H is the "tritium unit" (TU), which represents one ^{3}H atom in 10^{18} hydrogen atoms, or approximately 3.2 pCi/L of water. Various sources produce a background in our present environment of about 10 TU.

Range, Resolution, and Sample Size

AMS was developed for radiocarbon dating, which quantified the loss of ^{14}C in a sample compared to the natural level of 98 amol/mg carbon. AMS quantifies ^{14}C concentrations in materials as old as 50,000 yr (0.2% Mod) with precisions as high as 0.1%. The amount of ^{14}C in 1mg of 50,000-year-old material is 0.23 amol. A 0.1% precision on measurement of 1 mg, 1 Mod sample also quantifies ±0.1 amol. For several decades, earth scientists have routinely quantified atto- and even zepto- (10^{-21}) moles of ^{14}C in milligram-sized samples. The chemical methods for handling large numbers of biological samples are less rigorous than those for dating applications, raising the useful sensitivity limit of biological AMS to 1% Mod, or about 1 amol of isotope label in a milligram-sized sample. Samples can be as small as 100 μg, however, leading to quantitations in zmol of label. Sensitivity for tritium extends to similar concentrations at less than 100 TU in 200-μg samples for tracing limits less than 1 amol.

The upper limit of isotope concentration in AMS comes from the counting ability of the identifying detector. There are several kinds of detectors, and the electronic systems connected to them can correct for missed counts ("live-time" corrections) to some degree, but rates greater than 100,000 cps lead to larger uncertainties in quantitation (Vogel *et al.*, 2004). Typical counting rates for Modern material are 100 cps or more, and samples up to 1000 Mod are easily measured. Experiments should be designed to reliably provide samples below 100 Mod for ^{14}C, or 100 mega-TU for tritium. The ion source can become contaminated at concentrations a factor of 100 above these, and a time-consuming cleaning process is required to recover sensitivity for lower levels after such a sample.

The range of readily measured samples, thus, spans 5 orders of magnitude, with the highest precision available in the middle 2 orders, with reproducibility of 1–3% for ^{14}C and 2–5% for ^{3}H, as measured by repeated isotope ratio measurements each having less than 1% counting statistics. Uncertainties in quantitation expand to 5–10% for the high decade (≥100 Mod) and for the lowest decade (1–10% Mod). Natural biological hosts do contain Modern carbon, and the ability to quantify below this level may appear spurious. The extreme sensitivity of AMS makes extensive sample definition through highly selective fractionation possible. Chemical processing for AMS measurement needs more material than found in the fractions, and a well-defined mass of "carrier" compound is added to these isolated fractions that may contain only 1 ng or μg of natural carbon after solvents are removed. This carrier is chosen to have a low, but non-zero, ^{14}C or ^{3}H concentration. HPLC identification of metabolites has been done this way (Buchholz et al., 1999). Many separation media, such as acrylamide gel, are made from ^{14}C- and ^{3}H-free petroleum stock, in which isolated proteins that bind labeled compounds are easily distinguished at submicrogram amounts using the separation media directly as the carrier material (Vogel et al., 2001).

Measurement precision is highest at Modern ^{14}C concentrations, and tissue samples only a few percentages above natural levels can be confidently quantified using self-controls that are collected from a biological host just before dosing with the labeled compound. Resolution can be as little as 5% above the presumably Modern control, providing quantitation as low as 1 fmol/g of tissue (tissue $\approx$20% carbon):

$$(105 - 100)\% \times \text{Modern} \Rightarrow 0.05 \times 98\frac{\text{amol}}{\text{mgC}} \times 0.2\frac{\text{mgC}}{\text{mg}} \approx 1\frac{\text{amol}}{\text{mg}} = 1\frac{\text{fmol}}{\text{g}}$$

$$(4)$$

Even restricting intake to keep samples at less than 10 Mod results in a 200:1 dynamic range above minimal resolution:

$$10 \text{ Modern} \Rightarrow 10 \times 98\frac{\text{amol}}{\text{mgC}} \times 0.2\frac{\text{mgC}}{\text{mg}} \approx 200\frac{\text{amol}}{\text{mg}} = 200\frac{\text{fmol}}{\text{g}} \qquad (5)$$

Tissue concentration of the label is directly derived from an isotope ratio knowing only the carbon content of the measured tissue. The size of the individual sample does not enter the calculation, reducing the number of manipulations and increasing the precision of the determination.

Conversion of carbon to the solid fullerene (Vogel, 1992) or hydrogen to the titanium hydride (Chiarappa-Zucca et al., 2002) forms preferred in AMS quantitation work best if approximately 1 mg of the element is present. The carbon and hydrogen contents of some biological samples are listed in Table III, along with the amount of that sample that is useful for AMS sampling. Capillary collection using finger pricks for humans or tail punctures for rodents is possible for the occasional blood sample. These small blood volumes are also readily obtained during postmortem animal sampling without the use of skilled cardiac punctures. Indwelling catheters are useful when a kinetic study requires many closely spaced collections from humans or large animals (Dueker et al., 2000). The small AMS samples are a benefit to invasive collections such as blood or cells but

TABLE III

CARBON AND HYDROGEN CONTENTS OF VARIOUS BIOLOGICAL MATERIALS AND THE
APPROXIMATE AMOUNT OF EACH REQUIRED IN MAKING AN AMS SAMPLE

	%C	Sample	%H (wet)	Sample	%H (dry)	Sample	Collection
Plasma	4	25 μl	11	3 μl	5	3 mg	Capillary, syringe, catheter
RBC	17	5 μl	11	4 μl	8	3 mg	Capillary, syringe, catheter
Blood	10	10 μl	11	3 μl	6	3 mg	Capillary, syringe, catheter
Urine	0.3–1.5	200 μl	11	3 μl	2	6 mg	Cup
Feces	10	10 mg	7	4 mg	5	6 mg	Bag
Cells	22	5 mg	10	3 mg	6	5 mg	Various, e.g., buccal scrape
Organ	15	7 mg	10	3 mg	7	4 mg	Biopsy, dissection
Fat	64	1.5 mg	12	3 mg	12	3 mg	Biopsy, dissection
Protein	28	4 mg			5	6 mg	Biopsy, purification
DNA	29	4 mg			4	6 mg	Biopsy, purification
Nerve	13	8 mg	11	3 mg	9	3 mg	Biopsy, dissection
Muscle	11	8 mg	10	3 mg	6	5 mg	Biopsy, dissection
Breath	1	80 ml	1	60 ml			Tube, absorbent

complicate the use of the copious excreta samples that must be homogeneously sampled. Careful homogenization of urine and feces maintains relevance in the isotope ratio. Postmortem tissue or needle biopsy samples are 5–10 mg, because most tissue is 10–20% carbon. A few test samples can be weighed to provide a "feel" for the proper sample size, but a known sample size is not needed for quantitation. The elimination of sample weighing reduces the incidence of accidental contamination. Contamination control is of utmost importance when the instrument sensitivity reaches attomoles (Buchholz et al., 2000).

Tissue samples can be smaller than those outlined here, perhaps down to 10% of that listed, but special handling is required in preparing them. Even smaller samples are augmented with carrier compounds to obtain the desired process size. Two cases are most easily analyzed: either the carbon in the sample must arise from the natural tissue, or the added carrier carbon should dominate (>95%) the sample. Samples containing 10–50 μg of carbon or hydrogen fall between these limits and are problematic for solid sample AMS. Ion sources that accept combustion products directly are used for this mass range and allow more direct coupling of separation instruments (Liberman et al., 2004). A new technique of measuring sample masses between 0.1 and 100 μg permits quantitative isotope dilution by addition of a known amount of carrier carbon to the measured mass of the sample before combustion of the sample for AMS measurement (Grant *et al.*, 2003).

Dose Estimation

AMS quantitates most accurately for isotope ratios near 10^{-12}. Experiments designed for AMS quantitation need to keep samples within the range of high precision while introducing sufficient isotope label to have the needed dynamic range for the biological process being studied. In this section, we provide examples of dose estimations based on the data and constraints in the previous section. The calculations will use the data from the "reference human" listed in Table IV to allow for normalizations to other cases.

Decay counting experiments are designed to avoid a loss of signal at the low end. If there is an excess of isotope label, the high-end samples are just counted faster. AMS turns this around, because it has high precision for distinguishing signals near the natural background concentrations, but the instrument can be harmed by carelessly high isotope concentrations. Maximum concentration arises in plasma shortly after the absorption of a bioavailable or injected dose of a hydrophilic compound that remains isolated in the plasma. Such isolation in plasma is unlikely, and corrections are considered below. A labeled dose adding 10 Mod to the naturally Modern plasma leaves room for error and individual variation without approaching the upper limit of 100 Mod. The total dose for a 70-kg person is found from the 1mg of carbon available from 25 μl of the 3 liter plasma,

TABLE IV
VALUES FOR PHYSICAL PROPERTIES OF THE REFERENCE
HUMAN USED IN CALCULATING DOSES

Weight	70 kg
Body water	42 L
Intracellular water	24 L
Blood	5 L
Plasma	3 L
Soft tissue	45 kg
Fats	12 kg
Muscle	28 kg
Daily urine	2 L
Daily feces	1 kg

and Modern is expressed in any of the desired units (curie, becquerel, mole):

$$10 \text{ Mod} \Rightarrow 10 \times 6.1\frac{\text{fCi}}{\text{mgC}} \times \frac{1 \text{ mgC}}{25 \,\mu\text{l}} \times 31 = 7.3 \text{ nCi} \tag{6}$$

A total ^{14}C dose of about 10nCi is optimal if a chemical enters the plasma through complete gastrointestinal tract absorption or through intravenous injection and is expected to clear from plasma without extensive redistribution. Many compounds are labeled with ^{14}C at 10 Ci/mol (16.0% of the molecules containing a ^{14}C; pure ^{14}C = 62.4 Ci/mol, 2.3 teraBq/mol), yielding this ^{14}C dose from only 45 nmol, or 9 μg of a 200g/mol compound in the 70-kg human. Even physiological concentrations are greater than this 130 ng/kg, leading to the possibility of diluting labeled compounds to parts per hundred or more with unlabeled equivalents to make up the desired chemical dose at this radioactive content.

A similar calculation for tritium in plasma is given in Eq. (7), where the tritium concentration is found for a sample that yields 1 cps by AMS:

$$0.01 \text{ megaTU} \Rightarrow 10^4 \times \frac{0.99 \text{ zmol}}{\text{mgH}} \times \frac{1 \text{ mgH}}{9.5 \,\mu\text{l}} \times 31 = 3 \text{ nmol} \tag{7}$$

This represents 90 μCi of ^{3}H (pure ^{3}H = 29.1 kCi/mol, 1077 teraBq/mol), 3 μmol of a compound that is labeled at 1 in 1000, or 60 μg of a 200 g/mol compound for the 70-kg subject. A 0.01megaTU sample decays at 0.6 dpm, but AMS provides a dynamic range 3 orders of magnitude down from this concentration, and a 3% counting precision is available in 20 min rather than 30 h/sample.

The retention of a hydrophilic compound fully in the 3 liter of plasma is unrealistic, because such a compound is at least likely to expand into the interstitial water for a total distribution of 18 liters, increasing the estimated dose of ^{14}C to about 45 nCi and of ^{3}H to 560 μCi. The maximum distribution is limited to the full 42 liters of all body water, requiring a factor of 14 more isotope to obtain the same plasma concentrations as found in Eqs. (6) and (7). The human ^{14}C dose would then become 102 nCi, a dose that has been used in successfully tracing the hydrophilic folate vitamin within humans (Clifford et al., 1998; Lin et al., 2004). It is instructive to calculate the natural ^{14}C in our 70-kg reference human assuming a total body carbon abundance of 23%:

$$1 \text{ Mod} \Rightarrow 1 \times 6.1\frac{\text{fCi}}{\text{mgC}} \times \frac{23 \text{ mgC}}{100 \text{ mg}} \times 70 \text{ kg} = 98.2 \text{ nCi} \tag{8}$$

The ^{14}C dose for tracing the hypothetical hydrophile only doubles the ^{14}C in humans over the period that the compound requires for metabolism and elimination. These dose estimates contain sufficient leeway for variations in bioavailability and volume of distribution over factors of 5 or more. If bioavailability is expected to be less than 20%, the initial test dose should be increased. Label doses also increase for hydrophobic compounds that partition into the lipids and are less available in easily collected fractions. Beta-carotene was labeled at 200 nCi of ^{14}C per subject for studies that traced it and its derivative vitamin A for more than 6 months from a single dose (Dueker *et al.*, 2000; Hickenbottom *et al.*, 2002; Lemke *et al.*, 2003).

Minimum dose calculations ensure quantitation for extended kinetic studies. For example, minimum ^{14}C for detection at least 10% above natural levels in urine is calculated in Eq. (9) for 1 wk after a dermal exposure to a compound with 5% absorption whose biological mean life is 1 day. The urine is assumed to sample the 24 liters of interstitial body water, and the ^{14}C is found in approximately 1.5 mg carbon obtained in 200 μl:

$$(1.1 - 1) \text{ Mod} \Rightarrow 0.1 \times 6.1 \frac{\text{fCi}}{\text{mgC}} \times \frac{1.5 \text{ mgC}}{200 \text{ }\mu\text{l}} \times 24 \text{l} = 110 \text{ pCi} \qquad (9)$$

Because the urine is collected 7 mean lives after the exposure, the body water needs to be a factor of 1097 ($=e^7$) higher on the exposure day, giving a total interstitial water load of 120 nCi shortly after exposure. This absorbed amount arises from 2.4 μCi of ^{14}C applied dermally with a 5% chemical absorption ($=120$ nCi/0.05). This compares well with the actual human doses of [^{14}C]-atrazine whose metabolites were identifiable by HPLC-AMS 1 wk after dermal exposures (Buchholz *et al.*, 1999). Tritium does not have to overcome a large natural background, but a minimum concentration of about 100 TU will provide a readily quantifiable count rate in AMS:

$$100 \text{ TU} \Rightarrow 100 \times 29 \frac{\text{aCi}}{\text{mgH}} \times \frac{1 \text{ mgH}}{10 \text{ }\mu\text{l}} \times 24 \text{ l} = 7 \text{ nCi} \qquad (10)$$

The same factors apply for a 1-wk loss through excretion (7.7 μCi) and a 5% bioavailability to yield a minimum ^{3}H dermal dose of 150 μCi. Extrapolation of these simple examples of kinetics calculations to animals from human examples are usually done with a simple assumption of body mass scaling.

Biopsies or dissections provide tissues for quantitating molecular interactions with macromolecules such as DNA or proteins (Kautiainen *et al.*, 1997; Vogel *et al.*, 2002). Adjustments to the simple dose calculations must be made to allow for chemical partitioning into the cells or for binding probabilities. It is difficult to predict cellular uptake and interaction equilibria for chemicals at low doses within animals, and ranging experiments beginning with low doses are the best way to determine a final dose for an AMS experiment. Mauthe et al. (1999) compared tissue-available doses and DNA adduction of a labeled compound in humans, rats, and mice. All received approximately 60-pCi/g doses: 4.2 μCi total ^{14}C for the humans, 15 nCi for rats, and 1.5 nCi for mice. Colon tissue was sampled 4 h later by surgery of the human volunteers and through dissection of the rodents. The tissue was 8–30 Mod above natural ^{14}C, and the extracted purified DNA was 0.4–4.0 Mod above natural levels. DNA samples were small and augmented by carrier carbon before measurement. ^{3}H- and ^{14}C-labeled compounds were quantified in rat liver after doses containing 1.4 μCi ^{3}H and 50 pCi of ^{14}C, but quantitation of the same compounds bound to liver proteins required doses that were 10 times higher (Dingley et al., 1998b). Binding of a labeled compound to proteins available from simple blood samples, albumin and hemoglobin, was compared in humans and rats given 4.2 μCi and 0.10–120.00 μCi, respectively. ^{14}C concentrations in the isolated hemoglobin doubled at low doses and ranged to 20 Mod at high dose. Albumin, being more accessible to circulating chemicals, was measured at 1–300 Mod above natural ^{14}C (Dingley et al., 1998a). A starting dose for quantifying macromolecular interactions within living mammals is 50 pCi/g for ^{14}C and 5 nCi/g for ^{3}H. Similar uptake calculations and assumptions are needed in estimating exposures for cell cultures.

Data Interpretation

AMS measures an isotope ratio, expressed as ^{14}C per mass of carbon or ^{3}H per H. No other information is available. The isotope ratio is interpreted by knowing the sources of carbon or hydrogen isolated in the quantified sample. This is expressed mathematically in Eq. (11):

$$R_{meas} = \left(^{14}C \middle/ C\right)_{meas} = \frac{^{14}C_{trace} + ^{14}C_{natural} + ^{14}C_{carrier} + ^{14}C_{other}}{C_{trace} + C_{natural} + C_{carrier} + C_{other}} \quad (11)$$

Experimental design provides the data to unfold this equation into the desired quantitation. The $^{14}C_{other}$ represents unexpected isotope contamination that is assumed negligible until proven otherwise. C_{other} is also expected to be zero. C_{trace} is the mass of carbon due to the labeled compound within the isolated sample and is usually negligible, because 1 pmol or less of the compound is quantified in milligram-sized samples. These assumptions leave us with two easily calculated cases. The first comes from isolated samples that contain enough carbon for immediate chemical processing without addition of any carrier material:

$$R_{meas} = \frac{{}^{14}C_{trace} + {}^{14}C_{natural}}{C_{natural}} = \frac{{}^{14}C_{trace}}{C_{natural}} + \frac{{}^{14}C_{natural}}{C_{natural}} = \frac{{}^{14}C_{trace}}{C_{natural}} + R_{natural} \quad (12)$$

The tissue concentration of the labeled compound (grams of compound per gram of tissue) is quantified from the tissue sample, a control sample of similar tissue ($R_{natural}$), the carbon content of the tissue (C_t), the compound's label concentration (L), and its molecular weight (W):

$$Conc(g/g) = \frac{{}^{14}C_{trace}}{C_{natural}} \times C_t \times W/L = (R_{meas} - R_{natural}) \times C_t \times W/L \quad (13)$$

For example, the tissue dose of a 200-g compound, labeled at 10 Ci/mol, in liver ($C_t = 14.4\%$), measured as 1.6 Mod is

$$Conc = (1.6 - 1) \times 6.1 \frac{fCi}{mgC} \times \frac{\frac{14.4 \; mgC}{100 \; mg} \times 200 \frac{fg}{fmol}}{10 \frac{fCi}{fmol}} = 10.5 \frac{fg}{mg} = 10.5 \frac{pg}{g}$$

$$(14)$$

The mass of the measured sample is not required to calculate this tissue concentration, but an elemental analyzer is useful in checking the carbon and hydrogen content of specific animal tissues, because Table III is only adequate for dose estimations.

A second reduction of the isotope ratio is calculated for samples that contain very little "natural" carbon to which carrier compound has been added, such as HPLC eluent fractions. Equation (11) then reduces to Eq. (15):

$$R_{meas} = \frac{{}^{14}C_{trace} + {}^{14}C_{carrier}}{C_{carrier}} = \frac{{}^{14}C_{trace}}{C_{carrier}} + \frac{{}^{14}C_{carrier}}{C_{carrier}} = \frac{{}^{14}C_{trace}}{C_{carrier}} + R_{carrier} \quad (15)$$

The amount of labeled compound in the isolated fraction is found from the known amount of carrier added, the measured isotope ratio for control carrier aliquots, and the compound's label concentration:

$$Comp_{fraction} = {}^{14}C_{trace}/L = C_{carrier} \times (R_{meas} - R_{carrier})/L \quad (16)$$

Figure 2 shows HPLC aliquots around an amino acid peak from an AMS-Edman sequencing of a ^{14}C-labeled protein (Miyashita et al., 2001). Carrier carbon was added to each aliquot as 50 μl of a methanol solution at 40 mg/ml. Evaporation of the methanol left 1.19 mg of carbon from the 59.5% carbon material, tributyrin. Multiple aliquots of the carrier averaged to 8.3 ± 0.4% Mod, whereas HPLC eluents on either side of the glycine peak were 7.2% Mod. Loss of ^{14}C is unexpected, but the result could

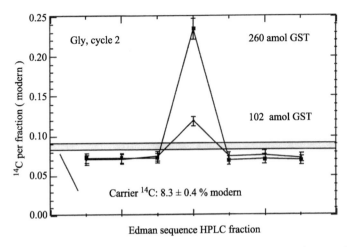

Fig. 2. [14]C content in high-performance liquid chromatography (HPLC) eluent fractions around a single amino acid peak. Three aliquots of carrier compound added to the fractions averaged 8.3 Mod, but the signal-free fractions averaged 7.2 Mod. Unexpected carbon from acetate buffer lowered the isotope ratio.

arise from about 100 μg of unaccounted ^{14}C-free carbon ("C_{other}") in each fraction. A review of the protocol found nonvolatile acetate in the HPLC solvent. Correcting for this oversight, the second Edman cycle of glutathione sulfur transferase produced 5.9 ± 1.1 amol of ^{14}C-glycine from 102 amol of protein and 20.4 ± 1.5 amol from 260 amol of protein with 1.29 mg of the 7.2% Mod effective carrier.

Radiation Exposure and Wastes

The radiation dose produced by an ingested radioisotope depends on the type and energy of the emitted decay product, electrons in the case of ^{3}H and ^{14}C, at average energies of 6.2 k and 52 k eV, respectively. The standard international unit of radiation dose equivalent is the sievert (Sv), which equals a joule of energy deposited in a kilogram of material by energetic electrons. Many people are still familiar with the radiation equivalent unit, the rem, which is 0.01 sieverts. An electronvolt is an amount of energy equal to the charge of the electron (1.6 × 10^{-19} coulomb [C]) brought across a 1 V (joule/coulomb) potential, so that the average decay energy deposited by ^{3}H is 992 aJoule and by ^{14}C is 8.32 fJoule. Because we all contain ^{14}C at about 1 Mod, we are constantly exposed to radiation energy from ^{14}C decays, d, within us:

$$1 \text{ Mod} \times 13.6 \frac{\text{dpm}}{\text{grC}} \times \frac{23 \text{ grC}}{100 \text{ gr}} \times 70 \text{ kg} \times 8.3 \frac{\text{fJoule}}{\text{d}} \times 60 \frac{\text{min}}{\text{hr}} = 110 \frac{\text{nJoule}}{\text{hr}}$$

(17)

This natural ^{14}C is spread throughout our bodies, so the energy deposition is converted to radiation dose equivalent by dividing by the body mass, 70 kg, to get 1.6 nSv/h or 160 nrem/h. We also contain other natural radioisotopes that produce greater radiation exposures than these from ^{14}C.

Isotope-labeled compounds do not provide a constant radiation exposure, because they are metabolized and excreted. Exact exposure calculation requires a full kinetic profile, but estimations are made using a model that incorporates instantaneous absorption followed by an exponential elimination characterized by a single time constant, the biological mean life, equal to biological half-life divided by ln(2). This exponential loss must be integrated over the expected life of the subject after dosing, perhaps 50 yr. However, the biological mean lives of drugs (<24 h) and nutrients (<60 days) are small compared to 50 yr, which is an effectively infinite time in comparison. Integrating the exposure to infinity allows a simple closed formula that overestimates the exposure by a small amount:

$$\text{Exposure} = \frac{E_d}{M_d} \times \text{Dose} \times \int_0^\infty e^{-t/\tau_{\text{bio}}} dt = \frac{E_d}{M_d} \times \tau_{\text{bio}} \times \text{Dose}$$

(18)

E_d is the energy deposited per decay in joules; M_d is the affected mass in kilograms; dose is the amount of radioactivity in dpm or becquerels (dps); and τ_{bio} is the biological mean life of the labeled compound in minutes or seconds. The absorbed dose decreases with radioactive decay and with chemical elimination, but the correction is negligible as long as the radioactive life of the isotope is much longer than the biological life of the compound, as is true in these cases. For cases of similar biological and radioactive mean lives, τ_{bio} in Eq. (18) is replaced by $\tau_{\text{bio}}\tau_{\text{rad}}/(\tau_{\text{bio}}+\tau_{\text{rad}})$, where τ_{rad} is the radioactive mean life of the isotope. The examples of Eqs. (9) and (10) assume a 1-day (1440 min) mean life of the compound with a total absorbed radioactive dose of 210 nCi of ^{14}C and 7.7 μCi of ^{3}H that would produce whole-body exposures of

$$\text{Exposure}(^{14}\text{C}) = \frac{8.3 \text{ fJ}}{70 \text{ kg}} \times 1440 \text{ min} \times 210 \text{ nCi} \times 2200 \frac{\text{dpm}}{\text{nCi}} = 79 \frac{\text{nJ}}{\text{kg}} = 79 \text{ nSv}$$

(19)

$$\text{Exposure}(^3\text{H}) = \frac{0.99 \text{ fJ}}{70 \text{ kg}} \times 1440 \text{ min} \times 7700 \text{ nCi} \times 2200\frac{\text{dpm}}{\text{nCi}} = 345\frac{\text{nJ}}{\text{kg}} = 345 \text{ nSv}$$

(20)

These exposures assume that the compounds spread evenly throughout the body. If these compounds concentrated in the 1.8-kg liver, the exposures are 3.1 and 13 μSv to that organ, replacing the 70-kg affected mass with the liver mass.

Human experimental subjects are not asked to accept risks to their health that are significantly greater than commonly accepted risks in everyday life, unless their health benefits. The lifetime exposures calculated in Eqs. 17 and 18 are equivalent to 2 and 9 days of the natural ^{14}C radiation and are acceptable risks to many. A common exposure to radiation occurs in aircraft at usual cruising altitudes, where the radiation dose is 5 μSv/hr (Friedberg et al., 2000). This radiation is primarily long-range energetic protons and muons that produce the same exposure to all tissues, compared to the electrons emitted from ^{3}H and ^{14}C, which stop within a few millimeters of their emittance. The liver and brain receive 5 μSv/hr in a plane, as does the leg or arm. Thus, even the exposures calculated for our example compound concentrated in the liver are lower than those obtained in 1–4 h of flying, a commonly accepted radiation risk, even for young children and pregnant women. Figure 3 shows the whole-body radiation dose from 100 nCi of ^{14}C and 1 μCi of ^{3}H in a 70-kg person as a function of biological mean life of the compounds incorporating the isotopes as labels. The left axis provides the exposure in nanosieverts, and the right axis converts the exposure to equivalent time in a cruising aircraft. Most drugs and toxins have short biological lives of less than a day, but nutrients can be recycled for many months.

Absorbed isotope doses to humans in AMS experiments range from 0.0014 nCi/g (100 nCi/70 kg) of ^{14}C to 0.14 nCi/g (10 μCi/70 kg) of tritium, with animal doses scaling proportionately to mass. Nuclear regulations allow disposal of wastes from biological experiments if the ^{14}C or ^{3}H content is less than 50 nCi/g, up to a total of 1 μCi/yr (U.S. Code of Federal Regulations, Title 10, Section 20.2005, 1991). Many experimenters can dispose of all their wastes from AMS experiments as nonradioactive. Others will reduce their radioactive waste streams by several orders of magnitude using AMS quantitation, for a significant savings in operation costs.

Conclusion

AMS increases the applicability of radioisotope labels for definitive molecular tracing in all biological systems, including humans. Chemicals are labeled with low specific activity, reducing synthesis costs and wastes.

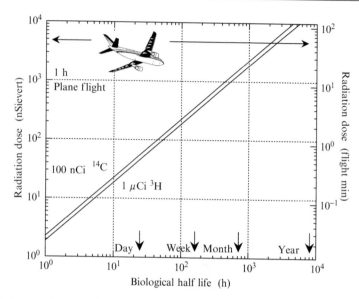

Fig. 3. Absorbed radiation-dose equivalents in a 70-kg human dosed with 100 nCi ^{14}C and 1 μCi ^{3}H are given as functions of the biological half-life of the compound containing the isotopic label. The radiation is expressed as nanosieverts and as equivalent minutes in an aircraft at typical flying altitude.

Laboratory safety is enhanced through low radioactivity usage. AMS provides high precision and throughput despite the greatly decreased isotope concentrations. Radiation exposures are below commonly accepted risks, even for children for whom many drug and nutrient studies can now be designed. AMS is available through an NIH Research Resource and from established academic and industrial facilities worldwide. A new class of smaller spectrometers is now available for institutions that can benefit from dedicated instruments.

References

Aitken, M. J. (1978). Archeological involvements of physics. *Phys. Rep. Rev. Section Phys. Lett.* **40**, 278–351.

Bennett, C. L., Beukens, R. P., Clover, M. R., Gove, H. E., Liebert, R. B., Litherland, A. E., Purser, K. H., and Sondheim, W. E. (1977). Radiocarbon dating using electrostatic accelerators: Negative-ions provide key. *Science* **198**, 508–510.

Brown, T. A., Roberts, M. L., and Southon, J. R. (2000). Ion-source modeling and improved performance of the CAMS high-intensity Cs-sputter ion source. *Nuclear Instr. Methods Phys. Res. Sect. B-Beam Interactions Materials Atoms* **172**, 344–349.

Buchholz, B. A., Freeman, S., Haack, K. W., and Vogel, J. S. (2000). Tips and traps in the C-14 bio-AMS preparation laboratory. *Nuclear Instr. Methods Phys. Res. Sect. B-Beam Interactions Materials Atoms* **172**, 404–408.

Buchholz, B. A., Fultz, E., Haack, K. W., Vogel, J. S., Gilman, S. D., Gee, S. J., Hammock, B. D., Hui, X. Y., Wester, R. C., and Maibach, H. I. (1999). HPLC-accelerator MS measurement of atrazine metabolites in human urine after dermal exposure. *Anal. Chem.* **71**, 3519–3525.

Chiarappa-Zucca, M. L., Dingley, K. H., Roberts, M. L., Velsko, C. A., and Love, A. H. (2002). Sample preparation for quantitation of tritium by accelerator mass spectrometry. *Anal. Chem.* **74**, 6285–6290.

Cleland, W. W. (2003). The use of isotope effects to determine enzyme mechanisms. *J. Biol. Chem.* **278**, 51975–51984.

Clifford, A. J., Arjomand, A., Dueker, S. R., Schneider, P. D., Buchholz, B. A., and Vogel, J. S. (1998). The dynamics of folic acid metabolism in an adult given a small tracer dose of C-14-folic acid. *In* "Mathematical Modelling in Experimental Nutrition," (A. J. Clifford and H. G. Müller, eds.), Vol. 445, pp. 239–251. Plenum Press, New York.

Dingley, K. H., Freeman, S., Nelson, D. O., Garner, R. C., and Turteltaub, K. W. (1998a). Covalent binding of 2-amino-3,8-dimethylimidazo 4,5-f quinoxaline to albumin and hemoglobin at environmentally relevant doses—comparison of human subjects and F344 rats. *Drug Metab. Dispos.* **26**, 825–828.

Dingley, K. H., Roberts, M. L., Velsko, C. A., and Turteltaub, K. W. (1998b). Attomole detection of ^{3}H in biological samples using accelerator mass spectrometry: Application in low-dose, dual-isotope tracer studies in conjunction with C-14 accelerator mass spectrometry. *Chem. Res. Toxicol.* **11**, 1217–1222.

Dueker, S. R., Lin, Y. M., Buchholz, B. A., Schneider, P. D., Lame, M. W., Segall, H. J., Vogel, J. S., and Clifford, A. J. (2000). Long-term kinetic study of beta-carotene, using accelerator mass spectrometry in an adult volunteer. *J. Lipid Res.* **41**, 1790–1800.

Friedberg, W., Copeland, K., Duke, F. E., O'Brien, K., and Darden, E. B. (2000). Radiation exposure during air travel: Guidance provided by the Federal Aviation Administration for air carrier crews. *Health Phys.* **79**, 591–595.

Grant, P. G., Palmblad, M., Murov, S., Hillegonds, D. J., Ueda, D. L., Vogel, J. S., and Bench, G. (2003). Alpha-particle energy loss measurement of microgram depositions of biomolecules. *Anal. Chem.* **75**, 4519–4524.

Hickenbottom, S. J., Lemke, S. L., Dueker, S. R., Lin, Y. M., Follett, J. R., Carkeet, C., Buchholz, B. A., Vogel, J. S., and Clifford, A. J. (2002). Dual isotope test for assessing beta-carotene cleavage to vitamin A in humans. *Eur. J. Nutr.* **41**, 141–147.

Hughey, B. J., Klinkowstein, R. E., Shefer, R. E., Skipper, P. L., Tannenbaum, S. R., and Wishnok, J. S. (1997). Design of a compact 1 MV AMS system for biomedical research. *Nuclear Instr. Methods Phys. Res. Sect. B-Beam Interactions Materials Atoms* **123**, 153–158.

Kautiainen, A., Vogel, J. S., and Turteltaub, K. W. (1997). Dose-dependent binding of trichloroethylene to hepatic DNA and protein at low doses in mice. *Chemico-Biol. Inter.* **106**, 109–121.

Lemke, S. L., Dueker, S. R., Follett, J. R., Lin, Y. M., Carkeet, C., Buchholz, B. A., Vogel, J. S., and Clifford, A. J. (2003). Absorption and retinol equivalence of beta-carotene in humans is influenced by dietary vitamin A intake. *J. Lipid Res.* **44**, 1591–1600.

Liberman, R. G., Tannenbaum, S. R., Hughey, B. J., Shefer, R. E., Klinkowstein, R. E., Prakash, C., Harriman, S. P., and Skipper, P. L. (2004). An interface for direct analysis of C-14 in nonvolatile samples by accelerator mass spectrometry. *Anal. Chem.* **76**, 328–334.

Lin, Y. M., Dueker, S. R., Follett, J. R., Fadel, J. G., Arjomand, A., Schneider, P. D., Miller, J. W., Green, R., Buchholz, B. A., Vogel, J. S., Phair, R. D., and Clifford, A. J. (2004). Quantitation of *in vivo* human folate metabolism. *Am. J. Clin. Nutr.* **80**, 680–691.

Loyd, D. L., Vogel, J. S., and Trumbore, S. (1991). Lithium contamination in AMS measurements of radiocarbon. *Radiocarbon* **33**, 297–301.

Mauthe, R. J., Dingley, K. H., Leveson, S. H., Freeman, S., Turesky, R. J., Garner, R. C., and Turteltaub, K. W. (1999). Comparison of DNA-adduct and tissue-available dose levels of MeIQx in human and rodent colon following administration of a very low dose. *Int. J. Cancer* **80**, 539–545.

Middleton, R. (1974). Survey of negative-ion sources for tandem accelerators. *Nucl. Instr. Methods* **122**, 35–43.

Miyashita, M., Presley, J. M., Buchholz, B. A., Lam, K. S., Lee, Y. M., Vogel, J. S., and Hammock, B. D. (2001). Attomole level protein sequencing by Edman degradation coupled with accelerator mass spectrometry. *Proc. Natl. Acad. Sci. USA* **98**, 4403–4408.

Nakamura, T., Kobayashi, K., Matsuzaki, H., Murayama, M., Nagashima, Y., Oda, H., Shibata, Y., Tanaka, Y., and Furukawa, M. (2004). Proceedings of the ninth International Conference on Accelerator Mass Spectrometry—Nagoya University, Nagoya, Japan, 9–13 September 2002. *Nuclear Instr. Methods Phys. Res. Sect. B-Beam Interactions Materials Atoms* **223–224**, VII–VII.

Nelson, D. E., Korteling, R. G., and Stott, W. R. (1977). Carbon-14: Direct detection at natural concentrations. *Science* **198**, 507–578.

Stuiver, M., and Polach, H. A. (1977). Reporting of C-14 data—discussion. *Radiocarbon* **19**, 355–363.

Suter, M., Jacob, S., and Synal, H. A. (1997). AMS of C-14 at low energies. *Nuclear Instr. Methods Phys. Res. Sect. B-Beam Interactions Materials Atoms* **123**, 148–152.

Turteltaub, K. W., Felton, J. S., Gledhill, B. L., Vogel, J. S., Southon, J. R., Caffee, M. W., Finkel, R. C., Nelson, D. E., Proctor, I. D., and Davis, J. C. (1990). Accelerator mass spectrometry in biomedical dosimetry: Relationship between low-level exposure and covalent binding of heterocyclic amine carcinogens to DNA. *Proc. Natl. Acad. Sci. USA* **87**, 5288–5292.

Vogel, J. S. (1992). Rapid production of graphite without contamination for biomedical AMS. *Radiocarbon* **34**, 344–350.

Vogel, J. S., Grant, P. G., Buchholz, B. A., Dingley, K., and Turteltaub, K. W. (2001). Attomole quantitation of protein separations with accelerator mass spectrometry. *Electrophoresis* **22**, 2037–2045.

Vogel, J. S., Keating, G. A., and Buchholz, B. A. (2002). Protein binding of isofluorphate *in vivo* after coexposure to multiple chemicals. *Env. Health Perspect.* **110**, 1031–1036.

Vogel, J. S., Nelson, D. E., and Southon, J. R. (1989). Accuracy and precision in dating microgram carbon samples. *Radiocarbon* **31**, 145–149.

Vogel, J. S., Ognibene, T., Palmblad, N. M., and Reimer, P. J. (2004). Counting statistics and ion interval density in AMS. *Radiocarbon* **46**, 1103–1109.

[14] Accelerator Mass Spectrometry for Biomedical Research

By KAREN BROWN, KAREN H. DINGLEY, and KENNETH W. TURTELTAUB

Abstract

Accelerator mass spectrometry (AMS) is the most sensitive method for detecting and quantifying rare long-lived isotopes with high precision. In this chapter, we review the principles underlying AMS-based biomedical studies, focusing on important practical considerations and experimental procedures needed for the detection and quantitation of ^{14}C- and ^{3}H-labeled compounds in various experiment types.

Introduction

AMS is the most sensitive method available for detecting and quantifying rare long-lived isotopes with high precision. This technique is widely employed in the earth and environmental sciences for purposes such as radiocarbon dating and studying the circulation of the world's oceans (Vogel *et al.*, 1995). It was not until the late 1980s that AMS was first used in biological research. Since then, it has been used primarily to investigate the absorption, distribution, metabolism, and excretion of radio-labeled drugs, chemicals, and nutrients, as well as in the detection of chemically modified DNA and proteins in animal models and humans. Newer applications of AMS include an isotope-labeled immunoassay (Shan *et al.*, 2000) and attomole-level sequencing of ^{14}C-labeled protein, achieved by coupling Edman degradation with AMS detection (Miyashita *et al.*, 2001). The high sensitivity of AMS measurements translates to the use of low chemical and radioisotope doses and relatively small sample sizes, which enables studies to be performed safely in humans, using exposures that are environmentally or therapeutically relevant while generating little radioactive waste.

Most biomedical AMS studies completed have employed carbon-14 as the radiolabel, although the capability exists for detecting other isotopes including ^{3}H, ^{26}Al, ^{41}Ca, ^{10}Be, ^{36}Cl, ^{59}Ni, ^{63}Ni, and ^{129}I. Both ^{14}C and ^{3}H are commonly used in tracing studies because they can be readily incorporated into organic molecules, either synthetically or biosynthetically. In this chapter, we review the principles underlying AMS-based biomedical studies, focusing on important practical considerations and experimental

METHODS IN ENZYMOLOGY, VOL. 402   0076-6879/05 $35.00
DOI: 10.1016/S0076-6879(05)02014-8

procedures needed for the detection and quantitation of ^{14}C- and ^{3}H-labeled compounds in various experiment types.

Methodology

AMS is 10^3–10^9 fold more sensitive than the decay counting methods routinely employed in biological studies involving radioisotopes (Turteltaub and Vogel, 2000). A detailed explanation of the differences between decay counting and AMS is presented in Chapter 13 by Vogel *et al.* Decay counting indirectly predicts the number of isotope nuclei present by measuring decay events. For ^{14}C and ^{3}H isotopes, this is an inefficient process, dependent on the length of time the sample is counted and the number of nuclei present. AMS directly quantifies each individual isotopic nucleus and is independent of these variables. This results in improved sensitivity, which means AMS biomedical studies can be performed with isotope doses 100–1000 times lower than those traditionally used and sample sizes can be 10^5–10^6 fold lower.

Instrumentation

A variety of instrument designs exist depending on the accelerating voltage and measurement requirements such as precision, resolution and the isotope range. At Lawrence Livermore National Laboratory (LLNL), a multipurpose 10 MV system (Fig. 2A) is used for the analysis of a wide range of isotopes from biomedical and earth and environmental science studies. This AMS machine includes two mass spectrometers separated by an electrostatic accelerator, through which negative ions are accelerated to high energies (Fig. 1). Individual graphite (for ^{14}C) or titanium hydride (for ^{3}H) pellets in aluminum holders, derived from the sample material, are bombarded with a large current of positive cesium ions (3–10 keV), which produces negatively charged elemental and molecular ions (Vogel, 1992). These undergo an initial separation through a low-energy mass spectrometer. Ions are then accelerated to the positive high-voltage terminal (3–10 MV) at the midpoint of the tandem accelerator where they pass through a thin carbon foil or gas, which strips off electrons, forming positive ions and causing dissociation of interfering molecular isobars. The charge state is dependent on ion velocity. With a 7 MV terminal, most C$^-$ ions take on a 4+ charge and are then further accelerated back to ground potential in the second half of the accelerator. After emerging from the accelerator, the rare ions are separated from the abundant ions by a high-energy mass spectrometer and crossed electric and magnetic fields (Wein filter).

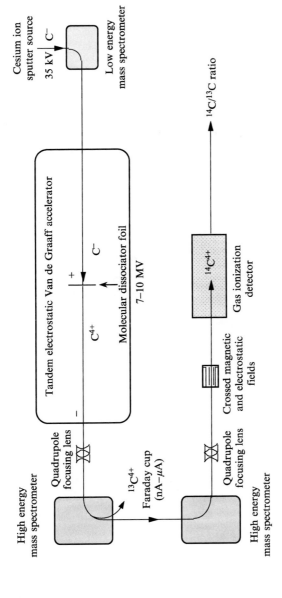

FIG. 1. Diagram of the 10 MV accelerator mass spectrometry (AMS) system at the Lawrence Livermore National Laboratory.

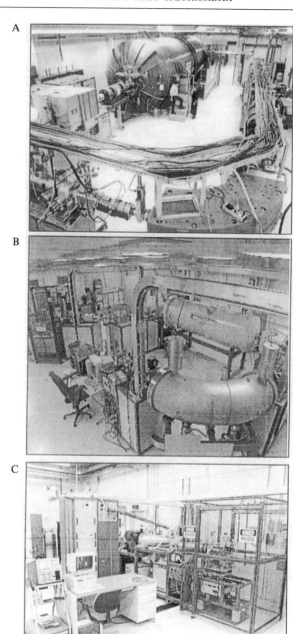

FIG. 2. Accelerator mass spectrometers in use at the Lawrence Livermore National Laboratory. (A) The high-energy 10 MV instrument used by multiple researchers for the measurement of various isotopes. (B) The 1 MV accelerator mass spectrometry (AMS) system dedicated to the analysis of ^{14}C in biomedical samples. (C) The compact tritium AMS system.

Isotopic ions are then individually identified through characteristic energy loss and counted in a gas ionization detector.

AMS does not determine absolute concentrations of radioisotopes; each measurement is an isotope ratio: the ratio of a rare isotope to a stable isotope of the same element. Expressing results as a ratio corrects for fluctuations in ion transport through the AMS system. For ^{14}C or ^{3}H analysis, ^{13}C or ^{1}H is measured sequentially with the radioisotope of interest (^{14}C or ^{3}H, respectively) as a current in a Faraday cup. The isotope ratios are then compared with a set of standards with known ^{14}C or ^{3}H content to determine the precise amount of ^{14}C or ^{3}H in the sample analyzed.

At the LLNL facility, a much smaller 1 MV spectrometer was built, designed specifically for measuring ^{14}C in samples from biomedical studies (Fig. 2B). This lower voltage spectrometer, which can measure more than 300 samples a day with precisions of 3%, is now used for quantifying ^{14}C in all biological samples at LLNL (Ognibene et al., 2002). A compact AMS system dedicated to the measurement of ^{3}H has also been developed (Roberts et al., 2000) (Fig. 2C). The system uses a radiofrequency quadrupole (RFQ) LINAC, which has a compact size (<1.5 m) and is able to accelerate all three hydrogen isotopic species to energies sufficient for measurement using a simple magnetic spectrometer. Other compact lower voltage AMS instruments are also being developed with variations in ion source and sample introduction (Hughey et al., 1997).

Sample Preparation: Methodology and Practical Considerations

The most important considerations in preparing samples for AMS analysis are the predicted amount of radioisotope in each sample, preventing contamination and knowing the sources and amounts of any carbon introduced during processing. Numerous precautions need to be in place throughout the procedure to ensure the amount of isotope present is within the dynamic range of the spectrometer and to minimize the potential for contamination, to ensure that the isotope detected in the sample is associated with the labeled compound under investigation.

The sample of interest, for example, blood, DNA, or high-performance liquid chromatography (HPLC) fractions, must be converted to a form that is compatible with the ion source of the instrument. Most ^{14}C samples are measured as graphite and ^{3}H samples as titanium hydride. The advantages of analyzing samples in a solid state rather than as a gas include a lower risk of cross-contamination in the ion source, shorter memory effects when a "hot" sample is encountered, and higher sample throughput due to more efficient ionization. Solid samples can be prepared at any location and sent to an AMS facility for analysis. This allows multiple researchers working on many experiments in various locations to use the instrument simultaneously. Solid samples can be left in the ion source for as long as necessary to accumulate sufficient events in the detector to attain the required measurement precision and can be saved for reanalysis if desired. However, the ability to directly couple an AMS instrument to an analytical HPLC, CE, or gas chromatograph (GC) system would provide a powerful and necessary technique for biomedical research, because it would enable online analysis of components in a sample without the need for prior separation and sample preparation. A prototype system has been described, which interfaces a GC to an AMS instrument (Hughey et al., 2000). Following GC separation, organic compounds eluting from the column are converted to CO_2 in a combustion chamber before entering the ion source. In addition, nonvolatile samples, including eluent from an HPLC system, can also be analyzed using a laser-induced combustion interface in which liquid is deposited into a bed of CuO powder (Liberman et al., 2004). Heating the matrix locally with a laser causes sample combustion and the resulting CO_2 is transported in a stream of carrier gas into the ion source of the AMS instrument. Ultimately, the authors intend this system to be used for the online separation and detection of both ^{14}C- and ^{3}H-labeled compounds using various chromatographic techniques.

The standard procedure for the production of graphite or titanium hydride involves oxidation of the crude sample to a gaseous mixture including carbon dioxide and water (Fig. 3). Carbon dioxide is then cryogenically extracted and reduced to form filamentous graphite, and water is reduced to hydrogen gas and adsorbed on to titanium, forming titanium hydride. The isolated sample, typically in the form of a DNA or protein solution, HPLC fraction, or wet tissue, is transferred to a quartz tube, which is baked before use to remove all bound carbon. This tube, which must be handled with disposable forceps to prevent contamination of the exterior, is placed in an outer borosilicate glass tube to facilitate handling, and the volatile components of the sample are removed by vacuum centrifugation. Copper oxide in wire form is added, and the inner quartz tube transferred to a quartz combustion tube, which is evacuated and sealed using an oxyacetylene torch. The sample is then oxidized by heating in a furnace at 900° for 2 h, which produces CO_2, N_2, and H_2O (Vogel, 1992). For production of graphite, the combustion tube is attached to a Y-shaped disposable plastic manifold, the other ends of which are connected to a reaction tube, containing reagents for the reduction of CO_2 to elemental carbon (titanium hydride, zinc, and cobalt powders) and a vacuum line, as shown in Fig. 4. The next step is to cryogenically separate CO_2 from the

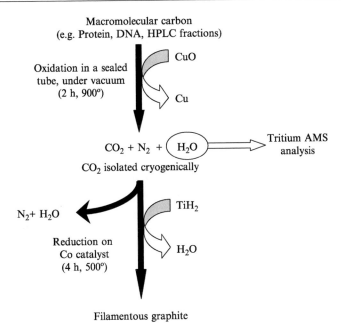

Fig. 3. Stages of sample combustion and graphitization.

other oxidation products and trap it in the reaction tube. This involves first evacuating the reaction tube, then clamping off the vacuum. The combustion tube is placed in a dry ice–isopropanol slurry, which condenses out the H_2O. The end of the combustion tube is broken open, allowing transfer of CO_2 over to the reaction tube. This tube is placed in liquid N_2, which freezes out and traps the CO_2. Finally the hose clamp is removed to evacuate N_2 gas, and the reaction tube is sealed. The samples are baked in a furnace for 4 h at 500°. The CO_2 is reduced to graphite by zinc and titanium hydride and deposits on the cobalt catalyst contained in a smaller inner tube of the reaction tube.

The sample-preparation stage is the rate-limiting step in AMS studies, but a method has been developed for preparing graphite from CO_2 gas in septa-sealed vials (Ognibene et al., 2003). This approach increases sample throughput and is less complex than the standard protocol, because it does not require the use of a plastic manifold or torch sealing of the transferred combustion products. Instead, the combusted sample is transferred to a septa-sealed vial containing zinc dust and an inner quartz tube with iron

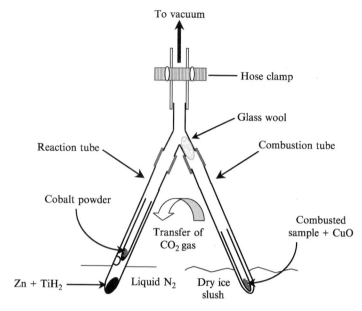

FIG. 4. Transfer of CO_2 from the combustion tube to the reaction tube in the preparation of graphite for ^{14}C analysis.

powder, via a needle connected to a Luer-Lok stopcock. The vial is kept in liquid nitrogen to cryogenically trap the CO_2 and H_2O while noncondensing gases are removed by vacuum pump. The needle is removed from the vial, which is then placed in a block heater and held at 500°, reducing the CO_2 to graphite. This newer protocol results in lower process backgrounds and allows for preparation of smaller sized samples. It has been adopted as the routine method at the LLNL for preparing biological samples for ^{14}C analysis by AMS and is expected to form the basis for an integrated automated system, aimed at considerably increasing sample throughput.

For analysis of ^3H, water from the combustion process is trapped using a dry-ice isopropanol bath and then reduced to hydrogen gas using zinc. The resulting hydrogen gas is adsorbed onto titanium to produce titanium hydride (Chiarappa-Zucca *et al.*, 2002; Roberts *et al.*, 1994). For AMS measurement, the prepared graphite or titanium hydride is packed into individual sample holders and placed in a sample wheel. Samples are measured at least three times, up to a maximum of seven, until the measurement precision is less than 3%.

The high sensitivity of AMS measurements means that a number of precautions must be taken during sample preparation to prevent contamination. Laboratory surfaces and equipment are routinely swiped using a glass filter wetted with ethanol, to monitor for removable contamination. Airborne ^{14}C contamination is detected using sorbent carbon, such as graphitized coal or fullerene soot mixed with fine metal powder and packed into AMS sample holders. These samples are left in the laboratory for several days or weeks and then measured by AMS to check for contamination during this period. These methods report airborne and surface contamination after the event, alerting users to a problem, and supplement the numerous procedures that are in place to prevent contamination of samples and the laboratory environment; each stage of the sample preparation process is physically segregated, according to the activity level being handled. To avoid sample contamination during handling, disposable plastic labware is used whenever possible, bench covering is changed frequently, and only specifically designated equipment (centrifuges, freezers, HPLC instruments, etc.) is used for AMS work (Buchholz *et al.*, 2000).

Data Manipulation

Living entities maintain a natural level of ^{14}C because of the relatively constant production of this isotope in the upper atmosphere. This level, referred to as "Modern," corresponds to 97.6 amol of ^{14}C/mg of total carbon, or 6.11 fCi ^{14}C/mg of carbon. AMS is used in biomedical studies to measure the increase in ^{14}C or ^3H above natural background in a particular sample, which is due to the presence of a ^{14}C- or ^3H-labeled parent compound or related derivative.

Samples that generate approximately 0.5–1.0 mg of total carbon or 2.0 mg of water upon combustion are optimal for AMS analysis using the methods described earlier. Samples containing sufficient natural carbon or ^1H are measured neat, whereas smaller samples are supplemented with a precise amount of carrier that provides sufficient carbon or water for optimal sample preparation. An ideal carrier has radioisotope concentrations well below the levels found naturally in living organisms (<1% contemporary isotope abundance). Tributyrin (25 μl of a 40 mg/ml solution in methanol or 1 μl neat tributyrin providing 0.6 mg of carbon) is often used as the carrier because it is a nonvolatile liquid and contains depleted levels of ^{14}C and ^3H. We have found measurements of tributyrin samples to be highly reproducible. With every series of carrier-supplemented samples submitted for AMS analysis, a set of three or four samples containing just the carrier is also measured.

To determine the exact amount of ^{14}C or ^3H in a sample by AMS measurement, the complete carbon or hydrogen inventory must be known.

For biological samples, this information can be acquired by elemental analysis or using published reference values for a particular tissue type. Calculation of ^{14}C or ^{3}H content is easiest for samples that can be obtained in quantities yielding in excess of 1 mg of carbon or 2 mg of water, such as blood, whole tissue, and isolated proteins. In this case, the excess level of ^{14}C or ^{3}H in a sample, above that of an undosed control, is due to the presence of ^{14}C- or ^{3}H-labeled compound. If carrier is added to the samples, this must also be accounted for in the calculations by subtracting the amount of radioisotope contributed by the carrier and that contributed by the natural radioisotopic abundance of the actual sample. To do this, the precise amount of carrier added and the exact mass of biological material in the sample being analyzed must be known. Values are then converted to an appropriate form, such as picograms of radiolabeled compound per gram of tissue or tissue isolate (e.g., DNA or protein) using the specific activity, molecular weight of the labeled compound, and the %w/w carbon present in the tissue. For a detailed account of data handling, including example calculations, readers are referred to Chapter 13 by Vogel *et al.*, in this volume.

Applications

AMS can be used to trace the fate of any molecule in *in vitro* systems or whole organisms if it is labeled with an isotope appropriate for AMS analysis. Studies have been undertaken with a variety of drugs, environmental carcinogens, and micronutrients, to determine the kinetics of absorption, distribution, and excretion, to identify and quantify metabolites, and to assess DNA or protein-binding capability.

The sensitivity of AMS measurement gives this technique a number of major advantages over other methods for the detection of isotopes. Importantly, because only low doses of chemical and radioactivity are required, studies can be performed with levels of chemicals equivalent to therapeutic or environmental exposures. This is a significant feature because the biological effects observed at high doses may not extrapolate to the low doses humans typically encounter. Furthermore, with safer, low radioisotope doses, it is possible to perform studies in humans. Doses used in human protocols range from 10 nCi/person for metabolism and mass balance studies up to 50 μCi/person for DNA binding studies. These are within the range that corresponds to natural background levels of radiation that individuals are exposed to during everyday life.

Although this chapter concentrates on ^{14}C- and ^{3}H-AMS, many other radioisotopes are increasingly being used in biological studies. Bone calcium metabolism has been studied in humans administered [41]C orally, as the carbonate (Freeman *et al.*, 2000), and this isotope has been investigated as a tracer for calcium uptake and deposition in cardiac ischemia (Southon *et al.*, 1994). A better understanding of aluminum absorption, distribution, and clearance in healthy patients and a group with Alzheimer's disease has been gained through AMS studies using [26]Al (Kislinger *et al.*, 1997; Moore *et al.*, 2000; Talbot *et al.*, 1995). In addition, uptake of silicic acid has been determined by [32]Si-AMS (Popplewell *et al.*, 1998). Other isotopes are also being developed for use in biomedical studies, including [10]Be and isotopes of plutonium.

AMS Determination of Radioisotope Concentrations: Pharmacokinetic Studies

One of the simplest and most common applications of biological AMS has been in the determination of absorption, distribution, metabolism, and excretion characteristics of isotope-labeled compounds in animal models and humans. Some unique aspects of AMS pharmacokinetic studies include the ability to determine long-term kinetics and metabolism using low doses, several months after isotope administration. Detailed pharmacokinetic data require frequent sampling, which is made possible with AMS detection, by virtue of the small sample sizes needed for analysis. AMS has been used to establish the kinetics of β-carotene uptake and plasma clearance in a human volunteer who received a single dose of ^{14}C-β-carotene obtained from ^{14}C-labeled spinach (Dueker *et al.*, 2000). Plasma concentrations of β-carotene and its metabolites were determined at intervals over a 7-mo period and required just 30 μl of plasma/analysis. Such complete investigations would not be possible using other methods that lack the necessary sensitivity to detect compounds and metabolites months after dosing.

To quantify concentrations of [14]C- or ^{3}H-labeled compounds and their derivatives, wet tissue samples (5–10 mg) or aliquots of fluid, such as blood, plasma, or urine (typically 10–200 μl), are prepared for analysis by placing them in quartz sample tubes and drying in a vacuum centrifuge to remove water before combustion. These are then processed as usual and measured by AMS. In these types of experiments, radioisotope concentrations can be relatively high, particularly with samples taken at early time points, so it is very important to screen samples before AMS analysis to ensure the samples do not contain too much radioisotope. This is done by liquid scintillation counting (LSC) an aliquot of each sample, using an amount equivalent to that intended for AMS analysis. The AMS measurement range is 0.01–100 Modern, which equates to approximately 0.1–1350 dpm/g of carbon. Consequently, if radioisotope can be detected above background levels using LSC, the sample must be diluted before AMS analysis. If the signal is below or at the limit of detection, analysis can be performed.

A new application of AMS, made possible by advances in tritium detection (Chiarappa-Zucca *et al.*, 2002; Love *et al.*, 2002), is dual isotope-labeling studies. Dingley *et al.* (1998, 2003) have demonstrated that two independent compounds can be traced if one is labeled with ^{14}C and the other [3]H. For example, the liver concentrations of two heterocyclic amines ^{3}H-PhIP and ^{14}C-MeIQx were determined when the compounds were administered to rats individually and in combination, to ascertain whether co-administration affected bioavailability (Fig. 5). The methodology permits liver samples from a single source to be assayed in parallel for the presence of either ^{14}C or [3]H. Furthermore, simultaneous extraction of CO_2 and H_2O from the same sample is also now possible so that both isotopes can be quantified using just one sample, thereby increasing sample throughput and decreasing the amount of material required (Chiarappa-Zucca *et al.*, 2002). This will enable animal and human studies to be carried out involving mixtures of compounds, to more closely mimic typical human exposures.

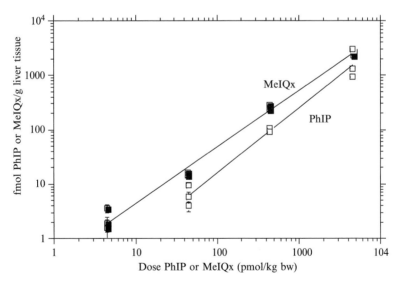

FIG. 5. Concentration of ^{3}H-PhIP and ^{14}C-MeIQx in the liver tissue of rats administered a dose range of either ^{3}H-PhIP or ^{14}C-MeIQx alone or in combination.

HPLC-AMS

AMS measurements provide no structural information, so any sample characterization or identification must be performed before sample preparation, using chromatographic (TLC, HPLC, size exclusion, and affinity) or electrophoresis (gel and capillary) separation techniques. Coupling an HPLC step to AMS detection affords both specificity and high sensitivity, in an approach that has been used for metabolite profiling and quantification (Buchholz et al., 1999; Dueker et al., 2000; Malfatti et al., 1999), as well as the identification of chemically modified nucleosides (Mauthe et al., 1999) in humans and animal models where other techniques are not sensitive enough to detect the low levels formed. Metabolites can be extracted and analyzed from a variety of biological matrices such as plasma, urine, tissues, and milk and a full metabolite profile can be obtained using just 100 μl of urine (Turteltaub and Vogel, 2000). Before beginning an experiment, aliquots of HPLC buffers, solvents, and any standards that might be run should be measured by AMS to check for contamination. Importantly, the HPLC solvents must be compatible with the graphitization process, in particular buffers containing sodium salts should not be used. Ideally, the buffer should not introduce an extra source of carbon, although it is sometimes possible to use such buffers if an isocratic system is employed and the amount of carbon in each fraction is constant throughout the run. In this situation, it is important that the samples do not contain too much carbon, because this can cause the tubes to explode during the combustion step. Samples are separated by HPLC and individual fractions collected at known intervals, typically every 20–60 s. In most cases, the fractions usually contain very little carbon after removal of the solvent, and tributyrin is added to each fraction to provide a fixed amount of carbon, sufficient for AMS measurement. The quartz sample tubes have a maximum capacity of approximately 300 μl, limiting the volume that can be submitted. Therefore, in situations in which the radioisotope content is expected to be low, fractions should be concentrated before this stage. Alternatively, 300-μl aliquots can be dried down repeatedly in the quartz tubes, but this can be time consuming. Performing and analyzing a control run of unlabeled material processed in an identical manner to the test sample is vital because the background ^{14}C or ^{3}H level of the system determines the detection limit of the method. The concentration of ^{14}C or ^{3}H in each fraction is determined from the isotope ratio and is used to reconstruct a chromatogram. Individual metabolites and compounds can then be identified based on retention times and comparison to authentic standards.

As has been demonstrated for folic acid, the high sensitivity of AMS analysis can reveal the formation of previously unidentified metabolites that are formed at levels below the limits of detection achievable with traditional techniques (Clifford et al., 1998). A further feature is the capability to detect metabolites formed after low-dose exposures, using doses of labeled chemicals comparable to therapeutic or environmental levels. This methodology was applied in a study designed to assess the distribution of PhIP and its metabolites in lactating female rats and their pups after administration of a single dose of ^{14}C-PhIP, equivalent to an average human daily dose of this compound (Mauthe et al., 1998). The detection of PhIP and PhIP metabolites in the breast milk, stomach contents, and liver tissue of the pups by AMS provided evidence that suckling pups are exposed to this carcinogen even at low doses. To further examine the types and concentration of metabolites excreted in breast milk, metabolites were extracted by adding methanol and centrifuging the milk samples to remove protein. The supernatant was evaporated to dryness, the residue dissolved in 0.1% trifluoroacetic acid, and subject to reverse-phase HPLC separation using a trifluoroacetic acid/acetonitrile mobile phase. This procedure achieved greater than 95% recovery of PhIP and its metabolites, based on the radiocarbon levels of samples measured before and after extraction. Fractions were collected at 1-min intervals throughout the run. These were then concentrated and redissolved in water/methanol before adding tributyrin and preparing for AMS analysis. Reconstruction of the HPLC chromatogram using the AMS data revealed the presence of three metabolite peaks in addition to the parent PhIP molecule (Fig. 6). These were identified by comparison to metabolites present in milk extracts from a rat treated with a high (10 mg/kg) dose of PhIP, which were fully characterized by HPLC-MS. This study demonstrates the importance of being able to investigate low-dose metabolism, because although the metabolite profile over the dose range examined was qualitatively similar, two glucuronide conjugates formed at high doses (10 mg/kg) of PhIP were not detectable at low doses.

As stated previously, other separation techniques can be used in conjunction with AMS. Binding of ^{14}C-labeled molecules to proteins has been quantified and specific protein targets identified, by combining polyacrylamide gel separations (one dimensional or two dimensional) with AMS detection (Vogel et al., 2001; Williams et al., 2002). Typically, gel bands are excised and after drying under vacuum can be directly analyzed, because the polyacrylamide, which is produced from petroleum-based chemicals, contains sufficient carbon (51%) to act as a carrier. The total amount of carrier carbon can be calculated if the volume and composition of the excised gel band is known.

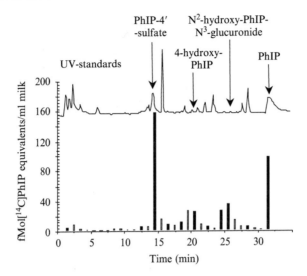

FIG. 6. High-performance liquid chromatography-accelerator mass spectrometry (HPLC-AMS) isotope chromatogram of ^{14}C-PhIP metabolites detected in breast milk from a female rat administered PhIP at 1000 ng/kg. The top chromatogram shows ultraviolet detection of metabolite standards.

AMS Detection of Chemically Modified DNA and Protein

AMS has demonstrated great value in determining whether chemicals bind to DNA or protein, forming covalent adducts. This is important because DNA adduct formation is considered an early initiating event in chemical-induced cancers. The presence of protein adducts, which may be formed at higher levels than DNA adducts, indicates a chemical is either itself reactive or can be converted to a reactive metabolite capable of binding to cellular macromolecules. Protein adducts can serve as surrogates for DNA adducts, giving a measure of an individual's level of exposure to the bioactive dose of a compound. Techniques used for detecting adducts must be very sensitive because adducts are usually formed at extremely low levels, particularly in humans exposed to environmental or therapeutic doses of carcinogens or drugs. AMS has been used to detect DNA adducts at levels of 1–10 adducts/10^{12} nucleotides, which is less than 1 modification/cell, following acute and chronic exposure to ^{14}C-labeled carcinogens (Turteltaub et al., 1993). This is at least 100-fold lower than the next most sensitive detection method, the ^{32}P-postlabeling assay.

One of the first biomedical applications of AMS was to determine whether the compound MeIQx, which is formed in cooked meat, binds to

DNA in rodent tissues at low doses and whether the relationship between adduct formation and dose is linear (Turteltaub et al., 1990). Since then, this technique has been used to answer similar questions for a number of chemical carcinogens in humans and animal models, including benzene (Mani et al., 1999), trichloroethylene (Kautiainen et al., 1997), and the breast cancer drugs tamoxifen and toremifene (Boocock et al., 2002; Martin et al., 2003; White et al., 1997).

In this type of study, DNA and protein are isolated from tissues using standard protocols. For measurement of DNA adduct levels, tissues are lysed and digested using the enzymes proteinase K, RNase T_1, and RNase A. The lysate is loaded on to a Qiagen anion exchange column, and the protein and other material, including non-covalently bound labeled compounds and metabolites, are washed from the column, while the DNA is retained (Frantz et al., 1995). The DNA is then eluted, precipitated, and washed with 70% ethanol before redissolving in a buffer appropriate for AMS analysis. It is important to demonstrate that all non-covalently bound compounds are removed by the isolation and purification procedure employed. For example, DNA could be enzymatically digested to nucleosides and separated by HPLC, ideally using a system that resolves adducted nucleosides from unmodified nucleosides and the free parent compound and metabolites (Mauthe et al., 1999). This method should also remove other potential ^{14}C-labeled contaminants that may be present, such as residual adducted protein or peptides. AMS analysis of the collected fractions will indicate what percentage of the total ^{14}C or ^{3}H signal can be attributed to covalently bound adducts. Because this will be compound specific, performing this type of experiment will identify situations in which it may be necessary to incorporate additional DNA purification steps.

To isolate protein (and other acid precipitable macromolecules), homogenized tissues are lysed overnight and then centrifuged. Perchloric acid (70%) is added to the supernatant precipitating the protein. After centrifuging, the resulting protein pellet is washed with 5% perchloric acid, followed by several organic solvents (a 50% methanol solution and a 1:1 mixture of ether:ethanol) to extract any residual non-covalently bound ^{14}C-labeled compounds (Dingley et al., 1999). The protein is then redissolved in 0.1 M of potassium hydroxide and aliquots of this solution are submitted for AMS analysis.

AMS can also be used to determine the ability of chemicals to bind to specific proteins, as was demonstrated by Dingley et al. (1999) in a study that reported PhIP binds to the blood proteins, hemoglobin, and albumin and to white blood cell DNA in humans administered a dietary relevant dose of [^{14}C]-PhIP. Adduct levels were monitored in five subjects over a

24 h period by taking blood samples (30 ml) at various times and separating into plasma, red blood cell, and buffy coat (containing white blood cells and platelets) fractions. Albumin and hemoglobin were isolated using standard protocols, with the addition of an initial dialysis step, and analyzed by AMS (Fig. 7). It is crucial when extracting albumin or hemoglobin that each sample of plasma or red blood cell lysate is first dialyzed extensively, in an individual beaker, for 48 h to remove non-covalently bound ^{14}C-labeled compounds and metabolites. This investigation demonstrated the value of AMS in providing information on adduct formation and

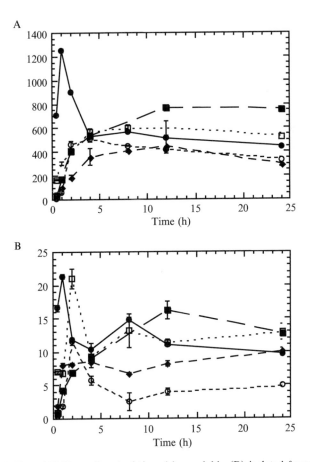

FIG. 7. Binding of PhIP to albumin (A) and hemoglobin (B) isolated from five human subjects administered a single dose of ^{14}C-PhIP.

kinetics in humans, which may be important in assessing an individual's susceptibility to specific carcinogens.

Protein samples, which are usually obtained in higher quantities than DNA, can normally be measured without the addition of carrier. Approximately 3 mg of protein will yield the required 1 mg of carbon, as will 3 mg of DNA when this amount is available, because both are composed of about 30% carbon, although this is dependent on tissue type and species. Smaller quantities can be analyzed (as little as 1 mg of DNA or protein), but there is greater potential for contamination and the sample may not produce sufficient graphite. In situations when the amount of tissue, and therefore, DNA is limited, as is often the case with human studies, the DNA is submitted with carrier. The amount of DNA analyzed in this way can be as little as 1 μg, although this will be influenced by the expected level of adduction, the compound specific activity, and availability. Material analyzed neat does not need to be accurately quantified, because the AMS measurement is a ratio, but for carrier-supplemented samples, the precise amount of DNA or protein must be determined. Consequently, when analyzing carrier-added samples, the precision and accuracy of the AMS data can be limited by the precision and accuracy of the methods used to quantify the mass of material. Alpha-particle energy loss measurement was developed as an accurate way to nondestructively quantify the mass of microgram quantities of macromolecules with high precision before analysis by AMS (Grant et al., 2003).

The Future

AMS methodology has advanced substantially over the last 10 years, resulting in its now routine use in the measurement of ^{14}C and to a lesser extent ^{3}H in biological samples. The technique offers unique capabilities in studies where low amounts of radioisotope-labeled compounds must be quantified with high precision. These qualities have led to the increasing use of AMS in areas of nutrition, biomarker detection, human metabolism studies, and drug development. Methods for the analysis of additional isotopes are being developed.

Over the coming years, we anticipate AMS applications will expand and become more diverse as smaller cheaper instruments designed specifically for biologists are constructed and the technology becomes more accessible. Furthermore, automation of the sample-preparation process will increase sample throughput and could allow for analysis of smaller samples, all of which will help expand AMS into all areas of biological and environmental research.

Acknowledgments

This work was performed under the auspices of the U.S. DOE (W-7405-ENG-48) with support from the National Institutes of Health/National Center for Research Resources (RR13461), DOD Prostate Cancer Program (DAMD17-03-1-0076), and the NCI (CA55861).

References

Boocock, D. J., Brown, K., Gibbs, A. H., Sanchez, E., Turteltaub, K. W., and White, I. N. H. (2002). Identification of human CYP forms involved in the activation of tamoxifen and irreversible binding to DNA. *Carcinogenesis* **23,** 1897–1901.

Buchholz, B. A., Freeman, S. P. H. T., Haack, K. W., and Vogel, J. S. (2000). Tips and traps in the ^{14}C bio-AMS preparation laboratory. *Nucl. Instr. Methods Phys. Res. B* **B172,** 404–408.

Buchholz, B. A., Fultz, E., Haack, K. W., Vogel, J. S., Gilman, S. D., Gee, S. J., Hammock, B. D., Hui, X., Wester, R. C., and Maibach, H. I. (1999). HPLC-accelerator MS measurements of atrazine metabolites in human urine after dermal exposure. *Anal. Chem.* **71,** 3519–3525.

Chiarappa-Zucca, M. L., Dingley, K. H., Roberts, M. L., Velsko, C. A., and Love, A. H. (2002). Sample preparation for quantitation of tritium by accelerator mass spectrometry. *Anal. Chem.* **74,** 6285–6290.

Clifford, A. J., Arjomand, A., Dueker, S. R., Schneider, P. D., Buchholz, B. A., and Vogel, J. S. (1998). The dynamics of folic acid metabolism in an adult given a small tracer dose of ^{14}C-folic acid. *Adv. Exp. Med. Biol.* **445,** 239–251.

Dingley, K. H., Curtis, K. D., Nowell, S., Felton, J. S., Lang, N. P., and Turteltaub, K. W. (1999). DNA and protein adduct formation in the colon and blood of humans after exposure to a dietary-relevant dose of 2-amino-1-methy-6-phenylimidazo[4,5-*b*]pyridine. *Cancer Epidemiol. Biomarkers* **8,** 507–512.

Dingley, K. H., Roberts, M. L., Velsko, C. A., and Turteltaub, K. W. (1998). Attomole detection of ^{3}H in biological samples using accelerator mass spectrometry: Application in low-dose, dual-isotope tracer studies in conjunction with ^{14}C accelerator mass spectrometry. *Chem. Res. Toxicol.* **11,** 1217–1222.

Dingley, K. H., Ubick, E. A., Chiarappa-Zucca, M. L., Nowell, S., Abel, S., Ebeler, S. E., Mitchell, A. E., Burns, S. A., Steinberg, F. M., and Clifford, A. J. (2003). Effect of dietary constituents with chemopreventive potential on adduct formation of a low dose of the heterocyclic amines PhIP and IQ and phase II hepatic enzymes. *Nutr. Cancer* **46,** 212–221.

Dueker, S. R., Lin, Y., Bucholz, B. A., Schneider, P. D., Lamé, M. W., Segall, H. J., Vogel, J. S., and Clifford, A. J. (2000). Long-term kinetic study of β-carotene, using accelerator mass spectrometry in an adult volunteer. *J. Lipid Res.* **41,** 1790–1800.

Frantz, C. E., Bangerter, C., Filtz, E., Mayer, K. M., Vogel, J. S., and Turteltaub, K. W. (1995). Dose–response studies of MeIQx in rat liver and liver DNA at low doses. *Carcinogenesis* **16,** 367–373.

Freeman, S. P. H. T., Beck, B., Bierman, J. M., Caffee, M. W., Heaney, R. P., Holloway, L., Marcus, R., Southon, J. R., and Vogel, J. S. (2000). The study of skeletal calcium metabolism with ^{41}Ca and ^{45}Ca. *Nuclear Instr. Methods Phys. Res.* **B172,** 930–933.

Grant, P. G., Palmblad, M., Murov, S., Hillegonds, D. J., Ueda, D. L., Vogel, J. S., and Bench, G. (2003). α-Particle energy loss measurement of microgram depositions of biomolecules. *Anal. Chem.* **17,** 4519–4524.

Hughey, B. J., Klinkowstein, R. E., Shefer, R. E., Skipper, P. L., Tannenbaum, S. R., and Wishnok, J. S. (1997). Design of a compact 1 MV AMS system for biomedical research. *Nuclear Instr. Methods Phys. Res.* **B123,** 153–158.

Hughey, B. J., Skipper, P. L., Klinkowstein, R. E., Shefer, R. E., Wishnok, J. S., and Tannenbaum, S. R. (2000). Low-energy biomedical GC-AMS system for ^{14}C and ^{3}H detection. *Nuclear Instr. Methods Phys. Res.* **B172,** 40–46.

Kautiainen, A., Vogel, J. S., and Turteltaub, K. W. (1997). Dose-dependent binding of trichloroethylene to hepatic DNA and protein at low doses in mice. *Chem. Biol. Interact.* **106,** 109–121.

Kislinger, G., Steinhausen, C., Alvarez-Brückmann, M., Winklhofer, C., Ittel, T.-H., and Nolte, E. (1997). Investigations of the human aluminium biokinetics with ^{26}Al and AMS. *Nuclear Instr. Methods Phys. Res. B* **123,** 259–265.

Liberman, R. G., Tannenbaum, S. R., Hughey, B. J., Shefer, R. E., Klinkowstein, R. E., Prakash, C., Harriman, S. P., and Skipper, P. L. (2004). An interface for direct analysis of ^{14}C in nonvolatile samples by accelerator mass spectrometry. *Anal. Chem.* **76,** 328–334.

Love, A. H., Hunt, J. R., Roberts, M. L., Southon, J. R., Chiarappa-Zucca, M. L., and Dingley, K. H. (2002). Use of tritium accelerator mass spectrometry for tree ring analysis. *Environ. Sci. Technol.* **36,** 2848–2852.

Malfatti, M. A., Kulp, K. S., Knize, M. G., Davis, C., Massengill, J. P., Williams, S., Nowell, S., MacLeod, S., Dingley, K. H., Turteltaub, K. W., Lang, N. P., and Felton, J. S. (1999). The identification of [2-^{14}C]2-amino-1-methyl-6-phenylimidazo[4,5-*b*]pyridine metabolites in humans. *Carcinogenesis* **20,** 705–713.

Mani, C., Freeman, S., Nelson, D. O., Vogel, J. S., and Turteltaub, K. W. (1999). Species and strain comparisons in the macromolecular binding of extremely low doses of [^{14}C]benzene in rodents, using accelerator mass spectrometry. *Toxicol. Applied Pharmacol.* **159,** 83–90.

Martin, E. A., Brown, K., Gaskell, M., Al-Azzawi, F., Garner, R. C., Boocock, D. J., Mattock, E., Pring, D. W., Dingley, K., Turteltaub, K. W., Smith, L. L., and White, I. N. H. (2003). Tamoxifen DNA damage detected in human endometrium using accelerator mass spectrometry. *Cancer Res.* **63,** 8461–8465.

Mauthe, R. J., Dingley, K. H., Leveson, S. H., Freeman, S. P. H. T., Turesky, R. J., Garner, R. C., and Turteltaub, K. W. (1999). Comparison of DNA-adduct and tissue-available dose levels of MeIQx in human and rodent colon following administration of a very low dose. *Int. J. Cancer* **80,** 539–545.

Mauthe, R. J., Snyderwine, E. G., Ghoshal, A., Freeman, S. P. H. T., and Turteltaub, K. W. (1998). Distribution and metabolism of 2-amino-1-methyl-6-phenylimidazo[4,5-*b*]pyridine (PhIP) in female rats and their pups at dietary doses. *Carcinogenesis* **19,** 919–924.

Miyashita, M., Presley, J. M., Buchholz, B. A., Lam, K. S., Lee, Y. M., Vogel, J. S., and Hammock, B. D. (2001). Attomole level protein sequencing by Edman degradation coupled with accelerator mass spectrometry. *Proc. Natl. Acad. Sci. USA* **98,** 4403–4408.

Moore, P. B., Day, J. P., Taylor, G. A., Ferrier, I. N., Fifield, L. K., and Edwardson, J. A. (2000). Absorption of aluminium-26 in Alzheimer's disease, measured using accelerator mass spectrometry. *Dementia Geriatr Cognitive Disord.* **11,** 66–69.

Ognibene, T. J., Bench, G., Brown, T. A., Peaslee, G. F., and Vogel, J. S. (2002). A new accelerator mass spectrometry system for ^{14}C-quantification of biochemical samples. *Int. J. Mass Spectrom.* **218,** 255–264.

Ognibene, T. J., Bench, G., Vogel, J. S., Peaslee, G. F., and Murov, S. (2003). A high-throughput method for the conversion of CO_2 obtained from biochemical samples to graphite in septa-sealed vials for quantification of ^{14}C via accelerator mass spectrometry. *Anal. Chem.* **75,** 2192–2196.

Popplewell, J. F., King, S. J., Day, J. P., Ackrill, P., Fifield, L. K., Cresswell, R. G., Di Tada, M. L., and Liu, K. (1998). Kinetics of uptake and elimination of silicic acid by a human subject: A novel application of Si-32 and accelerator mass spectrometry. *J. Inorg. Biochem.* **69,** 177–180.

Roberts, M. L., Hamm, R. W., Dingley, K. H., Chiarappa-Zucca, M. L., and Love, A. H. (2000). A compact tritium AMS system. *Nuclear Instr. Methods Phys. Res. B* **172,** 262–267.

Roberts, M. L., Velsko, C., and Turteltaub, K. W. (1994). Tritium AMS for biomedical applications. *Nuclear Instr. Methods Phys. Res.* **B92,** 459–462.

Shan, G., Huang, W., Gee, S. J., Buchholz, B. A., Vogel, J. S., and Hammock, B. D. (2000). Isotope-labeled immunoassays without radiation waste. *Proc. Natl. Acad. Sci. USA* **97,** 2445–2449.

Southon, J. R., Bishop, M. S., and Kost, G. J. (1994). [41]Ca as a tracer for calcium uptake and deposition in heart tissue during ischemia and reperfusion. *Nuclear Instr. Methods Phys. Res. B* **V92**(N1-4), 489–491.

Talbot, R. J., Newton, D., Priest, N. D., Austin, J. G., and Day, J. P. (1995). Inter-subject variability in the metabolism of aluminium following intravenous injection as citrate. *Human Exp. Toxicol.* **14,** 595–599.

Turteltaub, K. W., Felton, J. S., Gledhill, B. L., Vogel, J. S., Southon, J. R., Caffee, M. W., Finkel, R. C., Nelson, D. E., Proctor, I. D., and Davis, J. C. (1990). Accelerator mass spectrometry in biomedical dosimetry: Relationship between low-level exposure and covalent binding of heterocyclic amine carcinogens to DNA. *Proc. Natl. Acad. Sci.* **87,** 5288–5292.

Turteltaub, K. W., and Vogel, J. S. (2000). Bioanalytical applications of accelerator mass spectrometry for pharmaceutical research. *Curr. Pharmaceut. Design* **6,** 991–1007.

Turteltaub, K. W., Vogel, J. S., Frantz, C. E., and Fultz, E. (1993). Studies on DNA adduction with heterocyclic amines by accelerator mass spectrometry: A new technique for tracing isotope-labelled DNA adduction. *In* "Postlabelling Methods for Detection of DNA Adducts" (D. H. Phillips, M. Castegnaro, and H. Bartsch, eds.), Vol. 124, pp. 293–301. Lyon, International Agency for Cancer Research.

Vogel, J. S. (1992). Rapid production of graphite without contamination for biomedical AMS. *Radiocarbon* **34,** 344–350.

Vogel, J. S., Grant, P. G., Buchholz, B. A., Dingley, K. H., and Turteltaub, K. W. (2001). Attomole quantitation of protein separations with accelerator mass spectrometry. *Electrophoresis* **22,** 2037–2045.

Vogel, J. S., Turteltaub, K. W., Finkel, R., and Nelson, D. E. (1995). Accelerator mass spectrometry: Isotope quantification at attomole sensitivity. *Anal. Chem.* **67,** 353A–359A.

White, I. N. H., Martin, E. A., Mauthe, R. J., Vogel, J. S., Turteltaub, K. W., and Smith, L. L. (1997). Comparisons of the binding of [^{14}C]radiolabelled tamoxifen or toremifene to rat DNA using accelerator mass spectrometry. *Chem. Biol. Interact.* **106,** 149–160.

Williams, K. E., Carver, T. A., Miranda, J. L., Kautiainen, A., Vogel, J. S., Dingley, K., Baldwin, M. A., Turteltaub, K. W., and Burlingame, A. L. (2002). Attomole detection of *in vivo* protein targets of benzene in mice. *Mol. Cell. Proteom.* **1,** 885–895.

Author Index

A

Abel, S., 434
Abu-Threideh, J., 245
Ackloo, S., 59
Ackrill, P., 433
Adams, G. W., 35, 80, 210, 212, 233, 238
Adams, M. D., 245
Addona, T. A., 259, 301
Aebersold, R., 17, 79, 83, 149, 210, 223, 246, 247, 250, 252, 262, 263, 265, 266, 275, 294, 295, 304, 305, 307
Ahn, N. G., 307, 308, 342, 390
Aicher, L., 253, 272
Ainscough, R., 245
Aitchison, J. D., 249
Aitken, M. J., 404
Ajuh, P., 249
Akashi, S., 334
Akita, S., 186
Alaiya, A. A., 270
Al-Azzawi, F., 438
Alderdice, D. S., 161
Aleksandrov, M. L., 51
Alexander, A. J., 210
Alfred, R. L., 122
Alice, M. B., 317
Allemann, M., 135
Allen, M. H., 365
Allen, N., 40, 223, 300
Allen, N. P., 299, 306, 307
Allinson, E. T., 122
Allison, I., 52
Altschul, S. F., 302
Alvarez-Brückmann, M., 433
Alving, K., 238
Amanatides, P., 245
Ambulos, N. P., 233
Amess, B., 270, 271, 274
Amini, A., 267
Amster, I. J., 139
Amy, J. W., 110, 115, 116, 119, 120, 149, 159
Anderegg, R. J., 345, 346

Andersen, G. A., 303
Andersen, I., 271
Andersen, J. S., 235, 247, 269
Anderson, G. A., 34, 36, 37, 263, 308, 370, 371
Anderson, L., 252, 269
Anderson, N. G., 269, 272
Anderson, N. L., 269, 270, 272, 274
Anderson, S., 155
Anderson, U. N., 91
André, J., 122, 123
Andrews, L., 247, 292
Andrews, P., 262
Andriantomanga, V., 367
Andrien, B., 67, 73
Andrieu, J. P., 332
Andry, V., 375
Angeletti, R. H., 295, 342
Annan, R. S., 25, 250, 256
Anthony, R. M., 265
Aoi, S., 210
Aplin, R. T., 313, 317, 332, 339, 370, 383, 384, 392
Appel, R. D., 245, 246
Apweiler, R., 275, 295, 298
Arce, A., 272
Archer, D. B., 391, 392, 395, 396, 399, 400
Arcuri, F., 270
Ardekani, A. M., 271
Argo, E., 271
Ariyaratne, T. R., 62
Arjomand, A., 413, 436
Armstrong, R. N., 315
Arnott, D., 247, 257, 259, 260, 267, 270, 302
Arrington, C. B., 332, 337, 339, 342
Arrowsmith, C. H., 383
Asada, T., 80
Asano, K. G., 110, 152, 155, 169, 171, 173, 174, 178
Aswad, D. W., 245
Auberry, K. J., 308
Auer, G., 270
Austin, J. G., 433
Aveline-Wolf, L. D., 307, 308

445

M

Subject Index

A

B

C

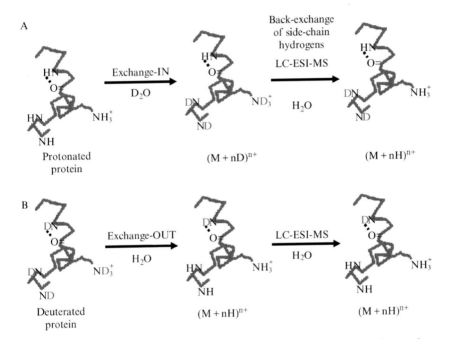

MAIER AND DEINZER, CHAPTER 10, FIG. 10. The two experimental approaches used to perform isotopic exchange studies: (A) the exchange-in and (B) the exchange-out approach.

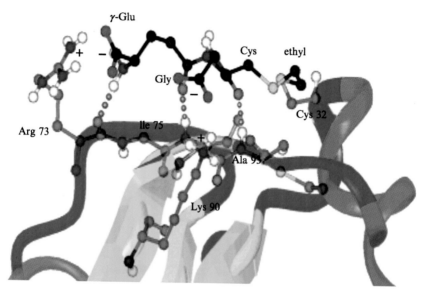

MAIER AND DEINZER, CHAPTER 10, FIG. 17. Part of the structural model of the ethylglutathionyl thioredoxin conjugate highlighting the interactions of ethylglutathionyl moiety with the thioredoxin backbone. Hydrogen bonding interactions are depicted as purple dots. The positively charged side chains of Arg-73 and Lys-93 are shown to form salt bridges to the negatively charged carboxylate groups of the glutathionyl γ-Glu and Gly residues. Carbons of the glutathione chain are shown as black spheres, while those in the thioredoxin protein are shown in green. Oxygen atoms are in red, sulfurs in yellow, and nitrogens in blue. The protein backbone is shown as blue ribbon, α-helical regions are colored in red and β-strands in yellow.

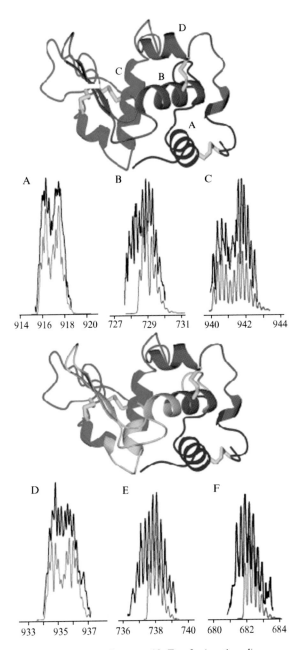

YAO *ET AL.*, CHAPTER 12, FIG. 3. Nano-electrospray mass spectra of the partially exchanged peptide fragments of the D67H variant. The multiply charged peptides are assigned as follows: (A) $(58–108)^{+6}$; (B) $([1–25]\text{-}[124–130])^{+4}$; (C) $([26–40]\text{-}[109–123])^{+5}$; (D) $([58–84]\text{-}[93–108])^{+5}$; (E) $([26–31]\text{-}[109–123])^{+4}$; and (F) $([1–19]\text{-}[124–130])^{+4}$. Before digestion, the protein was exposed to hydrogen exchange at pH 10 for 3 s. The location of the principal peptic digestion products is shown within the native structure of lysozyme in the upper panel. Similarly, the smaller peptide fragments, derived from three regions (D), (E), and (F), are indicated in the lysozyme structure in the lower panel. The lysozyme structure was generated from the X-ray structure (1REX) and produced using MOLMOL. The peaks in the spectra are compared with simulations of the isotope patterns color-coded for each peptide fragment. The peptide fragments (A), (C), and (D) fit well to an exchange model in which the total number of protected sites undergo exchange exclusively via an EX1 mechanism to give two distinct hydrogen exchange populations. The peptide fragments (B), (E), and (F) show a single hydrogen exchange population that undergoes exchange via an EX2 mechanism. (Reprinted, with permission, from Canet *et al.* [2002].)